COLLECTED PAPERS OF
SALOMON BOCHNER

Courtesy of The Friends of Fondren Library

Salomon Bochner

August 20, 1899–May 2, 1982

COLLECTED PAPERS OF SALOMON BOCHNER

Part 4

Robert C. Gunning
Editor

American Mathematical Society
Providence, Rhode Island

1991 *Mathematics Subject Classification.*
Primary 01A75, 00B60.

Library of Congress Cataloging-in-Publication Data

Bochner, S. (Salomon), 1899–1982.
[Works. 1991]
Collected papers of Salomon Bochner/Robert C. Gunning, editor.
p. cm.
Includes bibliographical references.
1. Mathematics. 2. Bochner, S. (Salomon), 1899–1982. I. Gunning, R. C. (Robert Clifford), 1931– . II. Title.
QA3.B5719 1991 91-31045
510—dc20 CIP
ISBN 0-8218-0161-9 (set: alk. paper)
ISBN 0-8218-0174-0 (pt. 1: alk. paper)
ISBN 0-8218-0175-9 (pt. 2: alk.paper)
ISBN 0-8218-0176-7 (pt. 3: alk. paper)
ISBN 0-8218-0177-5 (pt. 4: alk. paper)

Printed in the United States of America
A complete list of acknowledgments can be found at
the back of this publication.

The paper used in this book is acid-free and falls within the guidelines
established to ensure permanence and durability. ♾
10 9 8 7 6 5 4 3 2 1 97 96 95 94 93 92

Contents of Part 4

Preface vii

Chronological Bibliography ix

Section I: Lie Groups 1

Compact groups of differentiable transformations 3

Groups of differentiable and real or complex analytic transformations (with D. Montgomery) 13

Formal Lie groups 23

Locally compact groups of differentiable transformations (with D. Montgomery) 33

Groups on analytic manifolds (with D. Montgomery) 49

Section J: Miscellaneous Mathematical Papers 61

Über einen arithmetischen Satz der Rontgenkristallographie 63

Remark on the Euclidean algorithm 71

Harald Bohr, April 22, 1887–January 22, 1951 73

John von Neumann 77

Section K: History 97

Mathematics 99

The role of mathematics in the rise of mechanics 107

The significance of some basic mathematical conceptions for physics 125

Revolutions in physics and crises in mathematics 153

Aristotle's physics and today's physics 159

CONTENTS OF PART 4

Aristotle's notion of place (topos) in physics	187
Why mathematics grows	191
Plato and the One-ness of Knowledge	213
The size of the Universe in Greek thought	215
The rise of functions	237
Mathematical reflections	257
Mathematical background space in astronomy and cosmology	283
Note on Kepler's contribution to mathematical analysis	313
Commentary	315
Kepler: A personal Footnote	325
The emergence of analysis in the renaissance and after	327
Fourier series came first	373
Einstein between centuries	375
Vita	431
Theses Supervised	433
Acknowledgments	437

Preface

The present volumes contain the collected papers of Salomon Bochner. He worked in a variety of different areas of mathematics in a long and productive career. For the convenience of those using these volumes, the papers are grouped in eleven different categories: Almost Periodic Functions, Fourier Analysis, Real Analysis (topics in real analysis that do not easily fit into the categories of almost periodic functions and Fourier analysis), Complex Analysis (general and special analytic functions of one complex variable), Several Complex Variables, Differential Geometry, Lie Groups, Differential Equations (both ordinary and partial), Probability, History, and Miscellaneous (including a few papers that did not fit naturally anywhere else together with obituary articles). The choice of these categories and the assignment of papers to a particular category are of course both rather arbitrary. Bochner saw the interconnections between topics very clearly and was often led to results in one area by his work in other areas, so many papers properly fall between categories or alternatively within two or more categories.

This collection does not contain the many reviews of other papers and books that Bochner wrote. He was a long-time reviewer for *Mathematical Reviews* and wrote some review articles for the *Bulletin of the American Mathematical Society* as well. Although his reviews do convey his own opinions and evaluations of a good deal of the mathematics of his time, they contain little of his own characteristic contributions to the subject. However, not included in this collection are some entries he wrote for the *Dictionary of the History of Ideas* (Scribner, 1973), the entries on "Space," "Continuity and Discontinuity," "Infinity," "Symmetry and Asymmetry," "Mathematics in Cultural History," and the article "Analysis" in the *Encyclopedia Britannica* (1961). These are undoubtedly of interest to the general reader, but less so to those who are interested in Bochner's own work.

In addition to his mathematical papers, Bochner wrote seven books on mathematical and historical topics. The books are listed in the summaries at the beginning of the appropriate categories as well as in the chronological bibliography included in this collection, but are not reproduced here.

Robert C. Gunning

Chronological Bibliography

The notation to the right of a paper indicates the part and section where it is located in this collection. For example, 1:A means part 1, section A. The absence of notation or the word "book" means that the work has not been included.

Orthgonale Systeme analytischer und harmonischer Funktionen, Deutsche Math.-Ver. 30 (1921), 100–101. 1:A

Über orthogonale Systeme analytischer Funktionen, Math. Zeit. 14 (1922), 180–207. 1:A

Sur les fonctions presque périodiques de Bohr, C. R. Acad. Sci. Paris 180 (1925), 1156–1158. 2:C

Properties of Fourier series of almost periodic functions, Proc. London Math. Soc. 26 (1926), 433–452. 2:C

Remarks on the theorems of Picard-Landau an Picard-Schottky, J. London Math. Soc. 1 (1926), 100–103. 2:E

Properties of Fourier series of almost periodic functions, J. London Math. Soc. 1 (1926), 131–133. 2:C

Notes on two theorems of Norbert Wiener (with G. H. Hardy), J. London Math. Soc. 1 (1926), 240–244. 1:B

Über Fourier-Reihen von fastperiodischen Funktionen, Sitzungsberi chte Berliner Math. Ges. 26 (1927), 49–56. 2:C

Beiträge zur Theorie der fastperiodischen Funktionen I. *Teil. Funktionen einer Variablen,* Math. Ann. 96 (1927), 119–147. 2:C

Beiträge zur Theorie der fastperiodischen Funktionen I. *Teil. Funktionen mehrerer Variablen,* Math. Ann. 96 (1927), 383–409. 2:C

Darstellung reellvariabler und analytischer Funktionen durch verallgemeinerte Fourier- und Laplace-Integrale, Math. Ann. 97 (1927), 635–662. 1:A

Arithm. Satz der Röntgenkristallog (with H. Seyfarth), Probl. Mod. Physki (1928), 8 pp. 4:J

Hauptlösungen von Differenzengleichungen, Acta. Math. 51 (1928), 1–21. 1:B

Über die Struktur von Fourierreihen fastperiodischer Funktionen, Sitzungsberichte Math.-Nat. Abteil. Bayerischen Akad. Wiss. (1928), 181–190. 2:C

Konvergenzsäte für Fourierreihen grenzperiodischen Funktionen, Math. Zeit. 27 (1928), 187–211. 2:C

Fortsetzung Riemannscher Flächen, Math. Ann. 98 (1928), 406–421. 2:E

Über einen arithmetischen Satz der Rontgenkristallographie (with H. Seyfarth), Probleme der Moderne Physik, Leipzig, (1928), 88–94. 4:J

Über Factorenfolgen für Fouriersche Reihen, Acta Litt. Ac. Sci. Univ. Franc.-Josephinae, 4 (1929), 5 pp. 1:A

Über Sturm-Liouvillesche Polynomsysteme, Math. Zeit. 29 (1929), 730–736. 3:H

Ein Satz über lineare Operationen, Math. Zeit 29 (1929), 737–743. 1:B

Über gewisse Differential- und allgemeinere Gleichungen, deren lösungen fastperiodisch sind. I. *Existenzsatz.* Math. Ann. 102 (1929), 489–504. 2:C

Über eine Klasse singulärer Integralgleichungen, Sitzungsberichte Preuss. Akad. Wiss. (1930), 9 pp. 1:B

Beitrag zur absoluten Konvergenz fastperiodischer Funktionen, Jber. Deutschen Math. Ver. 39 (1930), 52–54. 2:C

Über gewisse Differential- und allgemeinere Gleichungen, deren lösungen fasterperiodisch sind. II. Beschranktheitsatz Math. Ann. 103 (1930), 588–597. 2:C

Der Dimensionsbegriff für Punktmengen in kartesischen Räumen, Math. Ann. 102 (1930), 768–784. 1:B

Kurven endlicher Länge, Jber. Dtsch. Math. Ver. 40 (1931), 245–249. 1:B

Über gewisse Differential- und allgemeinere Gleichungen, deren lösungen fastperiodisch sind. III. *Systeme von Gleichungen,* Math. Ann. 104 (1931), 579–587. 2:C

Allgemeine lineare Differenzengleichungen mit asymptotisch konstanten Koeffizienten, Math. Zeit. 33 (1931). 426–450. 1:B

Ein Konvergensatz für mehrvariablige Fouriersche Integrale, Math. Zeit. 34, 440–447. 1:A

Die Poissonsche Summationsformel in mehreren Veränderlichen, Math. Ann. 106 (1932), 56–63. 1:A

Limitierung mehrfacher Folgen nach dem Verfahren der aritmetischen Mittel, Math. Zeit. 35 (1932), 122–126. 1:B

Vorlesungen über Fouriersche Integrale, Leipzig, Akademische Verlagsgesellschaft, 1932. Book

Umkehrsätze für allgemeine Limitierungsverfahren, S.-B. preuss. Akad. Wiss. (1933), 126–144. 1:B

Spektralzerlegung linearer Scharen unitärer Operatoren, S.-B. preuss Akad. Wiss. (1933), 371–376. 1:B

Abstrakte Funktionen und die Besselsche Ungleichung, Nachr. Ges. Wiss. Göttingen (1933), 178–180. 1:A

Integration von Funktionen, deren Werte die Elemente eines Vektorraumes sind, Fundam. Math. 20 (1933), 262–276. 1:B

Eine Bemerkung zum Satz von Fubini, Fundam. Math. 20 (1933), 277–280. 1:B

Absolut-additiv abstrakte Mengenfunktionen, Fundam. Math. 21 (1933), 211–213. 1:B

Remark on the integration of almost periodic functions, J. London Math. Soc. 8 (1933), 250–254. 2:C

Homogeneous systems of differential equations with almost periodic coefficients, J. London Math. Soc. 8 (1933), 283–288. 2:C

Abstrakte fastperiodische Funktionen, Acta Math. 61 (1933), 149–184. 2:C

Monotone Funktionen, Stieltjessche Integrale und harmonische Analyse, Math. Ann. 108 (1933), 378–410. 1:A

Ein Satz von Landau und Ikehara, Math. Zeit 37 (1933), 1–90. 2:E

Simultankonvergenz von mehrfachen Fourierschen Reihen und Integralen, Math. Zeit 37 (1933), 566–571. 1:A

Average distribution of arbitrary masses under group translations, Proc. Nat. Acad. Sci. U.S.A. 20 (1934), 206–210. 1:B

A theorem on Fourier-Stieltjes integrals, Bull. Amer. Math. Soc. 40 (1934), 270–276. 1:A

Remark on the Euclidean algorithm, J. London Math. Soc. 9 (1934), 4. 4:J

An extension of a Tauberian theorem on series with positive terms, J. London Math. Soc. 9 (1934), 141–148. 1:A

Fastperiodische Lösungen der Wellengleichung, Acta. Math. 62 (1934), 227–237. 2:C

Inversion formulae and unitary transformations, Ann. of Math. 35 (1934), 111–115. 1:A

Analytic functions with almost periodic coefficients (with F. Bohnenblust), Ann. of Math. 35 (1934), 152–161. 2:C

Distribution functions and positive definite-functions (with B. Jessen), Ann. of Math. 35 (1934), 252–257. 1:A

Summation of multiple Fourier series by spherical means, Proc. Nat. Acad. Sci. U.S.A. 21 (1935), 353–355. 1:A

Almost periodic functions in groups, II Trans. Amer. Math. Soc. 37 (1935), 21–50. 2:C

On compact solutions of operational-differential equations, I (with J. v. Neumann), Ann. of Math. 36 (1935), 255–291. 2:C

Summation of double Fourier series by spherical means (abstract), Bull. Amer. Math. Soc. 41 (1935), 332. 1:A

Summation of multiple Fourier series by spherical means, Trans. Amer. Math. Soc. 40 (1936), 175–207. 1:A

Summation of derived Fourier series. An application to Fourier expansions on compact Lie groups, Ann. of Math. 37 (1936), 345–356. 1:A

A converse of Poisson's theorem in the theory of probability, Ann. of Math. 37 (1936), 816–822. 2:D

Some general convergence theorems (with S. Izumi), Tôhoku Math. J. 42 (1936), 191–194. 1:A

On general Fourier series with gaps, Prace mat.-fiz 43 (1936), 63–79. 1:A

Bounded analytic functions in several variables and multiple Laplace integrals, Amer. J. Math. 59 (1937), 732–738. 3:F

Singularities of composite functions in several variables (with W. T. Martin), Ann. of Math. 38 (1937), 298–302. 3:F

Remark on the theorem of Green, Duke Math. J. 3 (1937), 334–338. 3:G

Analytic mappings of compact Riemann spaces into Euclidean space, Duke Math. J. 3 (1937), 339–354. 3:G

Completely monotone functions of the Laplace operator for torus and sphere, Duke Math. J. 3 (1937), 488–502. 2:D

Stable laws of probability and completely monotone functions, Duke Math. J. 3 (1937), 726–728. 2:D

A theorem on analytic continuation of functions in several variables, Ann. of Math. 39 (1938), 14–19. 3:F

Closure theorems for translations (with R. P. Boas), Ann. of Math 39 (1938), 287–300. 2:C

Linear functionals on certain spaces of abstractly-valued functions (with A. E. Taylor), Ann. of Math. 39 (1938), 913–944. 1:B

Differentiable and Riemann metric, Duke Math. J. 4 (1938), 51–54. 3:G

Functions of integrable square in several complex variables, Duke Math. J. 4 (1938), 635–639. 3:F

Additive set functions on groups, Proc. Nat. Acad. Sci. U.S.A. 25 (1939), 158–160. 1:B

On a theorem of M. Riesz for Fourier series (with R. P. Boas), J. London Math. Soc. 14 (1939), 62–73. 1:A

Some theorems on quasi-analyticity for functions of several variables (with A. E. Taylor), Amer. J. Math. 61 (1939), 303–329. 3:F

Additive set functions on groups, Ann. of Math. 40 (1939), 769–799. 1:B

A uniqueness theorem for analytic almost-periodic functions, Duke Math. J. 5 (1939), 937–940. 2:C

Integration and differentiation in partially ordered spaces, Proc. Nat. Acad. Sci. U.S.A. 26 (1940), 29–31. 1:B

Harmonic surfaces in Riemann metric, Trans. Amer. Math. Soc. 47 (1940), 146–154. 3:G

Finitely additive integral, Ann. of Math. 41 (1940), 495–504. 1:B

Hadamard's theorem for Dirichlet series, Ann. of Math. 41 (1940), 711–714. 2:E

A generalization of Poisson's summation formula, Duke Math. J. 6 (1940), 229–234. 1:A

Additive set functions and vector lattices (with R. S. Phillips) Ann. of Math. 42 (1941), 316–324. 1:B

Hilbert distances and positive definite functions, Ann. of Math. 42 (1941), 647–656. 1:A

On a theorem on Tannaka and Krein, Ann. of Math. 43 (1942), 56–58. 2:C

Absolutely convergent Fourier expansions for non-commutative normed rings (with R. S. Phillips), Ann. of Math. 43 (1942), 409–418. 1:A

Completely monotone functions in partially ordered spaces, Duke Math. J. 9 (1942), 519–526. 1:B

Analytic and meromorphic continuation by means of Green's formula, Ann. of Math. 44 (1943), 652–673. 3:F

Group invariance of Cauchy's formula in several variables, Ann. of Math. 45 (1944), 686–707. 3:F

Boundary values of analytic functions in several variables and of almost periodic functions, Ann. of Math 45 (1944), 708–722. 3:F

Compact groups of differentiable transformations, Ann. of Math. 46 (1945), 372–381. 4:I

Groups of differentiable and real or complex analytic transformations (with D. Montgomery), Ann. of Math. 46 (1945), 685–694. 4:I

Finitely additive set functions and stochastic processes, Proc. Nat. Acad. Sci. U.S.A. 32 (1946), 259–261. 2:D

Bloch's theorem for real variables, Bull. Amer. Math. Soc. 52 (1946), 715–719. 3:G

Vector fields and Ricci curvature, Bull. Amer. Math. Soc. 52 (1946), 776–797. 3:G

Formal Lie groups, Ann. of Math. 47 (1946), 192–201. 4:I

Linear partial differential equations with constant coefficients, Ann. of Math. 46 (1946), 202–212. 3:H

Locally compact groups of differentiable transformations (with D. Montgomery), Ann. of Math. 47 (1946), 639–653. 4:I

Curvature in Hermitian metric, Bull. Amer. Math. Soc. 53 (1947), 179–195. 3:G

On compact complex manifolds, J. Indian Math. Soc. 11 (1947), 1–21. 3:G

Distributive order-preserving operations in partially ordered vector sets (with Ky Fan), Ann. of Math. 48 (1947), 168–179. 1:B

Groups on analytic manifolds (with D. Montgomery), Ann. of Math 48 (1947), 659–669. 4:I

Stochastic processes, Ann. of Math. 48 (1947), 1014–1061. 2:D

A homogeneous differentiable system of infinite order with nonvanishing solution (with D.V. Widder), Bull. Amer. Math. Soc. 54 (1948), 409–415. 3:H

On spherical partial sums of multiple Fourier series, Rev. Ci., Lima, 50 (1948), 85–104. 1:A

Curvature and Betti numbers, Ann. of Math. 49 (1948), 379–390. 3:G

On the localization property for multiple Fourier series (with K. Chandrasekharan), Ann. of Math. 49 (1948), 966–978. 1:A

Summations over lattice points in K-space, (with K. Chandrasekharan), Quart. J. Math. 19 (1948), 238–248. 1:A

Several complex variables, (with W. T. Martin). Princeton University Press, 1948. Book

Diffusion equation and stochastic processes, Proc. Nat. Acad. Sci. U.S.A. 35 (1949), 368–370. 2:D

Gauss summability of trigonometric integrals (with K. Chandrasekharan), Amer. J. Math. 71 (1949), 50–59. 1:A

Curvature and Betti numbers, II. Ann. of Math. 50 (1949), 77–93. 3:G

Fourier series of L_2-functions (with K. Chandrasekharan), Duke Math. J. 16 (1949), 579–583. 1:A

Fourier transforms (with K. Chandrasekharan), Princeton University Press, 1949; Annals of Math. Studies 19. Book

Partial ordering in theory of stochastic processes, Proc. Nat. Acad. Sci. U.S.A. 36 (1950), 439–443. 2:D

Su un teorema di Frobenius per le funzioni di Jacobi, Atti Accad. Naz. Lincei, Rend., Cl. Sci. fis. mat. natur. 8 (1950), 327–331. 2:E

Lattice points and Fourier expansions (with K. Chandrasekharan), Acta Sci. Math. 12B (1950), 1–15. 1:A

Quasi-analytic functions, Laplace operator, positive kernels, Ann. of Math. 51 (1950), 68–91. 1:A

Euler–Poincaré characteristic for locally homogeneous and complex spaces, Ann. of Math. 51 (1950), 241–261. 3:G

Vector fields on complex and real manifolds, Ann. of Math. 52 (1950), 642–649. 3:G

Summations over lattice points in K-space (A supplementary note) (with K. Chandrasekharan), Quart. J. Math. 1 (1950), 80. 1:A

Localization of best approximation, Contrib. Fourier Analysis, Ann. of Math. Studies 25 (1950), 3–23. 1:A

Dirichlet problem for domains bounded by spheres, Contrib. Fourier Analysis, Ann. of Math. Studies 25 (1950), 24–25. 1:A

Complex spaces with transitive commutative groups of transformations, Proc. Nat. Acad. Sci. U.S.A. 37 (1951), 356–359. 3:G

Tensor fields and Ricci curvature in Hermitian metric, Proc. Nat. Acad. Sci. U.S.A. 37 (1951), 704–706. 3:G

Theta relations with spherical harmonics, Proc. Nat. Acad. Sci. U.S.A. 37 (1951), 804–808. 2:E

Some properties of modular relations, Ann. of Math. 53 (1951), 332–363. 2:E

Tensor fields with finite bases, Ann. of Math. 53 (1951), 400–411. 3:G

A new viewpoint in differential geometry, Canadian J. Math. 3 (1951), 460–470. 3:G

Laplace operator on manifolds, Proc. Int. Congr. Math., Cambridge, Mass., 1950, 2 (1952), 189–201. 3:G

Partial differential equations and analytic continuations, Proc. Nat. Acad. Sci. U.S.A. 38 (1952), 227–230. 3:H

Local transformations with fixed points on complex spaces with singularities (with W.T. Martin), Proc. Nat. Acad. Sci. U.S.A. 38 (1952), 726–732. 3:F

Harald Bohr, April 22, 1887–January 22, 1951. Bull. Amer. Math. Soc. 58 (1952), 72–75. 4:J

On the addition theorem for multiply periodic functions, Proc. Amer. Math. Soc. 3 (1952), 99–106. 3:F

Remarks on Gaussian sums and Tauberian theorems, J. Indian Math. Soc. 15 (1952), 97–104. 1:A

Algebraic and linear dependence of automorphic functions in several variables, J. Indian Math. Soc. 16 (1952), 1–6. 3:F

Connection between functional equations and modular relations and functions of exponential type, J. Indian Math. Soc. 16 (1952), 99–102. 2:E

Hartogs' theorem in complex spaces with singularities (with W.T. Martin) J. Indian Math. Soc. 16 (1952), 137–146. 3:F

Tensor-fields in non-symmetric connections (with K. Yano), Ann. of Math. 56 (1952), 504–519. 3:G

Bessel functions and modular relations of higher type and hyperbolic differential equations, Meddel. Lunds Univ. Mat. Sem. Suppl.-band M. Riesz (1952), 12–20. 2:E

Stochastic processes with finite and non-finite variance, Proc. Nat. Acad. Sci. U.S.A. 39 (1953), 190–197. 2:D

Fourier transforms of time series, Proc. Nat. Acad. Sci. U.S.A. 39 (1953), 302–307. 2:D

Closure classes originating in the theory of probability, Proc. Nat. Acad. Sci. U.S.A. 39 (1953), 1082–1088. 2:D

The theorem of Morera in several variables, Ann. Mat. Pura Appl. 34 (1953), 27–39. 3:F

Zeta functions and Green's functions for linear partial differential operators of elliptic type with constant coefficients, Ann. of Math. 57 (1953), 32–56. 3:H

Length of random paths on general homogeneous spaces, Ann. of Math. 57 (1953), 309–313. 2:D

Complex spaces with singularities (with W.T. Martin), Ann. of Math. 57 (1953), 490–516. 3:F

Curvature and Betti numbers (with K. Yano), Princeton University Press, 1953; Annals of Math. Studies 32. Book

Structure of complex spaces, Ann. of Math. Studies 30 (1953), 189–201. 3:F

Boundedness and stationary of time series, Proc. Nat. Acad. Sci. U.S.A. 40 (1954), 289–294. 2:D

Limit theorems for homogeneous stochastic processes, Proc. Nat. Acad. Sci. U.S.A. 40 (1954), 699–703. 2:D

Positive zonal functions on spheres, Proc. Nat. Acad. Sci. U.S.A. 40 (1954), 1141–1147. 1:A

Green's formula and analytic continuation, Ann. of Math. Studies 33 (1954), 1–14. 3:F

Stationary, boundedness, almost periodicity of random-valued functions, Proc. Third Berkeley Symp. Math. Stat. and Prob. 2 (1954), 7–27. 2:D

Contributions to the theory of partial differential equations, (Edited with L. S. Bers and F. John), Ann. of Math. Studies 33, Princeton University Press, 257.

Stationary space-time in general relativity, Proc. Nat. Acad. Sci. U.S.A. 41 (1955), 485–490. 3:G

Existence of functionally independent automorphic functions (with R.C. Gunning), Proc. Nat. Acad. Sci. U.S.A. 41 (1955) 746–752. 3:F

Curvature and Betti numbers in real and complex vector bundles, Univ. e Politec. Torino, Rend. Sem. Mat. 15 (1955), 225–253. 3:G

Partial ordering in the theory of martingales, Ann. of Math. 62 (1955), 162–169. 2:D

Green-Goursat theorem, Math. Zeit. 63 (1955), 230–242. 3:F

Functions in one complex variable as viewed from the theory of functions in several variables, Lect. on functions of a complex variable, (University of Michigan, Ann Arbor, 1955), 315–333. 3:F

Harmonic analysis and the theory of probability, Berkeley, University of California Press, 1955. Book

Gamma factors in functional equations, Proc. Nat. Acad. Sci. U.S.A. 42 (1956), 86–89. 2:E

On Rieman's functional equation (with K. Chandrasekharan), Ann. of Math. 63 (1956), 336–360. 2:E

Weak solutions of linear partial differential equations, J. Math. Pures Appl. 35 (1956), 193–202. 3:H

Systèmes des équations aux dérivées partielles qui se rattachent aux formules de Green, Th. Eqns. aux Der. Part. Coll. Int. C.N.R.S. 71, Paris (1956), 19–27. 3:H

Sturm–Liouville and heat equations whose eigenfunctions are ultraspherical polynomials or associated Bessel functions, Proc. Conf. on Diff. Eqns., University of Maryland, 1956, 23–48. 3:H

Theorems on analytic continuation which occur in the study of Riemann's functional equation, J. Indian Math. Soc. 21 (1957), 127–147. 2:E

Krivizna i Chisla Betti (with K. Yano), Moscow, 1957, (translated from the English). (Translation of Curvature and Beth Numbers) Book

John von Neumann, Nat. Acad. Sci. U.S.A. Biographical Memoirs 32 (1958), 438–457. 4:J

On Riemann's functional equation with multiple Gamma factors, Ann. of Math. 67 (1958), 29–41. 2:E

A special integral transformation in Euclidean space (with T. Kawata), Ann. of Math. 68 (1958), 150–158. 2:D

A limit theorem for the periodogram (with T. Kawata), Ann. Math. Stat. 29 (1958), 1198–1208. 2:D

On identical vanishing of holomorphic functions in several complex variables, Proc. Nat. Acad. Sci. U.S.A. 45 (1959), 46–47. 3:F

Linear and algebraic dependance of functions on compact complex spaces with singularities, Proc. Nat. Acad. Sci. U.S.A. 45 (1959), 47–49. 3:F

Generalized conjugate and analytic functions without expansions, Proc. Nat. Acad. Sci. U.S.A. 45 (1959), 855–857. 2:E

Strong law of large numbers for sequences of almost periodic functions (with S. Izumi), Ann. of Math. 69 (1959), 718–732. 2:D

Lectures on Fourier integrals, with an Author's Supplement on Monotonic Functions, Stieltjes Integrals, and Harmonic Analysis. Princeton University Press, 1959; Annals of Math. Studies 42. (Translated from the original by Morris Tenenbaum and Harry Pollard.) (Translation of Vorlesungen über Fouriersche Integrale) Book

Classes of holomorphic functions of several variables in circular domains, Proc. Nat. Acad. Sci. U.S.A. 46 (1960), 721–723. 3:F

Almost periodic solutions of the inhomogeneous wave equation, Proc. Nat. Acad. Sci. U.S.A. 46 (1960), 1233–1236. 2:C

Mathematics, McGraw-Hill Encyc. of Sci. and Tech. 8 (1960), 175–180. 4:K

Hartogs' theorem in Euclidean space and a related Cousin the theorem on the torus, Contributions to Function Theory, Tata Institute of Fundamental Research, Bombay, 12–19 January 1960, 79–114. 3:F

Uniform convergence of monotone sequences of functions, Proc. Nat. Acad. Sci. U.S.A. 47 (1961), 582–585. 1:B

Vector fields on Riemannian spaces with boundary, Ann. Mat. Pura Appl. 53 (1961), 57–62. 3:G

General analytical setting for the central limit theorem of probability, Calcutta Math. Soc. Golden Jubilee Comm. Vol. part 1, Calcutta Math. Soc. (1961), 111–128. 2:D

Analysis. Encyc. Britannica 1 (1961), 864–864F.

Subordination of non-Gaussian stochastic processes, Proc. Nat. Acad. Sci. U.S.A. 48 (1962), 19–22. 2:D

A new approach to almost periodicity, Proc. Nat. Acad. Sci. U.S.A. 48 (1962), 2039–2043. 2:C

Entire functions in several variables with constant absolute values on a circular uniqueness set, Proc. Amer. Math. Soc. 13 (1962), 117–120. 3:F

Infinite linear pseudogroups of transformations (with R.C. Gunning), Ann. of Math. 75 (1962), 93–104. 3:G

The role of mathematics in the rise of mechanics, American Scientist 50 (1962), 294–311. 4:K

Lektsu ob integralakh Fur'l. Moscow, 1962 (translated from the English). (Translation of Vorlesungen über Fouriersche Integrale) Book

A generalization of bounded Haar measure, Proc. Nat. Acad. Sci. U.S.A. 50 (1963), 293–297. 1:B

Potential-theoretic approach to analytic Bohr almost periodic functions in half-planes, Math. Ann. 150 (1963), 150–155. 2:C

Function rings and analytic measures, J. de Math. Pures et Appl. 42 (1963), 209–221. 1:B

The significance of some basic mathematical conceptions for physics, Isis 54 (1963), 179–205. 4:K

Revolutions in physics and crises in mathematics, Science 141 (1963), 408–411. 4:K

Interpolation of general bounded and of almost periodic sequences by functions of stratified exponential type, Proc. Nat. Acad. Sci. U.S.A. 51 (1964), 164–168. 2:C

Continuous mappings of almost automorphic and almost periodic functions, Proc. Nat. Acad. Sci. U.S.A. 52 (1964), 907–910. 2:C

Analytic measures on compact Bohr groups, Scripta Math. 27 (1964), 5–26. 1:A

Aristotle's physics and today's physics, Inter. Phil. Quar. 4 (1964), 217–244. 4:K

Aristotle's notion of place (topos) in physics, Acts of the Tenth International Congress of the History of Science VIII, (Ithaca,1962), (1965), 471–474. 4:K

Why mathematics grows, Jour. Hist. Ideas 26 (1965), 3–24. 4:K

Intrinsic analytic continuation and envelopes of holomorphy, Proc. Nat. Acad. Sci. U.S.A. 53 (1965), 904–907. 3:F

Plato and the One-ness of Knowledge, University Magazine 32 (Spring 1967), Princeton University Press, p. 18. 4:K

The role of mathematics in the rise of science, Princeton University Press, 1966; Japanese edition, 1970. Book

The size of the Universe in Greek thought, Scientia 103 (1968), 511–531. 4:K

The rise of functions, Complex Analysis II: Proceedings of the Conference on Complex Analysis, Rice University, 26–29 March 1969, Rice University Studies, 56, No. 2 (1970), 3–23. 4:K

Space, Dictionary of the History of Ideas, Charles Scribner's Sons, 1973, New York.

Continuity and discontinuity in nature and knowledge, Dictionary of the History of Ideas, Charles Scribner's Sons, 1973, New York.

Infinity, Dictionary of the History of Ideas, Charles Scribner's Sons, 1973, New York.

Symmetry and asymmetry, Dictionary of the History of Ideas, Charles Scribner's Sons, 1973, New York.

Mathematics in Cultural History, Dictionary of the History of Ideas, Charles Scribner's Sons, 1973, New York.

Eclosion and synthesis, perspectives on the history of knowledge, W. A. Benjamin, Inc., 1969. Book

Singularities and discontinuities, Proceedings of the Conference on Complex Analysis, Rice University, March 1972, Rice University Studies, 59, No. 2 (1973). 3:F

Mathematical reflections, The American Mathematical Monthly 81, No. 8, (1974), 827–852. The first part of the article appeared as "Mathematical Americana" in Graduate Studies, Texas Tech University No. 13, October 1976, 87–97. 4:K

"Henry Bateman," Dictionary of American Biography, Supplement Four 1949–1950, 1974, 57–58.

A formal approach to stochastic stability, Zeitschrift für Wahrscheinlichkeitstheorie 31 (1975), 187–198. 2:D

Mathematical background space in astronomy and cosmology, Vistas in Astronomy 19, part 2 (1975), 133–161. 4:K

Note on Kepler's contribution to mathematical analysis (Kepler-Four Hundred Years), Vistas in Astronomy 18, (1975). 4:K

General almost automorphy, Proc. Nat. Acad. Sci. U.S.A. 72 (1975). 2:C

"Commentary" on Curtis A. Wilson's "Rheticus, Ravetz, and the 'necessity' of Copernicus' Innovation." In The Copernican Achievement (Robert S. Westman, Editor). Univ. of California Press (1975), 40–48. 4:K

Review of "Genevieve Guitel, Histoire Comparee des numeration écrites." Paris, Flammarion, 1975. American Scientist 65 (1977), 105.

Kepler: A personal Footnote, Vistas in Astronomy, 22, (1978), 19–20. 4:K

The Emergence of Analysis in the Renaissance and After, Rice University Studies, 64 Nos. 2 and 3 (1978), 11–56. 4:K

Fourier Series came first, The American Mathematical Monthly, 86 (1979), 197–199. 4:K

Einstein between Centuries, Rice University Studies, Vol. 65, No.3, Summer 1979, 1–54. 4:K

Positivity of the Heat Kernel for Ultraspherical Polynomials and Similar Functions, Archive of Rational Mechanics and Analysis, 70 (1979), 211–217. 1:A

Review of Gustar Herglotz, Gesammelte Schritten, 1979 in Bulletin of the American Mathematical Society, Vol. 1, 1020–1022.

Section I

Lie Groups

During the later 1940s, Bochner worked on Lie groups of transformations, in part, jointly with Deane Montgomery, and on some properties of formal Lie groups.

ANNALS OF MATHEMATICS
Vol. 46, No. 3, July, 1945

COMPACT GROUPS OF DIFFERENTIABLE TRANSFORMATIONS

BY S. BOCHNER

(Received February 26, 1945)

Introduction.

H. Cartan[1] has investigated automorphisms and general transformations into itself of a domain in several complex variables by means of complex analytic functions. His main results are as follows:

(i) If a compact group of automorphisms has a fix point then in suitably chosen local coordinates around the fix point all transformations are linear.

(ii) In a compact group of automorphisms all sufficiently small subgroups are the identity.[2]

(iii) If a transformation of a bounded domain into (part of) itself has an invariant point 0, and is of the special form

$$w_m = z_m + \text{(higher powers)}, \ m = 1, \cdots, n$$

around 0, then the transformation is the identity.

Finally, H. Cartan and C. Caratheodory[3] have proved the following theorem:

(iv) If a transformation of a bounded domain into itself with invariant point has a Jacobian of absolute value 1 at that point, then the transformation is an automorphism.

In connection with his proposition (ii), Cartan[4] observes that the analytic character of the automorphisms is no essential pre-requisite. Rather, it suffices to assume that the family of transforming functions is bounded and has equally continuous partial derivatives of the first order Now, the purpose of the

---

[1] For a systematic account which however does not include the second proposition, see W. T. MARTIN, Bulletin of the American Math. Society, 50 (1944), 5–19.

[2] H. CARTAN, *Sur les groupes de transformation analytiques*, Actualités Scientifiques et Industrielles 198, Paris, 1935.

[3] See C. CARATHEODORY, *Ueber die Abbildungen die durch Systeme von analytischen Funktionen von mehreren Veräuderlichen erzengt werden*, Math. Annalen, 34 (1932), p. 782.

[4] See the quotation under footnote 2; the proposition of Cartan, if generalized to uniformly differentiable transformations includes a theorem arrived at independently in S. B. MYERS and N. E. STEENROD, *The group of isometries of a Riemannian manifold*, Annals of Math., 40 (1939), 400–419. The theorem states that a (compact) group of isometries on a compact space with Riemannian metric is a Lie group. Specializing the transformations to be isometries is a restriction not as severe as one might first assume. In fact there is some metric on any locally Euclidean space. Given a compact group of transformation a new distance function can be derived for which the transformations are invariant. Now, if the latter distance function satisfies some differentiability conditions, then it corresponds locally to a Riemannian metric. See S. BOCHNER, *Differentiable and Riemann Metric*, Duke Journal, 4 (1938), 51–54.

present paper is to show that similar generalizations are available for the other propositions and to give a connected account of such generalizations.[5]

Compact families of functions.

We consider functions in a domain U of the Euclidean E_n. A family of functions is *compactifiable* C^k, if there exist a covering $\{U_0\}$ of U, such that on each neighborhood U_0, the functions and their derivatives of order $\leqq k$ are bounded and equally uniformly continuous, the derivatives existing. A sequence of functions converges *uniformly* C^k, if in each U_0 the sequence of the functions and of their derivatives (of order $\leqq k$) converges uniformly. If a family is compactifiable C^k, then each sequence contains a convergent sub-sequence, and each convergent sequence converges uniformly C^k. Also, adding the limits leaves the family compactifiable, and a family thus closed will be called *compact* C^k. We further note that the convex closure consisting of all linear combinations

$$\mu_1 f_1 + \cdots + \mu_p f_p$$

with $\mu_1 > 0, \cdots, \mu_p > 0$, $\mu_1 + \cdots + \mu_p = 1$, and their limits is again compact C^k.

In our context, compact families will occur in the following way. Let U be a Euclidean domain as before, let G be an arbitrary compact space (it will always be a compact group) and let A be a subset of G. For α in A and $x = (x_1, \cdots, x_n)$ in U we consider a function

$$f(\alpha; x_1, \cdots, x_n) \tag{1}$$

in $A \times U$. We will say that such a function is *uniformly* C^k if it has partial derivatives of order $\leqq k$ in the variables $x_1, \cdots, x_n$ and if the function and these derivatives are uniformly continuous in $A \times U_0$, for each U_0 from a covering $\{U_0\}$ of U. On the other hand, we will view (1) as a family of functions

$$f(\alpha; x) \tag{2}$$

in x, with α being a parameter to indicate membership in the family. If (1) is uniformy C^k, then the family (2) is compactifiable C^k. Conversely, if the family (2) is compactifiable C^k, and if (1) is continuous in α for each x, then (1) is uniformly C^k.

We will also deal with (real) analytic functions, but in a somewhat incidental manner. A family will be called *compactifiable analytic*, if on every neighborhood U_0 each function is given by a power series, these series being majorized by one series convergent in U_0. The (convex) closure will automatically have the

[5] In a forthcoming paper (*Topological groups of differentiable transformations*, Annals of Math.) Deane Montgomery has proved the remarkable result, that in a locally compact group of transformations each of which has continuous partial derivatives, those derivatives are automatically equally uniformly continuous. This removes as being redundant the most unnatural part of the assumptions which the present paper stipulates.

same property and will be termed *compact analytic*. The function (1) will be called *uniformly analytic* if the family (2) is compactifiable and if moreover (1) is continuous in α. The latter condition can also be replaced by the condition that in the power series expansion on U_0 the coefficients are continuous in α.

Compact groups of local transformations

We now take as U the sphere

$$U: x_1^2 + \cdots + x_n^2 < a^2 \tag{3}$$

and as A we take an entire compact group G. With each α in G we associate a system of functions

$$y_m = f_m(\alpha; x_1, \cdots, x_n), \qquad m = 1, \cdots, n, \tag{4}$$

which will be interpreted as a transformation from U to E_n. We will also write

$$y = f(\alpha; x) \tag{5}$$

and denote the transformation abstractly by $T(\alpha)$.

We will assume that the family $\{T(\alpha)\}$, that is each of the n functions (4), is uniformly C^k, $k \geqq 1$, and that the origin is invariant, $f(\alpha; 0) = 0$. This implies the existence of a sub-neighborhood

$$U_0: x_1^2 + \cdots + x_n^2 < a_0^2, \qquad a_0 < a_i,$$

such that for α in G we have

$$f(\alpha; U_0) \subset U. \tag{6}$$

Therefore, for any pair α, β in G we can define the iterated transformation

$$T(\alpha)T(\beta): y = f(\alpha; f(\beta; x)) \tag{7}$$

for x in U_0. We will say that our family (5) is a group of local transformations, if for x in U_0, (or some smaller concentric sphere) the iterated transformation (7) is identical with the transformation $T(\alpha\beta)$ where $\alpha\beta$ is the group product in G.

THEOREM 1. *If (4) is a compact group of local transformations with fix point, and if it is uniformly C^k, $k \geqq 1$, (or uniformly analytic), then in some neighborhood of the fix point there exists a change of coordinates of class C^k (or analytic) and of the special form*

$$y_m = x_m + o(|x|), \qquad m = 1, \cdots, n \tag{8}$$

where $|x| = (x_1^2 + \cdots + x_n^2)^{1/2}$, such that in the new coordinates all transformations are linear.

PROOF. The theorem states that if we write our transformations in the form

$$T(\alpha): y_m = \sum_{p=1}^{n} a_{mp}(\alpha)x_p + o(|x|) \tag{9}$$

then a change of coordinates will carry them into the truncated linear forms

(10) $$L(\alpha):y_m = \sum_{p=1}^{n} a_{mp}(\alpha)x_p,$$

about which, to start with, we can only say that they have the group property $L(\alpha)L(\beta) = L(\alpha\beta)$. For each α we introduce the transformation $L(\alpha^{-1})T(\alpha)$. It is given by functions $y = h(\alpha; x)$, where

(11) $$h_m(\alpha; x) = \sum_{p=1}^{n} a_{mp}(\alpha^{-1})f_p(\alpha; x),$$

and it is easily seen that

$$h_m(\alpha; x) = x_m + o(\mid x \mid).$$

The family (11) is compactifiable C^k (or compactifiable analytic). Finally we form the transformation

(12) $$R = \int_G L(\alpha^{-1})T(\alpha)\, d\alpha,$$

where "$d\alpha$" refers to the Haar measure on G, with total measure 1. The transformation R is defined in U_0 and if we write $y = g(x)$, then

$$g_m(x_1, \cdots, x_n) = \int_G h_m(\alpha; x_1 \cdots x_n)\, d\alpha.$$

It belongs to the convex closure of the family $h(\alpha; x)$ and it therefore is C^k (or analytic). Moreover it has the form $g_m = x_m + o(\mid x \mid)$ and thus there exists, an inverse R^{-1} in some neighborhood of the origin. If γ is any fixed element of G, we have

$$L(\gamma)R = L(\gamma) \int L(\alpha^{-1})T(\alpha)d\alpha = \int L(\gamma\alpha^{-1})T(\alpha)d\alpha.$$

If we put $\gamma\alpha^{-1} = \beta^{-1}$, that is $\alpha = \beta\gamma$ and if we recall that Haar measure is invariant under translations by γ, the last integral is

$$\int L(\beta^{-1})T(\beta\gamma)d\beta = \int L(\beta^{-1})T(\beta)d\beta\cdot T(\gamma) = RT(\gamma)$$

Thus $L(\gamma)R = RT(\gamma)$, or $L(\gamma) = RT(\gamma)R^{-1}$, and this completes the proof of our theorem.

Inverse Transformations

We will next derive a refinement of the classical theorem on implicit functions, the latter theorem being presupposed.

THEOREM 2. *If a family of transformations*

(13) $$T:y_m = f_m(x_1, \cdots, x_n), \qquad m = 1, \cdots, n$$

is defined and compact C^1 *in a neighborhood* (3), *if the Jacobian of the transformations is bounded away from zero at the origin, say*

$$\frac{\partial(y_1, \cdots, y_n)}{\partial(x_1, \cdots, x_n)} \geqq \lambda > 0 \tag{14}$$

for $x = 0$, *and if finally*

$$f(0) = 0, \tag{15}$$

then there exist concentric neighborhoods $U_0 \subset U' \subset U$ *such that each* T *is one-one on* U' *and that the image of* U' *covers* U_0: $f(U') \supset U_0$.

Proof. Owing to the compactness, relation (14) will hold not only at the origin but in all of U', for some U'. Now, let

$$(x_1, \cdots, x_n), \qquad (x_1 + h_1, \cdots, x_n + h_n),$$

be two points in U'. We have

$$f_m(x + h) - f_m(x) = \sum_{p=1}^{n} \frac{\partial f_m}{\partial x_p} h_p + o(|h|),$$

and therefore

$$\sum_{m=1}^{n} |f_m(x + h) - f_m(x)|^2 = \sum_{m=1}^{n} \left(\sum_{p=1}^{n} \frac{\partial f_m}{\partial x_p} h_p\right)^2 + o(|h|^2), \tag{16}$$

where the "little-o" estimates are meant to be understood uniformly in x. Furthermore, due to the compactness, in conjunction with (14), the quadratic form on the right side of (16) is $\geqq 2\rho^2h^2$ in some U', and thus the left side is $\geqq \rho^2h^2$ for some U'. Therefore T is one-to-one in U'. Furthermore if we put $x = 0$ in (16) and remember (15), we obtain

$$\sum_{m=1}^{n} |f_m(h)|^2 \geqq \rho^2 h^2$$

and thus $f(U')$ covers a fixed neighborhood U_0 as claimed.

Thus far we have assumed $f(0) = 0$. But we will also need a theorem in case $f(0)$ is only sufficiently near the origin. The following theorem can be easily derived from the previous one.

Theorem 3. *If the family* (13) *is compact* C^1 *in* (3), *and if the family contains the identity*

$$y_m = x_m, \qquad m = 1, \cdots, n,$$

then there exists an $\epsilon > 0$ *and a concentric sub-neighborhood* U_0 *such that any transformation* $y = g(x)$ *of the family for which*

$$\sum_{m=1}^{n} |g_m(x_1 \cdots x_n) - x_m|^2 \leqq \epsilon^2 \tag{17}$$

in U, *has an inverse in* U_0.

Compact groups of transformations

If S is an n-dimensional coordinate space of class C^k (or analytic); if G is compact; and if for α in G we consider a family of transformations

$$P = f(\alpha; P_0)$$

from S into (part of) itself; then we say that the family is uniformly C^k (analytic) if corresponding to any α and P_0 and to $P = f(\alpha; P_0)$ there exist a neighborhood $A = A(\alpha)$ in G and coordinate neighborhoods $U = U(P)$ and $U_0 = U_0(P_0)$ in S such that

$$f(\alpha; U_0) \subset U, \qquad \alpha \text{ in } A, \tag{18}$$

and that the corresponding functions $y = f(\alpha; x)$ from U_0 to U are uniformly C^k (analytic) according to the previous definition. Owing to the compactness of G, if P_0 is a fix point of the family, then the neighborhood A can be chosen as the entire space G.

Now, if S_0 is a subspace of S and if G is a compact group and if a family of transformations are defined in S, we will say that they are a group in S_0, if for points of S_0 the group property $T(\alpha)T(\beta) = T(\alpha\beta)$ holds.

THEOREM 4. *If the space S is C^k (or analytic) if the transformations $\{T(\alpha)\}$ are uniformly C^k (analytic) in S and a compact group G on a subspace S_0, and if a point P_0 of S_0 is a fix point, then in an allowable coordinate neighborhood of P_0 all transformations are linear.*

Also, if the set of fix points in S_0 contains an open set then every point of S_0 is a fix point.

PROOF. Around any point P_0 there exist neighborhoods U_0, U, such that (18) holds for all α. By Theorem 1, all transformations are linear in some suitable coordinate neighborhood U'. Furthermore, if all transformations are linear in U', and hence, analytic in U', and if the set of fixpoints in U' contains an inner point, all points of U' are fix points. Thus, the interior of the set of fix points in S_0, if not empty, must be all of S_0.

THEOREM 5. *If S is a coordinate space C^1, if the family of transformations $\{T(\alpha)\}$ is uniformly C^1 in S and forms a compact group G in a subspace S_0, then there exists a neighborhood $A_0 = A_0(1)$ of unity in G having the following property. If G_0 is a subgroup of G entirely contained in A, then for α in G_0 and P in S_0 we have $f(\alpha; P) = P$.*

In other words, a compact group of transformations C^1 has no small subgroups save identity.

PROOF. Let P_0 be a point in S_0. Since $f(1; P_0) = P_0$ there exists a neighborhood $A = A(1)$ in G, and a coordinate neighborhood U of P_0 in S_0, such that the family of transformations

$$y_m = f_m(\alpha; x); \qquad m = 1, \cdots, n \tag{19}$$
$$x \in U; \qquad \alpha \in A$$

is compactifiable C^1. The family contains in particular the identity

$$y_m = x_m, \qquad m = 1, \cdots, n.$$

We form the convex closure of (19), and then we chose an $\epsilon > 0$ and an U_0, as given by Theorem 3. Given $\epsilon > 0$ we select a subneighborhood A_0 of A such that for α in A_0, we have

$$\sum_{m=1}^{n} | f_m(\alpha; x) - x_m |^2 \leqq \epsilon^2; \quad x \in U.$$

Now, let G_0 be a subgroup of G which is contained in A_0. With the Haar measure on G_0 we form the transformation

$$R = \int_{G_0} T(\alpha)\, d\alpha,$$

which, if written, $y = g(x)$, has the form

$$g_m(x_1, \cdots, x_n) = \int_{G_0} f_m(\alpha; x_1, \cdots, x_n)\, d\alpha.$$

By Theorem 3, there exists an inverse R^{-1} in U_0. On the other hand, for α in G_0 we have

$$RT(\gamma) = \int T(\alpha\gamma)\, d\alpha = \int T(\beta)\, d\beta = R,$$

that is, $T(\gamma) = R^{-1}R = identity$, in some neighborhood of P_0.

For fixed γ, we form with $T = T(\gamma)$ the cyclic group $\{T^r\}$, $r = 0, \pm 1, \pm 2, \cdots$, and its closure. This leads to a compact group having the invariant points of T as fix points. By Theorem 4, the set of such fix points embraces all of S_0 and this proves our theorem.

A characterization of the identity

The remaining theorems will be stated for a Euclidean domain, although with some effort they could be extended to a coordinate space in general.

THEOREM 6. *If T is a transformation of a domain D into itself, if the family of iterates*

$$\{T, T^2, T^3, \cdots\} \tag{20}$$

is compactifiable C^1 in a sub-domain D_0, and if there is an invariant point P_0 in D_0 in the neighborhood of which T has the form

$$y_m = x_m + o(| x |), \qquad m = 1, \cdots, n \tag{21}$$

then T is the identity in D_0.

PROOF. We are aiming at proving first that T is the identity in some neighborhood of P_0. By Theorem 4, we may introduce the inverses

$$T^{-r} = (T^r)^{-1}$$

in a joint neighborhood of P_0, and we claim that in some smaller neighborhood we have

$$T^r T^s = T^{r+s}$$

for $-\infty < r, s < \infty$. In fact, if $p > 0, q > 0$, then $T^{-p}T^{-q} = T^{-(p+q)}$ follows from a general property of inverses; for $p > q > 0$ we have

$$T^p T^{-q} = T^{(p-q)+q}T^{-q} = T^{p-q}(T^q T^{-q}) = T^{p-q},$$

and for $q > p > 0$ we have

$$T^p T^{-q} = T^p(T^{-p}T^{-(q-p)}) = T^{p-q};$$

and similarly for $T^{-q}T^q$.

If we put $T(r) = T^r$, then the family

$$\{T(r)\}, \qquad -\infty < r < \infty \tag{22}$$

has the following local property. It is a compactifiable family of transformations, and it has the group property $T(r)T(s) = T(r + s)$. In order to apply Theorem 1 we would have to imbed the family (22) in a family $T(\alpha)$, where α is the element of a compact group G in which the group of integers is dense. The linear parts $L(\alpha)$ being the identity, the transformation T would be the identity in a neighborhood of P_0. In particular the change of coordinates R which occurs in the proof to Theorem 1 would at present be

$$R = \int_G T(\alpha)\, d\alpha.$$

We will not proceed in this way. Without enlarging the group of integers we will show that $T(r)$ is an almost periodic function of r, and the proof to Theorem 1 will remain valid if we put

$$R = \lim_{N\to\infty} \frac{1}{2N} \sum_{n=-N}^{N} T(n).$$

The function $T(r)$ is almost periodic if every sequence of integers $\{s_k\}$, $k = 1, 2, \cdots$, contains a subsequence $\{s_k'\}$, such that the sequence $T(r + s_k')$ converges as $k \to \infty$, uniformly in r. Since $T(r + s) = T(r)T(s)$ and since the family $\{T(r)\}$ is bounded, this will hold whenever $T(s_k')$ converges. The family (22) being compactifiable, there always exists a subsequence $\{s_k'\}$ for which the latter sequence is convergent, and therefore $T(r)$ is almost periodic. Thus T is the identity in a neighborhood of P_0. Let D' be the largest open subset of D_0 on which T is identity. In D' we have

$$\frac{\partial f_m}{\partial x_p} = \delta_{mp},$$

therefore T has the form (21) in every boundary point of D', and thus by the first part of the theorem no such boundary point can exist in D_0.

A characterization of automorphisms

The following theorem will involve a rather strict version of "compactness."

THEOREM 7. *If T is a transformation of D into itself; if the family* (20) *is compactifiable C^1 in D; if to each compact subset Δ_o of D there exists another compact subset Δ of D such that for all positive r we have*

$$T^r(\Delta_0) \subset \Delta; \tag{23}$$

and if there exists a fix point P_0 at which the Jacobian of T has absolute value 1; *then T is an automorphism.*

PROOF. On the basis of assumption (23), if (each component function of) a sequence of iterates

$$T^{r_1}, T^{r_2}, T^{r_3}, \cdots, \qquad r_1 < r_2 < r_3 < \cdots \tag{24}$$

is convergent, then the limit is again a transformation of D into D, and if we denote the limit by R, then all iterates R^s likewise belong to the compact closure of (20). Furthermore, if we write

$$T^r : y_m = \sum_{p=1}^{n} a_{mp}^{(r)} x_p + o(|x|),$$

then

$$\lim_{k\to\infty} a_{mp}^{(r_k)}$$

exists. On the other hand, by explicit assumption, the determinant

$$| a_{mp}^{(r)} | \qquad m, p = 1, \cdots, n$$

has value ± 1. If therefore we introduce the new sequence

$$T^{s_1}, T^{s_2}, T^{s_3}, \cdots \tag{25}$$

where $s_1 = r_1$, $s_{k+1} = r_{k+1} - r_k$, then the linear terms converge to identity. On the other hand, by what we have just stated, a subsequence of (25) converges to a transformation S of D into D. The iterates of S are a compactifiable family, and S has the form (21). Thus, by Theorem 6, S is the identity E. Altogether we obtain the following lemma.

LEMMA TO THEOREM 7. *A certain subsequence T^{m_k}, $k = 1, 2, \cdots$, converges to identity.*

We next prove that T is a one-one map of D into part of itself. Consider any two points P, Q which T maps into the same point,

$$T(P) = T(Q). \tag{26}$$

This implies $T^{m_k}(P) = T^{m_k}(Q)$ but by the lemma we have

$$T^{m_k}(P) \to P, \qquad T^{m_k}(Q) \to Q.$$

Thus (26) implies $P = Q$ and thus T is one-one.

Finally we will apply Theorem 3 to the closure of the family $\{T^{m_k}\}$. If P_0 is a point of D, then by Theorem 3, there exists an open sphere which contains P_0 and which in turn is contained in all $T^{m_k}(D)$ for $k \geqq k_0$. Since $T(D) \supset T^{m_k}(D)$, there exists an open sphere which contains P_0 and which in its turn is contained in $T(D)$. Therefore $T(D)$ includes every point of D, and thus T is an automorphism.

PRINCETON UNIVERSITY

ANNALS OF MATHEMATICS
Vol. 46, No. 4, October, 1945

GROUPS OF DIFFERENTIABLE AND REAL OR COMPLEX ANALYTIC TRANSFORMATIONS

BY SALOMON BOCHNER AND DEANE MONTGOMERY

(Received May 16, 1945)

INTRODUCTION

We will prove two loosely connected theorems.

(i) If a Lie group acts on a manifold, if the transforming functions $f(g; x)$ are continuous in the space coordinates x and the group coordinates g simultaneously, and if the functions $f(g; x)$ have derivatives in the space coordinates x, then they also have mixed derivatives in all group and space coordinates occurring. An analogous result is true in the analytic case.[1]

(ii) If a Lie group is complex analytic and if the group space is compact, or more generally, if the adjoint group is compact; then the group is Abelian.[2] This explains why torus spaces play such a preponderant role in the theory of functions of several complex variables.

From (i) and a known theorem[3] we obtain the following:

(iii) If a compact group G acts effectively on a connected manifold M which is analytic (of class C^k, $k \geqq 1$) and if for each g, $f(g; x)$ is analytic (C^k) in x, then G is a Lie group and $f(g; x)$ is analytic (C^k) in (g, x) simultaneously.

I. ANALYTIC TRANSFORMATIONS

If G is a (local) topological group and M is an n-dimensional coordinate space, then G is said to act on M if for each g in G there exists a homeomorphism of M, (which will be denoted by $g(x)$ or $f(g; x)$), if the group property

$$g_1[g_2(x)] = (g_1g_2)x$$

is fulfilled, and if $f(g; x)$ is continuous in (g, x) on $G \times M$. If P_0 is any point of M and g_0 any point of G and if $Q_0 = f(g_0; P_0)$, then there exists a coordinate neighborhood (x) of P_0, and a neighborhood (y) of Q_0, and a neighborhood U of g_0, such that for g in U, $y = g(x)$ is given by functions

$$y_i = f_i(g; x), \qquad i = 1, \cdots, n. \tag{1}$$

---

[1] For results related to (i) in the complex analytic case see H. CARTAN, *Sur les groupes de transformations analytiques*, Actualites Scien. et Industrielles, 198, Paris 1935. For results related to (iii) see MYERS AND STEENROD, *The group of isometries of a Riemannian manifold*, Annals of Math., vol. 40 (1939), pp. 400–416.

[2] This theorem is known and in particular it is familiar to C. Chevalley. However, its proof is included here since it was suggested during the study of the other questions treated here.

[3] MONTGOMERY, *Topological groups of differentiable transformations*, Annals of Math., vol. 46 (1945) pp. 382–387. Theorem 1 below was proved in this paper for locally compact groups but the proof as given is valid for locally compact group germs.

Whenever g is the unit element e of G we may and will assume that (x) and (y) are the same coordinate system.

We will require the following theorem that has been established before.[3]

THEOREM 1. *If G is locally compact, and M belongs to the differentiability class C^k, $k \geqq 1$, and if the functions* (1) *belong to class C^k for each g, then their partial derivatives of order $\leqq k$ in x are continuous in $(g; x)$.*

We will also require an analogue for analytic functions.

THEOREM 2. *If G is locally compact and M is (real) analytic, and if the functions* (1) *are analytic in x, then their power series expansions are dominated in sufficiently small neighborhoods by convergent power series which are independent of g.*

PROOF. Let U be a small neighborhood of the unit e in G, and (x) a fixed coordinate system in M. For each g in U we consider the expansions

$$f_i(g; x) = \sum a^i_{\lambda_1 \cdots \lambda_n}(g) x_1^{\lambda_1} \cdots x_n^{\lambda_n}, \tag{2}$$

and for each $k = 1, 2, \cdots$ we denote by E_k the subset of U for which

$$\sup_{i,\lambda_1 \cdots \lambda_n} \left| a^i_{\lambda_1 \cdots \lambda_n}(g) \frac{1}{k^{\lambda_1 + \cdots + \lambda_n}} \right| \leqq k. \tag{3}$$

Since (2) has some domain of absolute convergence for each g, U is the union of all E_k. Also the coefficients $a_\lambda(g)$ are values at $x = 0$ of partial derivatives of $f(g; x)$ with respect to x, and thus by Theorem 1 they are continuous. Therefore each E_k is closed in U and by a category argument, at least one E_k contains an inner point, call it g_0. Since, for fixed k, the inequalities (3) imply dominated convergence of the series (2) in some x-neighborhood, our theorem has been proved for some sub-neighborhood $U(g_0)$ of U.

Now, the group relations

$$f_1(g_0^{-1} g; x) = f_i(g_0^{-1}; f(g; x))$$

will permit the "transfer" of these majorants from the neighborhood of g_0 to some neighborhood of e, or indeed to the neighborhood of any group element, and the proof of the theorem is completed.

We next prove another type of theorem.

LEMMA 1. *If in a unitary transformation*

$$y_p = \sum_{q=1}^{n} g_{pq}(t) x q, \qquad p = 1, \cdots, n \tag{4}$$

the coefficients $a_{pq}(t)$ are analytic functions of complex variables, then they are constant in those variables.

More generally, if in a (real) orthogonal transformation (4) *the coefficients are harmonic functions of real variables, that is, solutions of the Laplacian*

$$\frac{\partial^2}{\partial t_1^2} + \cdots + \frac{\partial^2}{\partial t_r^2} = 0$$

or more generally, if they are sub-harmonic in those variables, then they are constants.

PROOF. The first part of the lemma is really contained in the second. In fact, separation into real and imaginary, will carry the unitary transformation into an orthogonal transformation (of $2n$ dimensions), and the real and imaginary parts of functions in complex variables are in particular solutions of the Laplacian in all real components of the parameters.

If $a_{pq}(t)$ is harmonic in $t_1, \cdots, t_r$ its value at the center of an r-dimensional Euclidean sphere is equal to its integrated mean over the $(r - 1)$-dimensional boundary of the sphere, in symbols,

$$a_{pq}(t) = \frac{1}{\omega} \int_S a_{pq}(t + \xi)\, d\omega_\xi . \tag{5}$$

For a sub-harmonic function we have

$$a_{pq}(t) \leqq \frac{1}{\omega} \int_S a_{pq}(t + \xi)\, d\omega_\xi . \tag{6}$$

By Schwarz inequality we hence obtain

$$a_{pq}(t)^2 \leqq \frac{1}{\omega} \int (a_{pq}(t + \xi))^2\, d\omega_\xi , \tag{7}$$

and, on summing with respect to q we will have

$$\sum_{q=1}^{n} a_{pq}(t)^2 \leqq \frac{1}{\omega} \int \sum_{q=1}^{n} (a_{pq}(t + \xi))^2\, d\omega_\xi . \tag{8}$$

However, orthogonality implies

$$\sum_{q=1}^{n} a_{pq}(t)^2 = 1$$

identically in t. Thus (8) must be a strict equality and so must (7). However, by the precise statement of Schwarz' inequality, the equality (7) will be consistent with (6) only if $a_{pq}(t + \xi)$ is constant in ξ, as asserted.

Since any compact linear group can be carried by a linear space transformation into an unitary or orthogonal group respectively, we therefore obtain the following theorem.

THEOREM 3. *If a Lie group G is compact or more generally if the adjoint group of a local Lie group is compact, and if the group is analytic in complex parameters, or more generally, if the coefficients in the adjoint group are harmonic functions for some choice of real group parameters; then the group G is Abelian.*

II. Differentiable Transformations

We consider a manifold M which is (at least) of class C^1 and a (local or global) group G acting on it such that for each g in G, the functions $f_i(g; x)$ belong to class C^1. We will work extensively with the derivatives

$$f_{ij}(g; x) = \partial/\partial x_j\, f_i(g; x) \tag{9}$$

and we recall the decisive fact that by Theorem 1 they are continuous in $(g; x)$.

LEMMA 2. *If G is the (local) one-parameter group of real numbers, then in each coordinate system (x) the derivative*

$$\partial/\partial s f_i(0; x) \tag{10}$$

exists and for small h we have

$$f_i(h; x) - f_i(0; x) = \textstyle\sum_j \left[\int_0^h f_{ij}(t; x)\, dt\right][\partial/\partial s\, f_j(0; x)]. \tag{11}$$

PROOF. We introduce the functions

$$T_i(h; x) = \int_0^h f_i(t; x)\, dt,$$

and apply the mean value formula

$$\begin{aligned} T_i(h; y) - T_i(h; x) &= \textstyle\sum_j \partial/\partial x_j\, T_i(h; \bar{x}(i))\cdot(y_j - x_j) \\ \bar{x}(i) &= x + \theta_i(y - x), \qquad 0 < \theta < 1. \end{aligned} \tag{12}$$

By Theorem 1 we have

$$\frac{\partial}{\partial x_j} T_i(h; x) = \int_0^h f_{ij}(t; x)\, dt \tag{13}$$

and if we put in particular $y = p(x) = f(p; x)$ where p is a small number we have the identity

$$\begin{aligned} T_i(h; p(x)) - T_i(h; x) &= \int_0^h f_i(t; p(x))\, dt - \int_0^h f_i(t; x)\, dt \\ &= \int_p^{p+h} f_i(t; x)\, dt - \int_0^h f_i(t; x)\, dt \\ &= \int_h^{h+p} f_i(t; x)\, dt - \int_0^p f_i(t; x)\, dt. \end{aligned} \tag{14}$$

On substituting (13) and (14) in (12) and dividing by p we obtain

$$\frac{1}{p}\int_0^p [f_i(t+h; x) - f_i(t; x)]\, dt = \textstyle\sum_j \left[\int_0^h f_{ij}(t; x)\, dt\right]\left[\frac{f_j(p; x) - x_j}{p}\right]. \tag{15}$$

Since $f_{ij}(t; x)$ is continuous in (t, x) and has the value δ_{ij} for $t = 0$, the matrix

$$\left\| \frac{1}{h}\int_0^h f_{ij}(t; \bar{x}(i))\, dt \right\| \qquad i, j = 1, \cdots, n$$

for h sufficiently small is near the identity matrix

$$\| \delta_{ij} \| \qquad i, j = 1, \cdots, n,$$

uniformly in x and p. Thus we can solve the system of relations (15) for the quantities

$$\frac{f_j(p; x) - x_j}{p} \tag{16}$$

and if the matrix

$$\| \varphi_{ij}(h; x) \|_{i,j=1,\cdots,n}$$

is the inverse to the matrix

$$\left\| \int_C^h f_{ij}(t; x)\, dt \right\|,$$

we find for (16) the value

(17) $$\sum_i \varphi_{ij}(h; \bar{x}(i)) \frac{1}{p} \int_{-0}^{p} [f_i(t + h; x) - f_i(t; x)]\, dt.$$

Now, (17) has a limit as $p \to 0$. But the limit of (16) is (10), and (15) converges to (11). This proves the lemma.

LEMMA 3. *If G has r parameters*

(18) $$g = (a_1, \cdots, a_r),$$

if the latter parameters are canonical coordinates such that the curves

$$\gamma_1 = e_1 t, \cdots, \gamma_r = e_r t$$
$$e_1^2 + \cdots + e_r^2 = 1$$

are one-parameter subgroups, and if we introduce the functions

(19) $$f_i(t; e; x) \equiv f_i(et; x)$$

pertaining to such a subgroup, then the derivatives

(20) $$F_i(e; x) = \frac{\partial}{\partial t} f_i(0; e; x)$$

are continuous in $(e_1, \cdots e_r; x_1, \cdots, x_n)$.

PROOF. If in (11) we replace $f_i(t; x)$ by (19) and invert, then $F_i(e; x)$ has the value

(21) $$\sum_i \varphi_{ij}(h; e; x)[f_i(h; e; x) - f_i(0; e; x)],$$

where the matrix $\| \varphi_{ij} \|$ is again inverse to

(22) $$\left\| \int_0^h f_{ij}(h; e; x)\, dt \right\|.$$

However, for a fixed small value of h, (21) is continuous in $(e; x)$ and so is (20).

LEMMA 4. *Moreover for such parameters* (18) *the partial derivatives*

(23) $$\frac{\partial}{\partial a_\rho} f_i(a_1, \cdots, a_r; x_1, \cdots, x_n), \qquad \rho = 1, \cdots, r,$$

exist and are continuous in (a, x).

PROOF. We will treat the index $\delta = 1$. We consider the group element (18), a nearby group element

$$g_h = (c_1 + h; a_2, \cdots, a_r)$$

and the group element

$$\gamma = g_h \cdot g^{-1} \equiv \gamma(a; h).$$

If we introduced the components

$$\gamma = (\gamma_1, \cdots, \gamma_r)$$

then they have the form

$$\gamma_1 = h + 0_2(h; a)$$

$$\gamma_\sigma = 0_2(h; a), \qquad \sigma = 2, \cdots, r,$$

where $0_2(h; a)$ is a power series in the variables $(h; a_1, \cdots, a_r)$; the latter series are divisible by h and contain only terms of degree two or more. It is not hard to see that we can write

$$\gamma_\rho = e_\rho(h; a) \cdot t(h; a), \ \rho = 1, \cdots, r,$$

with

$$\sum e_\rho^2 = 1;$$

where $t(h; a)$ and $e_\rho\ (h; a)$ are power series in $(h; a)$, and $e_1(h; a)$ is normalized to have the form

$$+1 + 0_1(h; a)$$

with $0_1(h; a)$ being a power series without a constant term.

Since $f(g_h; x) = f(\gamma; g(x)) = f(\gamma; f(g; x))$, the difference quotient

$$\frac{f(a_h; x) - f(a; x)}{h} \tag{24}$$

has the value

$$\frac{f(\gamma; a(x)) - f(0; a(x))}{h},$$

and by (17) this is

$$\sum_j \left[\frac{1}{h}\int_0^{t(h;a)} f_{ij}(\tau; e(h; a); a(x)\, d\tau\right] \cdot F_j(e(h; a); a(x)), \tag{25}$$

where F_j is defined by (20). Now

$$f_{ij}(\tau; e(h; a); a(x)) = f_{ij}(\tau e(h; a); f(a; x))$$

is continuous in $(\tau; h; a; x)$ and $F_j(e(h; a); a(x))$ is continuous in $(h; a; x)$ by Lemma 3. Therefore we may let $h \to 0$ in (25) and the limit is

$$\frac{\partial}{\partial h} t(0; a) \cdot \sum_j f_{ij}(0; e(0; a); a(x)) F_j(e(0; a); a(x)).$$

This however is continuous in $(a; x)$.

LEMMA 5. *Let G be a (local) Lie group with parameters chosen as in the preceding lemmas and let G act on an analytic (class C^k) manifold M in such a way that for each x the functions $f_i(g; x)$ are analytic (class C^k) in x for each fixed g. Then the functions $\partial/a_\rho f_i(0; x)$ are analytic (class C^{k-1}) functions of x.*

PROOF. For convenience we consider the case where the index ρ is one, and we let $a_\rho = 0$, for $\rho > 1$. We are therefore dealing with the case of a one parameter subgroup and rewriting (11) we have

$$f_i(h, 0, \cdots, 0; x) - f_i(0; x) = \sum_j \left[\int_0^h f_{ij}(t; 0; \cdots, 0; x)\, dt\right][(\partial/\partial a_1) f_j(0; x) .$$

Consider first the analytic case. The functions $f_i(h, 0, \cdots, 0; x)$ are analytic in x for a fixed h, and $f_i(0; x)$ is analytic. Furthermore

$$\int_0^h f_{ij}(t; 0, \cdots, 0; x)\, dt$$

is analytic in x because by Theorem 2 we know that the power series entering under the integral sign are majorized. As before these equations may be solved for $\partial/\partial a_1 f_j(0; x)$ and hence the conclusion is proved.

In the case of class C^k the conclusion follows because the function $f_i(h; x)$, $f_i(0, x)$ and

$$\int_0^h f_{ij}(t; 0, \cdots, 0; x)\, dt$$

are all of class C^{k-1} in x when k is fixed.

THEOREM 4. *If a (local) Lie group G acts on a manifold M of class C^k, $k \geqq 1$, and if for each g the functions $f_i(g; x)$ are of class C^k in x, then they are of class C^k in $(g; x)$ provided the parameters (18) are analytic, say.*

If M and f_i are real analytic in x, then f_i is analytic in $(g; x)$.

If M and f_i are complex analytic in x, and G has complex analytic parameters (g), then f_i is complex analytic in $(g; x)$.

PROOF. It is sufficient to establish the conclusion for a neighborhood of the identity; other neighborhoods of G can then be reached by translations. Also the group parameters will be chosen in a canonical form that will satisfy the preceding lemmas.

For $k = 1$, our theorem is nothing more than Lemma 4. Also we are in a position to derive Lie's differential equations by the following usual procedure. Consider the group elements a; b, c subject to the relation $ba = c$. Let $x' = f(a; x)$. Then we have the relation

$$f_i(b; x') = f_i(c; x)$$

in which we shall consider x, a, and c as independent variables. Differentiating the relation with respect to a_ρ we obtain

$$\sum_\sigma \frac{\partial f_i(b; x')}{\partial b_\sigma} \frac{\partial b_\sigma}{\partial a_\rho} + \sum_j \frac{\partial f_i(b; x')}{\partial x'_j} \frac{\partial x'_j}{\partial a_\rho} = 0 \tag{26}$$

the rules of differentiation being justified by Lemma 4. Since $\partial f_i/\partial x_j'$ is continuous in (g, x) and since for $g = e$

$$\frac{\partial f_i}{\partial x_j'} = \frac{\partial f_i(e; x')}{\partial x_j'} = \delta_{ij}$$

equations (22) can be solved for $\partial x_j'/\partial a_\rho$ and we obtain

$$\frac{\partial x_i'}{\partial a_\rho} = \sum_\sigma \xi_{i\sigma}(b; x')\Omega_{\sigma\rho}(a; b).$$

Letting $b = 0$

$$\frac{\partial x_i'}{\partial a_\rho} = \sum_\sigma \xi_{i\sigma}(0; x')\Omega_{\sigma\rho}(a; 0)$$

and we contract this to write the following classical equations

$$\frac{\partial f_i(a; x)}{\partial a_\rho} = \sum_\sigma \xi_{i\sigma}(f(a; x))\Omega_{\sigma\rho}(a). \tag{27}$$

In these the functions $\Omega_{\sigma\rho}(a)$ are analytic in a; the functions $\xi_{i\sigma}(x')$ are analytic in x if for each fixed g, $f_i(g; x)$ is analytic. This follows from our preceding results. The functions $\xi_{i\sigma}(x')$ are of class C^{k-1} in x' if $f(g; x)$ was of class C^k in x.

If we make the inductive assumption that $f(a; x)$ is of class C^{k-1} in $(a; x)$ we can take $(k - 1)$ successive mixed derivatives of the right side in (27), and therefore $f(a; x)$ is of class C^k in $(a; x)$.

If the functions

$$x_i' = f_i(a; x) \tag{28}$$

are (real) analytic in x, then the functions $\xi_{i\sigma}(x')$ are analytic. Also, for analytic group parameters the coefficients $\Omega_{\sigma\rho}(a)$ are analytic in a, and thus the functions (28) satisfy the analytic differential equations

$$\frac{\partial x_i'}{\partial a_\rho} = \sum_\sigma \xi_{i\sigma}(x')\Omega_{\sigma\rho}(a). \tag{29}$$

By our previous conclusion the functions (28) are at any rate of class C^k in $(a; x)$ for every k. Also they are a complete solution of the system (29). But if the system has such a complete solution, then the system is integrable and the complete solution is analytic in $(a; x)$. This proves our theorem in the real analytic case.

We next take the case of *real* parameters a, and complex analytic space variables $z_i = x_i + iy_i$. Thus we consider a group of transformations

$$z_i' = F_i(a; x), \tag{30}$$

that is of transformations

$$\begin{aligned} x_i' &= f_i(a; x, y) \\ y_i' &= g_i(a; x, y). \end{aligned} \tag{31}$$

By the result thus far established the functions (31) are power series in $(a;$ $x, y)$ and from this it follows easily that the functions

$$z_i' \equiv F_i(u; x) \equiv f_i(a; x, y) + \sqrt{-1}\, g_i(a; x, y),$$

since they are independent of the conjugate complex conjugate quantities $\bar{z}$, are power series in $(a; z)$. Also, the functions (31) satisfy certain differential equations

$$\frac{\partial x_i'}{\partial a_\rho} = \sum_\sigma \xi_{i\sigma}(x', y')\Omega_{\sigma\rho}(a)$$

$$\frac{\partial y_i'}{\partial a_\rho} = \sum_\sigma \eta_{i\sigma}(x', y')\Omega_{\sigma\rho}(a),$$

and thus we have

(32) $$\frac{\partial z_i'}{\partial a_\rho} = \sum_\sigma \zeta_{i\sigma}\Omega_{\sigma\rho}(a)$$

where the operators

$$\zeta_{i\sigma} = \xi_{i\sigma}(x', y') + \sqrt{-1}\eta_{i\sigma}(x', y')$$

are analytic in (x', y'). For fixed (a), $\zeta_{i\sigma}$ can be expressed in terms of the function

(33) $$\frac{\partial z_i'}{\partial a_\rho}$$

which are independent of the quantities $\bar{z}_i$. On solving (30) with respect to z, we can express (33) in terms of z', and thus $\zeta_{i\sigma}$ is complex analytic in z'.

Finally we will assume that the group parameters are also complex. If we denote them by

$$c_\rho = a_\rho + \sqrt{-1}b_\rho$$

we have a system of equations

$$\frac{\partial z_i'}{\partial a_\rho} = \sum_\sigma \zeta_{i\sigma}(z')\Omega_{\sigma\rho}(a, b)$$

$$\frac{\partial z_i'}{\partial b_\rho} = \sum_\sigma \zeta_{i\sigma}(z')\Omega^*_{\sigma\rho}(a, b)$$

and if we introduce the formal operator

$$\frac{\partial}{\partial c} = \frac{1}{2}\frac{\partial}{\partial a} + \frac{1}{2\sqrt{-1}}\frac{\partial}{\partial b}$$

we hence obtain

(34) $$\frac{\partial z_i'}{\partial c_\rho} = \sum_\sigma \zeta_{i\sigma}(z')\phi_{\sigma\rho}(a, b)$$

where

$$\Phi_{\sigma\rho} = \tfrac{1}{2}\Omega_{\sigma\rho} + \frac{1}{2\sqrt{-1}}\,\Omega^{*}_{\sigma\rho}\,.$$

We now consider the parameter group of G writing it in the form

$$z'_i = \varphi_i(c; z), \qquad i = 1, \cdots, r.$$

It satisfies a system of the form (34) in which the coefficients $\phi_{\sigma\rho}$ are the same as before. However for the parameter group the system (34) can be solved with respect to the quantities $\phi_{\sigma\rho}(a, b)$, and from this it can be easily inferred that the latter quantities are power series in the variables c alone. Thus, in the equations (34) as applied to the original functions (30) the right sides are power series in $(z'; c)$ and this proves that the functions (30) are power series in $(z; c)$.

We will terminate with a theorem on a (global) group G acting effectively on M. "Effectively" means that for an element g of G the relation $g(x) = x$ can be satisfied for all x only if g is the unit element e.

THEOREM 5. *If a compact group G acts effectively on a connected manifold of class C^k, $k \geqq 1$ (or analytic) and if the functions $f(g; x)$ are of class C^k (or analytic) then G is a Lie group and in suitable group parameters, the functions $x' = f_i(g; x)$ satisfy the Lie equations* (29) *and are of class C^k (or analytic) in $(a; x)$.*

This follows immediately from Theorem 4 because it is known[3] that such a group must be a Lie group.

PRINCETON UNIVERSITY.

ANNALS OF MATHEMATICS
Vol. 47, No. 2, April, 1946

FORMAL LIE GROUPS

BY S. BOCHNER

(Received August 25, 1945)

Introduction

In a local r-parametric group-space the law of multiplication $a \cdot b = \varphi(a; b)$ is given by a system of functions

$$\varphi_\rho(a_1, \cdots, a_r; b_1, \cdots, b_r), \qquad \rho = 1, \cdots, r. \tag{1}$$

We consider an analytic coordinate system around the group origin and we list the following properties of the functions:

I. The functions (1) are power series in the $2r$ variables

$$a_1, \cdots, a_r; b_1, \cdots, b_r, \tag{2}$$

and they have no constant terms.[1]

II. The linear part of $\varphi(a; b)$ is $a + b$, that is we have

$$\varphi_\rho(a; b) = a_\rho + b_\rho + \text{(higher powers)} \tag{3}$$

for $\rho = 1, \cdots, r$.

III. More precisely, we have

$$\varphi_\rho(0; b) = b_\rho, \qquad \rho = 1, \cdots, r \tag{4}$$

and

$$\varphi_\rho(a; 0) = a_\rho, \qquad \rho = 1, \cdots, r. \tag{5}$$

IV. The associative law

$$\varphi(\varphi(a; b); c) = \varphi(a; \varphi(b; c)) \tag{6}$$

holds identically in a, b, c.

Property III is more specific than property II. It states that except for $a_\rho + b_\rho$ each term in the expansion of (3) is divisible by at least one of the factors $a_1, \cdots, a_r$ and at least one of the factors $b_1, \cdots, b_r$, simultaneously.

In the present paper we will deal with formal power series. We will assume that the quantities (2) are algebraic variables and that the functions (1) are formal power series in these variables. Their coefficients and other occurring constants need not be ordinary numbers. As a matter of fact, we are making the assumption that they shall be elements of any fixed algebraic field of characteristic 0. Whenever explicitly referred to, the field will be denoted by K. We will view a system of functions (1) as a "group" whenever it satisfies the enumerated properties, subject to the familiar manner of operating with formal power series. And the purpose of the present paper is to point out that our

[1] That is, the constant terms have value 0.

artificial groups are subject to the classical theorems of Lie, provided we introduce formal solutions for differential equations in terms of formal functions. The formal treatment will render the classical proofs quite lucid, and we will be able to state the converse to Lie's third theorem even in the case in which the given constants of structure $\{e^{\alpha}_{\beta\gamma}\}$ are elements of our general field K.

Lie's equations

We take a formal transformation

(7) $$\eta_i = f_i(\xi_1, \cdots, \xi_n), \qquad i = 1, \cdots, n,$$

of the form

$$\eta_i = \sum_j c_{ij}\xi_j + \text{(higher powers)}$$

and we note the following lemma.

LEMMA 1. *If the determinant* $|c_{ij}|$ *is* $\neq 0$, *then there exists another transformation*

$$\xi_i = g_i(\eta_1, \cdots, \eta_n),$$

again without constant terms, which is the inverse to (7). *That is,*

$$f_i(g(\eta_1 \cdots \eta_n)) = \eta_i .$$

Now, given a group (1), we replace the letter a by x, and we introduce new variables z_ρ by the relations

(8) $$z_\rho = \varphi_\rho(x; b), \qquad \rho = 1, \cdots, r$$

If we add the relations

$$b'_\rho = b_\rho, \qquad \rho = 1, \cdots, r,$$

we obtain a transformation from $(x; b)$ to $(z; b')$, and on the basis of property II we may apply Lemma 1. In this way we obtain a transformation

(9) $$x_\rho = \psi_\rho(z; b)$$

which is inverse to (8), that is

(10) $$\varphi(\psi(x; b); b) = x.$$

We now form the partial derivatives

(11) $$\frac{\partial z_\rho}{\partial b_\sigma} = \frac{\partial \varphi_\rho(x;b)}{\partial b_\sigma}$$

(by formal term-by-term differentiation), and if we substitute (9) in the right side of (11) we obtain a system of equations

(12) $$\frac{\partial z_\rho}{\partial b_\sigma} = \Phi^\rho_\sigma(z; b) \qquad \rho, \sigma = 1, \cdots, r$$

of which the functions (8) are a solution. That is, we have

$$\frac{\partial \varphi_\rho(x;b)}{\partial b_\sigma} = \Phi^\rho_\sigma(\varphi(x;\, b);\, b) \tag{13}$$

identically in the independent variables $(x;\, b)$.

We will now make full use of properties III and IV. By property III, if we put $x = 0$, we obtain

$$\varphi_\rho(0;\, b) = b_\rho\,, \qquad \frac{\partial \varphi_\rho(0;\, b)}{\partial b_\sigma} = \delta^\rho_\sigma\,.$$

Therefore, we have

$$\Phi^\rho_\sigma(b;\, b) = \delta^\rho_\sigma \tag{14}$$

identically in b, and in particular

$$\Phi^\rho_\sigma(0;\, 0) = \delta^\rho_\sigma\,. \tag{15}$$

We next write property IV in the form

$$\varphi_\rho(\varphi(x;\, a);\, b) = \varphi_\rho(x;\, \varphi(a;\, b)) \tag{16}$$

for independent variables $(x,\, a,\, b)$. We will denote the common value of (16) by z_ρ, and we put

$$y_\alpha = \varphi_\alpha(x;\, a),\; c_\alpha = \varphi_\alpha(a;\, b).$$

On differentiating (16) with respect to b_σ we obtain

$$\frac{\partial \varphi_\rho(y;\, b)}{\partial b_\sigma} = \frac{\partial \varphi_\rho(x;\, c)}{\partial c_\tau}\,\frac{\partial \varphi_\tau(a;\, b)}{\partial b_\sigma} \tag{17}$$

[summation with respect to τ!], identically in $(x,\, a,\, b)$. Since (13) is true for independent variables $(x,\, b)$ it remains valid if in both sides we replace the variables $x_1\,, \cdots, x_r\,;\, b_1\,, \cdots, b_r$ by any $2r$ functions of these variables. Now, if we replace $(x;\, b)$ successively by $(y;\, b)$, $(x;\, c)$ and $(a;\, b)$, we obtain from (17) the relations

$$\Phi^\rho_\sigma(z;\, b) = \Phi^\rho_\tau(z;\, c)\Phi^\tau_\sigma(c;\, b). \tag{18}$$

We observe that these relations hold identically in $(x,\, a,\, b)$, if we put

$$c_\alpha = \varphi_\alpha(a;\, b),\; z_\alpha = \varphi_\alpha(x;\, c).$$

However, if we replace a_ρ by the previously defined function $\psi_\rho\,(c;\, b)$, see (9), relations (18) will be identities in $(x,\, c,\, b)$ with $z_\alpha = \varphi_\alpha(x;\, c)$; and if we finally replace x_α by $\psi_\alpha(z;\, c)$, relation (18) will be an identity in the independent variables $(z,\, c,\, b)$. In particular, we may put $c = 0$, and if we introduce the functions

$$A^\rho_\tau(z) = \Phi^\rho_\tau(z;\, 0), \qquad B^\tau_\sigma(b) = \Phi^\tau_\sigma(0;\, b)$$

we obtain the following theorem.

THEOREM 1. *If the functions*

$$c_\rho = \varphi_\rho(a; b), \qquad \rho = 1, \cdots, r \tag{19}$$

form a group, then they are solutions of a system of equations

$$\frac{\partial c_\rho}{\partial b_\sigma} = A^\rho_\tau(c)B^\tau_\sigma(b); \qquad \rho, \sigma = 1, \cdots, r \tag{20}$$

identically in (a, b) *and*

$$A^\rho_\tau(0) = \delta^\rho_\tau, \qquad B^\tau_\sigma(0) = \delta^\tau_\sigma \tag{21}$$

$$A^\rho_\tau(b)B^\tau_\sigma(b) = \delta^\rho_\sigma. \tag{22}$$

Integrability conditions

Consider a system of differential equations

$$\frac{\partial z_i}{\partial a_\rho} = F^i_\rho(z; a) \qquad \begin{matrix} i = 1, \cdots, n; \\ \rho = 1, \cdots, r, \end{matrix} \tag{23}$$

in which the functions F^i are formal series in the $n + r$ variables (z, a). If $x_1, \cdots, x_m$ are any additional variables, let

$$z_i = f_i(x; a), \qquad i = 1, \cdots, n, \tag{24}$$

be a system of series without constant terms in the $m + r$ variables (x, a). We call the functions (24) a solution of (23), if they satisfy the equations (23) identically in (x, a). It should be noted that we postulated no dependencies between the three integers m, n, r, and that we have not excluded the value $m = 0$. We first note a minor lemma.

LEMMA 2. *If a system* (23) *has two solutions* $z_i = f_i(x; a)$ *and* $z_i = g_i(x; a)$ *and if* $f_i(x; 0) = g_i(x; 0)$ *then* $f_i(x; a) = g_i(x; a)$.

In other words, any existing solution is determined by its values as the "origin" $a = 0$. If we write $f_i(x; a)$ in the form

$$\textstyle\sum_k \gamma^i_{k_1 \cdots k_r}(x) a_1^{k_1} \cdots a_r^{k_r},$$

and if we substitute these expansions in both sides of (23) and compare, then for $k = k_1 + \cdots + k_r \geqq 1$, γ^i_k can be expressed in terms of some γ^i_k with smaller values of k; provided the basic field K has characteristic 0, as we have assumed once for all. This proves the lemma.

A solution (24) of (23) will be called its complete solution, provided $m = n$ and

$$f_i(x; 0) = x_i, \qquad i = 1, \cdots, n. \tag{25}$$

For the complete solution we again have the following familiar theorem.

LEMMA 3. *The system* (23) *has a complete solution if and only if the expressions*

$$A^i_{\rho\sigma} = \frac{\partial F^i_\rho}{\partial z_j} F^j_\sigma + \frac{\partial F^i_\rho}{\partial a_\sigma} - \frac{\partial F^i_\sigma}{\partial z_j} F^j_\rho - \frac{\partial F^i_\sigma}{\partial a_\rho} \tag{26}$$

vanish identically in (z, a), *for* $i = 1, \cdots, n$; $\rho, \sigma = 1, \cdots, r$.

PROOF. The necessity of the condition is quite easy to establish. In fact, if (24) is a solution, then we must have

$$A^i_{\rho\sigma}(z; a) = 0 \tag{27}$$

for $z = f(x; a)$ identically in (x, a). However, due to (25), the system (24) can be inverted into $x_i = g_i(z; a)$ and thus (27) holds identically in (z, a).

The sufficiency can be proven by the familiar method of Adolph Mayer.[2] In the prospective solution (24) we put

$$a_\rho = \lambda_\rho s, \qquad \rho = 1, \cdots, r, \tag{28}$$

where $\lambda_1, \cdots, \lambda_r$; s are at first independent variables. The resulting functions

$$\varphi_i(x; \lambda; s) = f_i(x; \lambda s) \tag{29}$$

would have to satisfy the equations

$$\frac{\partial \varphi_i}{\partial s} = F^i_\rho(\varphi; \lambda s)\lambda, \qquad i = 1, \cdots, n, \tag{30}$$

[summation with regard to ρ]. Now, for the actual proof. We start with the differential equations (30). It follows, as usual, that there exists a system of solutions

$$\varphi_i = \varphi_i(x; \lambda; s) \tag{31}$$

for which

$$\varphi_i(x; \lambda; 0) = x_i, \qquad i = 1, \cdots, n, \tag{32}$$

The next steps are to show that the functions (31) satisfy the relations

$$\lambda_\rho \frac{\partial \varphi_i}{\partial \lambda_\rho} - s \frac{\partial \varphi_i}{\partial s} = 0 \tag{33}$$

$$\frac{\partial \varphi_i}{\partial \lambda_\rho} = sF^i_\rho(\varphi; \lambda s). \tag{34}$$

In order to prove (33) we form the derivative

$$\frac{\partial}{\partial s}\left\{\lambda_\rho \frac{\partial \varphi_i}{\partial \lambda_\rho} - s \frac{\partial \varphi_i}{\partial s}\right\}.$$

By using (30), a direct calculation will show that the derivative is 0. Hence the left side of (33) is independent of s. However on the basis of (32) it is 0 for $s = 0$, and thus (33) holds identically. In order to prove (34) we set up the differences

$$w^i_\rho = \frac{\partial \varphi_i}{\partial \lambda_\rho} - sF^i_\rho \tag{35}$$

---

[2] See C. CARATHEODORY, *Variationsrechung*, p. 26.

as functions in (x, λ, s), and by a direct calculation[2] it can be shown that they satisfy the relations

$$\frac{\partial w_\rho^i}{\partial s} - \lambda_\sigma \frac{\partial F_\sigma^i}{\partial x_j} w_\rho^j = s\lambda_\sigma A_{\sigma\rho}^i .$$

Therefore, if (27) holds, we obtain for the system of functions $\{w_\rho^i\}$ the linear homogeneous differential equations

$$\frac{\partial w_\rho^i}{\partial s} = \lambda_\sigma \frac{\partial F_\sigma^i}{\partial x_j} w_\rho^j .$$

Also, for $s = 0$, we have $w_\rho^i = 0$, identically in (x, λ), and therefore the functions (35) vanish identically in (x, λ, s).

Now, if we write $\varphi_i(x; \lambda; s)$ as a series

$$\sum \gamma_{k_1 \cdots k_r, k}^i (x) \lambda_1^{k_1} \cdots \lambda_r^{k_r} s^k,$$

then on the basis of (33) it will effectively contain only those terms for which

$$k_1 + \cdots + k_r - k = 0.$$

Therefore, for each i, there exists a function $f_i(x; a)$ in independent variables $x_1, \cdots, x_n; a_1, \cdots, a_r$, such that

$$\varphi_i(x; \lambda; s) = f_i(x; \lambda s).$$

However, for such functions (24) the relations (34) imply relations (23), and (32) implies (25). This proves the lemma.

Lie's Theorems

THEOREM 2. *If a system*

(36) $$\frac{\partial c_\rho}{\partial b_\sigma} = A_\tau^\rho(c) B_\sigma^\tau(b)$$

with the properties (21) *and* (22), *has a complete solution*

(37) $$c_\rho = \varphi_\rho(a; b)$$

(38) $$\varphi_\rho(a; 0) = a_\rho ,$$

then the functions (37) *are a group.*

The system (36) *with the properties* (21), (22) *is completely integrable if and only if the functions* B_σ^τ *satisfy the Maurer relations*

(39) $$\frac{\partial B_\sigma^\tau}{\partial b_\rho} - \frac{\partial B_\rho^\tau}{\partial b_\sigma} = c_{\alpha\beta}^\tau B_\sigma^\alpha B_\rho^\beta$$

in which the quantities $c_{\alpha\beta}^\tau$ *are constants* (*in* K) *with the familiar properties*

(40) $$c_{\beta\gamma}^\alpha = c_{\gamma\beta}^\alpha$$

(41) $$c_{\alpha\beta}^\tau c_{\tau\gamma}^\delta + c_{\beta\gamma}^\tau c_{\tau\alpha}^\delta + c_{\gamma\alpha}^\tau c_{\tau\beta}^\delta = 0.$$

PROOF. We consider the functions

(42) $$z_\rho = \varphi_\rho(x; \varphi(a; b)).$$

On the one hand they satisfy the relations

$$\frac{\partial z_\rho}{\partial b_\sigma} = \frac{\partial \varphi_\rho(x; c)}{\partial c_\alpha} \frac{\partial c_\alpha}{\partial b_\sigma}$$
$$= A^\rho_\tau(z) B^\tau_c(c) B^\alpha_\gamma(c) B^\lambda_\sigma(b) = A^\rho_\tau(z) B^\tau_\sigma(b).$$

On the other hand we have

$$\varphi(x; \varphi(a; 0)) = \varphi(x; a) = \varphi(\varphi(x; a); 0).$$

By a judicious application of Lemma 2 we hence conclude that

(43) $$z_\rho = \varphi_\rho(\varphi(x; a); b),$$

and this proves the all-important property IV. We now set up $\varphi_\rho(a; b)$ as a series in b, thus writing

$$\varphi_\rho(a; b) = a_\rho + \gamma^\rho_\sigma(a) b_\sigma + \cdots,$$

see (38). If we substitute this in (36), and then put $a = 0$, $b = 0$, then on the basis of (21) we obtain $\gamma^\rho_\sigma(0) = \delta^\rho_\sigma$. Therefore we have

$$\varphi_\rho(a; b) = a_\rho + b_\rho + \text{(higher powers)},$$

and this proves property II. On the basis of this, relations (37) have inverses

(44) $$b_\rho = \chi_\rho(a; c).$$

Next, in the general relations

$$\varphi_\rho(a; \varphi(c; b)) = \varphi_\rho(\varphi(a; c); b)$$

we put $c = 0$, and due to (38) we therefore have

$$\varphi_\rho(a; \varphi(0; b)) = \varphi_\rho(a; b) = c_\rho .$$

Since the inverses (44) exist, we thus have

$$\varphi_\rho(0; b) = \chi_\rho(a; c) = b_\rho ,$$

and this proves relation (4). Since relation (5) has been stipulated explicitly, we have thus verified all properties of a group.

The proof of the second half of Theorem 2 is much the same as in the classical case, and this completes the proof of Theorem 2.

THEOREM 3. *Any set of constants $c^\alpha_{\beta\gamma}$ which satisfy relations* (40) *and* (41) *are the constants of structure of a group.*

Proof. We will imitate a classical proof in its main steps.[3] We start with the system of differential equations

$$\frac{\partial \theta_\sigma^\rho}{\partial s} = \delta_\sigma^\rho + c_{\alpha\beta}^\rho \, \theta_\sigma^\alpha \, \lambda_\beta \,, \tag{45}$$

in which the variables $\lambda_1, \cdots, \lambda_r$ are parameters. If the functions

$$\theta_\sigma^\rho = \theta_\sigma^\rho(\lambda; s), \qquad \rho, \sigma = 1, \cdots, r \tag{46}$$

are an arbitrary solution of (45), then the functions

$$\Delta_\sigma^\rho = \frac{\partial \theta_\sigma^\rho}{\partial \lambda_\gamma} \lambda_\gamma - s \frac{\partial \theta_\sigma^\rho}{\partial s} + \theta_\sigma^\rho \tag{47}$$

will satisfy the system of equations

$$\frac{\partial \Delta_\sigma^\rho}{\partial s} - c_{\alpha\beta}^\rho \, \Delta_\sigma^\alpha \, \lambda_\beta = 0. \tag{48}$$

If in particular,

$$\theta_\sigma^\rho(\lambda; 0) = 0, \tag{49}$$

then $\Delta_\sigma^\rho(\lambda; 0) = 0$, and on the basis of (48) we have

$$\Delta_\sigma^\rho(\lambda; s) = 0.$$

Thus if $\{\theta_\sigma^\rho(\lambda; s)\}$ is the solution of (45) which vanishes for $s = 0$, then we have

$$\frac{\partial \theta_\sigma^\rho}{\partial \lambda_\gamma} \lambda_\gamma - s \frac{\partial \theta_\sigma^\rho}{\partial s} + \theta_\sigma^\rho = 0. \tag{50}$$

From this it follows that there are functions (that is, formal power series)

$$B_\sigma^\rho(b_1, \cdots, b_r) \tag{51}$$

such that

$$\theta_\sigma^\rho(\lambda; s) = s B_\sigma^\rho(\lambda_1 s, \cdots, \lambda_r s). \tag{52}$$

On the other hand, if we make use of the relations (40) and (41), then for an arbitrary solution (46) of (45) we have

$$\frac{\partial}{\partial s} \left\{ \frac{\partial \theta_\sigma^\tau}{\partial \lambda_\rho} - \frac{\partial \theta_\rho^\tau}{\partial \lambda_\sigma} - c_{\alpha\beta}^\tau \, \theta_\rho^\alpha \, \theta_\rho^\beta \right\} = c_{\gamma\delta}^\tau \lambda_\delta \left(\frac{\partial \theta_\sigma^\gamma}{\partial \lambda_\rho} - \frac{\partial \theta_\rho^\gamma}{\partial \lambda_\sigma} - c_{\alpha\beta}^\gamma \, \theta_\sigma^\alpha \, \theta_\rho^\beta \right).$$

If we again add the assumption (49), we hence derive the relations

$$\frac{\partial \theta_\sigma^\tau}{\partial \lambda_\rho} - \frac{\partial \theta_\rho^\tau}{\partial \lambda_\sigma} = c_{\alpha\beta}^\tau \, \theta_\sigma^\alpha \, \theta_\rho^\beta$$

[3] L. P. Eisenhart, *Continous groups*, p. 53.

identically in (λ, s). It now follows that the functions B^ρ_σ as defined by (52) have the properties

$$\frac{\partial B^\tau_\sigma}{\partial b_\rho} - \frac{\partial B^\tau_\rho}{\partial b_\sigma} = c^\tau_{\alpha\beta} B^\alpha_\sigma B^\beta_\rho \tag{53}$$

$$B^\rho_\sigma(0) = \delta^\rho_\sigma . \tag{54}$$

Finally, if we define functions $A^\rho_\sigma(b)$ by the relations (22), we will obtain a completely integrable system (20), and the resulting group has the given coefficients $c^\alpha_{\beta\gamma}$ as constants of structure.

Remarks

Equations (20) are equivalent with the Cartan equations

$$B^\rho_\alpha(c) dc_\alpha = B^\rho_\alpha(b) db_\alpha .$$

If we introduce a change of coordinates $b = Tb'$ of the form

$$b_\rho = b'_\rho + \text{(higher powers)}, \tag{55}$$

then

$$B^\rho_\alpha(b)\, db_\alpha = B^\rho_\alpha \frac{\partial b_\alpha}{\partial b'_\beta} db'_\beta = \bar{B}^\rho_\beta(b')\, db_\beta$$

where

$$\bar{B}^\rho_\beta(b') = B^\rho_\alpha(b) \frac{\partial b_\alpha}{\partial b'_\beta} ; \bar{B}^\rho_\beta(0) = \delta^\rho_\beta .$$

If we also put $c = Tc'$, we obtain

$$\bar{B}^\rho_\alpha(c')dc'_\alpha = \bar{B}^\rho_\alpha(b')db'_\alpha ,$$

and the functions $\bar{B}^\rho_\alpha(b')$ are another solution of the equations (53), (54) for the same constants $c^\tau_{\alpha\beta}$. We remark without proof that conversely any two solutions of (53), (54) can be obtained, the one from the other, by a change of coordinates of type (55). The special solution, as was constructed explicitly in the proof to Theorem 3, was the one corresponding to canonical coordinates. Also the change of coordinates which leads from any coordinates $(b_1, \cdots, b_r)$ to canonical coordinates $(u_1, \cdots, u_r)$ can be obtained in the familiar manner from formal solutions of the equations

$$\frac{\partial b_\rho}{\partial s} = A^\rho_\tau(b)\, u_\tau .$$

If (1) is a group space as before, then we will define formal functions

$$z_i = f_i(x_1, \cdots, x_n; b_1, \cdots, b_r) \qquad i = 1, \cdots, n \tag{56}$$

to be a group of transformations whenever

$$f(f(x; a); b) = f(x; \varphi(a; b))$$
$$f_i(x; a) = x_i .$$

We remark without proof that the functions (56) are the complete solution of a system of equations of the form

(57)
$$\frac{\partial z_i}{\partial b_\rho} = \zeta^i_\tau(z)\, B^\tau_\rho(b),$$

where the functions B^τ_ρ belong to the group space. Also

(58)
$$\zeta^j_\alpha \frac{\partial \zeta^i_\beta}{\partial z_j} - \zeta^j_\beta \frac{\partial \zeta^i_\alpha}{\partial z_j} = c^\tau_{\alpha\beta}\zeta^i_\tau .$$

Also, if for a given group space, the function $\zeta^i_\alpha(z)$ are solutions of (58), then the equations (57) are completely integrable, and the complete integral is a group of transformations.

PRINCETON UNIVERSITY

Annals of Mathematics
Vol. 47, No. 4, October, 1946

LOCALLY COMPACT GROUPS OF DIFFERENTIABLE TRANSFORMATIONS

By Salomon Bochner and Deane Montgomery

(Received January 3, 1946)

1. Introduction

There are several theorems which show under various light restrictions that certain groups of transformations not originally assumed to be Lie groups must nevertheless actually be Lie groups. For example, H. Cartan [4] has shown that the group of all complex analytic homeomorphisms of a bounded domain in several complex variables into itself must be a Lie group. Myers and Steenrod [7] have proved a similar result for the group of all isometries of a Riemannian manifold into itself. Recently it has been shown [6] that any compact effective group of transformations of class C^1 of a manifold of class C^1 must be a Lie group.

In this paper we shall prove an extension of the theorem of H. Cartan to non-analytic transformations, and our theorem will also include the main result of Myers and Steenrod. In the light of our previous results [2], [3], [6] the methods used by Cartan have become applicable to general differentiable transformations, and in large part our reasoning will closely parallel his. However, in one respect our argument will be more elaborate inasmuch as we must show that our group is finite dimensional. In Cartan's case the finite dimensionality of the group follows at an early stage from the analyticity, but this approach is not open to us.

Our theorem may be stated as follows:

Theorem A. *Let G be a topological group which acts on a manifold M in such a way that the identity is the only element whose set of fixed points contains interior points. If G is locally compact, if the manifold M is of class C^2, and if each transformation*

$$y_i = f_i(g; x)$$

is of class C^2, then G is a Lie group.

From a previous paper by the authors [3] it follows that the transforming functions

$$f_i(g; x)$$

are of class C^2 in the variables g and x simultaneously, this statement having become meaningful as a result of Theorem A. The coordinates g in G are of course taken so the group operations are analytic. If the transforming functions are of class C^k (k at least 2) or analytic, then our earlier result shows that these functions are of class C^k or analytic in g and x simultaneously.

2. Definitions

For the sake of completeness we recall that a topological group G is said to be a transformation group acting on a space M if there is a continuous function

$g(x) = f(g; x)$ which is defined on the product of G and M and which satisfies the following conditions:

1) For each fixed $g, f(g; x) = g(x)$ is a homeomorphism of M onto itself.

2) If g_1 and g_2 are any two elements of G, then $g_1[g_2(x)] = (g_1g_2)(x)$ or in the other notation

$$f[g_1; f(g_2; x)] = f(g_1g_2; x).$$

It follows that

3) If e is the identity of G, then $e(x) = x$ for all x in M.

The group G is called effective provided

4) The only element which leaves all of M fixed is e.

In the hypothesis of Theorem A we make the more stringent requirement that e is the only element which leaves all the points of any open set fixed.

Also in the case at hand G is locally compact and M is a manifold of class C^2. We assume that these terms are known and we make the same assumption for the idea of a homeomorphism of class C^2.

We remark explicitly that in the hypothesis of Theorem A it is not assumed that G is locally Euclidean, nor even that it is finite dimensional. It must therefore be shown in the course of the proof that G is locally Euclidean and that coordinates in G may be so chosen that the group operations have the proper differentiability properties.

3. Non-existence of small subgroups

In this section we derive a property of groups of differentiable transformations which will be of use in the sequel.

THEOREM 1. *Let G be a locally compact effective transformation group of a connected manifold M of class C^1 and let each element of G be of class C^1. Then there exists a neighborhood of the identity in G which contains no subgroup except for the trivial subgroup whose only element is the identity.*

It is known by hypothesis that when expressed in terms of the local coordinates the transforming functions $f_i(g; x)$ have, for each fixed g, partial derivatives with respect to the x's which are continuous in x. It follows [6] that these partial derivatives are simultaneously continuous in g and x. From this it follows [4], [2] that G "can not have arbitrarily small subgroups." The phrase quoted means that, given a coordinate neighborhood V in M, there is a neighborhood Q of e in G such that any subgroup in $\bar{Q}$ leaves all of V fixed. Now assume that $\bar{Q}$ is compact. We need consider only closed subgroups of $\bar{Q}$ and we assume that H is such a group. Then H leaves all of V fixed. Near any fixed point coordinates may be chosen in such a way that H becomes a group of orthogonal transformations [2]. When such an H leaves an open set fixed, it must leave fixed every point of the neighborhood on which it operates. Therefore the interior of the set of all fixed points of H must be identical with all of M. Thus every element of H leaves all of M fixed and because G is effective H can contain only the identity element. It has now been shown that $\bar{Q}$ contains no subgroup except the trivial one.

THEOREM 2. *Let G be a locally compact effective transformation group of a connected manifold of class C^1 and let each element be of class C^1. Then there exists a neighborhood of the identity in G having the property that every element in it, except e, has some positive iterate not in it.*

From the preceding theorem we know there exists a neighborhood containing no non-trivial subgroup. If g is not e and is in this neighborhood, then some positive or negative iterate of g is not in the neighborhood. By choosing, if necessary, a smaller neighborhood of e which is invariant under the operation of taking inverses we obtain a neighborhood having the desired property.

4. Local metric in G

So far we have only required each individual transformation to be of class C^1. We shall be able to proceed somewhat farther without increasing the order of differentiability, but we shall find it convenient to begin to use the additional hypothesis in Theorem A.

To be precise, *we assume from this point forward and without further remark that G is a locally compact group acting on a manifold M of class C^1 and that no element of G except the identity leaves every point of an open set fixed.* Additional assumptions will be made as they are needed.

We also assume from now on that V, V_1, and V_2 are coordinate neighborhoods in M whose closures are homeomorphic to the solid unit sphere in n space, and which are such that

$$\bar{V} \subset V_1 \text{ and } \bar{V}_1 \subset V_2 .$$

There exists a neighborhood of the identity in G, call it Q, which is such that $\bar{Q}$ is compact and

$$\bar{Q}(\bar{V}_1) \subset V_2 \text{ and } \bar{Q}(\bar{V}) \subset V_1 .$$

The topology in $\bar{V}_2$ may be given by its Euclidean metric $\rho(x, y)$ and we assume it is so given.

THEOREM 3. *If in $\bar{Q}$ we introduce a new topology by means of the distance function*

$$d(h_1, h_2) = \max_{x \in V_1} \rho[h_1(x), h_2(x)],$$

then the topology given by this metric is equivalent to the original topology.

Because $\bar{Q}$ is a topological group germ in each topology it is sufficient to prove that a neighborhood of the identity in either topology includes a neighborhood of the identity in the other.

PROOF. First let X_2 be any neighborhood of the identity in the new topology. Since $\bar{V}_1$ is compact we may, for every ϵ, choose a neighborhood of the identity X_1 in the old topology so that for every x in $\bar{V}_1$

$$\text{diameter } \bar{X}_1(x) < \epsilon.$$

This is because of the simultaneous continuity of $f(g; x)$. Hence for a proper choice of ϵ we see that X_1 is in X_2.

Next let X_1 be any neighborhood of the identity in the old topology. Let X_{2n} be the set of elements g of $\bar{Q}$ which are such that for all x in $\bar{V}_1$

$$\rho(x, g(x)) \leqq 1/n.$$

The set X_{2n} is closed in both topologies. Since it is closed in the old topology, it is compact in the old topology. As n increases X_{2n} decreases. It will be sufficient to show that for some n

$$X_{2n} \subset X_1 .$$

In order to do this, notice first that $X_{2n} - X_1$ is closed and hence compact in the old topology. Assuming that no $X_{2n} - X_1$ is null, then

$$A = \bigcap_n (X_{2n} - X_1)$$

contains at least one point g. Now g has the property that for all x in $\bar{V}_1$

$$\rho(x, g(x)) = 0$$

and hence g is the identity, but on the other hand g is not in X_1. Hence the assumption that no $X_{2n} - X_1$ is null has led to a contradiction and therefore some $X_{2n} - X_1$ is null. This proves the theorem.

This theorem proves that $\bar{Q}$ is a compact metric space which is useful later for various reasons. We assume henceforth that the topology for $\bar{Q}$ is given in this way.

5. Existence of one-parameter groups

Let $S(e; r)$ be a closed sphere in $\bar{Q}$ with center at the identity e and radius r. On the basis of Theorem 2 we choose r in such a way that every element in $S(e; r)$ has a positive iterate outside $S(e, r)$. For each g in $S(e, r)$ there must, therefore, exist a point x in $\bar{V}_1$ and a positive integer q such that

$$\rho[x, g^q(x)] > r.$$

We now define an integer valued function $q(g)$ for each element g in $S(e; r)$ as follows: for all integers m, $1 \leqq m \leqq q$ and for all x in $\bar{V}_1$

$$\rho(x, g^m(x)) \leqq r$$

while there exists an x in $\bar{V}_1$ such that

$$\rho(x, g^{q+1})) > r.$$

THEOREM 4. *For each positive number* ϵ *there exists a positive number* δ *such that for any* g *in* $S(e; r)$ *and any positive integer* p *if*

$$\frac{p}{q(g)} < \delta,$$

then for all x *in* $\bar{V}_1$

$$\rho(x, g^p(x)) < \epsilon.$$

PROOF. Let g be any element of $\bar{Q}$ and assume that q is an integer such that

$$g(x),\ g^2(x),\ \cdots,\ g^q(x)$$

are in V_2 for any x in $\bar{V}_1$ or in $g(\bar{V}_1)$ and that

$$g,\ g^2,\ \cdots,\ g^{q+1}$$

are in $\bar{Q}$.

As in [4], [3], let

$$T_i(g; x) = x_i + f_i(g; x) + \cdots + f_i(g^{q-1}; x) \tag{1}$$

and let

$$\xi_i(g; x) = f_i(g; x) - x_i.$$

Then if we let

$$T_{ij}(g; x) = (\partial/\partial x_j)T_i(g; x)$$

we have

$$T_i(g; x + \xi) - T_i(g; x) = \sum_j \xi_j \int_0^1 T_{ij}(g; x + t\xi)dt. \tag{2}$$

Now

$$\begin{aligned} T_i(g; x + \xi) &= T_i(g; f(g; x)) \\ &= f_i(g; x) + f_i(g^2; x) + \cdots\cdots + f_i(g^q; x) \end{aligned} \tag{3}$$

so that

$$T_i(g; x + \xi) - T_i(g; x) = f_i(g^q; x) - x_i. \tag{4}$$

Hence

$$f_i(g^q; x) - x_i = \sum_j \left[\int_0^1 \{\delta_{ij} + f_{ij}(g; x + t\xi) + \cdots + f_{ij}(g^{q-1}; x + t\xi)\} dt\right]. \\ [f_j(g; x) - x_j]. \tag{5}$$

Multiplying and dividing by q on the right gives

$$f_i(g^q; x) - x_i = \sum_j [\delta_{ij}(g; x; q)][q\{f_j(g; x) - x_j\}] \tag{6}$$

where

$$\delta_{ij}(g; x; q) = (1/q) \int_0^1 \delta_{ij} + f_{ij}(g; x + t\xi) + \cdots + f_{ij}(g^{q-1}; x + t\xi)dt. \tag{7}$$

If $\bar{Q}$ is sufficiently small, then, because $f_{ij}(g; x)$ is continuous simultaneously in g and x [6] it follows that $\delta_{ij}(g; x; q)$ is near δ_{ij} and this uniformly in q. We may therefore let α_{ji} be the matrix inverse to the matrix $\delta_{ij}(g; x; q)$ and write the relation (6) as follows:

$$q\{f_j(g; x) - x_j\} = \sum {}_i \alpha_{ji}(g; x; q)\{f_i(g^q; x) - x_i\}. \tag{8}$$

We point out that both $\alpha_{ij}(g; x; q)$ and $\delta_{ij}(g; x; q)$ approach δ_{ij} uniformly as the diameter of the set $e, g, g^2, \cdots g^q$ approaches zero. More precisely for any positive ϵ there exists a positive δ such that, if

$$\text{diam } \{e, g, \cdots g^q\} < \delta,$$

then

$$| \delta_{ij}(g; x; q) - \delta_{ij} | < \epsilon$$

and

$$| \alpha_{ij}(g; x; q) - \delta_{ij} | < \epsilon.$$

For any g, x, and q let

$$\beta(g; x; q)$$

be the smallest positive number such that any vector operated upon by $\delta_{ij}(g; x; q)$ or $\alpha_{ij}(g; x; q)$ assumes a norm which is between $1 - \beta(g; x; q)$ and $1 + \beta(g; x; q)$ times its original norm. The number $\beta(g; x; q)$ approaches zero uniformly with diameter $\{e, g, \cdots, g^q\}$.

For the present we are not interested in β tending to zero, but only in β being less than one. *We make now a final choice of $\bar{Q}$ which is such that*

$$\beta(g; x; q) < \beta < 1.$$

Having developed these useful formulas we turn to the actual proof of Theorem 4. From (8), we have

$$f_j(g; x) - x_j = (1/q(g)) \sum {}_i\alpha_{ji}(g; x; q(g)) \{f_i(g^{q(g)}; x) - x_i\}.$$

We know that

$$| f(g^{q(g)}; x) - x | \leqq r.$$

Hence

$$| f(g; x) - x | \leqq \frac{(1 + \beta)r}{q(g)}.$$

From (6)

$$f_i(g^p; x) - x_i = \textstyle\sum_j \delta_{ij}(g; x; p)[p\{f_j(g; x) - x_j\}1$$

and as a result

$$| f(g^p; x) - x | \leqq p(1 + \beta) | f(g; x) - x |.$$

Thus

$$| f(g^p; x) - x | \leqq r(1 + \beta)^2 p/q(g)$$

and if

$$p/q(g) < \epsilon/r(1 + \beta)^2$$

then

$$| f(g^p; x) - x | < \epsilon.$$

This proves Theorem 4 with $\delta = \epsilon / r(1 + \beta)^2$.

As an immediate consequence we have the following result.

THEOREM 5. *For each positive number ϵ there exists a positive number δ such that for any g in $S(e; r)$, if*

$$p/q(g) < \delta,$$

then

$$d(e, g^p) \subset \epsilon.$$

We now define other integral valued functions of g for $g(\neq e)$ in $S(e; r)$. For convenience we understand g^0 to be equal to e. For any positive integers m and s where $m \leqq s$ let

$$q(g; m/s)$$

be the largest non-negative integer which is such that

$$q(g; m/s) \leqq (m/s) \cdot q(g).$$

We notice that

$$q(g; 1) = q(g).$$

Let g_k be a sequence of elements in $S(e; r)$ which approach e. We may select a subsequence (we denote it with no change of subscripts) such that for each fixed m and s the sequence

$$g_k^{q(g_k; m/s)}$$

converges to an element of $S(e; r)$ which will be denoted by $h(m/s)$.

THEOREM 6. *If m_α and s_α are such that m_α / s_α converges to t, then $h(m_\alpha / s_\alpha)$ converges to a point of $S(e; r)$. Furthermore if m'_α and s'_α are such that m'_α / s'_α converges to t, then $h(m'_\alpha / s'_\alpha)$ converges to the same element as does $h(m_\alpha / s_\alpha)$. It is therefore permissible to denote this element by $h(t)$.*

PROOF. We know that

$$| q(g_k; m_\alpha / s_\alpha) - (m_\alpha / s_\alpha) q(g_k) | < (1/s_\alpha) q(g_k)$$

and

$$| q(g_k; m'_\alpha / s'_\alpha) - (m'_\alpha / s'_\alpha) q(g_k) | < (1/s)'_\alpha q(g_k).$$

Assume that

$$| (m_\alpha / s_\alpha) - (m'_\alpha / s'_\alpha) | < \eta.$$

Then

$$| q(g_k; m_\alpha/s_\alpha) - q(g_k; m'_\alpha/s'_\alpha) | \leqq (1/s_\alpha + 1/s'_\alpha)q(g_k) + \eta.$$
$$\leqq (1/s_\alpha + 1/s'_\alpha + \eta)q(g_k).$$

If $\epsilon > 0$ is given and if s_α, s'_α and η are such that for the δ of Theorem 5

$$1/s_\alpha + 1/s'_\alpha + \eta < \delta,$$

then (assuming $q(g_k; m_\alpha/s_\alpha) \geqq q(g_k; m'_\alpha/s'_\alpha)$ which implies no loss of generality)

$$d[e, g_k^{q(g_k;m_\alpha/s_\alpha)-q(g_k;m'_\alpha/s'_\alpha)}] < \epsilon.$$

This shows, that, uniformly in k,

$$g_k^{q(g_k;m_\alpha/s_\alpha)} \quad \text{and} \quad g_k^{q(g_k;m'_\alpha/s'_\alpha)}$$

are near each other, and this is sufficient to prove the theorem.

THEOREM 7. *The set of elements $h(t)$, $0 \leqq t \leqq 1$, defined above forms the positive part of a local one parameter group.*

That this set of elements is a continuous image of the unit interval follows from the methods of the preceding theorem. It remains only to see that for small numbers t and t' we have

$$h(t)h(t') = h(t + t').$$

But this follows from the fact that, if m_α/s_α approaches t and m'_α/s'_α approaches t', then $m_\alpha/s_\alpha + m'_\alpha/s'_\alpha$ approaches $t + t'$ and therefore

$$g_k^{q(g_k;m_\alpha/s_\alpha+m'_\alpha/s'_\alpha)}$$

approaches $h(t + t')$.

THEOREM 8. *If G contains a sequence of elements approaching the identity, then G contains a non-trivial one parameter group.*

This is an immediate corollary of the preceding theorems.

6. Vector fields

The preceding work has applied to transformations of class C^1, but *from now on we shall draw on all the hypotheses of Theorem A which means in particular that the manifold and transformations are of class C^2.*

We prove next a theorem which is one of the main points in the method of H. Cartan. We speak of a vector as being of class C^1 if each of its components is of class C^1.

THEOREM 9. *If g_k is a sequence of elements ($\neq e$) in $S(e; r)$ which approach e, then we may choose a subsequence (denoted by the same subscripts) such that for $1 \leqq i \leqq n$*

$$q(g_k)[f(g_k; x) - x]$$

approaches uniformly a vector which is of class C^1 in $\bar{V}_1$ and which is not identically zero.

PROOF. By formula (8)

(9) $$q(g_k)\{f_j(g_k; x) - x_j\} = \sum_i \alpha_{ji}(g; x; q(g_k))\{f_i(g_k^{q(g_k)}; x) - x_i\}.$$

For k sufficiently large there must be an x in $\bar{V}_1$ such that

$$|f(g_k^{q(g_k)}; x) - x| \geqq r/2.$$

Hence for all sufficiently large k there is an x in $\bar{V}_1$ such that

(10) $$|q(g_k)[f(g_k; x) - x]| \geqq (1 - \beta)r/2.$$

All the quantities appearing on the right of (9) are uniformly bounded and have first partial derivatives which are uniformly bounded and equicontinuous. Therefore, replacing the sequence in k by a subsequence, we may assume that

$$q(g_k)\{f_j(g_k; x) - x_j\}$$

converges uniformly for $1 \leqq j \leqq n$ to functions $\alpha_j(x)$ which are of class C^1. Furthermore because of (10) we have some x in $\bar{V}_1$ such that

$$|\alpha(x)| \geqq (1 - \beta)r/2,$$

and this completes the proof.

THEOREM 10. *Let g_k be a sequence of elements in $S(e; r)$ approaching e and let m_k and q_k be two sequences of real numbers such that the vectors*

$$m_k[f(g_k; x) - x] \text{ and } q_k[f(g_k; x) - x]$$

converge uniformly to vectors $\alpha^1(x)$ and $\alpha^2(x)$. If $\alpha^2(x)$ does not vanish identically in $\bar{V}_1$, then m_k/q_k approaches a limit A and

(10.1) $$\alpha^1(x) = A\alpha^2(x).$$

PROOF. In a certain compact region of $\bar{V}_1$ at least one component of $\alpha^2(x)$ does not vanish. Call this component $\alpha_j^2(x)$. In this component we have (as everywhere) uniformly

(10.2) $$m_k[f_j(g_k; x) - x] \to \alpha_j^1(x)$$

and

(10.3) $$q_k[f_j(g_k; x) - x] \to \alpha_j^2(x).$$

Hence by dividing we obtain

(10.4) $$\frac{m_k}{q_k} \to \frac{\alpha_j^1(x)}{\alpha_j^2(x)}.$$

This proves that in the given domain $\alpha_j^1(x)/\alpha_j^2(x)$ is constant and that m_k/q_k approaches this constant. Denoting this constant by A and remembering that (10.2) and 10.3) hold for all j we obtain

$$\alpha_j^1(x) = A\alpha_j^2(x)$$

for all j and this completes the proof.

THEOREM 11. *If g_k is a sequence of elements in $S(e; r)$ which approach e and if m_k is a sequence of positive integers such that*

$$m_k[f(g_k\,; x) - x]$$

converges uniformly on $\bar{V}_1$ to a vector $\psi(x)$ of class C^1, then for each t sufficiently near to zero there is a transformation in G coinciding in $\bar{V}$ with the local one parameter group defined by the following system of differential equations:

$$\frac{dx_i}{dt} = \psi_i(x_1, \cdots, x_n). \tag{11}$$

PROOF. Let $h(t)$ be the one parameter group determined by a certain subsequence of the g_k's as in Theorem 7. We know [3] that $h(t)(x)$ has a derivative with respect to t, that is if

$$F(t; x) = f(h(t); x)$$

then the functions $F(t; x)$ have derivatives with respect to t which are continuous functions of t and x. Since these derivatives exist we may find them by considering the limit of the difference quotient for positive rational values of the increment. Let m/s be a rational number between 0 and 1 and put

$$F_i(m/s; x) = f_i(h(m/s); x).$$

Then

$$\begin{aligned} F_i(m/s; x) - x_i &= \lim_{k\to\infty} f_i(g_k^{q(g_k;m/s)}; x) - x_i \\ &= \lim_{k\to\infty} [\textstyle\sum_j \delta_{ij}(g_k; x; q(g_k; m/s))] \\ &\qquad \cdot [i(g_k; m/s)\{f_j(g_k; x) - x_j\}] \cdot \end{aligned}$$

We choose a subsequence of the k's so that

$$q(g_k)\{f_j(g_k\,; x) - x_j\}$$

converges uniformly to a vector on $\bar{V}_1$, say $\alpha(x)$, where $\alpha(x)$ is of class C^1 and does not vanish identically. With this choice we see, since $q(g_k\,; m/s)$ is approximately $(m/s)q(g_k)$, that

$$q(g_k\,; m/s)\{f(g_k\,; x) - x\}$$

approaches $(m/s)\alpha(x)$. Hence replacing part of the formula above by its known limit we have

$$\frac{F_i(m/s; x) - x_i}{m/s} = \lim_{k\to\infty} [\textstyle\sum_j \delta_{ij}(g_k\,; x; q(g_k\,; m/s))][\alpha_j(x)].$$

The elements $e, g, g^2, \cdots, g^{q(g_k;m/s)}$ have a diameter which is approximately $(m/s)r$ so that this diameter approaches zero with (m/s). We may assume, by choosing a subsequence if necessary, that

$$\lim_{k\to\infty} \delta_{ij}(g_k; x; q(g_k; m/s))$$

exists and is a function of class C^1, call it $\delta_{ij}(x; m/s)$. We see also that $\delta_{ij}(x; m/s)$ approaches δ_{ij} uniformly with m/s. Thus

$$\frac{F_i(m/s; x) - x_i}{m/s} = \sum_j \delta_{ij}(x; m/s)\alpha_j(x).$$

Hence the limit of the difference quotient exists when m/s approaches zero and the limit is $\alpha_i(x)$.

But from a preceding theorem the vectors $\psi(x)$ and $\alpha(x)$ can not differ except by a constant multiple and therefore the transformations defined by (11) must coincide with $F(t; x)$ where t may have to be changed by a certain scale factor.

7. The vector space

For each one parameter group in G, say $h(t)$, there is associated a vector field $\psi(x)$ defined on $\bar{V}_1$.

THEOREM 12. *The vector fields $\psi(x)$ defined on $\bar{V}_1$ and determined by one parameter groups in G from a linear vector space B which is closed under the bracket operation.*

Let $\psi_1(x)$ and $\psi_2(x)$ be two elements of B which are determined by two one-parameter groups $h_1(t)$ and $h_2(t)$. In the first place, if $\psi_1(x)$ is multiplied by a constant A it is associated with $h_1(At)$, so that B is closed under multiplication by real numbers.

Let t_k approach zero and consider the elements of G given as follows:

$$h_1(t_k)h_2(t_k).$$

Now

$$h_1(t_k)(x) = x + t_k\psi_1(x) + \eta_1(x, t_k)t_k$$

$$h_2(t_k)(x) = x + t_k\psi_2(x) + \eta_2(x, t_k)t_k$$

where $\eta_1(x, t_k)$ and $\eta_2(x, t_k)$ approach zero uniformly with t_k. Then

$$h_1(t_k)h_2(t_k)(x) = x - t_k\psi_1(x) + t_k\psi_2(x) + \eta(x, t_k)t_k \tag{12}$$

where $\eta(x, t_k)$ approaches zero uniformly with t_k. If we put $t_k = 1/k$, then

$$k[h_1(t_k)h_2(t_k)(x) - x]$$

tends uniformly to

$$\psi^1(x) + \psi^2(x).$$

Hence there is a one parameter group in G associated with $\psi_1(x) + \psi_2(x)$.

It remains to show that B is closed under the bracket operation. We know [3] that $h_1(t)(x)$ and $h_2(t)(x)$ have continuous second derivatives with respect to t. Hence we may make use of standard information (see for example [5, p. 437]) to conclude that

$$h_1(1/k)h_2(1/k)h_1(-1/k)h_2(-1/k)(x) - x = \left(\frac{1}{k}\right)^2 \{[\psi^1, \psi^2] + \eta(x, k)\}$$

where $\eta(x, k)$ approaches zero with $1/k$. Hence

$$k^2\{h_1(1/k)h_2(1/k)h_1(-1/k)h_2(-1/k)(x) - x\}$$

approaches uniformly

$$[\psi^1(x), \psi^2(x)]$$

which proves that B is closed under the bracket operation.

THEOREM 13. *If $h_1(t)$, $h_2(t)$, $\cdots$ is a sequence of one parameter subgroups such that for $1 \leqq t \leqq 1$, $h_k(t)$ is in $S(e; r)$, and if for each rational m/s it is true that $h_k(m/s)$ converges to a point of $S(e; r)$, then for any real t, $h_k(t)$ converges to a point of $S(e; r)$, say $h(t)$. Furthermore $h(t)$ is a one parameter group.*

PROOF. If $h_k(1)$ approaches e, then, for all t, $h_k(t)$ approaches e uniformly in t, and the theorem is proved. Hence we consider only the case where $h_k(1)$ approaches $a \neq e$. In this case however the argument is exactly the same as in Theorems 6 and 7 and need not be repeated.

Consider now the linear space B of vector fields defined on $\bar{V}_1$ determined by one parameter groups in G. If $\psi(x)$ is in B we define the norm of $\psi(x)$ to be

$$\max_x [\psi_1^2(x) + \cdots + \psi_n^2(x)]^{1/2}.$$

THEOREM 14. *The vector space B is locally compact and hence finite dimensional.*

PROOF. If B is locally compact it is well known that it is finite dimensional [1]. We proceed now to show that B is locally compact.

Let B_1 denote the subset of B coming from one parameter groups $h(t)$ in G which have the property that for $0 \leqq t \leqq 1$ $h(t)$ is in $S(e; r)$. The set B_1 clearly includes the zero element of B as an interior point. We shall prove the theorem by showing that B_1 is compact.

Let $\psi^1(x)$, $\psi^2(x)$, $\cdots$ be a countable subset of B_1 and let $h_1(t), h_2(t), \cdots$ be the associated one parameter subgroups in G. From a formula in a previous paper [3]

$$(13)\quad f_i(h_k(1); x) - f_i(0; x) = \sum_j \left[\int_0^1 f_{ij}(h_k(t); x)\, dt\right] \cdot [(\partial/\partial s) f_j(h_k(0); x)],$$

that is,

$$(14)\quad f_i(h_k(1); x) - f_i(0; x) = \sum_j \left[\int_0^1 f_{ij}(h_k(t); x)\, dt\right] \cdot \psi_j^k(x)].$$

These equations may be solved for the functions $\psi_j^k(x)$ and from the formula we see that all these functions and their first partial derivatives are uniformly bounded. Hence from $\psi^k(x)$ we may select a subsequence which converges uniformly in $\bar{V}_1$ to a vector $\psi(x)$. We need only show that $\psi(x)$ is in B_1.

By choosing, if necessary, another subsequence of the k's we have for each rational m/s that

$$h_k(m/s)$$

converges. Then by the preceding theorem, for all t,

$$h_k(t) \longrightarrow h(t)$$

where $h(t)$ is a one parameter subgroup. We shall prove that $h(t)$ and $\psi(x)$ are associated which will prove that $\psi(x)$ is in B_1. We know that the derivative of $f_i(h(t); x)$ with respect to t exists and we need only prove that when $t = 0$ it coincides with $\psi_i(x)$. From the same formula used in deriving (14) we have

$$(15)\qquad f_i(h(1); x) - f_i(h(0); x) = \sum_j \left[\int_0^1 f_{ij}(h(t); x)\, dt\right] [(\partial/\partial s) f_j(h(0); x)].$$

By solving (14) and (15) for

$$\psi_j^k(x)$$

and

$$(\partial/\partial s) f_j(h(0); x)$$

respectively we see that

$$\lim_{k\to\infty} \psi_j^k(x) = (\partial/\partial s) f_j(h(0); x).$$

Since it is also true that

$$\lim_{k\to\infty} \psi_j^k(x) = \psi_j(x),$$

it follows that

$$\psi_j(x) = (\partial/\partial s) f_j(h(0); x)$$

which concludes the proof of the theorem.

8. Proof of Theorem A

In this section we conclude the proof of Theorem A, by proving first that G contains a local Lie group and second that this is a complete neighborhood of the identity.

THEOREM 15. *The space B determines a local group Γ which operates on $\bar{V}$, in symbols,*

$$\Gamma(t_1, \cdots, t_r; x)$$

where B is r-dimensional; that is, these transformations are defined for $x \epsilon \bar{V}$ and $| t_i | \leqq \eta$. For every γ in Γ there is a unique g in G such that $g(x) = \gamma(x)$ for x in $\bar{V}$.

PROOF. The fact that B determines a local group follows from standard existence theorems since we know that B is linear and is closed under the bracket operation [8]. If γ is in Γ there is a one parameter subgroup $\gamma(t)$ such that γ is on $\gamma(t)$. Now $\gamma(t)$ is determined by a vector field $\psi(x)$ in B and hence, for small values of t, $\gamma(t)$ and $h(t)$ coincide where $h(t)$ and $\Omega(x)$ are associated as above. The elements $h(t)$ are in G and as transformations are defined on all of M. It is important to notice that the element of G corresponding to each $\gamma(t)$ is unique.

The topologies of the elements as given by Γ and as in G are equivalent. We shall therefore regard Γ for $| t_i | \leqq \eta$ as a local subgroup of G. The following theorem will then complete the proof of Theorem A.

THEOREM 16. *The set Γ is a complete neighborhood of the identity in G.*

PROOF. The proof will be by means of a contradiction paralleling the argument of H. Cartan.

Assuming the theorem to be false, let g_k be a sequence of elements in $S(e; r)$ which approach e, with no g_k being in Γ. Let γ_k be an element of Γ which makes

$$d(e, g_k)$$

a minimum for γ in Γ, and introduce the element

$$g_k' = \gamma_k g_k .$$

As in Cartan's case, the element g_k' is not in Γ and the sequence g_k' approaches e. The positive integers $q(g_k')$ have the property that for $1 \leqq i \leqq n$

$$q(g_k')\{f(g_k' ; x) - x\} \tag{16}$$

converges uniformly to a vector $\psi(x)$ of class C^1 which does not vanish identically on $\bar{V}_1$. Let $h(t)$ be the one parameter subgroup associated with $\psi(x)$; the elements of this group are in Γ, by definition of Γ. Consider the elements

$$g_k'' = g_k' h(- 1/q(g_k')).$$

We see that

$$h(- 1/q(g_k')(x) = x - [1/q(g_k')][\psi(x) + \eta(q(g_k'); x))] \tag{17}$$

where

$$\eta(q(g_k'); x)$$

approaches zero with $1/k$. From (16)

$$f_j(g_k' ; x) - x_j = [1/q(g_k')][\psi_j(x) + \eta_j'(g_k' ; x)]$$

where $\eta'(g_k' ; x)$ approaches zero with $1/k$.
We have also

$$\begin{aligned} f_j(g_k' ; h(- 1/q(g_k')(x)) &- [h(- 1/q(g_k')(x)]_j \\ &= [1/q(g_k')][\psi_j(h(- 1/q(g_k')(x)) + \eta'(g_k' ; h(- 1/q(g_k')(x) \end{aligned}$$

and

$$\begin{aligned} f_j(g_k'' ; x) - x_j = [(1/q(g_k')][\psi_j(x) + \eta_j(q(g_k'); x)] \\ + (1/q(g_k'))[_j(h(- 1/q(g_k')(x) + '(g_k' ; h(- 1/q(g_k')(x). \end{aligned}$$

We know that

$$\max_x | f(g_k'' ; x) - x | \geqq \max_x | f(g_k' ; x) - x |.$$

But the above

$$\lim_{k \to \infty} q(g_k')[f(g_k''; x) - x] = 0$$

which is a contradiction. This completes the proof of Theorem 16, which in turn completes the proof of Theorem A.

9. Compact Complex Manifolds (May 17, 1946)

Let M be a connected compact complex analytic manifold, and let G be the set of all complex analytic homeomorphisms taking M onto itself. The set G is a group, and this group may be topologized, for example by regarding M as a metric space and defining the distance between two transformations to be the maximum distance between corresponding images. With this topology G becomes locally compact. In order to see this, choose a neighborhood U of the identity in G so small that all transformations in U move any point within a single bounded coordinate neighborhood. From properties of complex functions it then follows that from any sequence of transformations in U we may select a uniformly convergent subsequence. This subsequence converges to a complex analytic transformation. The limit cannot be degenerate, as can be seen from topological considerations, and for complex variables it can therefore be concluded that it is a homeomorphism (for an outline of this conclusion see p. 369 of Caratheodory, Uber die Abbildungen, die durch Systeme von analytischen Funktionen von mehreren Veränderlichen erzeugt werden, Math. Zeit., vol. 34 (1931) pp. 758–792). From these considerations and Theorem A, we may infer the following corollary which, in the light of the work of H. Cartan, is not a novel statement.

COROLLARY: *The group of all complex analytic homeomorphisms of a compact complex analytic manifold is a Lie group.*

Perhaps it should be remarked that we use the term Lie group to include the case where the group is countable and discrete. The result of this corollary is not true without compactness.

PRINCETON UNIVERSITY
THE INSTITUTE FOR ADVANCED STUDY

BIBLIOGRAPHY

1. BANACH, *Théorie des óperations linéaires*, Warsaw, 1932.
2. BOCHNER, *Compact groups of differentiable transformations*, Annals of Math., vol. 46 (1945), pp. 372–381.
3. BOCHNER AND MONTGOMERY, *Groups of differentiable and real or complex analytic transformations*, Annals of Math., vol. 46 (1945), pp. 685–694.
4. H. CARTAN, *Sur les groupes de transformations analytiques*, Actualities Scien. et Industrielles 198, Paris, 1935.
5. LIE-SCHEFFERS, Vorlesungen über continuierliche gruppen, Leipzig, 1893.
6. MONTGOMERY, *Topological groups of differentiable transformations*, Annals of Math., vol. 46 (1945), pp. 382–387.
7. MYERS AND STEENROD, *The group of isometries of a Riemannian manifold*, Annals of Math., vol. 40 (1939), pp. 400–416.
8. P. SMITH, *Foundations of the theory of Lie groups with real parameters*, Annals of Math., vol. 44 (1943), pp. 481–513.

ANNALS OF MATHEMATICS
Vol. 48, No. 3, July, 1947

GROUPS ON ANALYTIC MANIFOLDS

BY SALOMON BOCHNER AND DEANE MONTGOMERY

(Received September 20, 1946)

1. Introduction

This paper is largely concerned with groups on complex analytic manifolds but in part also with groups on real analytic manifolds. It has been shown [3] that the group of all complex analytic homeomorphisms of a compact complex manifold into itself is a Lie group. The present paper will improve on this result by showing that this Lie group is a complex Lie group For the validity of this improvement the compactness (or something very near to compactness) of the manifold is essential. Indeed we point out the counter theorem that on a bounded domain in the space of several complex variables no complex Lie group can act effectively, although here again [4] the total group is a real Lie group. Furthermore we try to explain this difference between a compact manifold and a bounded domain as a contrast between elliptic and hyperbolic geometry.

Another type of theorem discussed here may be described as follows. If M is a real analytic manifold with a positive definite Riemannian metric which is homeomorphic (not necessarily analytically homeomorphic) to E_n and if G is a compact connected group of analytic transformations with at least one n-dimensional orbit, then there is an analytic homeomorphism from M onto E_n which takes each element of G into an orthogonal transformation. There is an analogous theorem for complex analytic manifolds.

2. Compact Complex Manifolds

Let M be a compact complex manifold. The group of homeomorphisms of M has a natural topology which can be defined as follows: Observe that M can be considered a metric space and define the distance between any two homeomorphisms h_1 and h_2, in the usual way, as

$$\sup_{x \in M} \text{distance } [h_1(x), h_2(x)].$$

When we speak of the group G of all complex analytic homeomorphisms of M, we mean the topological group obtained by giving G a topology equivalent to the one stated.

A complex Lie group is a Lie group whose identity component may be made into a complex analytic manifold in such a way that the group operations are given by complex analytic functions. A given Lie group is complex, if a neighborhood of the identity may be specified by complex coordinates in such a way, that, for elements sufficiently near the identity, the group operations are given by complex analytic functions. In this connection a useful tool is the concept of a complex Lie algebra and its interrelations with complex Lie groups.

With these preliminaries the theorem which interests us in this section may be stated in the following manner.

THEOREM 1. *Let M be a compact complex analytic manifold, and let G be the group of all complex analytic homeomorphisms of M into itself where G has been topologized as above. Then G is a complex Lie group.*

It is known that G is a Lie group so that its points may be specified locally by a certain number, say r, of real parameters. If the element g in G takes a point x of M to $f(g; x)$ then by a known result [2, p. 693], $f(g; x)$ may be expressed locally in terms of coordinates as

$$f_i(g; x) = f_i(g_1, \cdots, g_r; x_1, \cdots, x_n), \qquad i = 1, \cdots, n \tag{1}$$

where $g_1, \cdots, g_r$ are analytic real parameters for G and $x_1, \cdots, x_n$ are complex analytic parameters for M, and the functions f_i are convergent power series in the real quantities $g_1, \cdots, g_r$ and the complex quantities $x_1, \cdots, x_n$ simultaneously.

Let V and W be two coordinate neighborhoods of M with $\bar{V}$ in W, and let Q be a neighborhood of the identity in G which is so small that

$$f(Q; \bar{V}) \subset W.$$

We assume that V and Q have been chosen so small that for g in Q and x in $\bar{V}$ the functions f_i are given by convergent power series as mentioned above, that is, for $i = 1, \cdots, n$,

$$f_i(g; x) = \sum a_{j_1 \cdots j_r k_1 \cdots k_n} g^{j_1} \cdots g_r^{j_r} x_1^{k_1} \cdots x_n^{k_n}. \tag{2}$$

Also

$$f[g; f(h; x)] = f(gh; x) \tag{3}$$

for x in $\bar{V}$ and g, h, gh in Q. However the power series (2) will continue to be convergent if $g_1, \cdots, g_n$ are any complex numbers such that

$$| g | = (g_1\bar{g}_1 + \cdots + g_r\bar{g}_r)^{\frac{1}{2}} < \delta \tag{4}$$

for δ sufficiently small, and (3) will then be valid for g and h both complex.

For any real value of g satisfying (4) the Jacobian of the transformation (2) denoted by

$$J(g; x)$$

is different from zero for all x in $\bar{V}$ and the same will be true for complex g if δ is sufficiently small.

Suppose now that W, which is covered by an allowable coordinate system x, is also covered by an allowable coordinate system y. These two coordinate systems are related by

$$y_i = H_i(x) \tag{5}$$

where the functions $H_i(x)$ define a transformation which together with its inverse is complex analytic and has a non-vanishing Jacobian in W. Assume that for this set of coordinates the transformation group is given by

$$f_i^*(g; y) \tag{6}$$

with properties similar to those of (2). Since the transformations expressed by $f(g; x)$ and $f^*(g; y)$ are the same it follows that

$$f_i^*[g; H(x)] = H_i[f(g; x)] \tag{7}$$

holds for all x in $\bar{V}$ and all real g satisfying (4). It will then also hold for all complex g satisfying (4), and this proves that the transformations of $\bar{V}$ which have been defined for complex values of g subject to (4) are independent of the local coordinate system.

For each pont of the compact manifold M there are two neighborhoods V_x and W_x subject to the conditions on V and W which have previously been discussed. Since M is compact there exists a finite subfamily of the sets V_x, the members of which cover M. Let these with their associated W's be denoted by

$$V_1, W_1; \cdots; V_p, W_p.$$

Assume that in V_k, W_k the transformations are given by

$$f_{ki}(g; x^k) = \sum a_{j_1 \cdots j_r k_1 \cdots k_n} g_1^{j_1} \cdots g_r^{rj} (x_1^k)^{k1} \cdots (x_n^k)^{k_n} \tag{8}$$

where the coordinates $x_1^k, \cdots, x_n^k$ refer, of course, to W_k. There are only a finite number of neighborhoods under consideration and therefore a positive δ may be selected so that (8) converges for $i = 1, \cdots, n$, provided g is complex with

$$|g| < \delta \tag{9}$$

and provided x^k is in $\bar{V}_k$, $k = 1, \cdots, p$.

Because of the fact that the transformations determined by (8) are independent of the local coordinate system, we see that for every g in G^*, G^* denoting the set of complex g's satisfying (9), there is defined a complex analytic transformation of M into itself denoted by $T(g; x)$. It may also be assumed that δ of (9) is so chosen that

$$f_{ki}(g; x^k) \qquad i = 1, \cdots, n$$

has a Jacobian

$$J_k(g; x^k) \qquad k = 1, \cdots, p$$

which does not vanish in $\bar{V}_k$.

From topological considerations it can be seen that for G^* sufficiently small the transformation $T(g; x)$ must be a continuous transformation of M *onto* itself. It has also been shown previously that for g, h, and gh in G^*

$$T[g; T(h; x)] = T[gh; x]. \tag{10}$$

The set G^* is a complex local Lie group so that if g is sufficiently small it has an inverse g^{-1} which is also in the set G^*. This implies that for sufficiently small g, $T(g; x)$ is one-one, since

$$T(g; x) = T(g; y)$$

implies by (10)

$$x = T[g^{-1}; T(g; x)] = T[g^{-1}; T(g; y)] = y.$$

This, together with the fact that M is compact and $T(g; x)$ is continuous, shows that $T(g; x)$ is a homeomorphism.

To sum up what has been obtained thus far, we may say that G^* is a complex local Lie group obtained by complex extension of a nucleus of G and that for each g in G^* the following conditions are satisfied:

(a) $T(g; x)$ is a complex analytic homeomorphism of M onto itself.

(b) If g, h and gh are in G^*, then

$$T[g; T(h; x)] = T(gh; x).$$

(c) If g is real and in G^* then

$$T(g;(x) = f(g; x)$$

where $f(g; x)$ is the original transforming function of G

(d) The function $T(g; x)$ is simultaneously complex analytic in g and x.

These facts justify calling G^* a local transformation group of the manifold M. However, it should be noticed that this local group is not effective. A transformation group or a local transformation group is called effective provided the only element leaving every point of M fixed is the identity. In the case at hand this is not true and the local group G^* is not effective.

Let H denote the set of elements h in G^* which have the property that for all x in M,

$$T(h; x) = x.$$

For any fixed x in M let G_x^* denote the set of all elements g in G^* such that

$$T(g; x) = x.$$

We observe that H and G_x^* are local subgroups of G^*, that H is an invariant local subgroup, and that

$$\bigcap_x G_x^* = H.$$

(For definitions regarding local subgroups see Pontrjagin [9, pp. 83–86].)

It is known (see for instance [9, p. 196]) that H is locally a real analytic manifold of some dimension, because this is true for any closed local subgroup of a local Lie group. We wish to see that H is a complex subgroup which is imbedded in a complex analytic way in G^*. In order to see this it is most convenient to make use of the complex Lie algebra, say R, which is associated with G^*. The space R is a complex vector space and with each vector v in R there is associated a complex analytic vector field in M. In any local coordinate system this field may be written in coordinate form as

$$\psi_i(v; x) \qquad i = 1, \cdots, n$$

where these n functions are complex analytic simultaneously in v and x (see Pontrjagin [9, p. 287]. Let R_1 be the subset of R consisting of all vectors v which are such that

$$\psi_i(v; x) = 0 \qquad i = 1, \cdots, n$$

for all x in M.

The vector fields associated with R, themselves form a vector space isomorphic to R [9, pp. 287–289] in which vector addition and the bracket operation correspond to the same operations in R. The set R_1 can be seen to be a linear subspace of R which is a Lie algebra and which furthermore is an invariant Lie algebra. Therefore R_1 gives rise to an invariant local subgroup of G^* and this local subgroup must be H. It must be complex analytic because it is determined by R_1.

But then the factor local group G^*/H must also be complex analytic and this complex analytic factor group acts as a transformation group of M and the action is effective. Since G contains all complex analytic homeomorphisms of M, the group G must be locally isomorphic to G^*/H, and we see in this way that G can be given a complex analytic structure. This completes the proof of Theorem 1. It is worth remarking that when G is given a complex analytic structure, the transforming functions become simultaneously analytic in g and x [2].

We point out also that the above proof could be broken into two parts each one of which might have some interest for its own sake. The first part shows that any real Lie group nucleus which acts analytically on a compact complex analytic manifold can be extended to a complex Lie group nucleus which acts in a complex way. The second part shows that if a complex group acts in a complex way on a complex manifold, then the subgroup leaving every point fixed is a complex group so that the effective factor group is also complex.

3. Bounded Domains and Non-compact Manifolds

We shall now make a few observations about complex groups which act on non-compact manifolds.

THEOREM 2. *Let M be a bounded domain in the space of several complex variables. If G is a complex connected group it cannot act effectively and in a complex way on M. In fact it can only act, in a complex way, in the trivial manner of leaving every point stationary.*

In order to prove the theorem, consider the transforming functions

$$f(g; x).$$

Since G is a complex group there must be local one-parameter complex groups through the orgin. This means (to consider one such group) that there is a complex analytic mapping with values in G

$$g(t)$$

defined for complex numbers t with

$$|t| < d$$

when d is a positive number, and where if

$$|t_1| < d, \qquad |t_2| < d, \qquad |t_1 + t_2| < d,$$

then

$$g(t_1 + t_2) = g(t_1)\, g(t_2). \tag{11}$$

However the mapping $g(t)$ can be extended so as to be defined and analytic for all t and so as to satisfy (11) for all t_1 and t_2. These extensions can be carried out by using (11).

Now let x be a definite point of M. Since G operates on a bounded domain, it follows that

$$f_i[g(t); x] \tag{12}$$

is an analytic bounded function defined for every complex number t. This function must be a constant and since for $t = 0$ its value is x_i it follows that for all t

$$f_i[g(t); x] = x_i. \tag{13}$$

But x was arbitrary and therefore (13) holds for all t and all x in M. Furthermore every point in the neighborhood of e in G is on a one-parameter subgroup. Hence for all x in M and all g in a neighborhood of e

$$f_i(g; x) = x. \tag{14}$$

But if (14) holds in a neighborhood it holds everywhere and it follows that, as stated in the theorem, G must act in a trivial way and certainly is not effective.

THEOREM. 3. *Let G be a complex connected group which acts in a complex way on a complex manifold M. If $F(x)$ is a bounded analytic function defined everywhere on M, then $F(x)$ is constant on each orbit of G.*

The method of proof of the preceding theorem carries over immediately to this theorem.

THEOREM 4. *If G is a semi-simple complex group and if G acts in a simply transitive way on a complex manifold, then M cannot be of the topological type of a cell. More generally, the homology groups of M* mod p *(p any prime) cannot all vanish.*

The semi-simple complex group G must contain a real compact group H [9] and H must contain an element g of order p, that is $g \neq e$, and

$$g^p = e.$$

Consider the homeomorphism $T(x)$ of M given by

$$T(x) = g(x).$$

This homeomorphism, being a group translation, can have no fixed point. On the other hand $T(x)$ is periodic of period p and would have to have a fixed point if all homology groups of M (mod p) vanished [10].

4. Non-Euclidean Geometry

Take a compact complex manifold of $2n$ real dimensions and let G be a complex group of $2r$ real dimensions which acts effectively on M. If $r > n$ there can exist no metric which is invariant under G. In fact, for $r > n$ there eixsts for any x in M a non-trivial complex group G_x which is the subgroup leaving x fixed. If there is an invariant metric, G_x would act on points y near x so that the orbits $G_x(y)$ lie in a coordinate neighborhood and this contradicts Theorem 2.

If G is semi-simple it is generated locally by complex extension of a compact real subgroup H of r real dimensions. Since H is compact, the topology of M can always be given by a metric invariant under H, for this is true of any compact group acting on M.

As an example, let M be complex projective space and let G be the projective group so that $r = n(n + 2)$. The group G contains the unitary unimodular group H of $n(n + 2)$ real dimensions and H has as one invariant metric the elliptic distance of Fubini [6] and Study. This metric of course cannot be invariant under G. As in the case of projective space then so in many other cases in the literature the invariant metric is elliptic in the sense of having positive Riemannian curvature.

However hyperbolic geometry, negative Riemannian curvature, arises in the theory of automorphic functions in several complex variables on bounded domains. In most cases, in particular in the case of matrix spaces, Hua [7], there exists a metric with negative curvature, the metric being invariant under the entire group of complex homeomorphisms, the latter group being in all interesting cases non-compact.

Thus we may loosely state that elliptic geometry is invariant under a compact group whose non-compact complex completion does not leave the metric invariant; whereas in hyperbolic geometry the metric is invariant under a maximal non-compact group but this group is not a complex Lie group.

5. Real Analytic Space homeomorphic to E_n with $(n - 1)$-dimensional Orbits

Let M be a real analytic manifold which is topologically homeomorphic to E_n. We use the expression "topologically homeomorphic" to emphasize that it is not assumed that M is analytically homeomorphic to E_n, by which we would mean that there is a homeomorphism between M and E_n which is analytic in both directions. Whether the existence of a topological homeomorphism between M and E_n implies the existence of an analytic homeomorphism is unknown, and appears to be a difficult question. As a matter of fact one of the conclusions of Theorem 5 below is that under an additional strong assumption there is such a homeomorphism.

Let G be a compact topological transformation group of M which acts effectively and which has at least one $n - 1$ dimensional orbit. It follows that [8]

(a) G is a Lie group.

(b) with a single exception all orbits are $(n - 1)$-dimensional.

(c) the exceptional orbit is a stationary point a.

(d) every two $(n - 1)$-dimensional orbits are homeomorphic, that is, all these orbits are homeomorphic to a fixed $(n - 1)$-dimensional manifold P.

(e) there is a topological ray Q beginning at a which pierces each orbit precisely once.

(f) the space $M - a$ is the topological product of $Q - a$ and P.

(g) the action of G on $M - a$ is such that if (p, q) is a point of the product of P and $Q - a$, then

$$g(p, q) = [g(p), q]$$

where $g(p)$ indicate the action of G in the homogeneous space P.

This topological information will be utilized in proving the following theorem.

THEOREM 5. *Let M be a real analytic manifold which has a positive definite analytic Riemannian metric and which is topologically homeomorphic to E_n. Let G be a compact effective group af analytic homeomorphisms of M onto itself which has at least one $(n - 1)$-dimensional orbit. Then there exists a reversibly analytic homeomorphism H of M onto E_n such that for any g in G, the transformation*

$$Hg H^{-1}$$

is an orthogonal transformation of E_n.

The functions $g_{ij}(x)$ which are analytic and define the Riemannian metric may be assumed to be such that the metric is invariant under the action of the group. For this situation can, if necessary, be obtained by integrating over the group G.

It is known [1] that in the vicinity of the stationary point there is an analytic change of coordinates under which all transformations g become orthogonal. We shall therefore assume that coordinates in the neighborhood of a have been chosen in this way. It follows from this that the $(n-1)$-dimensional orbits are homeomorphic to $(n-1)$-dimensional spheres and from (g) above it is evident that there is a topological homeomorphism taking M onto E_n under which each transformation g becomes orthogonal. We remark in passing that for the case of differentiable transformations and a differentiable manifold M, it would not be difficult to choose this homeomorphism in a differentiable manner. However for the analytic case which interests us a little more care is necessary.

Let x be any point of $M - a$. There is a geodesic through x in the direction orthogonal to the $(n-1)$-dimensional manifold $G(x)$. This geodesic (see Eisenhart [5, p. 210]) must be orthogonal to each $(n-1)$-dimensional orbit through which it passes. Starting at x and following this geodesic outwardly from $G(x)$ we see that no orbit is intersected more than once for if so the geodesic, at some point g between two points of intersection with a single orbit, would not be orthogonal to $G(y)$. Furthermore this geodesic cannot remain in any bounded domain for in order to do so it would also be necessary for it to fail to be orthogonal to certain orbits which it crossed. It follows, then, that this outward geodesic pierces each orbit outside $G(x)$ in one and only one point.

Starting along this geodesic in the inward direction from $G(x)$ it can be seen in a similar way that the inward geodesic must finally touch the stationary point a and that the analytic arc ax pierces each orbit between a and x in one and only one point.

Hence, if R denotes the geodesic ray beginning at a, continuing through x, and indefinitely beyond, as has just been described, we observe that

(1) R is an analytic ray.

(2) R touches each orbit in M in precisely one point.

The fact that R is an analytic ray means that R is the reversibly analytic image of the ray $0 \leqq s < d$ where d is a positive real number or infinity. The mapping from R to the ray as just described is made by means of the arc length.

Any two images of R under G either coincide or have only the point a in common. Near a, each image of R is an actual linear ray in the local coordinate system, and furthermore each point of $M - a$ has one and only one image of R passing through it.

In order to construct the mapping H from M to E_n we first construct a mapping H^* from M to the interior $s(d)$ of a sphere of radius d. If d is infinity the mapping H can then be chosen as coinciding with H^* and if d is not infinite there is an elementary and reversibly analytic homeomorphism carrying $s(d)$ onto E_n. Hence in any case the construction of the mapping H^* is sufficient to solve the problem.

The mapping H^* is constructed as follows. The stationary point a is mapped into the origin 0 of E_n. Any other point y of M has uniquely associated with it a geodesic distance from the origin along an image of R, as well as the angles at which this image of R approaches a (we recall that this image is a straight line segment in the local coordinate system around a). The point y is mapped to $H^*(y)$ where $H^*(y)$ is such that the distance from 0 to $H^*(y)$ is equal to the geodesic distance from a to y and also such that the segment $OH^*(y)$ makes the same angles as the geodesic ay. This defines $H^*(y)$ for all y in M and we see that this mapping is a reversibly analytic homeomorphism. By definition this homeomorphism is such that for all g in G

$$H^*gH^{*-1}$$

is an orthogonal transformation of the interior of the unit sphere. It has already been remarked that the required H can easily be constructed from H^*, so that the proof of the Theorem 3 is now complete.

6. Complex analytic space topologically homeomorphic to E_{2n} with $(2n-1)$-dimensional orbits

THEOREM. *Let M be a complex analytic manifold with a positive definite Hermitian metric which is topologically homeomorphic to E_{2n}, and let G be a compact group of complex analytic homeomorphisms of M onto itself which has at least one $(2n-1)$-dimensional orbit. Then there is a reversibly complex analytic homeomorphism H which maps M onto the interior of a sphere $S(d)$ (finite or infinite radius d) in the space C_n of n complex variables and which is such that for each g in G,*

$$HgH^{-1}$$

is a unitary map of $S(d)$ onto itself.

PROOF: This theorem follows essentially as a corollary of the preceding theorem. Consider the mapping H^* which has already been described. This is a real analytic mapping from M onto the interior of a sphere $S(d)$ and is given by $2n$ functions

$$u_p = \phi_p(x, y) \qquad v_p = \psi_p(x, y) \tag{15}$$

where u_p, v_p, $(p = 1, \cdots, n)$ are the real and imaginary parts of the complex $w_p = u_p + iv_p$ on $S(d)$ and x_p, y_p are the real and imaginary parts of local coordinates $z_p = x_p + iy_p$. If we consider the neighborhood of the fixed point in M which is being mapped on the neighborhood of the origin in the w-space then in that neighborhood there exists a complex analytic transformation

$$w_p = f_p(z_1, \cdots, z_n) \tag{16}$$

such that (16) takes each element of G to a unitary transformation [2]. However, it may be assumed that the mapping H^* is given by (16) in the vicinity of the stationary point. In terms of the real components (15) the special form (16) means

$$\frac{\partial u_p}{\partial x_q} - \frac{\partial v_p}{\partial x_q} = 0, \qquad \frac{\partial u_p}{\partial y_q} + \frac{\partial v_p}{\partial x_q} = 0 \tag{17}$$

for $p, q = 1, \cdots, n$. However the left sides of these relations are certain tensors in all of M, and if (17) holds in one neighborhood it holds throughout M. Therefore (15) has the special form (16) throughout M.

PRINCETON UNIVERSITY
INSTITUTE FOR ADVANCED STUDY, YALE UNIVERSITY

BIBLIOGRAPHY

1. BOCHNER, *Compact groups of differentiable transformations*, Annals of Math., vol. 46 (1945) pp. 372–381.
2. BOCHNER AND MONTGOMERY, *Groups of differentiable and real or complex analytic transformations*, Annals of Math., vol. 46 (1945), pp. 685–694.

3. BOCHNER AND MONTGOMERY, *Locally Compact groups of differentiable transformations*, Annals of Math., vol. 47 (1946), pp. 639–653.
4. H. CARTAN, *Sur les groupes de transformations analytiques*, Actualites Scien. et Industrielles 198, Paris, 1935.
5. EISENHART, Continuous groups of transformations, Princeton, 1933.
6. FUBINI, *Sulle metriche definite du una forma Hermitiona*, Istituto Veneto, vol. 63 (1904) pp. 502–513.
7. HUA, *On the theory of Fuchsian functions of several variables*, Annals of Math., vol. 47 (1946) pp. 167–191.
8. MONTGOMERY AND ZIPPIN, *A class of transformation groups in* E_n , Amer. Jour. of Math., vol. 63 (1941) pp. 1–8.
9. PONTRJAGIN, Topological Groups, Princeton, 1939.
10. P. A. SMITH, *Fixed point theorems for periodic transformations*, Amer. Jour. of Math., vol. 63 (1941) pp. 1–8.

Section J

Miscellaneous Mathematical Papers

Grouped together here are two papers on crystallography, one on the Euclidean algorithm, and obituary articles on Harald Bohr and John von Neumann.

Sonderdruck aus der Festschrift zum 60. Geburtstage „Arnold Sommerfelds"
Verlag von S. Hirzel in Leipzig

Über einen arithmetischen Satz der Röntgenkristallographie

Von
S. Bochner und **H. Seyfarth** in München

In der vorliegenden Note werden wir einen arithmetischen Satz beweisen, der die exakte Behandlung einer physikalischen Frage gestattet. Der Satz selbst und wesentliche Bestandteile des Beweises sind einer Arbeit des an zweiter Stelle genannten Verfassers entnommen[1]). Das Physikalische werden wir am Schlusse angeben.

I.

Wir bezeichnen in diesem Abschnitte mit m und n positive ganze Zahlen, mit p und r nicht-negative ganze Zahlen, mit a und b irgendwelche (reelle oder komplexe) Zahlen und mit ζ, η, φ irgendwelche Zahlen vom absoluten Betrage eins ($|\zeta| = |\eta| = |\varphi| = 1$). Weiterhin bezeichnen wir mit $\Lambda = (\lambda_1, \lambda_2, \lambda_3)$ irgendein Tripel ganzer rationaler Zahlen, mit anderen Worten irgendeinen Punkt des dreidimensionalen Zahlengitters. Für diese Gitterzahlen verabreden wir die Schreibweise

$$\Lambda_\sigma = (\lambda_{\sigma 1}, \lambda_{\sigma 2}, \lambda_{\sigma 3})$$

$$\Lambda_\sigma + \Lambda_\tau = (\lambda_{\sigma 1} + \lambda_{\tau 1}, \lambda_{\sigma 2} + \lambda_{\tau 2}, \lambda_{\sigma 3} + \lambda_{\tau 3}).$$

Nunmehr betrachten wir Ausdrücke der Gestalt

$$S(\Lambda) = \sum_{k=1}^{n} \zeta_{k1}^{r_1 + \lambda_1 p_1} \cdot \zeta_{k2}^{r_2 + \lambda_2 p_2} \cdot \zeta_{k3}^{r_3 + \lambda_3 p_3} = \sum_{k=1}^{n} \prod_{l=1}^{3} \zeta_{kl}^{r_l + \lambda_l p_l},$$

die wir als Funktionen der Gittervariablen Λ ansehen wollen, und befassen uns mit der Frage, für welche Wertekombinationen der Zahlen ζ, r, p und n die Funktion $S(\Lambda)$ identisch verschwindet,

$$S(\Lambda) = 0. \tag{1}$$

Wir ziehen eine Hilfsfunktion

$$D(\Lambda) = \sum_{k=1}^{n} a_k \prod_{l=1}^{3} \eta_{kl}^{\lambda_l} \tag{2}$$

heran, von der gleichfalls vorausgesetzt sei

$$D(\Lambda) = 0. \tag{3}$$

---

[1]) Ztschr. f. Krist. **67**, 131, 1928.

Für n voneinander unabhängige Gitterzahlen $\Lambda_1, \Lambda_2, \cdots \Lambda_n$ bilden wir successive die Ausdrücke.

$$\left.\begin{aligned}
&D(\Lambda_1),\\
&D(\Lambda_1, \Lambda_2) = D(\Lambda_1) - \prod_{l=1}^{3} \eta_{1l}^{-\lambda_{2l}} \cdot D(\Lambda_1 + \Lambda_2),\\
&D(\Lambda_1, \Lambda_2, \Lambda_3) = D(\Lambda_1, \Lambda_2) - \prod_{l=1}^{3} \eta_{2l}^{-\lambda_{3l}} \cdot D(\Lambda_1 + \Lambda_3, \Lambda_2),\\
&\cdots\cdots\cdots\cdots\cdots\cdots\cdots\cdots\\
&D(\Lambda_1, \Lambda_2, \cdots, \Lambda_{\varrho+1}) = D(\Lambda_1, \Lambda_2, \cdots, \Lambda_\varrho) - \prod_{l=1}^{3} \eta_{\varrho l}^{-\lambda_{\varrho+1,l}} \cdot D(\Lambda_1 + \Lambda_{\varrho+1}, \Lambda_2, \cdots, \Lambda_\varrho).
\end{aligned}\right\} \quad (4)$$

Man findet induktiv die Gleichungen

$$\left.\begin{aligned}
&D(\Lambda_1, \Lambda_2) = \sum_{k=2}^{n} a_k \prod_{l=1}^{3} \eta_{kl}^{\lambda_{1l}} \left(1 - \prod_{l=1}^{3} (\eta_{kl}\eta_{1l}^{-1})^{\lambda_{2l}}\right),\\
&D(\Lambda_1, \Lambda_2, \Lambda_3) = \sum_{k=3}^{n} a_k \prod_{l=1}^{3} \eta_{kl}^{\lambda_{1l}} \left(1 - \prod_{l=1}^{3} (\eta_{kl}\eta_{1l}^{-1})^{\lambda_{2l}}\right)\left(1 - \prod_{l=1}^{3} (\eta_{kl}\eta_{2l}^{-1})^{\lambda_{3l}}\right),\\
&\cdots\cdots\cdots\cdots\cdots\cdots\cdots\cdots\\
&D(\Lambda_1, \Lambda_2, \cdots, \Lambda_n) = a_n \prod_{l=1}^{3} \eta_{nl}^{\lambda_1} \prod_{\varrho=1}^{n-1} \left(1 - \prod_{l=1}^{3} (\eta_{nl}\eta_{\varrho l}^{-1})\, \lambda_{\varrho+1,l}\right).
\end{aligned}\right\} \quad (5)$$

Aus (3) folgt, nach dem Bildungsgesetz (4), daß für alle Wertekombinationen $\Lambda_1, \Lambda_2, \cdots, \Lambda_n$

$$D(_1, \Lambda_2, \cdots, \Lambda_n) = 0. \tag{6}$$

Machen wir die Voraussetzung $a_n \neq 0$, also auch $a_n \prod_{l=1}^{3} \eta_{nl}^{\lambda_{1l}} \neq 0$, so kann (6) nur in der Weise zustandekommen, daß für einen passenden Wert ϱ_0, $1 \leqq \varrho_0 \leqq n-1$,

$$\eta_{nl} = \eta_{\varrho_0 l} \quad \text{für} \quad l = 1, 2, 3.$$

Gäbe es nämlich zu jedem Index ϱ einen Index l_ϱ, so daß

$$\eta_{n l_\varrho} \neq \eta_{\varrho l_\varrho},$$

so wäre $D(\Lambda_1, \Lambda_2, \cdots, \Lambda_n) \neq 0$, falls man die Gitterzahlen folgendermaßen definiert: Λ_1 ist beliebig und $\Lambda_{\varrho+1}$, $\varrho = 1, 2, \cdots, n-1$, hat die „Komponenten"

$$\lambda_{\varrho+1,\mu} = \begin{cases} 1 \text{ für } \mu = l_\varrho, \\ 0 \text{ für } \mu \neq l_\varrho. \end{cases}$$

In der die Funktion $D(\Lambda)$ definierenden Summe (2) fassen wir diejenigen Summanden zusammen, für welche die Zahlen $\eta_{k1}, \eta_{2k}, \eta_{k3}$ je einander gleich sind. Genauer formuliert machen wir folgendes: Wir verteilen die Gesamtheit der Indexreihe $k = 1, 2, \ldots, n$ auf ein Schema

$$\left.\begin{matrix}
k_{11}, & k_{12}, & \ldots\\
k_{21}, & k_{22}, & \ldots\\
\cdots & \cdots & \cdots\\
k_{m1}, & k_{m2}, & \ldots
\end{matrix}\right\} \quad (7)$$

von der folgenden Beschaffenheit: Für je zwei Indices k_1 und k_2, aus der gleichen Zeile:

$$k_1 = k_{\mu\varrho}, \qquad k_2 = k_{\mu\sigma}$$

ist

$$\eta_{k_1,l} = \eta_{k_2,l} \quad \text{für} \quad l = 1, 2, 3\,;$$

sind hingegen die Indices k_1 und k_2 verschiedenen Zeilen entnommen, so gibt es ein $l_0 = l(k_1, k_2)$, so daß

$$\eta_{k_1,l_0} \neq \eta_{k_2,l_0}.$$

Bezeichnet man nunmehr den gemeinsamen Wert derjenigen Zahlen η_{kl}, deren Indices k der μ-ten Zeile von (7) entnommen sind, mit φ_{kl}, und bezeichnet man die Summe der entsprechenden Zahlen a mit b_μ,

$$b_\mu = a_{k_{\mu_1}} + a_{k_{\mu_2}} + a_{k_{\mu_3}} + \cdots,$$

so nimmt die Funktion $D(\Lambda)$ die Gestalt an

$$D(\Lambda) = \sum_{\mu=1}^{m} b_\mu \prod_{l=1}^{3} \varphi_{\mu l}.$$

Auf Grund von (3) schließen wir nunmehr, daß alle b_μ verschwinden müssen:

$$b_\mu = 0, \quad \mu = 1, 2, \ldots, m.$$

Denn wäre nur ein b_μ von Null verschieden, etwa $b_m \neq 0$, so müßte es auf Grund der obigen Überlegungen mindestens einen Index $\varrho, \varrho < m$, geben. so daß

$$\varphi_{\varrho l} = \varphi_{ml} \quad \text{für} \quad l = 1, 2, 3,$$

und dies widerspricht der Definition der Größen φ.

Wir kehren zu unserer eigentlichen Funktion $S(\Lambda)$ zurück. Durch die Substitution

$$\eta_{kl} = \zeta_{kl}^{p_l}$$

$$a_k = \prod_{l=1}^{3} \zeta_{kl}^{r_l}$$

geht die Funktion $S(\Lambda)$ in eine Funktion $D(\Lambda)$ und unsere Betrachtung für die Funktion $D(\Lambda)$ in eine Aussage für die Funktion $S(\Lambda)$ über. Wir schreiben

$$S(\Lambda) = \sum_{k=1}^{n} A_k(\Lambda), \qquad A_k = \prod_{l=1}^{3} \zeta_{kl}^{r_l + \lambda_l p_l}$$

und interpretieren jede Zahl $A_k(\Lambda)$ als eine Vektorstrecke in einer Gaussschen Ebene. Jede Strecke A_k hat die absolute Länge 1. Den Wert $S(\Lambda)$ denken wir uns, bei fest gegebenem Λ, so versinnbildlicht, daß wir von einem festen Punkte, etwa dem Anfangspunkte, der Gaussschen Ebene ausgehend, zuerst die Strecke A_1 auftragen, an den „positiven" Endpunkt von A_1 die Strecke A_2 antragen, an A_2 wieder A_3 anschließen usw. Auf diese Weise erhalten wir einen zusammenhängenden Polygonzug, dessen Gestalt offenbar von Λ abhängt. Der Polygonzug ist auch bei festem Λ nicht eindeutig, hängt vielmehr von der Reihenfolge der Summanden ab. Wenn wir aber Polygonzüge für verschiedene Werte von Λ miteinander vergleichen, setzen wir stillschweigend voraus, daß in allen die Anordnung der Summanden die gleiche ist.

Die Voraussetzung $S(\Lambda) = 0$ ist damit gleichbedeutend, daß alle Polygonzüge geschlossen sind, und wir haben bewiesen, daß man unter dieser Voraussetzung die Strecken $A_1, A_2, \ldots, A_n$ derart auf ein Schema

$$\begin{array}{l} A_{11}, A_{12}, \ldots \\ A_{21}, A_{22}, \ldots \\ \ldots\ldots\ldots\ldots \\ A_{m1}, A_{m2}, \ldots \end{array}$$

aufteilen kann, daß das „Polygon" $S(\Lambda)$ in m Teilpolygone zerfällt.

$$S(\Lambda) = S_1(\Lambda) + S_2(\Lambda) + \cdots + S_m(\Lambda),$$

$$S_\mu = A_{\mu_1} + A_{\mu_2} + \cdots,$$

$$S_\mu = \varphi_{\mu_1}^{\lambda_1}\varphi_{\mu_2}^{\lambda_3}\varphi_{\mu_2}^{\lambda_3}(a_{\mu_1} + a_{\mu_2} + \cdots), \tag{8}$$

von denen jedes geschlossen ist, und bei Variation der Gitterzahl Λ nur Drehungen erleidet, also sich immer gleichsinnig kongruent bleibt; denn die Summe $(a_{\mu_1} + a_{\mu 2} + \cdots)$ ist von Λ unabhängig, und der in (8) davorstehende Ausdruck ist zwar mit Λ variabel, aber vom absoluten Betrage eins und demnach nur ein Richtungsfaktor.

II.

Die Zerlegung in drehungsinvariante Polygone ist nicht eindeutig. In vielen Fällen kann man jedes der $S_\mu(\Lambda)$ weiter unterteilen. Man kann insbesondere eine derartige Unterteilung geben, bei der jedes Polygon im euklidischen Sinne regulär ist. Das beruht auf folgendem algebraischen Hilfssatz.

Gegeben seien eine positive ganze Zahl α und ein von α verschiedener positiver Teiler d von α. Unter der Funktion $\varphi(x;\ \alpha, d)$ verstehen wir die Funktion

$$\varphi(x;\alpha,d) = \sum_{\nu=0}^{\alpha/d-1} x^{\nu d} = \frac{x^\alpha - 1}{x^d - 1}.$$

Unter ε verstehen wir irgendeine feste primitive α-te Einheitswurzel. Es ist

$$\varphi(\varepsilon;\alpha,d) = 0. \tag{9}$$

Man betrachte irgendeine

$$f(x) = \sum_A^B a_k x^k, \qquad A < B, \tag{10}$$

in welcher A, B und die a_k irgendwelche (positive oder negative) ganze Zahlen sind. Falls $x = \varepsilon$ eine Nullstelle von $f(x)$ ist,

$$f(\varepsilon) = 0, \tag{11}$$

kann man Teiler $d_1, d_2, \ldots, d_j$ von α und dazugehörige ganze Zahlen $c_1, c_2, \ldots, c_j$ und $q_1, q_2, \ldots q_j$ finden, so daß

$$f(x) \equiv \sum_{\nu=1}^{j} c_\nu x^{q_\nu}(x;\ \alpha,\ d_\nu). \tag{12}$$

Ist insbesondere α eine Primzahlpotenz, $\alpha = \pi^u$, so kann man alle d_ν einander gleich wählen, nämlich $d_\nu = \pi^{u-1}$.

Beweis: Wir können für den Beweis annehmen, daß $A \geqq 0$. Sonst betrachte man die Funktion $f(x) = x^{-A} \cdot f(x)$. Aus jeder Aufspaltung

$$f_1(x) = \sum_{\nu=1}^{j} c_\nu x^{q_\nu} \varphi(x;\, \alpha,\, d_\nu)$$

geht dann, wenn man q_ν durch $q_\nu + A$ ersetzt, eine Aufspaltung für $f(x)$ selbst hervor.

Es sei vorerst α eine Primzahlpotenz, $\alpha = \pi^u$. Die Funktion

$$\varphi(x;\, \pi^u,\, \pi^{u-1})$$

enthält genau alle primitiven αten Einheitswurzeln als Nullstellen und ist demnach irreduzibel[1]). Wegen (11) gibt es daher ein Polynom $g(x) = \sum_{\nu=1}^{j} c_\nu x^\nu$ mit ganzzahligen Koeffizienten c, so daß

$$f(x) = g(x) \cdot \varphi(x)\,,$$

womit unser Satz bewiesen ist.

Im allgemeinen Falle $\alpha = \pi_1^{u_1} \cdot \pi_2^{u_2} \ldots \pi_\varrho^{u_\varrho}$ kommen die Funktionen

$$\varphi_\sigma(x) = \varphi\left(x;\, \alpha,\, \frac{\alpha}{\pi_\sigma}\right)$$

in Frage. Man kann leicht die Nullstellen dieser Funktionen angeben und auf Grund dessen feststellen, daß ihre gemeinsamen Nullstellen genau die primitiven αten Einheitswurzeln sind. Von ihrem größten gemeinsamen Teiler $\Delta(x)$ gilt daher wiederum

$$f(x) = g(x) \cdot \Delta(x)\,,$$

und andererseits gilt von ihm, daß er in der Form

$$\Delta(x) = \sum_{\sigma=1}^{\varrho} \Delta\sigma(x) \cdot \varphi_\sigma(x)$$

dargestellt werden kann. Die Funktionen $g(x)$ und $\Delta_\sigma(x)$ sind hierbei Polynome mit ganzzahligen Koeffizienten; also gibt es tatsächlich eine Aufspaltung der Form (12).

Wir kehren jetzt zur Funktion $S(\Delta)$ zurück. Unter $\bar{S}(\Delta)$ verstehen wir irgendeine festgehaltene unter den Funktionen $S_\mu(\Delta)$. Unter Fortlassung des Index μ ist also definitionsgemäß

$$\bar{S}(\Delta) = \prod_{l=1}^{3} \varphi_l^{\lambda_l} \cdot \sum_{k=1}^{N} b_k$$

mit

$$b_k = \prod_{l=1}^{3} \zeta_{kl}^{r_l}\,, \qquad (k = 1, 2, \cdots, N)$$

und

$$\varphi_l = \zeta_{\varrho l}^{p_l} = \zeta_{\sigma l}^{p_l} \qquad (l = 1, 2, 3;\, \varrho, \sigma = 1, 2 \cdots, N).$$

---

[1]) Wegen der in diesem Abschnitte verwandten Sätze über Einheitswurzeln und Teilbarkeit von Polynomen vergleiche man: O. Perron, Lehrbuch der Algebra, 1927, Bd. I, Satz 78; Bd. II, Satz 59 und § 35. — Die vorliegende einfache Fassung des Beweises verdanken wir einer Mitteilung von Herrn Perron.

Wenn wir die Größen ω_{kl} durch die Relationen

$$\zeta_{kl} = \zeta_{1l}\,\omega_{kl} \quad (l = 1, 2, 3; k = 1, 2 \cdots, N)$$

einführen, ist

$$\bar{S}(\Lambda) = \prod_{l=1}^{3} \zeta_{1l}^{r_l+\lambda_l p_l} \cdot \sum_{k=1}^{N} \beta_k$$

mit

$$\beta_k = \sum_{l=1}^{3} \omega_{kl}^{r_l},$$

wobei die ω_{kl} Einheitswurzeln sind,

$$\omega_{kl}^{p_l} = 1.$$

Setzt man

$$\alpha = p_1\, p_2\, p_3,$$

so ist offenbar jedes ω_{kl}, und daher jedes Produkt $\prod_{l=1}^{3} \omega_{kl}^{r_l}$, eine αte Einheitswurzel. Mit Hilfe einer primitiven αten Einheitswurzel ε kann man daher setzen

$$\sum_{k=1}^{N} b_k = f(\varepsilon) = \sum_{k=1}^{N} a_k\, \varepsilon^k.$$

Wegen der Geschlossenheit des Polygons $\bar{S}(\Lambda)$ ist $f(\varepsilon) = 0$. Auf Grund des voraufgehenden Hilfssatzes kann man daher $\bar{S}(\Lambda)$ und somit auch das Gesamtpolygon $S(\Lambda)$ in eine Summe von regulären Polygonen zerlegen. Wir bemerken, daß man durch eine geringfügige Modifikation unserer Überlegung einen schärferen Zahlenwert für den Exponenten α erhalten kann. Bezeichnet man mit (a, b) den größten gemeinsamen Teiler der ganzen Zahlen a und b, so kommt man bereits mit dem Exponenten

$$\alpha = \prod_{l=1}^{3} \frac{p_l}{(p_l, r_l)} \tag{13}$$

aus.

III.

Für einen vorgelegten Kristall bezeichnen wir die Koordinaten der n Basisatome in einem kristallfesten, affinen Koordinatensysteme mit

$$\varrho_l^k (l = 1, 2, 3; \quad k = 1, 2, \ldots, n).$$

Wenn wir auf den Kristall einen festgerichteten Röntgenstrahl auffallen lassen, so wird zu jeder Wellenlänge die Gesamtheit der Kristallagen, für welche eine Braggsche Reflexion in Frage kommt, durch die Gesamtheit aller Tripel von ganzen Zahlen

$$H = (h_1, h_2, h_3), \qquad -\infty < h_1, h_2, h_3 < +\infty$$

beschrieben (es sind dies die bekannten Millerschen Indices). Die Intensität der Reflexionsstrahlen ist dem Quadrat des absoluten Betrages des „Strukturfaktors"

$$S_H = \sum_{k=1}^{n} e^{2\pi i \sum_{l=1}^{3} \varrho_l^k h_l}$$

proportional. Es kann vorkommen, daß für ein Teilgitter

$$h_l = r_l + \lambda_l p_l, \quad (l = 1, 2, 3; \quad -\infty < \lambda_l < +\infty) \tag{14}$$

der Strukturfaktor S_H null wird, daß also die Gitterfunktion

$$S(\Lambda) = \sum_{k=1}^{n} \prod_{l=1}^{3} \zeta_{kl}^{r_l + \lambda_l p_l}$$

mit

$$\zeta_{kl} = e^{2\pi i \varrho_l^k}$$

identisch verschwindet.

Wenn nun für einen Kristall ein Auslöschungsgesetz (14), d. h. Konstanten r_l und p_l beobachtet sind, so kann man nach obigem auf folgende Relationen zwischen den Zahlen ϱ_l^k und n schließen.

Die Gesamtheit der Indices κ verteilt sich auf ein Schema (7), das in jeder Zeile mindestens zwei Zahlen enthält von der folgenden Beschaffenheit. Zu je zwei Indices μ und ν aus der gleichen Zeile gibt es ganze Zahlen $g_{\mu,\nu}^l$, so daß

$$\varrho_l^\mu - \varrho_l^\nu = \frac{g_{\mu,\nu}^l}{p_l}.$$

Diese Beziehungen bleiben zu Recht bestehen, wenn in dem Strukturfaktor den einzelnen Basisatomen verschiedene Gewichte ψ_k zukommen,

$$S_H = \sum_{k-1}^{n} \psi_k e^{2\pi i \sum_{l=1}^{3} \varrho_l^k h_l}.$$

Im speziellen Falle, daß alle ψ_k untereinander gleich sind, bestehen für die ϱ_l^k aus einer und derselben Zeile unseres Schemas (7) weitere Beziehungen. Bezeichnen wir mit N die Anzahl der Indices der vorgelegten Zeile, und bringen wir die durch (13) definierte Zahl α in die Form

$$\alpha = \pi_1^{u_1} \cdot \pi_2^{u_2} \cdots \pi_s^{u_s},$$

so kann man die Zahl N in der Gestalt

$$N = N_1 \pi_1 + N_2 \pi_2 + \cdots + N_s \pi_s$$

mit nicht-negativen Zahlen N_σ schreiben und die zugehörigen Indices k (und dementsprechend die Koordinaten ϱ_l^k) in N_1-Gruppen zu π_1, N_2-Gruppen zu $\pi_2, \ldots, N_s$-Gruppen zu π_s-Elementen anordnen, so daß folgende Relationen bestehen: $\sum_{l=1}^{3} (\varrho_l^\tau - \varrho_l^1) r_l = \frac{\tau}{\pi_\sigma} +$ eine beliebige ganze Zahl ($\tau = 1, 2, \ldots, \pi_\sigma - 1$). Hierin sind: π_σ eine der in Frage kommenden Primzahlen $\pi_1, \pi_2, \ldots, \pi_\sigma$ und, bei festgehaltenem π_σ, die ϱ_l^τ und ϱ_l^1 die Koordinaten aus einer der N_σ-Gruppen zu π_σ-Elementen. Demnach ergeben sich für jede Zeile unseres Schemas (7) $N_1 + N_2 + \cdots + N_s$ und, über alle Zeilen summiert, alles in allem n Relationen. Ähnliche Relationen gibt es auch dann noch, wenn die ψ_k nicht alle einander gleich sind, sondern in rationalen Verhältnissen zueinander stehen.

Alle unsere Überlegungen sind durchaus davon unabhängig, daß die Dimensionszahl 3 beträgt, d. h. daß dem Index l die Werte 1, 2, 3 zukommen. Vielmehr behalten sie für irgendeine feste Dimensionszahl Gültigkeit. Davon könnte man unter Umständen Gebrauch machen bei Anwendung auf Vierindices-Symbole hexagonaler Systeme.

REMARK ON THE EUCLIDEAN ALGORITHM

S. Bochner*.

[*Extracted from the Journal of the London Mathematical Society, Vol.* 9, *Part* 1.]

For polynomials of one variable (with, say, real coefficients) the proof of the existence of a greatest common divisor and, subsequently, of a unique resolution into prime factors is generally based on the Euclidean algorithm, or some reasoning equivalent to it. For polynomials of more variables the theorems remain true, but they cannot be proved in the same straightforward manner, because the Euclidean algorithm "fails" to be applicable. Text-books on algebra, as a rule, do not sufficiently discuss the true nature of that "failure". The following discussion is, perhaps, the simplest one.

Let there be given two polynomials f_0, f_1 in a fixed order. We say that the Euclidean algorithm is applicable to f_0, f_1, if there exist, for an integer n, polynomials $q_1, q_2, \ldots, q_n; f_2, \ldots, f_n$, *of arbitrary degrees*, such that

$$\text{(A)} \qquad \begin{cases} f_0 = q_1 f_1 + f_2, \\ f_1 = q_2 f_2 + f_3, \\ \ldots \quad \ldots \quad \ldots \\ f_{n-2} = q_{n-1} f_{n-1} + f_n, \\ f_{n-1} = q_n f_n. \end{cases}$$

Now we shall show that there are polynomials f_0, f_1 of two variables to which the algorithm is not applicable. If the set of relations (A) is satisfied, then, by a familiar argument,

(1) f_n is a common divisor of f_0 and f_1, and

(2) there exist polynomials p_0, p_1 such that

$$f_n = f_0 p_0 + f_1 p_1.$$

But if $f_0 \equiv x, f_1 \equiv y$, where x, y are independent variables, there, obviously, exists no common divisor other than a constant, and, on the other hand, the expression

$$x p_0(x, y) + y p_1(x, y)$$

cannot be a non-vanishing constant.

* Received and read 15 June, 1933.

Printed by C. F. Hodgson & Son, Ltd., Newton St., London W.C.2.

Reprinted from the
BULLETIN OF THE AMERICAN MATHEMATICAL SOCIETY
Vol. 58, No. 1, pp. 72-75
January, 1952

HARALD BOHR

April 22, 1887–January 22, 1951

Born in Copenhagen, Denmark, lived, learned and taught there, and died there; but liked to visit near and far. His father was Christian Bohr, Professor of physiology at the medical school; his mother was Jewish, and he had to flee the country in 1943, going then to Sweden for two years. During his last years he was Master of the one resident college in Copenhagen and he greatly enjoyed being surrounded by young people. Bohr was a very friendly person, kind and good-natured, chatty but not indiscreet, and he made and kept to his friends all his life. For all that, he was a shade melancholy, but this side of him was brightened by being married exceedingly happily. In his younger years he had been a football (or rather soccer) player of some renown, a rare enough occurrence among budding savants on the Continent. In after-years he would trounce a ball if he saw one, but he did not look much of an athlete.

Bohr's fame is his almost periodic functions. He did the theory in his mid-thirties, and very little in other topics afterwards. He wrote afterwards many papers in his theory, though, and rather substantial ones too. But the vim of the first papers did not return, and by now Bohr lives forth as much in the work of others as in his own. Privately, he gloried in the interest his theory has evoked and he would help to propagate it. He loved to lecture, and he lectured extremely well. In an individual lecture, no matter how learned the audience, he would first recite the definition and basic properties and make introductory comments; and he would do the same in print even when full references were included.

Whether he had an opinion as to future potentialities of the theory we do not know. Right now it is rather quiet about the theory, except for some distant rumbling in the field of automorphic functions, especially of several variables, which might precipitate in a new shower of almost periodicity. However that may be, it may be said that Harald Bohr has had the good fortune of having a very distinctive theory of magnitude very clearly and originally attached to his name, and this, after all, is as much as a mathematician may expect to secure for himself in a lifetime.

When Bohr was a student the study of mathematics at Copenhagen was not organized, and Bohr learned in a somewhat haphazard way from persons like the geometer Zeuthen, the astronomer Thiele, and the analyst Niels Nielsen (cylinder functions) who made a fierce play for mathematical rigor but not always very successfully; but he could not learn from J. L. W. V. Jensen, probably the greatest of them all, because he was an electrical engineer by profession and a self-taught mathematician, and had no connection with academic life. The course fare was generously supplemented by reading in books such as the second edition of C. Jordan's *Cours d'analyse* (of Jordan theorem fame) and the *Zahlentheorie* by Dirichlet and Dedekind.

Even before earning his masters degree in 1909 (the Ph.D. degree, which was quite a serious affair, followed in 1910) Bohr was led to the study of the theory of summability, mainly of Dirichlet series, and he dwelt on this topic quite a number of years. Summability, even in its pre-Tauberian era, used to be an important topic of study which even algebraists of the stature of Frobenius and Schur would dignify with their attention, and the reasons for this were manifold. First of all, to the uninitiated it exuded a certain mystique of esoterically conjuring up a limit where none had been before. In the second place (and this led to the most far-reaching consequences), Fejér had succeeded in "summing" the Fourier series of a continuous function towards its value, where DuBois Reymond had distressingly shown that the partial sums themselves need not so converge. And finally, and this must have been decisive for Bohr for the time being, it held out some hope of "moving-in" on the zeta function in its critical strip (or at least on L-series without a pole) where other means of intrusion had not been successful. This purpose proper was not to be attained, really, but a new view on summability did remain. Actually, Bohr's contribution was eventually absorbed in M. Riesz' more felicitous approach, and the resulting summability is named after Riesz, and not Bohr; but Bohr himself described his share in the theory in his article written jointly with H. Cramér, *Die neuere Entwicklung der analytischen Zahlentheorie* in Encyklopädie der Math. Wiss. Bd. II-3, Heft. 6, pp. 722–849.

We might mention that this early work of his brought Bohr soon in contact with E. Landau, and he spent 1909–1910 in Göttingen, which place he used to revisit frequently. He also became a very close friend of Hardy's, and also knew Littlewood very well, and stayed at Oxford and Cambridge in 1913–1914; and finally he went for several months in 1914 to Lebesgue in Paris to study his measure from him first-hand.

We think that the dominant interest of Harald Bohr was diophantine approximation and (equi-) distribution of numbers, but in a certain broad sense that would not lead him to a study of Weyl sums and the Waring problem, say. Somehow, to Bohr, the entire theory of almost periodic functions was an emanation from these sources, and this reflects itself best in his own proof for his "completeness theorem" which is so unlike any given afterwards and, in a nontechnical sense, is the most arithmetical among them all. Bohr was also infatuated with the classical theorem of Kronecker itself, and he repeatedly would give a new proof for it, which after all was not such an urgent task when compared with all the new problems suggesting themselves. Also, several years before arriving at the new theory, and repeatedly afterward, Bohr studied the zeta function and related functions as regards average frequency with which values are attained in the critical strip.

Bohr was the creator of the almost periodic functions, and he gave them such an appropriate name that no need arose for turning them into "fonctions bohriennes." Actually Bohr had a forerunner in the person of P. Bohl, who came within sight of the theory but missed it, although he has to his credit at least one nontrivial theorem which Bohr could take over virtually without change. One might even claim that Lagrange had anticipated a part of the theory by his study of mean motion. But after all this is conceded, the fact remains that almost periodic functions were Bohr's in the same sense as Lie groups were Lie's and in a stronger and much more significant sense than Banach spaces were Banach's, say. Bohr's high points of achievement consist of his having broken through towards a general definition for his functions such as he has found it, and of having then succeeded in proving his completeness theorem in his own fussy manner, such as he did; and as regards the genesis of his definition it is worth looking up a little known paper of his in which a tentative version is sought after (*Über eine quasi-periodische Eigenschaft Dirichletschen Reihen mit Anwendung auf die Dirichletschen L-funktionen*, Math. Ann. vol. 85, pp. 115–122). Bohr's personal triumph in attaining these achievements must be assessed against the fact that Bohr had been an arithmetico-analyst only, and always remained one only, and that nevertheless the true orientation of his achievements was towards topology, general group theory, homogeneous spaces, and such like directions, all of which had been rather alien to most arithmetico-analysts of his generation. We might perhaps point out for what it is worth that Hardy and Landau, both very prolific writers and close friends to Bohr, never took up the new topic themselves, or as much as published a note about it.

The brisk development of almost periodic functions on groups, as initiated by von Neumann, has by the very rapidity of its progress very quickly led to the sobering realization that after inherent compactifications have been introduced, the general theory is by actual scope, though not by initial intent, not larger than a theory of functions on compact groups only, or compact homogeneous spaces at best; and theories of functions on genuinely open groups are yet to be developed. In Euclidean space the current body of Fourier analysis including the theory of generalized Fourier integrals (alias "distributions tempérées") is a partial answer to the problem, perhaps. But on open semi-simple group spaces, say, the developments are not yet foreseeable, unless the recent work of Gelfand and Neumark and others is an indication of them.

If we contemplate Bohr's work the thought suggests itself that some of what is best in algebra, group theory, and topology has been coming forth from the fullness and fertility of analysis and has been stimulated if not always brought to surface by analysts themselves; and that perhaps analysis is the stuff of which our mathematics is made of, after all.

SALOMON BOCHNER

JOHN VON NEUMANN

December 28, 1903—February 8, 1957

BY S. BOCHNER

JOHN VON NEUMANN's academic career was in mathematics, even though he was very much associated with theoretical physics and to an extent with theoretical economics. He began his career in earnest with axiomatic set theory and logic and pursued this topic actively for several years. In the closing years of his life, while busily filling many positions as scientific executive on a national level, he was laying plans for a treatise in the "logics" underlying the mathematical foundations of games and strategies with applications to economics, of the theory of codifying instructions to computing machines, and of the theory of networks both electrical and nervous. The activities mentioned did not complete his professional life by any means. But in a sense they were very characteristic of the man and his basic style.

The contemporary mathematical activities we have just enumerated, in which von Neumann so excelled, whatever their implications for mid-century's modern man and whatever their fascination for the philosopher in search of modernity, are not yet too well established with the old-line mathematician, and the density of their mathematical content is as yet not always deemed to be optimal. In fact, von Neumann's mathematician's mathematics was of stronger texture by its very nature, and his prowess in theoretical physics also had to maintain itself rather more competitively. But his lifelong commitment to questions of logics, axiomatization, and formalization was expressive of his mathematical personality, nonetheless.

In his more traditional achievements he had his peers, and in his work in theoretical physics he did not originate a basic verity as several of his contemporaries in the 1920s did. But if to his preponderance in mathematics and physics, and his familiarity with many other problems of science, one adds his momentous excursion into the theory of rational behavior—as the implications of games and strategies are named by some—then it can be said that, among mathematicians at any rate, he was an exceptional contemporary indeed. By intellectual faculty he was a mathematician, but by species he was a scientific man within all the major connotations of today, from esoteric ones to those of popular appeal.

His effectiveness was largely due to his ever-present mental manipulatory quickness. He could literally "think on his feet," and much of his best work may have received its initial impulse in just this way. He had a prodigious memory, and legend has it that he knew all the facts and dates from many volumes of standard histories by heart.

Apart from achievements in the areas enumerated above, the high points of his work are as follows: (1) the extension of operators in Hilbert space from bounded to unbounded ones (# 30 of the Bibliography), (2) almost periodic functions on groups (# 52), and (3) his masterwork, algebras of bounded operators in Hilbert space (beginning with # 62). To these one might add (4) his definition of a "state" in quantum theory as an element in a Hilbert space (beginning with # 9) and (5) the introduction of analytic parameters in a locally Euclidean group in the compact case (# 48).

1. His theory of non-bounded operators in Hilbert space was arresting. The naturalness with which he axiomatized the Hilbert space and introduced concepts like maximal operator, hypermaximal operator, deficiency of an operator, etc., was breathtaking to specialist and nonspecialist alike. And yet there was nothing greatly contrived in the analysis, nor was there anything that some of the other mathematicians who were familiar with the topic could not have really done. As a matter of fact, Frederic Riesz, who had known the theory

as well as anybody, after catching his breath wrote a comment on the new theory, in the introduction to which he asks himself somewhat plaintively what he, Frederic Riesz, had been doing with himself all these years.

2. His theory of almost periodic functions on groups was glamorous. It soon turned out that syllogistically the theory can be put together much more simply than von Neumann had done it. But the effect of the theory, especially on the so-called Abstract Harmonic Analysis which was soon to emerge, was a decisive one, and he fully deserved obtaining the Bôcher prize for it.

3. The theory of algebras of operators, written in important phases in collaboration with F. J. Murray, has not the ease and naturalness of much of his earlier work and it is somewhat heavy in the manner of composition, but in strength, tenacity, and importance it is not outdone by anything in its generation. Any representation of a group by bounded operators (in an infinitely dimensional space) can be enlarged to such a representation of the enveloping group algebra and thus subsumes under the Murray-von Neumann theory. Now it is a fact that the theory of representations in the particular case of a group algebra has been and continues to be strongly guided by the more general theory, and in certain parts must still rely on taking over results directly from it.

4. Dr. von Neumann's papers in the 1920s on the foundations of quantum mechanics and their effect on thermodynamics were extremely mature. His formal definition of a physical "state" as being a point in a Hilbert space representing the system was as readily accepted by physicists as if it had been the most obvious thing since creation, and this concept of a "state" has had some repercussions in a purely mathematical context as well.

5. His solution of "Hilbert's fifth problem" in the compact case had no bearing on the subsequent solution by others in the non-compact case, but it is nonetheless interesting as an achievement of von Neumann's. He shows in it his mastery of finite linear representations, and he employs in it the group invariant measure which had

just been found by Haar. It is somewhat surprising that von Neumann did not construct the group invariant measure himself, since measure theory, as a bridge between Boolean algebras and linear functionals on vector spaces, was very close to his thinking. But at any rate, soon after this measure had been introduced, he studied it extensively and contributed much to its propagation.

Dr. von Neumann also attached major standing to his work on the ergodic theorem, 1931–32 (#41). On one occasion, he listed his essential work under the following three headings: the papers on quantum mechanics in the 1920s; his total work in operator theory, 1930–39; his version of the ergodic theorem.

Most of von Neumann's published work involves in a decisive phase linear spaces, finite or infinitely dimensional ones, or functionals and linear operations between such. The approach via linear spaces covers, or can be made to cover, a large part of the total mathematical area nowadays. Still, an important part of the landscape is not included in it. It leaves out almost anything which involves "geometry" in a proper global sense, such as topology, differential geometry and harmonic integrals, general relativity globally, algebraic geometry, etc. Into this kind of mathematics von Neumann ventured rarely, if at all, and he actually had no affinity with it in his thinking. An incident in his work may illustrate this. In 1929 he published the important result (# 24) that a group of linear transformations in a finite dimensional space, if closed, is a Lie group. Élie Cartan, who was a geometer, immediately recognized that the theorem is a special case of the general proposition that a closed subgroup of a Lie group is again a Lie group, and he proved it by the same procedure, although in an abbreviated form. There is hardly a young mathematician today who realizes that this familiar general proposition was von Neumann's discovery originally.

The various parts of mathematics which involved geometry have of late been tending to coalesce into one, and thus to restratify the classical traditional body of mathematics into a new pattern. This is a significant development, and to a historical observer it would almost appear as if this mathematics were harking back to its Pythagorean origin, when all mathematics was geometry and "structure" in dissoluble union, by deliberate design, and less naively so than is sometimes assumed. On the other hand, the mathematics of logical and operational approach, to which von Neumann was gravitating by natural disposition, is also aware of its origin genetically, but in a rather more negating manner on the whole, and in a much more consciously modern and even antagonistic spirit. Many a chapter in mathematical logic or semantics is like a page from Aristotle's *Organon* or *Rhetoric,* but with all sentences crossed out and each replaced by an entirely different one on the margin.

In short, there are two trends of modernism discernible in mathematics today. One is incisive, emphatic, frequently stirring, and always provocative. The other is a kind of regeneration with an element of traditionalism operating as a positive motive power in it. Von Neumann did not share in the latter.

John von Neumann always had the natural air of someone who had been reared from childhood in material comfort. He was born on December 28, 1903, in Budapest, Hungary, the eldest of three sons, and first came to the United States when he was twenty-seven years old. His brothers and his widowed mother also came here, and his mother died during von Neumann's illness, only six months before his death. His father, Max von Neumann, was a banker. First manager and later vice-president of a major bank, he finally was made one of the partners of a private banking firm, all in Budapest. According to John, it was his father who encouraged him to choose a career in science. John showed a pronounced interest in mathematics from the age of thirteen. He was also a great reader of books on history throughout his life, and in both science and history his retentive memory was most remarkable.

He attended grammar school and high school in Budapest. As a youth he had no bent towards physical or sporting activities. And even when later on he drove some very dashing looking cars with the maximum velocity for which they were designed, he never acquired the mien of a "sporting type." He did claim, however, that he had once learned to ride a bicycle, but admitted that a pair of trousers was part of the "tuition price." In his teens he was somewhat gauche and not quite the type of "leader." But there was nothing awkward about his manner in later life, and he gave the impression of considerable collectedness and self-confidence. He was courteous and could be quite jovial. In company he was an excellent raconteur. He never smoked, and would drink but rarely, although he would sometime simulate gaiety as if he had imbibed.

Even before entering on college studies, he was co-author of a paper (# 1 of the biblography) at the age of eighteen. He studied two years at the University of Berlin, 1921–23, and also took courses in chemistry. Fritz Haber, the then academic leader of chemistry in Germany, soon met him, and is reported to have said to friends that he expected or wished that von Neumann would take up chemistry as a career. John went from Berlin to the Institute of Technology in Zürich, Switzerland, where Hermann Weyl was in the mathematics department at the time. He left the Institute in 1925 with an engineering degree in chemistry. In the meantime, he had begun publishing very good mathematical papers pertaining to axiomatic set theory, and in 1926 he was awarded a doctorate for such a thesis at the University of Budapest. This enabled him to secure immediately a position as Privatdozent at the University of Berlin. One of the ranking mathematicians there was Erhard Schmidt, known mainly for his work in integral equations, who also had a strong interest in set theory, fostered by a lifelong friendship with Ernst Zermelo, of "Zermelo-axiom" fame.

John's interest soon turned to the recently created quantum theory, which was produced, packaged, and delivered by several "youngsters" in their twenties, barely older than himself. He began to think through the mathematical scheme of the theory in his own terms and, in the process, learned theoretical physics, and readied mathematical tools for future use. He wrote some lengthy articles on quantum theory, and there are indications in the first ones that it was Max Born's statistical interpretation of the Schroedinger wave function and also Born's attempt at rigorization of approach that put von Neumann initially on the track of becoming one of the busiest and most sought-after scientific consultants to military and other organizations in this country.

In 1928, between papers on mathematical logic and his first papers in physics, he "squeezed in" his first and basic paper on the theory of games. There had been forerunners in the field (E. Zermelo, H. Borel, and H. Steinhaus, among others) by whom most of the concepts like play and counterplay, chance and deception, pure strategy and mixed strategy had been anticipated in one form or another. But von Neumann brought the results of others and his own new ones together, clearly and axiomatically, into twenty-five pages. In the briefest of footnotes he made reference to the analogy with some problems in economics, and in another footnote, added in proof, he made reference to a very recent note of Borel's in which for the two-person zero-sum game the underlying minimax theorem was proposed, although not proven, as in von Neumann's paper. Eventually the paper grew into the treatise with Oscar Morgenstern published by the Princeton University Press in 1944 (# 90). When deciding to print the book, the Press was cheerfully resigned, for the sake of scholarship, to face the financial deficit they expected. But, in fact, they had to reprint the book in 1947 and again in 1953.

The quantum theory led von Neumann naturally to a re-examination of the theory of operators in Hilbert space and, while still in Berlin, he published in 1929 his paper # 30 on unbounded operators. It was soon followed by his paper # 31 on algebras of operators, the first in that direction. He discussed this material with Erhard Schmidt, who had some influence on clarifying the distinction between maximal (symmetric) and hypermaximal (selfadjoint) operators. This last work won for him the admiration of mathematicians. In 1929 he transferred to the University of Hamburg as Privatdozent, but remained only one year. In 1930 he came to Princeton, N. J., which remained his academic address afterwards.

About that time the Springer publishing firm solicited him to write for their so-called Yellow Collection (this was not the official name) a volume on the mathematical foundations of quantum mechanics (# 47a). While negotiations for the book were still in progress, von Neumann became dissatisfied with the galley proofs of an article in print, and wanted to make very costly revisions in it. The journal in which the article was to be published was also controlled by Springer, and in order to humor him they let him make the revisions without charge.

Just before coming to Princeton in 1930, von Neumann married Marietta Kovesi in Budapest. They were divorced in 1937, shortly after their daughter Marina was born. She is his only child, and has been Mrs. Robert Whitman since June 1956. In 1938, von Neumann married Klara Dan, also originally from Budapest, who later worked on his Computer Project.

From 1930 until 1933 he was Visiting Professor at Princeton University, and in 1933 became a permanent member of the Institute for Advanced Study, which academic post he held till his death. He soon became widely known to mathematicians in this country, at first chiefly to senior ones, but after his paper on almost periodic functions (# 55) to the "rank-and-file" as well. Until 1940, when the war began to change the pattern of his activities, he gave lecture courses fairly regularly, and he had large audiences. Furthermore, his papers were written in a uniform orderly style and were accessible for study by younger research workers, even though the subject was a difficult one.

In 1935, when F. J. Murray was a National Research Fellow in residence at the Institute for Advanced Study, von Neumann began the study of algebras of operators in earnest. Several of the papers were written in collaboration (# 62, # 70, # 87) with Murray, and Murray's partnership was not a nominal one. There were many discussions between them with a brisk give-and-take throughout. An offshoot of these studies was von Neumann's "continuous geometries." They are typified by a kind of projective space in which the hierarchy of linear subspaces of increasing dimension is no longer specified by discrete dimension numbers (that is, integer numbers, or multiples of integer numbers with some common factor) but by continuous real numbers out of an entire interval. This was the topic of his Colloquium Lectures before the American Mathematical Society in 1937. He did not publish the lectures, as most lecturers do, but several sets of notes found among his papers, written as always in his legible round hand, may have been intended to be used for this purpose.

Dr. von Neumann was elected to the National Academy of Sciences in 1937.

After completing the book with Morgenstern in 1944, von Neumann's attention turned to computing machines and, somewhat surprisingly, he decided to build his own. As the years progressed, he appeared to thrive on the multitudinousness of his tasks. It has been stated that von Neumann's electronic computer hastened the hydrogen bomb explosion of November 1, 1952.

When the war came, von Neumann was called to serve on an increasing number of upper echelon boards and committees concerned with armament programs or the utilization of nuclear energy. As most of the research work involved was classified, it is not easy to assay what kind of original work he did. But all published citations which he received are unanimous and convincing in praising him for outstanding contributions in many areas of vital importance.

In the summer of 1955, only a few months after his appointment to the Atomic Energy Commission had been approved by the Senate, he became ill. Cancer was soon suspected, but he continued to work fairly regularly for several more months. His last public appearance was early in 1956 when, in a wheelchair at the White House, he received the Medal of Freedom from President Eisenhower. About that time it began to be known that he was not expected to recover. In April, 1956, he was taken to the Walter Reed Hospital, where he remained till the end. He lingered on longer than was expected, dying on February 8, 1957.

Von Neumann was baptized a Roman Catholic in 1930, and he died in the faith. He is buried in Princeton, N. J.

CHRONOLOGY

1903 Born, Budapest, Hungary, December 28.
1930–33 Visiting Professor, Princeton University.
1933–57 Professor of Mathematics, Institute for Advanced Study, Princeton, N. J.
1937 Gibbs Lecturer, Colloquium Lecturer, Bôcher Prize, all in American Mathematical Society.
1940–57 Scientific Advisory Committee, Ballistics Research Laboratories, Aberdeen Proving Ground, Maryland.
1941–55 Navy Bureau of Ordnance, Washington, D. C.
1943–55 Los Alamos Scientific Laboratory (AEC), Los Alamos, N. M.
1945–57 Director of Electronic Computer Project, Institute for Advanced Study, Princeton, N. J.
1947 D. Sc. (hon.), Princeton University; Medal for Merit (Presidential Award); Distinguished Civilian Service Award, U. S. Navy.
1947–55 Naval Ordnance Laboratory, Silver Spring, Maryland.
1949–53 Research and Development Board, Washington, D. C.
1949–54 Oak Ridge National Laboratory, Oak Ridge, Tennessee.
1950 D. Sc. (hon.), University of Pennsylvania and Harvard University.
1950–55 Armed Forces Special Weapons Project, Washington, D. C.; Weapons System Evaluation Group, Washington, D. C.
1950–57 Member Board of Advisors, Universidad de los Andes, Colombia, South America.
1951–53 President, American Mathematical Society.
1951–57 Scientific Advisory Board, U. S. Air Force, Washington, D. C.
1952 D. Sc. (hon.), University of Instanbul, Case Institute of Technology, and University of Maryland.
1952–54 Member, General Advisory Committee, U. S. Atomic Energy Commission, Washington, D. C. (Presidential appointment).
1953 D. Sc. (hon.), Institute of Polytechnics, Munich; Vanuxem Lecturer, Princeton University.
1953–57 Technical Advisory Panel on Atomic Energy, Washington, D. C.
1955–57 U. S. Atomic Energy Commissioner (Presidential appointment).
1956 Medal of Freedom (Presidential award); Albert Einstein Commemorative Award; Enrico Fermi Award.
1957 Died, Washington, D. C., February 8.
Academy memberships:
Academia Nacional de Ciencias Exactas, Lima, Peru.
Academia Nazionale dei Lincei, Rome, Italy.
American Academy of Arts and Sciences.
American Philosophical Society.
Instituto Lombardo di Scienze e Lettere, Milan, Italy.
National Academy of Sciences.
Royal Netherlands Academy of Sciences and Letters, Amsterdam, Netherlands.

KEY TO ABBREVIATIONS

Acta Szeged. = Szeged. Egyetem. Acta universitatis szegediensis. Acta scientarum mathematicarum.
A. M. S. Bull. = American Mathematical Society Bulletin
A. M. S. Trans. = American Mathematical Society Transactions
Ann. Math. = Annals of Mathematics
Ann. Math. Stat. = Annals of Mathematical Statistics
Ann. Math. Studies = Annals of Mathematics Studies
Astrophys. J. = Astrophysical Journal
Commun. Pure Appl. Math. = Communications on Pure and Applied Mathematics
Compos. Math. = Compositio Mathematica
Erg. eines Math. Coll. = Ergebnisse eines mathematischen Kolloquiums
Fund. Math. = Fundamenta Mathematicae
Gött. Nach. = Gesellschaft der Wissenschaften zu Göttingen Nachrichten. Mathematisch-physikalische Klasse.
Hamb. Abh. = Hamburg Abhandlungen
Jahresb. = Deutsche Mathematiker-Vereinigung, Leipzig, Jahresbericht
J. Appl. Phys. = Journal of Applied Physics
J. f. Math = Journal für Mathematik
Math. Ann. = Mathematische Annalen
Math. Nach. = Mathematische Nachrichten
Math. Tables and Other Aids to Comp. = Mathematical Tables and Other Aids to Computation
Math. Zschr. = Mathematische Zeitschrift
Math. Phys. Lapok = Mathematikaies physikai Lapok
Math. Sbornik = Matica slovenska, Turócz-Szent-Márton. Sbornik.
N. A. S. Proc. = National Academy of Sciences Proceedings
Nat. Bur. Standards Appl. Math. Series = National Bureau of Standards Applied Mathematics Series
Phys. Zschr. = Physikalische Zeitschrift
Portugaliae Math. = Portugaliae Mathematica
Proc. Comp. Sem. = Proceedings, Comptroller's Seminar
Rev. Econ. Studies = Review of Economic Studies
Sitzungsber. d. Preuss. Akad. = Sitzungsberichte der Preussischen Akademie
Tomsk. Univ. Rev. = Tomsk Universitet Review
Zschr. f. Phys. = Zeitschrift für Physik

BIBLIOGRAPHY

1922

1. With M. Fekete. Über die Lage der Nullstellen gewisser Minimumpolynome. Jahresb., 31:125–38.

1923

2. Zur Einführung der transfiniten Ordnungszahlen. Acta Szeged., 1:199–208.

1925

3. Eine Axiomatisierung der Mengenlehre. Jour. f. Math., 154:219–40.
4. Egyenletesen sürü számsorozatok. Math. Phys. Lapok, 32:32–40.

1926

5. Zur Prüferschen Theorie der idealen Zahlen. Acta Szeged., 2:193–227.
6. Az általános Nalmazelmélet axiomatikus folépitése. (Doctor's thesis, Univ. of Budapest.) See also #18.

1927

7. With D. Hilbert and L. Nordheim. Über die Grundlagen der Quantenmechanik. Math. Ann., 98:1–30.
8. Zur Theorie der Darstellungen kontinuierlicher Gruppen. Sitzungsber. d. Preuss. Akad., pp. 76–90.
9. Mathematische Begründung der Quantenmechanik. Gött. Nach., pp. 1–57.
10. Wahrscheinlichkeitstheoretischer Aufbau der Quantenmechanik. Gött. Nach., pp. 245–72.
11. Thermodynamik quantenmechanischer Gesamtheiten. Gött. Nach., pp. 273–91.
12. Zur Hilbertschen Beweistheorie. Math. Zschr., 26:1–46.

1928

13. Allgemeine Eigenwerttheorie symmetrischer Funktionaloperatoren. "Habilitationsschrift," Univ. of Berlin. See also #30.
14. Zerlegung des Intervalles in abzählbar viele kongruente Teilmengen. Fund. Math., 11:230–38.
15. Ein System algebraisch unabhängiger Zahlen. Math. Ann., 99:134–41.
16. Über die Definition durch transfinite Induktion, und verwandte Fragen der allgemeinen Mengenlehre. Math. Ann., 99:373–91.
17. Zur Theorie der Gesellschaftsspiele. Math. Ann., 100:295–320.
18. Die Axiomatisierung der Mengenlehre. Math. Zschr., 27:669–752.
19. With E. Wigner. Zur Erklärung einiger Eigenschaften der Spektren aus der Quantenmechanik des Drehelektrons, I. Zschr. f. Phys., 47:203–20.

20. Einige Bemerkungen zur Diracschen Theorie des Drehelektrons. Zschr. f. Phys., 48:868–81.
21. With E. Wigner. Zur Erklärung einiger Eigenschaften der Spektren aus der Quantenmechanik des Drehelektrons, II. Zschr. f. Phys., 49:73–94.
22. With E. Wigner. Zur Erklärung einiger Eigenschaften der Spektren aus der Quantenmechanik des Drehelektrons, III. Zschr. f. Phys., 51:844–58.

1929

23. Über eine Widerspruchfreiheitsfrage der axiomatischen Mengenlehre. J. f. Math., 160:227–41.
24. Über die analytischen Eigenschaften von Gruppen linearer Transformationen und ihrer Darstellungen. Math. Zschr., 30:3–42.
25. With E. Wigner. Über merkwürdige diskrete Eigenwerte. Phys. Zschr., 30:465–67.
26. With E. Wigner. Über das Verhalten von Eigenwerten bei adiabatischen Prozessen. Phys. Zschr., 30:467–70.
27. Beweis des Ergodensatzes und des H-Theorems in der neuen Mechanik. Zschr. f. Phys., 57:30–70.
28. Zur allgemeinen Theorie des Maasses. Fund. Math., 13:73–116.
29. Zusatz zur Arbeit "Zur allgemeinen. . . ." Fund. Math., 13:333.
30. Allgemeine Eigenwerttheorie Hermitescher Funktionaloperatoren. Math. Ann., 102:49–131.
31. Zur Albegra der Funktionaloperatoren und Theorie der normalen Operatoren. Math. Ann., 102:370–427.
32. Zur Theorie der unbeschränkten Matrizen. J. f. Math., 161:208–36.

1930

33. Über einen Hilfssatz der Variationsrechnung. Hamb. Abh., 8:28–31.

1931

34. Über Funktionen von Funktionaloperatoren. Ann. Math., 32:191–226.
35. Algebraische Repräsentanten der Funktionen "bis auf eine Menge vom Maasse Null." J. f. Math., 161:109–15.
36. Die Eindeutigkeit der Schrödingerschen Operatoren. Math. Ann., 104:570–78.
37. Bemerkungen zu den Ausführungen von Herrn St. Leśniewski über meine Arbeit "Zur Hilbertschen Beweistheorie." Fund. Math., 17:331–34.

38. Die formalistische Grundlegung der Mathematik. Report to the Congress, Königsberg, Sept. 1931. Erkenntniss, 2:116–21.

1932

39. Zum Beweise des Minkowskischen Statzes über Linearformen. Math. Zschr., 30:1–2.
40. Über adjungierte Funktionaloperatoren. Ann. Math., 33:294–310.
41. Proof of the Quasi-ergodic Hypothesis. N. A. S. Proc., 18:70–82.
42. Physical Applications of the Quasi-ergodic Hypothesis. N. A. S. Proc., 18:263–66.
43. With B. O. Koopman. Dynamical Systems of Continuous Spectra. N. A. S. Proc., 18:255–63.
44. Über einen Satz von Herrn M. H. Stone. Ann. Math., 33:567–73.
45. Einige Sätze über massbare Abbildungen. Ann. Math., 33:574–86.
46. Zur Operatorenmethode in der klassischen Mechanik. Ann. Math., 33:587–642.
47. Zusätze zur Arbeit "Zur Operatorenmethode. . . ." Ann. Math., 33:789–91. See also #46.
47a. Mathematische Grundlagen der Quantenmechanik. Berlin, J. Springer; New York, Dover Publications (1943); Presses Universitaires de France (1947); Madrid, Instituto de Mathematicas "Jorge Juan" (1949); trans. from German by Robert T. Beyer, Princeton University Press (1955).

1933

48. Die Einführung analytischer Parameter in topologischen Gruppen. Ann. Math., 34:170–90.
49. A koordináta-mérés pontosságának határai az elektrón Dirac-féle elméletében (Über die Grenzen der Koordinatenmessungs-Genauigkeit in der Diracschen Theorie des Elektrons). Mat. és Természettud . . . Értesitó, 50:366–85.

1934

50. With P. Jordan and E. Wigner. On an Algebraic Generalization of the Quantum Mechanicalformalism. Ann. Math., 35:29–64.
51. Zum Haarschen Maass in topologischen Gruppen. Compos. Math., 1:106–14.
52. Almost Periodic Functions in a Group. I. A. M. S. Trans., 36:445–92.
53. With A. H. Taub and O. Veblen. The Dirac Equation in Projective Relativity. N. A. S. Proc., 20:383–88.

1935

54. On Complete Topological Spaces. A. M. S. Trans., 37:1–20.
55. With S. Bochner. Almost Periodic Functions in Groups. II. A. M. S. Trans., 37:21–50.
56. With S. Bochner. On Compact Solutions of Operational-Differential Equations. I. Ann. Math., 36:255–91.
57. Charakterisierung des Spektrums eines Integraloperators. Actualités Scient. et Ind. Series, No. 229. Exposés Math., publiés à la mémoire de J. Herbrand, No. 13. Paris. 20 pp.
58. On Normal Operators. N. A. S. Proc., 21:366–69.
59. With P. Jordan. On Inner Products in Linear, Metric Spaces. Ann. Math., 36:719–23.
60. With M. H. Stone. The Determination of Representative Elements in the Residual Classes of a Boolean Algebra. Fund. Math., 25:353–78.

1936

61. On a Certain Topology for Rings of Operators. Ann. Math., 37: 111–15.
62. With F. J. Murray. On Rings of Operators. Ann. Math., 37:116–229.
63. On an Algebraic Generalization of the Quantum Mechanical Formalism (Part I). Mat. Sbornik, 1:415–84.
64. The Uniqueness of Haar's Measure. Mat. Sbornik, 1:721–34.
65. With G. Birkhoff. The Logic of Quantum Mechanics. Ann. Math. 34:823–43.
66. Continuous Geometry. N. A. S. Proc., 22:92–100.
67. Examples of Continuous Geometries. N. A. S. Proc., 22:101–8.
68. On Regular Rings. N. A. S. Proc., 22:707–13.

1937

69. With C. Kuratowski. On Some Analytic Sets Defined by Transfinite Induction. Ann. Math., 38:521–25.
70. With F. J. Murray. On Rings of Operators, II. A. M. S. Trans., 41:208–48.
71. Some Matrix-Inequalities and Metrization of Matrix-Space. Tomsk. Univ. Rev., 1:286–300.
72. Über ein ökonomisches Gleichungssystem und eine Verallgemeinerung des Brouwerschen Fixpunktsatzes. Erg. eines Math. Coll., Vienna, ed. by K. Menger, 8:73–83.

73. Algebraic Theory of Continuous Geometries. N. A. S. Proc., 23:16–22.
74. Continuous Rings and Their Arithmetics. N. A. S. Proc., 23:341–49.

1938

75. On Infinite Direct Products. Compos. Math., 6:1–77.

1940

76. With I. Halperin. On the Transitivity of Perspective Mappings. Ann. Math., 41:87–93.
77. On Rings of Operators, III. Ann. Math., 41:94–161.
78. With E. Wigner. Minimally Almost Periodic Groups. Ann. Math., 41:746–50.
78a. With R. H. Kent. The Estimation of the Probable Error from Successive Differences. Aberdeen Proving Ground, Md., Report No. 175. 19 pp.

1941

79. With R. H. Kent, H. R. Bellinson, and B. I. Hart. The Mean Square Successive Difference. Ann. Math. Stat., 12:153–62.
80. With I. J. Schoenberg. Fourier Integrals and Metric Geometry. A. M. S. Trans., 50:226–51.
81. Distribution of the Ratio of the Mean Square Successive Difference to the Variance. Ann. Math. Stat., 12:367–95.

1942

82. A Further Remark Concerning the Distribution of the Ratio of the Mean Square Successive Difference to the Variance. Ann. Math. Stat., 13:86–88.
83. With P. R. Halmos. Operator Methods in Classical Mechanics, II. Ann. Math., 43:332–50.
84. With S. Chandrasekhar. The Statistics of the Gravitational Field Arising from a Random Distribution of Stars, I. Astrophys. J., 95: 489–531.
85. Note to: Tabulation of the Probabilities for the Ratio of the Mean Square Successive Difference to the Variance, by B. I. Hart. Ann. Math. Stat., 13:207–14.
86. Approximative Properties of Matrices of High Finite Order. Portugaliae Math., 3:1–62.

1943

87. With F. J. Murray. On Rings of Operators, IV. Ann. Math., 44:716–808.
88. With S. Chandrasekhar. The Statistics of the Gravitational Field Arising from a Random Distribution of Stars. II. The Speed of Fluctuations; Dynamical Friction; Spatial Correlations. Astrophys. J., 97:1–27.
89. On Some Algebraical Properties of Operator Rings. Ann. Math., 44:709–15.

1944

90. With O. Morgenstern. Theory of Games and Economic Behavior. Princeton Univ. Press (1944, 1947, 1953). 625 pp.

1945

90a. A Model of General Economic Equilibrium. Rev. Econ. Studies, 13(1):1–9.

1946

91. With V. Bargmann and D. Montgomery. Solution of Linear Systems of High Order. Report Prepared for Navy BuOrd under Contract Nord–9596. 85 pp.
92. With A. W. Burks and H. H. Goldstine. Preliminary Discussion of the Logical Design of an Electronic Computing Instrument. Part I, Vol. I. Report Prepared for U. S. Army Ord. Dept. under Contract W–36–034–ORD–7481. 42 pp.

1947

93. The Mathematician. In: *The Works of the Mind,* ed. by R. B. Heywood (Univ. of Chicago Press), pp. 180–96.
94. With H. H. Goldstine. Numerical Inverting of Matrices of High Order. A. M. S. Bull., 53:1021–99.
95. With H. H. Goldstine. Planning and Coding of Problems for an Electronic Computing Instrument. Part II, Vol. I. Report Prepared for U. S. Army Ord. Dept. under Contract W–36–034–ORD–7481. 69 pp.

1948

96. With H. H. Goldstine. Planning and Coding of Problems for an Electronic Computing Instrument. Part II, Vol. II. Report Prepared

for U. S. Army Ord. Dept. under Contract W-36-034-ORD-7481 68 pp.

97. With H. H. Goldstine. Planning and Coding of Problems for an Electronic Computing Instrument. Part II, Vol. III. Report Prepared for U. S. Army Ord. Dept. under Contract W-36-034-ORD-7481 23 pp.
98. On the Theory of Stationary Detonation Waves, File No. X122. BRL, Aberdeen Proving Ground, Md. 26 pp.

1949

99. On Rings of Operators. Reduction Theory. Ann. Math., 50:401-85.

1950

100. With R. D. Richtmyer. A Method for the Numerical Calculation of Hydrodynamic Shocks. J. Appl. Phys., 21:232-37.
101. Functional Operators. Vol. I: Measures and Integrals (Ann. Math. Studies, No. 21, 261 pp.); Vol. II: The Geometry of Orthogonal Spaces (Ann. Math. Studies, No. 22, 107 pp.). Princeton Univ. Press.
102. With G. W. Brown. Solutions of Games by Differential Equations. In: *Contributions to the Theory of Games* (Ann. Math. Studies No. 24, Princeton Univ. Press), pp. 73-79.
103. With N. C. Metropolis and G. Reitwiesner. Statistical Treatment of Values of First 2000 Decimal Digits of *e* and of Pi Calculated on the ENIAC. Math. Tables and Other Aids to Comp., 4:109-11.
104. With J. G. Charney and R. Fjørtoft. Numerical Integration of the Barotropic Vorticity Equation. Tellus, 2:237-54.
105. With I. E. Segal. A Theorem on Unitary Representations of Semisimple Lie Groups. Ann. Math., 52:509-17.

1951

106. Eine Spektraltheorie für allgemeine Operatoren eines unitären Raumes. Math. Nach., 4:258-81.
107. The Future of High-Speed Computing. Proc., Comp. Sem., Dec., 1949, published and copyrighted by IBM (1951), p. 13.
108. Discussion on the Existence and Uniqueness or Multiplicity of Solutions of the Aerodynamical Equations. Chapter 10 of *Problems of Cosmical Aerodynamics,* Proceedings of the Symposium on the Motion of Gaseous Masses of Cosmical Dimensions held at Paris, Aug. 16-19, 1949, Central Air Doc. Office, pp. 75-84.

109. With H. H. Goldstine. Numerical Inverting of Matrices of High Order, II. A. M. S. Proc. 2:188–202.

110. Various Techniques Used in Connection with Random Digits. Chapter 13 of Proceedings of Symposium on "Monte Carlo Method" Held June–July 1949 in Los Angeles. Nat. Bur. Standards, Appl. Math. Series, 12:36–38.

111. The General and Logical Theory of Automata. In: *Cerebral Mechanisms in Behavior—The Hixon Symposium* (Sept. 1948, Pasadena), ed. by L. A. Jeffress (New York, Wiley), pp. 1–31.

1952

112. Five Notes on Continuous Geometry, prepared by W. Givens for Use of Participants in Seminar at University of Tennessee, Summer, 1952. Includes items 66, 67, 68, 73, and 74.

1953

113. A Certain Zero-Sum Two-Person Game Equivalent to the Optimal Assignment Problem. In: *Contributions to the Theory of Games,* Vol. II (Ann. Math. Studies, No. 28, Princeton Univ. Press), pp. 5–12.

114. With D. B. Gillies and J. P. Mayberry. Two variants of Poker. In: *Contributions to the Theory of Games,* Vol. II (Ann. Math. Studies, No. 28, Princeton Univ. Press), pp. 13–50.

1954

115. Wtih E. P. Wigner. Significance of Loewner's Theorem in the Quantum Theory of Collisions. Ann. Math., 59:418–33.

116. The Role of Mathematics in the Sciences and in Society. Address before Association of Princeton Graduate Alumni, June 16, 1954.

117. A Numerical Method to Determine Optimum Strategy. Naval Res. Logistics Quarterly, 1:109–15.

118. The NORC and Problems in High Speed Computing. Speech at First Public Showing of IBM Naval Ordnance Research Calculator, Dec. 2, 1954.

119. Entwicklung und Ausnutzung neuerer mathematischer Maschinen. Arbeitsgemeinschaft für Forschung des Landes Nordrhein-Westfalen, Heft 45. Düsseldorf.

1955

120. Can We Survive Technology? Fortune, June.
121. With A. Devinatz and A. E. Nussbaum. On the Permutability of Self-Adjoint Operators. Ann. Math., 62:199–203.
122. With Bryant Tuckerman. Continued Fraction Expansion of $2^{1/3}$. Math. Tables & Other Aids to Comp., IX, No. 49.
123. With H. H. Goldstine. Blast Wave Calculation. Comm. Pure & Applied Math., 8:327–53.

1956

124. Probabilistic Logics and the Synthesis of Reliable Organisms from Unreliable Components. In: *Automata Studies,* ed. by C. E. Shannon and J. McCarthy (Princeton Univ. Press), pp. 43–98.

Section K

History

From the 1960s onward, Bochner wrote extensively on various topics in the history of mathematics and, more generally, in the history of science. This interest occupied most of his attention in his last years. His final books *The Role of Mathematics in the Rise of Science* (Princeton, 1966) and *Eclosion and Synthesis: Perspectives on the History of Knowledge* (W. A. Benjamin, 1969) dealt with these topics. He contributed a number of articles to the *Dictionary of the History of Ideas* (Scribner, 1973), and these are included in this collection.

Mathematics

Mathematics is frequently encountered in association and interaction with astronomy, physics, and other branches of natural science, and it also has deep-rooted affinities to what are called humanities. Actually, it is a realm of knowledge entirely by itself, and one of considerable scope; the word mathematics stems from a root which means learnable knowledge. Mathematical knowledge is commonly deemed to have a high degree of validity, irrespective of man's cultural conditioning and predilection, although it can be argued that, in the past, cultural settings have affected its development noticeably.

Relation to science. As far as the scientist is concerned, mathematics is not a branch of natural science itself. It does not deal with phenomena and objects of the external world and their relations to each other, but, strictly speaking, only with objects and relations of its own imagery. Mathematical figures in two- or three-dimensional geometry are largely idealizations of objects occurring in the physical world, but figures in n-dimensional space for general n no longer are. Integer numbers 1, 2, 3, . . . , and even real numbers in general, can be claimed to be abstractions from quantities occurring in the physical world, but the "imaginary" number $i = \sqrt{-1}$ has received its name from the very fact that it has no basis in the material world, even though the use of complex numbers $a + bi$ is indispensable to science. If a laboratory experiment idealizes a physical system in order to eliminate secondary features not essential to the study at hand, its ultimate objective is an understanding of the unidealized physical system. But even if it were true that the five regular solids as investigated in the 13th book of Euclid's *Elements* (cube, tetrahedron, octahedron, icosahedron, and dodecahedron) had been found because of occurrence of approximating crystals in nature, nevertheless, once found, the idealized geometric figures would still have become primary objects of consideration in mathematics.

Mathematics is not subordinate to natural science by being a handmaiden of it, and one can practice competently meaningful mathematics without being concerned with science at all. Especially, philosophical attempts to reduce all origin of mathematics to utilitarian motives are wholly unconvincing. However, it is fair to say that mathematics is the language of science in a deep sense. Mathematics is an indispensable medium by which and within which science expresses, formulates, continues, and communicates itself. And just as the language of true literacy not only specifies and expresses thoughts and processes of thinking but also creates them in turn, so does mathematics not only specify, clarify, and make rigorously workable concepts and laws of science but also at certain crucial instances becomes an indispensable constituent of their creation and emergence as well. In Isaac Newton's formula for the motion of a particle on a straight line

$$m\frac{d^2x}{dt^2} = F$$

the mass m and the force F are nonmathematical objects, perhaps. But the instantaneous velocity $v = dx/dt$ and the instantaneous acceleration $a = dv/dt = d^2x/dt^2$ are wholly mathematical, and without a mathematical theory of the infinitesimal calculus, they are not conceivable. Newton, the physicist, was driven to creating his version of the calculus because of this. Also, Newton had to have not only the process of differentiation but also the concept of a mathematical function, because only a function can be differentiated. He required not only the path function $x = x(t)$, but, for the second derivative, he had to envisage the velocity $v = dx/dt$ itself again as a function depending on t, even though in the definition of v this dependence had been reduced to 'instantaneousness." This concept of function was given to Newton by the then new theory of analytic geometry of René Descartes, and it is a fact that after Archimedes' work on the motion of a lever, theoretical mechanics stood virtually still for almost 2000 years until the twin mathematical concepts of function and derivative were ready to emerge. *See* CALCULUS, DIFFERENTIAL AND INTEGRAL.

A striking case of the manner in which mathematics may shape physics has occurred in the theory of relativity. When Albert Einstein was pondering the transition from special to general relativity, his attention was drawn to G. Ricci's work on tensor calculus. Einstein found it an excellent medium for his thoughts, adopted it, and made it widely known to mathematicians. In the 1920s the quantum theory could move as rapidly as it did only because certain prefabricated parts from the

S. BOCHNER

theory of matrices and differential equations were lying about somewhat idly. See DIFFERENTIAL EQUATION; MATRIX THEORY.

Creative formulas. A formula is a string of mathematical symbols subject only to certain general rules of composition. To a working mathematician a string of symbols is a formula if it is something worth remembering. Much mathematics is concentrated in and propelled by certain formulas of unusual import.

The oldest such is the Pythagorean theorem $c^2 = a^2 + b^2$, where a, b are the sides and c is the hypotenuse in a right-angled triangle. If a, b are rational numbers p/q, then c need not be so; for instance, when $a = b = 1$, $c = \sqrt{2}$ and is not a rational number. This compelled the Greeks to study ratios of incommensurable quantities and also quadratic irrationalities. In modern times the formula was extended to $c^2 = a_1{}^2 + \cdots + a_n{}^2$ for the diagonal of a rectangular parallelepiped in n-dimensional euclidean space, and gave the expression

$$s^2 = (x^1 - y^1)^2 + \cdots + (x^n - y^n)^2 \qquad (1)$$

for the distance between two points $x = (x^1, \ldots, x^n)$, $y = (y^1, \ldots, y^n)$ in cartesian coordinates there. For infinitesimally near points this becomes the line element $ds^2 = (dx^1)^2 + \cdots + (dx^n)^2$ which in the more general affine version

$$ds^2 = \sum_{i,j=1}^{n} g_{ij}\, dx^i\, dx^j \qquad (2)$$

is the cornerstone of Riemannian geometry. On the other hand, analytic developments demanded the extension of Eq. (1) to a space of infinitely many dimensions and then even to the case in which coordinates are complex numbers; thus

$$s^2 = |z^1 - w^1|^2 + |z^2 - w^2|^2 + \cdots$$

With this definition of distance the space becomes a Hilbert space, the basis of operator theory and quantum mechanics. Finally, again for finite dimension but still complex coordinates, it became significant, in analogy to Eq. (2), to introduce the line element

$$ds^2 = \sum_{\alpha,\beta=1}^{n} g_{\alpha\bar{\beta}}\, dz^\alpha\, d\bar{z}^\beta \qquad (2')$$

It is the basis for the study of Riemannian geometry on complex manifolds, and the study of such manifolds is penetrating even into relativity and field theory. See GEOMETRY, RIEMANNIAN; OPERATOR THEORY; QUANTUM MECHANICS.

The simple relation $(1 + x)^2 = 1 + 2x + x^2$ was generalized by Newton to

$$(1 + x)^n = 1 + \binom{n}{1} x + \binom{n}{2} x^2 + \cdots \qquad (3)$$

The coefficients are the binomial coefficients

$$\binom{n}{k} = \frac{n(n-1)\cdots(n-k+1)}{1 \cdot 2 \cdots k} \qquad (4)$$

and the formula is a generating relation for them. The expression in Eq. (4) can also be set up if n is any nonintegral number, real or complex, and it was indeed established that Eq. (3) holds for any n for $|x| < 1$. The expansion is usually an infinite series, and it is in fact the Taylor series of $(1 + x)^n$. By studying, for general n, the behavior of the series as $|x| \to 1$, H. Abel laid the foundation for the theories of summability of nonconvergent series. See SERIES.

Euclid's theorem that any integer is a unique product of primes (stated in Euclid only implicitly) can be expressed as a formula

$$a = p_1{}^{n_1} \times p_2{}^{n_2} \times \cdots \times p_k{}^{n_k}$$

The concept of a prime factor and the quest for a theorem on unique decomposition into prime factors has been a leading principle of arithmetic and algebra on all levels ever since. See NUMBER THEORY.

The constant π, widely known from antiquity through the formulas $l = 2\pi r$, $A = \pi r^2$ for a circle and $A = 4\pi r^2$, $V = 4/3\pi r^3$ for a sphere, also enters the sophisticated formula

$$\frac{\pi}{4} = 1 - \tfrac{1}{3} + \tfrac{1}{5} - \tfrac{1}{7} + \cdots$$

derived by G. W. Leibnitz. Both kinds of formulas were much used for the approximate computation of this transcendental number, and a great deal of computational technique (including the use of computing machines) was initiated in the course.

The transcendental number e, the basis of natural logarithms, appears in the eighteenth century in the formulas

$$e^{2\pi\sqrt{-1}} = 1 \qquad \cos\varphi + \sqrt{-1}\sin\varphi = e^{\varphi\sqrt{-1}}$$

which have been of immeasurable consequence to all of mathematics. Even the age-old plane trigonometry was greatly streamlined by them. See NUMERICAL ANALYSIS.

Near the end of the eighteenth century, L. Euler gave for convex polyhedrons the formula $e_0 - e_1 + e_2 = 2$ in which e_0 is the number of vertices, e_1 of edges, and e_2 of surfaces. This was a first step in function theory of closed Riemann surfaces and in algebraic geometry, and it was a starting point for H. Poincaré's combinatorial topology. Poincaré's own formula $B_p = B_{n-p}$ for Betti numbers of compact manifolds is even today one of the most arresting statements of the theory.

Cauchy's formula

$$f(z) = \frac{1}{2\pi i} \oint_C \frac{f(\zeta)\, d\zeta}{\zeta - z}$$

is a centerpoint from which the theory of complex functions radiates. See COMPLEX NUMBERS AND COMPLEX VARIABLES.

Galileo's main achievement is embodied in the formula $s = \tfrac{1}{2}gt^2$ for a falling body, and Newton's gravitational law in the formula $F = m_1 m_2 / r^2$ which also represents Coulomb's law of electricity.

Albert Einstein's daring formula $E = mc^2$ from special relativity created the atomic bomb, and W. Heisenberg's interpretation (1927) of the formula $pq - qp = h/2\pi i$ from quantum theory as an indeterminacy relation for physical observables has been an irritant to philosophers ever since. *See* RELATIVITY; UNCERTAINTY PRINCIPLE.

Foundations—mathematical logic. A prime demand on mathematics is that it be deductively rigorous, and a traditional model for intended rigor is Euclid's presentation of mathematical assertions in theorems. For the philosopher G. Spinoza in the seventeenth century, "more geometrico" is a synonym for "deductively rigorous." A theorem is a proposition which has been proved, excepting certain first theorems called axioms which are admitted without proof; and to prove a theorem means to obtain it from other theorems by certain procedures of deduction or inference. It had been a commonplace for long that each branch of mathematics is based on its own axioms, but during the nineteenth century, mathematicians arrived at the insight that even the same branch might have alternate axioms. Specifically, there were envisaged alternate versions of two- and three-dimensional geometry, the axiom varied being the axiom on parallels. It was also recognized that a set of axioms becomes mathematically possible if it is logically consistent, that is, if one cannot deduce from the axioms two theorems one of which, as a proposition, is the negation of the other. *See* GEOMETRY, EUCLIDEAN; GEOMETRY, NONEUCLIDEAN.

At the same time certain developments led to the realization that not only the axioms but the rules of inference themselves might be, and even ought to be, subject to variations. Now if axioms and rules of inferences are both viewed as subject to change, it is customary to speak of a mathematical system or also a formal system, and, of course, an irreducible first requirement is that the system shall be consistent after the manner just stated. Consistency alone is a somewhat negative property. There is a further property, called completeness, which is more positive, and which, if present, is very welcome. A system is complete if for any proposition which can be formulated it either can be proved that it holds or that its negation holds. If a theorem holds in a system and if the system is altered, then the same proposition, or what corresponds to it in the new system, may become doubtful if not outright false; and if it remains valid it may require a new proof because certain axioms or rules of inference are no longer available.

Some of the developments which led to doubt as to whether the traditional rules of inference are inviolate were the following:

1. G. Boole had found (1854) that the classical Aristotelian connectives "and," "or," "negation of" for propositions follow rules similar to those which the operations addition, multiplication, "the negative of" obey in ordinary algebra (Boole's algebra of propositions) and this gradually took from rules of inference the privileged status of untouchability.

2. G. Cantor, the founder of the theory of sets and operations between them, defined a set (intuitively or naively) as the collection of all objects having a certain property which is verbally expressible. Especially "the set of all sets" is again a set and it has the peculiarity of being a set which contains itself as one of its elements. But this leads to the following contradictory situation (Russell's paradox). Divide the totality of all possible sets into two categories. A set shall belong to category I if it does not contain itself as an element, and to category II if it does contain itself as an element. Now form the set M whose elements are the sets of category I. It can now be reasoned by deductive steps admissible in Cantor's own theory that the set M cannot belong to either of the two categories, although the original division into categories did assign each set to one of them.

3. In 1904, E. Zermelo formulated the following axiom of choice. Given any family of nonvacuous sets $\{S\}$, no matter how (infinitely) large the family may be, it is possible to choose simultaneously an element $x = x_S$ from each given set S and thus to consider the set M consisting of precisely these elements. By the use of this axiom, some striking theorems in classical mathematics could be proven, which, without the use of the axiom, seemed to be logically out of reach entirely. Mathematicians began to wonder whether a theorem based on the axiom of choice is indeed valid or, at any rate, whether it has the same level of validity as one without it, and as a consequence of this, theorems employing the axiom of choice were being frequently labeled as such. *See* SET THEORY.

Some of the doubts were resolved eventually; the most striking results to date are the following ones of K. Gödel. (1) Any consistent mathematical system which is sufficient for classical arithmetic must be incomplete. In other words, in a formal system which is expressible within the arithmetical language taught to a student of a secondary school in any country it is possible to formulate assertions which can be neither proved nor disproved. (2) Any such system remains consistent if one adds to it the axiom of choice, and it also remains consistent if one adds the negation of the axiom of choice. In other words, the system of working mathematics can neither disprove nor prove the axiom of choice. (3) The so-called general continuum hypothesis is also consistent with ordinary mathematics, and, in fact, ordinary mathematics remains consistent if the axiom of choice and the general continuum hypothesis are added simultaneously.

G. Cantor says that two sets A_1, A_2 have the same cardinal number if there exists a one-to-one correspondence between the elements of A_1 and A_2. Any infinite set has the same cardinal number as some proper subset of itself. For instance, the set of all integers $A_1 = \{1,2,3, \ldots\}$ has the same cardinal number as the subset of all even integers $A_2 = \{2,4,6,8, \ldots\}$ because the association of the number n in A_1 to the number $2n$ in A_2 is one-to-one. However, for a finite set this cannot happen, and the cardinal number of a finite set is the ordinary number of its elements. Next, Cantor says that a set A_2 has a larger cardinal number than A_1 if there is a one-to-one correspondence from A_1 to a subset of A_2 but not one from A_2 to A_1 or a subset of A_1. He then shows that if A is any set, and S is the set of its subsets including the empty set, the cardinal number of S is greater than that of A. If A is finite and has n elements, then S has 2^n elements. Now 2^n is indeed greater than n, and for $n \geqq 2$ there are even other integer numbers between n and 2^n. This is true for finite sets. However, for infinite sets the situation seems to change radically, and the generalized continuum hypothesis asserts that there are no other cardinal numbers between those of A and S. See LOGIC.

Counterexamples. It follows easily from the definition of a derivative that a differentiable function $y = f(x)$ in $a < x < b$ is also continuous. K. Weierstrass deepened considerably this statement by demonstrating that the converse is not true. He constructed a function $f(x)$ which is continuous in $a < x < b$ but does not have a derivative at any point (but his function does have right and left derivatives at all points). A construction of this kind is called a counterexample. Systematic preoccupation with counterexamples, which sometimes require considerable ingenuity, did not arise before the middle of the nineteenth century. Another early counterexample of considerable consequence referred to Fourier series. If $f(x)$ is periodic with period 2π then its Fourier series is

$$\tfrac{1}{2}a_0 + \sum_{n=1}^{\infty} (a_n \cos nx + b_n \sin nx) \qquad (5)$$

where, by definition,

$$a_n = \frac{1}{\pi}\int_0^{2\pi} f(t) \cos nt\, dt \quad b_n = \frac{1}{\pi}\int_0^{2\pi} f(t) \sin nt \qquad (6)$$

Fourier himself stated somewhat vaguely that for "any" function $f(x)$ the partial sums of the Fourier series are convergent to $f(x)$. P. G. L. Dirichlet was the first to give a specific criterion of convergence. In his criterion the function $f(x)$ may have certain simple discontinuities. But in the present discussion these discontinuities are excluded and $f(x)$ must be continuous, first of all. Dirichlet proved that the Fourier series converges everywhere to $f(x)$, if $f(x)$ is piecewise monotone, that is, if the interval of periodicity $0 \leqq x < 2\pi$ may be divided into a finite number of smaller intervals such that in each of these the function is either monotonely increasing or monotonely decreasing. Then, P. du Bois Reymond demonstrated that this requirement of monotoneity cannot be dispensed with entirely. He constructed a continuous periodic function $f(x)$ for which the Fourier series fails to converge in at least one point. By a general principle developed by A. Harnack (condensation of singularities) it is then possible to construct a continuous function whose Fourier series fails to converge at countably many points even if they are everywhere dense, for instance, at all rational points $x = p/q$. But to this day it is not known whether there exists a continuous function whose Fourier series fails to converge at all points, or at least at all points with the only exception of a set of Lebesgue measure 0. See FOURIER SERIES AND INTEGRALS.

Counterexamples frequently fall into patterns, meaning that the same underlying construction is used over again for different but related purposes. A construction used in many counterexamples is Cantor's ternary set. Divide the closed interval $0 \leqq x \leqq 1$ into three equal parts $(0, \tfrac{1}{3})$, $(\tfrac{1}{3}, \tfrac{2}{3})$, $(\tfrac{2}{3}, 1)$ and remove the middle one, but only the open inner part of it $\tfrac{1}{3} < x < \tfrac{2}{3}$. This leaves two closed intervals $0 \leqq x \leqq \tfrac{1}{3}$, $\tfrac{2}{3} \leqq x \leqq 1$. In each of these remove the open inner third, that is, $\tfrac{1}{9} < x < \tfrac{2}{9}$ and $\tfrac{2}{3} + \tfrac{1}{9} < x < \tfrac{2}{3} + \tfrac{2}{9}$. This leaves four closed intervals altogether. From each, remove the open inner third, and so on. After n steps the number of intervals removed is $1 + 2 + 2^2 + \cdots + 2^{n-1}$ and the sum of the length of the removed intervals is

$$\frac{1}{3} + \frac{2}{3^2} + \frac{2^2}{3^3} + \cdots + \frac{2^{n-1}}{3^n} = 1 - \left(\frac{2}{3}\right)^n$$

This value is less than 1, which must be so, because the total interval has length 1 and the removed intervals do not overlap. However if one lets n go to ∞, then $(\tfrac{2}{3})^n$ tends to 0, and the aggregate length of all infinitely many intervals which have been removed has value 1. Thus, the pointset which is left over ought somehow to have a one-dimensional measure which has value 0. This is so indeed in the theory of Lebesgue. Also the left-over set is nowhere dense. However, it is a very big set still, by a crude counting of its points, at any rate. More precisely, it turns out that its cardinal number is the same as for the entire interval $0 \leqq x \leqq 1$. This can be deduced from the following fact which is interesting for many purposes. Any number $0 \leqq x \leqq 1$ can be represented, instead of by a decimal expansion, by a ternary expansion, that is, by an expansion $0.a_1a_2a_3 \ldots$ in which each symbol a_1, a_2, a_3, . . . , has one of the values 0, 1, 2. Now, the left-over set consists of precisely those points in whose expansion the symbols a_1, a_2, a_3, . . . , assume only the values 0, 2 and not 1.

Constructiveness and approximations. Some mathematicians object to mere existence proofs, and they demand that any proof shall also be constructive. The interpretations of this demand differ widely. Some proofs closely approach what a practical mathematician welcomes; if, for instance, a theorem asserts the existence of a number or a function, then the proof must also embody a procedure for actual computation of the solution, approximately, at least. Other versions are little more than the negative requirement that certain combinations of inference shall be avoided. There are also views which combine both; the best known among the last is the intuitionist view. It firmly demands a certain kind of constructiveness, which, however, does not necessarily guarantee the calculation by present-day computing machines. However, the actual stricture by which intuitionism became widely known is that proof by contradiction is not admissible. It is also called proof by double negation, and it is equivalent to the Aristotelian *tertium non datur*. It assumes tentatively that the proposition to be proved is false and from this assumption deduces a contradiction to a previously established theorem.

There is one difference between practical and theoretical demands on constructiveness. Sometimes it is possible to prove the existence of a function nonconstructively and after that, based on the existence statement, to devise a procedure for a manageable construction. To practical mathematics this is satisfactory; to theoretical, it is not.

A demand superficially related to constructiveness, and without any overtones of "logic," is a desire to estimate the degree of approximation, whenever a task of approximation is being performed. For instance, a classical result of J. L. Lagrange for the Taylor series with remainder term states that if $f(x)$ has n derivatives in the interval $0 \leqq x \leqq 1$, then $f(x) = P_n(x) + R_n(x)$ where $P_n(x)$ is the polynomial

$$P_n(x) = \sum_{m=0}^{n} \frac{f^{(m)}(0)}{m!} x^m \qquad (7)$$

and the "error term" $R_n(x)$ is in numerical value bounded by $M_n/n!$, where M_n is the maximum of $|f^{(n)}(x)|$ in $0 \leqq x \leqq 1$. This result leads to the following topic. The polynomial (7) can be formed only if $f(x)$ has n derivatives. However, there is a theorem of Weierstrass that any function $f(x)$ which is continuous in an interval $a \leqq x \leqq b$ (and nothing more) can be approximated uniformly by polynomials

$$P_n(x) = a_0x^n + a_1x^{n-1} + \cdots + a_n \qquad (8)$$

Each proof for this theorem produces its polynomials, and there is an extensive theory for securing those for which the approximation is a best possible one.

The parallel question of approximating not by ordinary polynomials (8) but by trigonometric polynomials

$$s_n(x) = \tfrac{1}{2}a_0 + \sum_{m=1}^{n} (a_m \cos mx + b_m \sin mx) \qquad (9)$$

leads to mathematical procedures which at present pervade all of science. If one wishes to make the difference

$$\mathrm{Max}_{0 \leqq x < 2\pi} |f(x) - s_n(x)| \qquad (10)$$

rather small, then the partial sums of the Fourier series (5) are not too good, meaning that it is not very advantageous to use the polynomials (9) in which the coefficients a_m, b_m are the expressions (6). However, it improves the degree of approximation if one replaces the partial sums by their arithmetic means (Fejer sums)

$$\sigma_n = \frac{s_0 + s_1 + \cdots + s_{n-1}}{n}$$

which are again exponential polynomials, namely

$$\sigma_n(x) = \tfrac{1}{2}a_0 + \sum_{m=1}^{n} \left(1 - \frac{m}{n}\right)(a_m \cos mx + b_m \sin mx)$$

There are other averages which approximate even better, and one is led to studying procedures of averaging and smoothing which are of importance to analysis, probability, and statistics. But a major development begins if, following F. W. Bessel and M. A. Parseval (nineteenth century), the expression (10) is replaced by the expression

$$\left[\frac{1}{2\pi}\int_0^{2\pi} |f(x) - s_n(x)|^2\, dx\right]^{1/2} \qquad (11)$$

as a measure for the degree of approximation. In this case the partial sums of the Fourier series are indeed the best approximating sums for any order n, not only for ordinary Fourier series but for so-called orthogonal series in general. Pertinent application to the theory of accoustical and then optical waves led to an interpretation of the Fourier series (5) as a spectral resolution of its function $f(x)$, and this interpretation has been gradually extended to a host of related and analogous expansions in analysis, algebra, theory of probability, quantum theory, and other parts of physics. The measuring of approximation, deviation, and dispersion by a square of integral mean—which is Eq. (11)—has become a strong and pervasive influence on scientific thinking.

Space in mathematics. If geometry is the mathematics of space, then, in a superficial sense, all mathematics began with geometry, because apparently it began with measurements of figures: length, area, volume, size of angles. It did not concern itself with questions of shape but with clarifying and deciding when figures are equal or substantially equal with regard to form. The first true theory of geometry was the great theory of the Greeks, whose primary concern was study of the basic concept of equality of figures—their congruence and similarity—and the Greeks were so determined to dissociate their theory from the preceding phase of merely making measurements that Euclid's extensive work, for instance, avoids to a fault any kind of actual measurements. In the case of the Pythagorean theorem $c^2 = a^2 + b^2$, there is no hint that it might be an equality between numbers. It is only an equality between areas based on congruences, two of the squares being cut up into pieces and then put together into the third square after the manner of a jigsaw puzzle. But, for all its lofty purposes, Greek geometry was too rigid and circumscribed to be able really to cope with the mathematical problem of space. It came to a standstill and geometry did not progress further until, with the advent of coordinate systems, introduced by Descartes and his predecessors, a better mathematics of space could be initiated.

If a cartesian coordinate system is etched into two- or three-dimensional euclidean space, then the space becomes a pointset, each point being a pair

(x^1,x^2) or a triple (x^1,x^2,x^3) of real numbers, and any figure a suitable subset of it. This is a deliberate process of arithmetization of space which unifies space and number at the base. This does not hamper geometry in its task of pursuing problems of shape but instead aids it. In the cartesian plane, two figures are similar if the points of one can be obtained from the points of the other by means of a transformation

$$\begin{aligned} y^1 &= a^1 + \alpha_1{}^1 x^1 + \alpha_2{}^1 x^2 \\ y^2 &= a^2 + \alpha_1{}^2 x^1 + \alpha_2{}^2 x^2 \end{aligned} \tag{12}$$

where

$$\alpha_1{}^1 \cdot \alpha_1{}^2 + \alpha_2{}^1 \cdot \alpha_2{}^2 = 0 \tag{13}$$

and for some $\rho > 0$

$$(\alpha_1{}^1)^2 + (\alpha_2{}^1)^2 = (\alpha_1{}^2)^2 + (\alpha_2{}^2)^2 = \rho^2 \tag{14}$$

The similarity is a congruence if, and only if, $\rho = 1$ (orthogonal transformation). Now this analytic representation of congruence and similarity suggests a geometric examination of the most general linear transformations (12) which are nonsingular, that is, for which the determinant $|\alpha_q{}^p| \neq 0$. They were virtually unknown to the Greeks, although they highlight the axiom of parallels of Euclid's geometry. A one-to-one transformation of the cartesian plane is such a linear transformation if, and only if, it carries a straight line into a straight line and parallel straight lines into parallel straight lines. Thus, a parallelogram goes into a parallelogram, and, in fact, given any two parallelograms there is a linear transformation which carries the one into the other no matter what the angles and ratios of sides in the two figures are. There is a geometry, the so-called affine geometry, in which any two parallelograms are considered equal. It cannot measure angles, and nonparallel segments cannot be compared for length. This geometry does have conics, however, and it can separate them into ellipses, parabolas, and hyperbolas. *See* CONFORMAL MAPPING.

The family of all linear transformations constitutes a transitive group and the subfamily of orthogonal transformations is already a transitive group. F. Klein has made the pronouncement, which is generally accepted, that there arises a geometry on a space if on the space there is given a transitive group of transformations; two figures are considered equal whenever one figure can be carried into the other figure by one of the transformations. For the noneuclidean geometries of Bolyai-Lobachevski-Gauss various models have been exhibited which conform to this view and perhaps the most interesting one is the following. In the plane (x^1,x^2) introduce the complex variable $z = x^1 + ix^2$ and consider the family of transformations

$$w = e^{i\theta}\frac{z-a}{1-\bar{a}z}$$

for all constant complex numbers a for which $|a| < 1$ and all real numbers θ. They give rise to the following noneuclidean geometry. The space is not the entire z plane but only the unit disk: $|z| < 1$. A "point" is an ordinary point in it and a "straight line" of the new geometry is in the euclidean geometry either a diameter of the disk or any circular arc inside the disk whose endpoints are on the boundary line of the disk and which meets the boundary line of the disk at right angles. Through a point outside a straight line of the geometry there go infinitely many straight lines which are "parallel" to it; that is, they do not meet it inside the disk. This makes the geometry a hyperbolic one. If, however, in a geometry any two straight lines always intersect, the geometry is called elliptic. The prototype of an elliptic geometry is a geometry on a surface of a sphere (for example, the surface of the earth) in which the straight lines are the great circles. As a matter of fact, the hyperbolic geometry just described is in a very precise sense a counterpart to the elliptic geometry of great circles on a sphere.

The arithmetization of space led to a purely mathematical creation of n-dimensional space, euclidean and other, for any integer dimension n, by defining its points generally as n-tuples of real numbers $(x^1, \ldots, x^n)$ with suitable definitions for various geometrical relations between such points. The most eye-catching consequence in science was the four-dimensional space of the theory of relativity, but actually multidimensional geometry had been playing a part in physics before that. If a mechanical system involves M mass points, it was customary in effect to introduce the space of dimension $n = 3M$ whose points are the states of the system, that is, the n-tuples of coordinates $\{x_m{}^1, x_m{}^2, x_m{}^3\}$, $m = 1, \ldots, M$, at any one time point. Also, if there are restraints operative in the system, then the Lagrange-Hamilton theory suitably reduced the dimension of the space by the use of the free parameters of the system instead of the original n coordinates themselves. The use of free parameters spread from mechanical systems to other systems in physics and chemistry, and all so-called equations of state are geared to such. Finally, in quantum theory a state of a system has infinitely many coordinates and the infinitely dimensional space representing it is a Hilbert space. Also, partly under the influence of Hilbert space, mathematicians have become fascinated with infinitely dimensional spaces in general. They are being studied intensively, and large parts of mathematics are being pressed into these new frames of reference.

The arithmetization of space also is reflected in the ever-widening use of graphs and charts. Any tabulated dependence of a number y on a number x is a function $y = f(x)$ and hence representable as a curve on graph paper; a great amount of information is thus illustrated and stored. This is nothing else but a practical application of the concept of a manifold in modern topology and differential geometry, and, as frequently, the abstract formulation of a mathematical concept is but a circumlocution of what common sense dictates. See CALCULUS OF VECTORS; COORDINATE SYSTEMS, GRAPHICAL; *see also* ALGEBRA; ANALYTIC GEOMETRY; PROBABILITY; STATISTICS; TOPOLOGY.

[S.B.]

Bibliography: N. J. Achieser, *Theory of Approximation*, 1956; A. Ambrose and M. Lazerowitz, *Fundamentals of Symbolic Logic*, 1948; A. A. Fraenkel, *Abstract Set Theory*, 1953; D. Jackson, *The Theory of Approximation*, 1930; P. C. Rosenbloom, *The Elements of Mathematical Logic*, 1951; P. Suppes, *Introduction to Logic*, 1957.

Reprinted from AMERICAN SCIENTIST, Vol. 50, No. 2, June 1962

THE ROLE OF MATHEMATICS IN THE RISE OF MECHANICS*

By SALOMON BOCHNER

IN THE present article we will make some observations on the mathematical nature of mechanics during some phases of its development, and we will be concerned with the inward mathematical structure rather than with the outward mathematical setting. But we will not be able to go essentially beyond the beginning of the nineteenth century. After that time, mechanics and paradigms from mechanics began to penetrate into many other areas of theoretical physics, so that observations on the mathematical nature of mechanics after the early nineteenth century would have to take large parts of theoretical physics into account. But we are not prepared to offer a commentary of such a scope.—Also, in the seventeenth and eighteenth centuries mathematics and mechanics were developing not only in constant interactions but also in close intimacy. But, in the nineteenth century, relations between mathematics and mechanics or rather between mathematics and theoretical physics began to be different. The interactions continued to be no less in importance, but they began to manifest themselves in parallelisms of development rather than in direct interpenetrations. An analysis of their new relations would require categories of description other than those applicable to previous stages.

Modern mechanics came into being soon after the Renaissance towards the end of the sixteenth century and its stages of development, taken by centuries, have been as follows. The seventeenth century in mechanics extended from Stevin to Newton and culminated in Newton's *Principia*. The eighteenth century in mechanics extended from Newton and Leibniz to Lagrange and culminated in his *Mécanique analytique*. The nineteenth century in mechanics extended from Lagrange and Laplace up to the first stirrings of quantum theory in 1900, and in a certain qualified sense it culminated firstly in Maxwell's *Electrodynamics*, inasmuch as the latter theory imitated hydrodynamic wave theory, and secondly also in leading treatises on statistical mechanics. In the twentieth century the leading new theories of mechanics have become, more than before, direct and full representatives of the new physical theories of which they are a part and can hardly be isolated from them for separate analyses.

We venture to characterize the past centuries of modern mechanics by the following headings. The seventeenth century was an age of revelation, the eighteenth century was an age of patristic organization and the

* This research was supported by the Air Force Office of Scientific Research.

nineteenth century was an age of canonical legislation. We dare not continue and suggest that the twentieth century is an age of reformation, or perhaps an age of reformation and counter-reformation in one.

Before the Seventeenth Century

In Greek antiquity, hellenic and hellenistic, if viewed in retrospect and if compared with the seventeenth century, say, there seemingly was an imbalance between mathematics and rational mechanics. The Greeks had a systematic mathematical astronomy but outside of this they seemingly did not arrive at a theoretical mechanics or theoretical physics that was equal in status to their mathematics. Their mathematics, whatever its limitations, had the size and organization of a system. Their mechanics however was only an assortment of achievements, whatever their eminence.

Some of the difficulties that kept Greek mechanics from developing into a system may have been linked to peculiarities which kept Greek mathematics from advancing beyond the stage at which it was arrested. Anybody who appraises the mechanical work on the balancing of a lever in Archimedes and in his successors, hellenistic and medieval, cannot fail to notice that in some covert and tacit manner each of them had the concept of the statical moment in his reasoning, but that none of them reached the point of formulating the concept expressly as the mathematical product $P \cdot L$ in which the factor L represents the length of an arm of the lever and the factor P represents the size of the weight which is suspended from it. Now the concept of statical moment and of other moments became a great leaven in modern mechanics, and one might be tempted to ask how it came about that the concept of statical moment was not expressly conceptualized long before the seventeenth century and especially why Archimedes himself, great intellect that he was, was not forcibly led to introducing it. Now, a historical question like this may be deemed to be methodologically not admissible. But, in fact, if it be admitted, then a specific answer to it quickly suggests itself. The answer is this that even if Archimedes the physicist would have "wanted" to introduce the statical moment expressly, Archimedes the mathematician would not have been able to do so. If L_1 and L_2 are two quantities of a certain kind, say both lengths, and P_1 and P_2 are two quantities of another kind, say both weights, then Greek mathematics could envisage the proportion

$$P_1 : P_2 = L_2 : L_1$$

and decide whether it does or does not hold for special values of the quantities at hand. But it could not convert the proportion into an equality

$$P_1 \cdot L_1 = P_2 \cdot L_2$$

and verify its validity in this form. Greek mathematics could not or "would" not form a product $P \cdot L$. If P and L are general real numbers it could not form the product because it simply did not have real numbers. And if P and L were two general magnitudes, even if it did "have" the magnitudes separately, it still would not conceive of forming their product, especially if the factors P and L represented magnitudes of heterogeneous kind as weights and lengths are. And so Archimedes the physicist and generations of physicists after him could not have a statical moment in mechanics as they should have had.

In a certain sense, and again as viewed from the seventeenth century, the quick answer we have just given does no more than raise another question in the place of the first. If all that was wanting was the mathematical conceptualization of a product like $P \cdot L$ then why was Archimedes, the great mathematician, not driven to doing just that? To this question, if it be not ruled out as inadmissible, there is no quick answer. In fact, in the seventeenth and eighteenth centuries, if mechanics would edge up to a conceptualization which the then mathematics could not articulate then mechanics would simply bid mathematics to learn to do so; and mathematics, after having learned to do so, would react on mechanics in return by suggesting to it more than originally required of it. In other words, mathematics and mechanics were then erecting and exploiting parallelisms and correspondences between mathematical and mechanical entities, conceptions and articulations. The opening pages of Newton's *Principia* are impregnated with such parallelisms. They exasperate expounders of the *Principia* who insist on hierarchical orderliness and unequivocality of meaning in depth, but they also generate the vigor and the speed with which the stream of assertions and implications rushes by.

S. Sambursky in his recent book on *Physics of the Stoics* (1959) rightly observes in passing that Stoic attempts at building up a mechanics of continuous media ultimately failed because, among other reasons, mathematics did not aid them. In this Professor Sambursky is right. But we wish to qualify his finding by adding that, to a Stoic physicist, Archimedes was a mathematician, and a mathematician only, and not somebody who could also be a mathematician and a "theoretical" physicist in one.

The Greeks did not themselves introduce mathematics as a technique of mastering problems that arise in the consciousness and thinking of man. This had been done by others before, as the Greeks themselves fully acknowledged. But they did recognize in mathematics a universal arché of knowing. They so proclaimed it to be and they would not have allowed this universality to be qualified by differences of eras and of philosophical settings. We think therefore that it is legitimate to probe, even in retrospect and even against the different philosophical spirit of the seventeenth century, the extent to which the Greeks themselves succeeded in realizing and exploiting the force and effectiveness of this universality.

Archimedes also accomplished basic work, perhaps his most famous one, in the mechanics of floating bodies. Here again he did not introduce the physical concept which is central to the subject matter, namely, the concept of hydrostatic pressure. But in this case Archimedes may be "excused." Modern mechanics had great difficulties in conceptualizing the notion of pressure, although Stevin immediately mapped out the task of doing so and almost everybody after him was pursuing it. Even Newton was not yet quite certain of it. In a sense the clear-cut mathematization of the concept of pressure was arrived at only in the course of the nineteenth century beginning perhaps with work by A. Cauchy on equations of motion for a continuous medium in general. In the nineteenth century the mathematical "image" of pressure became a tensor, albeit a very special one, and the actual formalization of the concept of tensor and a full realization of its mathematical status took a long time to emerge.

The Schoolmen of the thirteenth and fourteenth centuries were inspired to attempt to conceptualize notions from mechanics like velocity, acceleration and force, by introducing corresponding elements of the infinitesimal calculus and by even trying to formulate coordinate geometry. It has not yet been fully established to what extent these medieval efforts were a truly anticipatory phase of seventeenth century developments. The fact is that in the fifteenth and sixteenth centuries the pace of mechanics slackened whereas mathematics was quite active but in directions of its own. By the end of the sixteenth century mechanics somehow became very vital again. It showed great vigor, acted as if it made a fresh start, and suddenly there existed a Rational Mechanics in the full sense of the word. Mathematics was quite active during the sixteenth century, but after 1600 it quickened its tempo even more and also entered on developments which were different from those of the preceding century.

During the Renaissance period, mathematics was preoccupied mainly with what is commonly called arithmetic and algebra. The initiative for this arithmetic and algebra came from Northern Italy and a most distinguished very early representative of it was Leonardo da Pisa (Fibonacci) around 1200. It is not quite clear how the great interest in these subjects came about and then sustained itself so unflaggingly. Economic determinists claim that the Rise of Commerce was a great stimulant. Now, whatever the origins of this mathematics may have been, about the effects of it there can be no doubt. It brought about momentous innovations which gradually wrought a change-over from the inertness of traditional syllogistic schemata of Greek mathematics to the mobility of symbols and functions and mathematical relations. These symbols and functions and relations have penetrated into many areas of systematic thinking, much more than sometimes realized, and they have become the syllables of the language of science. What mathematical symbols and functions are nobody can tell, other than in tautologies, that is. The Greeks have found that mathematical thinking has a universality *sui generis*, and investigations into foundations of mathematics are not much more than systematic explications of what this implies.

The Renaissance age also initiated the theory of probability. From being a gambler's calculus of finitely many combinatorial possibilities it eventually developed into an analytical theory of laws of large numbers and of infinitely many possibilities in discrete and continuous distributions. During the past 100 years, Probability entered Mechanics in a major way, and it did so twice. In the second half of the nineteenth century it created statistical mechanics for the kinetic theory of matter; and in the twentieth century it was called upon to mediate between two conflicting approaches to quantum mechanics, corpuscular and undulatory. The mediator's judgment was unprecedented and at first quite disturbing.—Also, as if in return, a phenomenon from mechanics which was first observed by a naturalist in the early 19th century, the so-called Brownian motion, has in the twentieth century stimulated the emergence of the probabilistic concept of a general stochastic process. The concept is of undetermined definitory scope, but it has been in the center of probabilistic thinking for several decades past.—Except perhaps for combinatorial speculations, there are no records of quantitative probabilistic pre-concepts in Greek antiquity. There is a provocative section on Chance and Fortune (or rather Chance and Fortuitous) in Aristotle's Physics (Book II, Ch. 4, 5, 6) but regrettably it has no relevance to Mathematics or Physics as we understand them. There seem to be no antecedents to modern Probability in earlier eras and civilizations, and in view of Probability's link to Chance and Fortune this is most remarkable.

The Seventeenth Century

The seventeenth century produced the greatest single statement on the relation between mathematics and physics. It is Galileo's dictum that mathematics is the language of science. It is more than an aphorism. It asserts that although mathematics and physics are inextricably linked they nevertheless have separate identities and are hierarchically not comparable. Attempts to bridge the separateness of their identities and to put them into hierarchical dependence have been undertaken frequently.

Attempts to subordinate mathematics to physics (or even mechanics) usually center in the statement that "geometry is a branch of physics." On the other hand, certain contemporary trends, especially those which trace themselves to Leibniz' plan for a universal mathesis, may be interpreted as giving to mathematical knowledge and to logico-mathematical formulations a supremacy over the realm of science, not only natural science but even social science including perhaps history too. Leibniz' plan had some antecedents in the fourteenth and fifteenth centuries. Also, some Schoolmen of the thirteenth century, above all Roger Bacon, have become known for their esteem for mathematics, but it would not be easy to attribute to them a reasoned metaphysical judgment on the relation between mathematics and physics.

In Greek antiquity, according to Aristotle, Pythagorean circles maintained that in nature "all is number" meaning apparently that "all is mathematics." Some of them also maintained that space, meaning apparently the space of physical events, is made up of discrete points. Before evaluating the degree of sophistication of this alleged Pythagorean view, one should realize that over 100 years ago the mathematician, Bernhard Riemann, in a famous address on differential geometry, also suggested that space might turn out to be discrete. Plato's doctrine or doctrines also incline in general to give to mathematics a supremacy over much of other knowledge. However, Aristotle also reports, with some emphasis, that Plato assigned to objects of mathematics a hierarchical position intermediate between sensibles and pure Forms (Ideas), which obviously circumscribes the supremacy of mathematics, and, what is more important, gives it an identity of its own. This was a very fine glimpse of the true nature of mathematics, but Aristotle, within his general condemnation of Plato's theory of Ideas, battles this platonic view and advances instead his own that mathematics is an abstraction from nature. Aristotle's view clearly does not fit modern mathematics. If π is, as usual, the area of the circle of radius 1, and e is the basis of natural logarithms then the number e^π is hardly an abstraction from anything in nature, although π so is, and e by an expenditure of casuistry, could be made to be so too. Also the complex number $i = \sqrt{-1}$, although a solution of the equation $x^2 + 1 = 0$, is certainly not an abstraction from anything in nature, and yet complex numbers have been used successfully by A. J. Fresnel in the theory of refraction in 1818 and have spread into all parts of physics for many purposes.

In the seventeenth century almost everybody excelled in more than one area. Galileo the physicist had a standing as philosopher. Pascal was scientist and "penseur." Leibniz was philosopher, mathematician, logician. Kepler was astronomer and mathematician. Descartes was evenly divided between philosophy and mathematics. And Newton was equally great in mathematics and in physics. A notable exception was Huygens. He was only a physicist. But he was a physicist's physicist, although he did not even write a treatise.

Some leading works of the seventeenth century give the impression of continuing, by direct succession, works of a distant past, thus spanning gaps of long stretches of time. For instance, Galileo does not take much notice of fourteenth century Schoolmen but he acts as if he were a direct successor to Aristotle and he "corrects" Aristotle on specific statements made in the *Physics*. Descartes in his geometry acts as if he were a direct successor to Apollonius (210 B.C.) and Pappus (320 A.D.) and in a certain sense he even taunts these "predecessors" of his for not having achieved themselves what he is about to do now. And it is indeed inexplicable why Greek mathematics which lasted over 1000 years, namely, from 600 B.C. to 400 A.D. and beyond, and which had a geometric tenor did not arrive at such simple things as placing figures into preintroduced coordinate systems and plotting curves on graph paper. Fermat who was mainly a number theorist considered himself to be direct successor and disciple to the ancient number theorist Diophantus (250 A.D.). He revered him, and as a gesture of reverence he even quoted Diophantus in a non-arithmetical context in a paper on maxima and minima (Fermat, *Oeuvres*, Vol. I, p 133). But the most notable instances of such "leaping backward" are in the achievements of Kepler and Newton.

Kepler was an assistant to Tycho Brahe, a disciple to Copernicus, and a spiritual descendant of Nicolas of Cusa (fifteenth century) who is commonly considered to have been the harbinger of an important phase of modernism. These are straight facts and could not be gainsaid. But if, from a distance, and in retrospect, one elects to look at the bare mathematics and physics only, and only at the sharp lines in the contours of development, then one may also find that Kepler's discipleship goes back directly to Apollonius and also Ptolemy (150 A.D.) and that Kepler was a successor directly to them. To generations of astronomers, from Eudoxus (370 B.C.) to Copernicus and Tycho Brahe, the basic celestial figure was a circle and they took it apparently for granted that a celestial trajectory is a figure composed of circles kinematically. Kepler broke with this supposition. He introduced ellipses instead, and, guided by astronomy, he grasped the true meaning of foci of conics which others before him had no occasion to recognize. Conics, and their theory, were in no way Kepler's private mathematical invention and discovery. They had been in the public domain for nearly 2000 years for anybody to find and use. Also, in point of literal fact, Kepler did not find his ellipses in the tables of Tycho or the "revolutionary" writings of Copernicus or the "learnedly ignorant" utterances of Cusanus. He found them by searching, untiringly, in the work of Apollonius all his life, and by being, apparently, also thoroughly familiar with the second half of Apollonius' treatise which by then had not yet been translated from the Arabic into Latin, as far as is known. Apollonius knew astronomy and Ptolemy undoubtedly knew conics, but they did not use conics as celestial orbits. Kepler did so, and the mathematics which he pre-required should have been readily available to Apollonius and Ptolemy. Therefore, from a certain pragmatic approach, Kepler was a direct successor to Apollonius and Ptolemy.

The Isaac Newton of the *Principia* is by the principal subject matter of the treatise a direct successor to Kepler and by the manner and style of its conceptualizations a direct successor to Archimedes. What makes the *Principia* the book that it is are not its Definitions and its Laws and its views on time and space and not even its central forces if taken by themselves and on their own merit. The *Principia* are what they are because with this congeries of definitions, laws, concepts of time and space, and his gravitational force, and by mathematically controlled reasoning, Newton constructed and deduced what Kepler had only divined and posited, and because into this design of Kepler's celestial orbits, and from the identical blueprints, Newton also fitted the (then) earthbound trajectories of Galileo and Toricelli. The twentieth century has apparently verified that even artificial celestial bodies conform to the Newtonian theory of the seventeenth century, and, what is most remarkable, nobody in the twentieth century doubted that they so would (except of course for "relativistic" adjustments if such should be called for, eventually).

Now, outwardly the *Principia* are written in a pure idiom of Archimedes and Apollonius, and outwardly there is hardly any mathematics invoked or presupposed which should not have been quickly accessible to them. Newton was well familiar with Descartes' work, and in his book on plane curves Newton masters the method of Descartes in analytical detail much better than Descartes himself. Also Newton was quite skilled in the use of symbols and functions from the beginnings of his mathematical career, as his early work on power series and also on fluxions shows. But in the *Principia* no attempt is made to modify the Greek idiom by the use of Descartes' innovations or to employ symbols and functions analytically; even the inviting trait of discursiveness that animates the *Opticks* is hardly present in the *Principia*. Furthermore, and this we would like to stress, Newton's definitions of limit and derivative ("ultimate ratio") as presented in the *Principia* seem to us to be such, in syntax and in content, that a personal disciple to Archimedes should have been able to compose them. At any rate, it would be a major task to substantiate philosophically that a student of Archimedes could not have possibly arrived at them because of their being removed from and alien to the hellenic-hellenistic milieu of natural philosophy within which the developments from Eudoxus to Archimedes and Apollonius had taken place.

And yet, inwardly, the Euclidean Space that underlies the *Principia* is mathematically not quite the same as the Euclidean Space that underlies Greek mathematics (and physics) from Thales to Apollonius. The geometry of the Greeks emphasized congruences and similarities between figures, that is, in analytical parlance, orthogonal and homothetic transformations of the underlying space. The Greeks themselves were half-aware of this. Late Greek traditions about the origin of their mathematics, whether true or legendary, ascribe to Thales mathematical exploits which involve congruences and similarities emphatically. The Euclidean Space of the *Principia* continued to be all this, but it was also something new in addition. Several significant physical entities of the *Principia*, namely, velocities, momenta, and forces are, by mathematical structure, vectors, that is, elements of vector fields, and vectorial composition and decomposition of these entities constitute an innermost scheme of the entire theory. This means that the mathematical space of the *Principia*, in addition to being the Greek Euclidean substratum, also carries a so-called affine structure, in the sense that with each point of the space there is associated a three-dimensional vector space over real coefficients, and that parallelism and equality between vectors which emanate from different points are also envisaged. In the course of the eighteenth century the vectorial statements of Newton and others were gradually transferred and re-interpreted into analytical statements. This involved Euclidean coordinates and eventually also "generalized" coordinates of Lagrange. In these analytical developments the vectorial character of the space of Newton ceased to be featured. But in the nineteenth century, partly within the concepts of Hamilton's theory of quaternions, the underlying vectorial trait was gradually recalled, and even taken account of by a symbolism of its own which has remained in use. The twentieth century widened the concept of a vector into the broader concept of a tensor. This called forth readjustments in mechanics and physics and affected pure mathematics even more profoundly.

Newton asserts the vectorial character of mechanical force in "Corollary II to the second law" in Book I of the *Principia*, and his assertion consists of two parts. The first part states that any two forces give rise to a resultant force which is the vectorial composition of the two. The second part is in the nature of a converse to the first part and it states that in a plane a force can be decomposed into two component forces along any two non-parallel lines. The second part of the assertion is much more recondite than the first, even if applied to the simpler case of velocities rather than forces. It is a simple observational fact that if two persons pull at a load in different directions then the load moves along some suitable diagonal, and this fact was undoubtedly known to Aristotle as commonly asserted. But it is an entirely different assertion to say that if in a plane a force is given "concretely" in a well-determined line of action, then purely speculatively it can be imagined to be a vectorial sum of two "fictitious" component forces along "fictitious" lines of action of arbitary choice; and there is nothing to suggest that such a possibility was in the minds of natural philosophers before the seventeenth century, say.

The Eighteenth Century

The eighteenth century was an age of organization, for mechanics and even for mathematics. It produced several so-called Principles of mechanics and some special theories and methods. Books from the eighteenth century are frequently still readable and many of their notations have maintained themselves into current or relatively recent use. The organizing activity was not all spent on a continuation and elaboration of what the seventeenth century had achieved, but a large part of it also on a recreation and re-establishment of the subject matter from its roots.

The eighteenth century introduced, more or less formally, the concept of a mass particle in the sense of a geometric mass point, and of a finite system of mass particles with general forces and constraints acting between them, and it recognized that the theory of such finite systems is a guide and pilot theory for rational mechanics as a whole. But it also studied extensively the so-called "mechanics of continua" which comprises large and detailed theories of its own, chief among which are hydrodynamics, acoustics, and the general theory of elasticity. Acoustics on a very elementary level is as old as the Pythagoreans, a hydrostatistical theorem is one of Archimedes' great claims to fame, and rudimentary beginnings of elasticity have been ascribed to Stoics.

The organization of mechanics necessitated the creation of a considerable amount of mathematics, mainly in Analysis. Of this mathematics, the calculus of variation was instigated largely by mechanics of particles, but other theories were instigated largely by mechanics of continua. Virtually all of partial differential equations started in this way, especially the mathematical theory of waves which eventually became the hallmark of theoretical physics in all its parts. Mechanics of continua also created half of Fourier analysis. The other half was created by the theory of heat which also contributed the name of the theory. Mechanics of continua also had a fair share in the promotion of potential theory. But it should be said that the concept of potential energy originated in the Lagrangean theory of finite systems of particles, and that, on the other hand, through the work of Poisson and others, the concept of potential acquired the status of a commanding physical entity not within mechanics but within electrostatics.

Finally, it appears that mechanics of continua had a share in the emergence of tensor theory which after slow stages of nascency began to take shape towards the end of the nineteenth century, and in the twentieth century has been penetrating various areas of physics and mathematics. Important entities in the mechanics of continua, at whose mathematical conceptualization the emergent theory of tensors was aiming, were what are nowadays the pressure and stress tensor and the accompanying deformation tensor. Pressures, tensions, stresses and such like entities are species of "force" but phenomenologically they seem not to subordinate themselves to the concept of force as introduced in Book I of the *Principia* the mathematical substratum of which is a vector of three components issuing at a point. In fact they seem to be entities which associate themselves with the various surface elements through a point in space rather than with a point of space by itself. In the outgoing eighteenth and beginning nineteenth centuries efforts were made to overcome these disparaties by reducing all surface forces to purely vectorial forces through appropriate molecular approximations; but there were also parallel efforts to conceptualize things as they present themselves phenomenologically. In this conceptualization there are two types of forces, point forces and surface forces and they are linked by certain basic laws which are modeled on Newton-Lagrange prototypes. A surface force is a space tensor of two indices. Such a tensor in general has 9 components. In surface forces of the customary kind a symmetry property reduces the number of components "effectively" to 6. Special symmetry requirements may reduce it further. The prime instance of this arises in the case of an incompressible fluid. In this case the pressure tensor reduces effectively to only one component, and this is the so-called pressure. Although it is a one-component quantity, it is definitely not a scalar. In fact, as a mathematical object it is far removed from anything which Archimedes might have been able to conceptualize in his work on floating bodies against the background of the mathematics then alive.

A particular situation arises in the case of rigid bodies. They are continuous media, but tensorial forces need not be envisaged although a tensorial setting does illumine the context. The theory of rigid bodies is a doctrine by itself. Large parts of it are geometrical, but it is a "modern" geometry of analytical variability, and not of Greek stationarity. The theory was fully started in the eighteenth century by L. Euler and is a showpiece among the many masterpieces of this great master. In retrospect, his kinematics is a first meaningful study of groups, namely, of one-dimensional continuous groups of orthogonal transformations in Euclidean space, well over 100 years before Sophus Lie, Erlanger Programs and such like. His dynamics was a transfer, with suitable elaborations, of appropriate statements for finite systems of points to the continuous case, without a trace of brooding over the melancholy fact that there are no truly rigid bodies in nature.

All told, the eighteenth century achieved very much for the theory of continuous media and mathematics involved in it. And yet, the historical "image" of eighteenth century mechanics is composed of achievements which relate to finite systems of particles primarily.

Of great originality are two so-called "differential" principles, the principle of Bernoulli and the principle of d'Alembert. They can be stated and appreciated separately. But what has given them their status is the fact that jointly they are the foundation for the general Lagrange equations of motions for finite systems of particles. In any rationalization of these equations the two principles are involved deeply and inseparably.

The principle of d'Alembert mathematically "transforms" a dynamic system of particles into a static one, and it views the so-called Newtonian equations of motion as if they were conditions for equilibrium. If a single point moves on a straight line under the influence of an "external" or "impressed" force F, then the principle of d'Alembert puts the equation of motion in the form

$$-m\frac{d^2x}{dt^2} + F = 0$$

and, as many in the nineteenth century understood the principle, it "interprets" the term

$$-m\frac{d^2x}{dt^2}$$

as representing also a force, the so-called inertial force. Thus it interprets the equation of motion to mean that the "total" force, that is the sum of external force and of inertial force, is zero. In a general system of points it asserts that suitable vectorial sums of external forces and of inertial forces are each zero. Since the notion of an external force is usually an "abstraction," for instance if it is a gravitational force, the inertial force of d'Alembert is in a sense an abstraction from an abstraction, or rather an abstraction vertically imposed on a previous one. Such "multifold" abstractions, that is, abstractions from abstractions, abstractions from abstractions from abstractions, etc., if performed in cognitive depth, generate much of the motive power of modern mathematics and also modern science; and we think that not many of such will be found among all the "ideations" from Plato to Plotinus to Abélard to Ockham, especially not from Plato to Plotinus.

The principle of Bernoulli is more technical, and in a sense even more daring. For our purposes we will present it in two stages. The first stage will be seemingly a mathematical tautology in preparation of the second stage. But we should point out that we will follow the presentation of the principle by Lagrange rather than Bernoulli's own.

If in a finite system of mass points we denote the totality of all components of all mass points in a uniform lettering by

$$x_1, x_2, \ldots, x_N \tag{1}$$

and the corresponding components of the external forces by

$$F_1, F_2, \ldots, F_N$$

then in the case of equilibrium all these components of force are zero. That is, we have

$$F_1 = 0, F_2 = 0, \ldots, F_N = 0. \tag{2}$$

Purely mathematically this set of N conditions can be replaced by the one condition

$$F_1 \cdot \lambda_1 + F_2 \cdot \lambda_2 + \ldots + F_N \cdot \lambda_N = 0 \tag{3}$$

which is assumed to hold for all possible choices of real numbers

$$\lambda_1, \lambda_2, \ldots, \lambda_N.$$

Now the principle interprets each number λ_n as a "virtual" displacement of the corresponding space component x_n. Such a displacement is usually denoted by δx_n, and thus condition (3) is now written as

$$F_1 \cdot \delta x_1 + F_2 \cdot \delta x_2 + \ldots + F_N \cdot \delta x_N = 0. \tag{4}$$

The first interpretative gain of this formulation was as follows. The expression on the left side, that is the sum

$$F_1 \cdot \delta x_1 + F_2 \cdot \delta x_2 + \ldots + F_N \cdot \delta_N, \tag{5}$$

was conceived as a certain "action" of the given forces along the given virtual displacement—which the nineteenth century eventually called "work"— and equation (4) is now interpreted to state that the system of forces is in equilibrium if and only if this action is 0 along any virtual displacement whatsoever.

This is the first stage of the Bernoulli principle. For the second stage we reconsider matters from the following approach. Thus far we have tacitly assumed that our system is a "free" system of particles, that is that the presence of external forces is all that influences the motion of the particles. But now we take a system of particles in which so-called "constraints" are present, where constraints are certain primary conditions which prescribe that only certain geometrical configurations of the mass points can occur. Also the constraints are assumed to be independent of the external forces and to be superimposed on these. Now, the second stage of the Bernoulli principle states that for a system with constraints the necessary and sufficient conditions for equilibrium continue to be the same as in a free system. Thus, it is again equation (4), and $F_1, \ldots, F_N$ are again the components of external forces, and $\delta x_1, \ldots, \delta x_N$ are again virtual displacements. But, and this is decisive, *only such virtual displacements are to be considered as are compatible with the constraints.*

It took a long time before it was generally understood that the principle of Bernoulli if applied to systems with constraints is a far-reaching hypothesis of axiomatic nature. L. Poinsot in an angry note in 1838 (*Journal de Mathématiques pures et appliquées*, Vol. 3, pp. 244–248) castigated Laplace for misrepresenting this point in his *Mécanique Céleste.* And C. G. Jacobi when stating matters very clearly in his *Vorlesungen über Dynamik* (1842–43, published 1865) still deemed it appropriate to invoke the great authority of Gauss in support of his statement.

Lagrange obtained his equations of motion in the following manner. He superimposed the principle of d'Alembert on the principle of Bernoulli, that is he applied the second stage of the principle of Bernoulli to forces which are sums of external and inertial forces. Then, by a purely mathematical argument, which is one of his leading mathematical achievements, he transformed the constraining *conditions* also into constraining *forces*. As a result, his equations of motion for a finite system with constraints have the same analytical form as for a system without constraints except that in addition to the external forces there are some further "ideal" forces which account for the constraints. These ideal forces are perpendicular to those virtual displacements which are compatible with the constraints so that they make no contribution to the expression (5). Also, the actual sizes of the constraining forces are measured by the values of so-called "Lagrange multipliers."

This was only half of what Lagrange did in this area, and the second half is the one that really matters. Lagrange subjects the original space coordinates (1) to a general analytic transformation of variables. This gives rise to new variables, again of the same total number N, and they are frequently denoted by the letters

$$q_1, q_2, \ldots, q_N. \tag{6}$$

These are his so-called "general coordinates," and Lagrange translates the previous equations of motion into new equations for the new coordinates. Now, if there are constraints in the system, and if they are expressible by s "independent" equations between the original N coordinates (1) then it is possible to choose the general coordinates (6) in such a way that the constraining conditions become the requirement that s among the new coordinates shall be zero, say

$$q_{N-s+1} = 0, q_{N-s+2} = 0, \ldots, q_N = 0. \tag{7}$$

The remaining N—s coordinates

$$q_1, q_2, \ldots, q_{N-s} \tag{8}$$

are then unconstrained and these are Lagrangean "free parameters" of the system. If general parameters have been so chosen, then the system of equations of motion decomposes into two parts, a primary part and a secondary part. The primary part involves only the free parameters (8), does *not* make express reference to constraining forces, and completely determines the solutions of the problem. The secondary part, on the other hand, only determines the constraining forces and nothing else. Thus the secondary part may be ignored altogether, and textbooks usually omit it.

Constraints and constraining forces, and attempts to understand their nature and effects, are as old as mechanics itself and a full history of the genesis of these concepts would be virtually coextensive with a history of the genesis of mechanics itself, at least of the mechanics of weights and ordinary bodies. One would certainly have to take into account the work of Jordanus (thirteenth century), and perhaps even go back to Heron of Alexandria. Stevin and Galileo operate with constraints and constraining forces in the open, and Newton does so even more expressly. And yet their analytical formulations in the eighteenth century, apparently more-or-less due to Lagrange, constituted a major novelty nonetheless.

The emergence of general coordinates and free parameters in the eighteenth century was intellectually a prelude to the rise of multi-dimensional geometries, Euclidean and non-Euclidean, in the nineteenth century. Some of the creators of these geometries were not only experts in rational mechanics but even leaders in it. We think that, by evolution of ideas, the geometries of the nineteenth centuries were, in good part, descendant from eighteenth century mechanics, and not only in the general sense that any period of mathematics is descendant from the preceding one. The total development was an evolutionary process. In the seventeenth and eighteenth centuries the structure of space and place was Euclidean as a matter of course. In the nineteenth century the possibilities of non-Euclidean structure were being explored for geometry and for geometry alone. By the end of the century, when the mathematics was ready for it, first steps were taken towards inserting non-Euclidean features into mechanics also. And in the course of the twentieth century some non-Euclidean (that is "relativistic") features of mechanics have already become basic and commonplace.

Some philosophers and mathematicians unduly single out the emergence of non-Euclidean geometry as the decisive mathematical event of the nineteenth century. Non-Euclidean geometry is all important for any confrontation between modern mathematics and Greek mathematics. But it was not the only great achievement of the nineteenth century, far from it; and against the background of what the eighteenth century had been preparing for, it was not a startling innovation but rather more a "natural" event, albeit a very notable one.

It was providential that the seventeenth and eighteenth centuries, especially the seventeenth, took the Euclidean structure of space for granted and knew of no other. Newton did find the law of gravitation for Euclidean space. But if, by some inspiration, he would have known about non-Euclidean possibilities and tried to find the law of gravitation directly for a space with curvature, he might have been led woefully astray. Greek mathematics, viewed in retrospect, undertook to develop things out of order and came to grief. Instead of developing a concept of real number as a common mathematical measure for physical quantities, Greek mathematics undertook to develop directly a mathematical theory of general physical quantities. They seemingly succeeded, as the fifth book of Euclid and the monumental works of Archimedes and Apollonius testify. But the resulting mathematics was too inflexible to adjust itself to needs of mechanics and physics, and too cumbersome and heavyweight to propel even itself far beyond the platform on which it had gained its crown of laurels.

In the evolution of modern mathematics and mechanics a contingent dis-harmonious feature was the timing of the rise of the great Kantian philosophical system. Having come into being in the eighteenth century it tied itself to Euclidism, and in fact to a doctrinaire seventeenth century version of it, and adherents in later generations wished it were not so.

The Principle of Least Action

A lively activity in the field of eighteenth century mechanics and mathematics was the search for a suitable version of the principle of least action. The search was beset with difficulties and misunderstandings because, as seen in retrospect, it involved the search for the concepts of both kinetic and potential energy on a level of quantitative precision. These concepts however were apparently difficult and recondite, and, in fact, only the nineteenth century was equipped to master them.

The search for the concept of energy began early. A notion of energy, or perhaps a set of companion notions, was a dominant and ever recurrent feature of Aristotle's philosophical system. One may even claim for him that the distinction between kinetic and potential energy was already foreshadowed in his work, especially in his famous statement that "motion (that is the Aristotlean kinesis) is the actualization of that which is potentially, as such." But in cold fact it must be said that there is a long distance between his notion of energy and a concept, or even preconcept, of energy that fits into mechanics as we understand it. Some of his medieval followers may have come closer to it, but the important researches of Anneliese Maier suggest that the Schoolmen had great difficulties in keeping apart force, energy, and momentum.

Almost everybody in the seventeenth and eighteenth centuries had something to say about *vis viva,* that is kinetic energy, except Isaac Newton who most prudently avoided getting involved in it, perhaps because, unlike force, energy could not be spatially exhibited. Huygens, on the other hand, was keenly interested in it and made an important application of it. He reasoned that, in an elastic collision of two masses, not only that total translational momentum remains continuous throughout the collision, which fact had already been known before, but the total kinetic energy too.

Lagrange was the first to introduce potential energy. He also gave a satisfactory version of the principle of least action, and in it he also uses the concept of total energy as the sum of kinetic and potential energy. The Lagrange version of the principle has become overshadowed by the

Hamiltonian principle, but Lagrange can claim credit for the fact that his version, which he found much before Hamilton, is in basic mathematical structure the more difficult one.

Although Lagrange introduced potential energy and applied it, the theory of it was elaborated by others (Laplace, Poisson, Gauss, Green, etc.). Also the actual interest in potential energy came not from mechanics but from electrostatics in which it is the identifying concept, almost. Similarly the concept of total energy of a system, although originating in mechanics, acquired its status in the theory of heat to which, again, it is the dominant concept.

Theoretical Physics, as against rational mechanics, fully began only with the concept of energy, after mechanics had completed its main phase without yet concentrating on it. The (prequantum) energy was measured by arbitrary real numbers, and it is perhaps more than a coincidence that the proper conceptualizations of real numbers and of energy were taking place in the nineteenth century virtually simultaneously.

Simple cases of the Lagrange version of the principle of least action are rarely presented in textbooks in analytical detail and we are going to append one.

Let T denote the kinetic energy of a conservative system, U its potential energy and let t denote time. The principle states that the equations of motion are obtained by the variation

$$\delta' \int_{t_0}^{t_1} 2T\, dt = 0 \tag{9}$$

under the side condition

$$\delta'(T + U) = 0, \tag{10}$$

where δ' denotes the variation not only of the space variables but also of the time scale, both between fixed endpoints.

Consider the case of a mass point m moving on the x-axis. Its kinetic energy is

$$T = \frac{m}{2}\left(\frac{dx}{dt}\right)^2$$

and we have to vary the integral

$$m \int_{t_0}^{t_1} \left(\frac{dx}{dt}\right)^2 dt. \tag{11}$$

Since t is also to be varied we must introduce a function $t = t(u)$ over an interval $u_0 \leq u \leq u_1$. The expression (11) is then

$$m \int_{u_0}^{u_1} \left(\frac{dx}{du}\right)^2 \left(\frac{dt}{du}\right)^{-1} du. \tag{12}$$

We must now replace x by $x + \epsilon\xi$ and t by $t + \epsilon\tau$ where ξ and τ are finite displacements and ϵ is a small real parameter.

If we substitute this in (12) we obtain a function

$$A(\epsilon) = m \int_{u_0}^{u_1} \left(\frac{dx}{du} + \epsilon \frac{d\xi}{du}\right)^2 \left(\frac{dt}{du} + \epsilon \frac{d\tau}{du}\right)^{-1} du \tag{13}$$

and the requirement (9) is

$$\left.\frac{dA(\epsilon)}{d\epsilon}\right|_{\epsilon=0} = 0.$$

By carrying out the differentiation in (13) and then putting $\epsilon = 0$ we obtain

$$2m \int_{u_0}^{u_1} \frac{dx}{du}\frac{d\xi}{du}\left(\frac{dt}{du}\right)^{-1} du - m \int_{u_0}^{u_1} \left(\frac{dx}{du}\right)^2 \left(\frac{dt}{du}\right)^{-2} \frac{d\tau}{du} du = 0. \tag{14}$$

Next, for the requirement (10) we have to introduce the function

$$B(\epsilon) = \frac{m}{2}\left(\frac{dx}{du} + \epsilon \frac{d\xi}{du}\right)^2 \left(\frac{dt}{du} + \epsilon \frac{d\tau}{du}\right)^{-2} + U(x + \epsilon\xi),$$

and the requirement itself is

$$\left.\frac{dB(\epsilon)}{d\epsilon}\right|_{\epsilon=0} = 0.$$

By carrying out the differentiation and putting $\epsilon = 0$ we obtain

$$m \frac{dx}{du} \cdot \frac{d\xi}{du}\left(\frac{dt}{du}\right)^{-2} - m\left(\frac{dx}{du}\right)^2 \left(\frac{dt}{du}\right)^{-3} \frac{d\tau}{du} + \frac{\partial U}{\partial x}\xi = 0.$$

From this we obtain

$$m\left(\frac{dx}{du}\right)^2 \left(\frac{dt}{du}\right)^{-2} \frac{d\tau}{du} = m \frac{dx}{du}\frac{d\xi}{du}\left(\frac{dt}{du}\right)^{-1} + \frac{\partial U}{\partial x}\xi \frac{dt}{du};$$

and if we substitute this into (14) the result is

$$\int_{u_0}^{u_1} \left[m \frac{dx}{du}\frac{d\xi}{du}\left(\frac{dt}{du}\right)^{-1} - \frac{\partial U}{\partial x}\xi \frac{dt}{du}\right] du = 0. \tag{15}$$

Thus, the time displacement τ has been eliminated and we can now return from the variable u to the original variable t. Then, (15) becomes

$$\int_{t_0}^{t_1} \left[m \frac{dx}{dt}\frac{d\xi}{dt} - \frac{\partial U}{\partial x}\xi\right] dt = 0. \tag{16}$$

Now, by partial integration we have

$$\int_{t_0}^{t_1} \frac{dx}{dt}\frac{d\xi}{dt} dt = \left.\frac{dx}{dt}\xi\right|_{t_0}^{t_1} - \int_{t_0}^{t_1} \frac{d^2x}{dt^2}\xi\, dt. \tag{17}$$

Next, we assume that the displacement ξ is 0 at the time endpoints t_0, t_1, and we substitute (17) into (16). The result is, after changing the sign,

$$\int_{t_0}^{t_1} \left[m \frac{d^2x}{dt^2} + \frac{\partial U}{\partial x}\right]\xi\, dt = 0.$$

Now, for a given function $x(t)$ between the limits t_0, t_1 this must hold for any displacement $\xi(t)$ which is 0 at t_0 and t_1. This can only occur if the factor of ξ under the integral is 0, that is

$$m \frac{d^2x}{dt^2} + \frac{\partial U}{\partial x} = 0.$$

But this is indeed the equation of motion of a mass point m on the x-axis subject to a force which is derived from a potential $U(x)$.

The Significance of Some Basic Mathematical Conceptions for Physics

By *Salomon Bochner* *

THE present paper supplements and continues a preceding one,[1] and we will again make observations on the relations between mathematics and physics on the level of abstractions and conceptualizations. In the preceding paper we emphasized mechanics, and at present we will be more concerned with physics in general; but we will find it convenient to begin with the same problem about Greek mechanics as before.

The Greeks created rational philosophy and a general conception of rational philosophy, and they created a general conception of mathematics and a general conception of physics. Their mathematics is justly renowned; the Greeks imparted to it from the beginning a certain metaphysical accent which has become a distinguishing mark of mathematics in general. They developed an astronomy and geometrical optics which were quite mathematical, and the Greeks themselves wondered whether, and in what sense, these could be distinguished from mathematics proper. Also the *Physica* of Aristotle alone contains manifold reflections which, if suitably interpreted, are flashes of insight of stirring profundity.

And yet if one undertakes to assess in retrospection and pragmatically the outcome of Greek mathematics and physics, and if one confronts the mathematics and physics of Greek antiquity with modern mathematics and physics, which came into being during and after the Renaissance, then one may nevertheless find that Greek physics as a whole, and in its inward structure, never developed into a " theoretical " system, that the system of Greek mathematics had severe limitations and shortcomings, and that these inadequacies made Greek mathematics unsuited to promote the rise of theoretical physics as we have it or to mold scientific thinking in our sense. A mathematics that was much more suited to do this began only after the Middle Ages, and its first impact on physics and other science

* Princeton University. This research was supported by the United States Air Force Office of Scientific Research.

ISIS, 1963, VOL. 54, PART 2, NO. 176.

[1] S. Bochner, " The Role of Mathematics in the Rise of Mechanics," *American Scientist*, 1962, *50*: 294-311.

was to promote the rise of a Rational Mechanics, which held the stage of science during the seventeenth and eighteenth centuries and without which the general mathematization of physics and then, progressively, of ever larger areas of so-called " exact " science and also of other (even social) science would not have come about.

Of course, the difference between Greek and modern mathematics cannot be fully located in one single indivisible attribute. It is a complex phenomenon, or rather a single phenomenon of many aspects, and in the course of the present paper we will refer to several aspects of it as they become manifest in different contexts. But there is one aspect of it which is very overt, and perhaps the most overt of all. It is this aspect with which we began our analysis previously,[1] and we are going to begin again by first of all recapitulating this analysis and expanding it into a larger context.

I. The Significance of Multiplication

1. Greek mathematics did not create real numbers, and it did not arrive at a conceptualization of general products of quantities. Instead, it created a calculus of ratios and proportions with which it could operate competently and skillfully. But, as seen in retrospect, this calculus was restrictive and stultifying, and it did not have an innate dynamism of evolution. Even Archimedes (250 B. C.) could not break out of the constrictions of the mathematics of his era.

Archimedes established two timeless propositions of mathematical physics. One was on the equilibrium of a lever, and the other on the equilibrium of a floating body. The proposition on the lever was as follows. If L_1 and L_2 are the lengths of the two arms of the lever and P_1 and P_2 are two weights suspended from them, then the lever is in equilibrium if and only if

$$P_1 : P_2 = L_2 : L_1. \tag{1}$$

This was the criterion of Archimedes, and he established it by a reasoning which has become a standard of mathematical rigor for all times. But Archimedes did not transform the proportion (1) into an equality

$$P_1 \cdot L_1 = P_2 \cdot L_2,$$

because a product like $P \cdot L$ did not exist in his mathematics.

The Greeks did not have a concept of real number, but they had a notion of magnitude (= μέγεθος) which was their substitute for it.[2] For instance, lengths were one kind of magnitude, areas were another kind of magnitude, weights were still another kind of magnitude, etc. They could envisage the ratio $P_1 : P_2$ for any two values of the same magnitude, and the ratio $L_2 : L_1$ for any two values of any other magnitude, and they could envisage the proportion (1) between them; they could verify, by a famous criterion

[2] There is need of a comprehensive study of the Greek notion of *megethos* in its various occurrences and usages. There is nothing about it in the many works of T. L. Heath on the history of Greek mathematics. Scattered remarks about *megethos* will be found in this author's article " Analysis " in the 1961 edition of the *Encyclopaedia Britannica*, Vol. I, p. 864A-F, especially 864A b, c.

which was a summit achievement of Greek mathematics, whether the proportion does or does not hold for the four specific values given. But the "classical" Greek mathematics of Euclid, Archimedes, and Apollonius did not form the conceptual product

$$P \cdot L$$

for two magnitudes P and L in general. In the very special case in which P and L were both lengths, a product did "exist" for them, as a matter of course, and was an area; and similarly if one of the factors was a length and the other an area, a product did exist and was a volume.

In modern physics, if P is a weight and L is a length, then P·L is the so-called statical momentum. The express formation of this momentum and of other mechanical and then electrodynamical momenta has been a most creative aspect in the unfolding of modern physics. Archimedes had this momentum in his context, but something in the metaphysical background and ambient of his thinking barred him from conceptualizing it overtly and "operationally." Generations of physicists after him, Heron (60 A.D.), Pappus (320 A.D.), Jordanus (thirteenth century), Leonardo da Vinci (fifteenth century), and many others were groping for the statical momentum, but it kept eluding them for over nineteen hundred years. The first clear-cut formulation of a momentum occurs in Newton's *Principia*. It is definition II of Book I, and it reads as follows: "The quantity of motion is the measure of the same arising from the velocity and quantity of matter conjointly."

We note that "quantity of motion" is an inertial (translational) momentum rather than a "static" one, but a clearly defined momentum it is. Before Newton, Galileo frequently uses the word *momento* in his writings, but its usage is not consistent. And it is not easy to decide whether Galileo did or did not have a concept of momentum in a fixed sense.[3] It should be remarked however that Huygens expressly defined *vis viva* as the scalar quantity mv^2, and he probably did so even before Newton arrived at defining his "quantity of motion" as the vector quantity mv; which fact may tend to lessen the originality of Newton's achievement in this respect.[4]

2. The product P·L not only multiplies the numerical values of P and L but it also forms a new "magnitude" out of the two magnitudes P and L. The nineteenth century extended this process of multiplication from magnitudes in mechanics to magnitudes in all of physics, especially to those which originate in electricity and magnetism; in fact it created a certain doctrine of "physical dimensions," where "physical dimension" is the same as "physical magnitude" except that no quantitative value is attached to it. J. Clerk Maxwell and Fleeming Jenkin introduced in 1863[5] the symbol [P] for the "dimension" of any magnitude P, whether the magni-

[3] See, for instance, E. Jouquet, *Lectures de mécanique*, Première partie, 1908, pp. 24 ff., and Max Jammer, *Concepts of Force*, 1957, p. 98.

[4] We are indebted for this remark to a communication from C. B. Boyer.

[5] It was published ten years later in: *Reports*

tude be " simple " or " composite," and they stipulated expressly that in a composite dimension

$$(2) \qquad [P_1^{a_1} \cdot P_2^{a_2} \ldots P_k^{a_k}]$$

the exponents $a_1, a_2, \ldots, a_k$ may be arbitrary real numbers. In 1822, in a short but systematic essay on physical dimension, which had been given by J. Fourier in Chapter II, Section 9, of the *Théorie de la chaleur,* the exponents had still been presumed to be integers only, but they already were positive or negative integers without further restriction.

For instance, if M, L, and T are mass, length, and time, then

$$[\text{energy}] = [\text{work}] = [M \cdot L^2 \cdot T^{-2}]$$

and

$$(3) \qquad [\text{" action "}] = [M \cdot L^2 \cdot T^{-1}].$$

We mention that the absolute constant h of quantum theory has the dimension (3) of an " action " and that Max Planck (1858–1947), the creator of the constant (1900), was in the habit of pointing this out with some emphasis.

The possibility of a twin conceptualization of a " physical dimension " and of a formal product (2) by which to symbolize it is no trifling matter. Not even the separate notion of a " physical dimension " by itself, without its involvement with a symbolism, will be found anywhere in Greek philosophy of nature, although the total course of this philosophy extended over more than a thousand years, namely from Thales (585 B. C.) to Proclus (fifth century A. D.) and Simplicius (sixth century A. D.) and included Heraclitus, Parmenides, Plato, Democritus, and Aristotle; and only some blurred and passing hints of it might be detected in utterances of some of the most forward-thinking philosophers in the Middle Ages.

The over-all intellectual development within which the twin conceptualization came about is a certain comprehensive process of algebraization and symbolization of mathematics as a whole and also of rational thinking in general. The process began in the Middle Ages and in a sense is still continuing; a particular manifestation of it is the gradual rise of algebra proper as a main division of mathematics and perhaps also the emergence of a symbolic logic in the nineteenth century.

3. A peculiar application of the Greek calculus of proportions occurs in the *Timaeus* of Plato.[6] If a cube has a side of length c, then its volume is c^3. The Greeks would state this by saying, in our notation, that if two cubes have sides a, b, then the ratio of their volumes is the triplicate of the ratio a : b; which means that the quantities a^3, b^3 are the extreme terms in the chain proportion

of the Committee on Electrical Standards, British Association for the Advancement of Science, 1873, pp. 59 ff.

[6] To the following, see the commentary of F. M. Cornford, *Plato's Cosmology,* 1937, pp. 43 ff.

$$a^3 : a^2b = a^2b : ab^2 = ab^2 : b^3.$$

This proportion contains four terms, namely

$$a^3, a^2b, ab^2, b^3,$$

and therefore, according to a Pythagorean view,[7] spatiality is associated with the enumerative role of the integer 4. Now, the *Timaeus* asserts (31B-34A) that this integer 4 also enumerates the number of space-filling elements of the cosmos, which are earth, water, air, and fire. Before disparaging this rationalization, one ought to consider that contemporary combinatorial topology shares the Pythagorean view that a three-dimensional simplex (that is tetrahedron) is determined by its four vertices, and we note that the physical and cosmological work of Arthur Eddington (1882–1944) also had a numerological slant.[8]

4. The Greek " horror of multiplication " also left a mark on the arithmetical theory of prime numbers. The ninth book of Euclid's *Elements* (300 B. C.) is a celebrated treatise on the theory of numbers, mainly prime numbers. But it does not have the so-called " fundamental theorem " which states that every integer can be represented as a finite product of prime numbers in which some factors may repeat themselves, and that the representation is unique. The product of equal factors is a power p^n, and therefore the theorem states in symbols that every integer N has a unique representation as a product

$$p_1^{n_1} \cdot p_2^{n_2} \cdots p_k^{n_k}, \tag{4}$$

in which $p_1, p_2, \ldots, p_k$ are different prime numbers, say

$$p_1 < p_2 < \cdots < p_k,$$

and the exponents $n_1, n_2, \ldots, n_k$ are some integers. What makes the omission of the fundamental theorem from Euclid's treatise noticeable is the fact that Euclid presents other theorems which, in modern accounts, are syllogistically very close to it and are deemed to be auxiliary to it.

We do not assert that Euclid and the later Greeks were not able to " find " the fundamental theorem simply because it involves the formation of products. In fact, in the case of numbers the Greeks did form products, of two or more factors, even if the number of factors was unspecified. But, to Greek mathematics our theorem would have offered another very serious difficulty; and this difficulty was interlocked with the difficulty of formation of products in such a manner and to such an extent that the overcoming of this difficulty would have taxed the Greek capacity greatly. The fundamental theorem is only an existence theorem. It only asserts the existence of a representation (4) and it is not concerned with the question of finding or constructing the integers $p_1, p_2, \ldots, p_k$ and $n_1, n_2, \ldots, n_k$ if N is given.

[7] See, for instance, G. S. Kirk and J. E. Raven, *The Presocratic Philosophers*, 1957, pp. 253-254. Also Aristotle, *Metaphysics*, 1090b 20-25.

[8] See, for instance, his posthumous work, *Fundamental Theory*, 1946.

At first sight one might think that this fact would make the theorem easier and less complicated, but actually to a Greek this would have made the theorem only harder to find. The Greeks have never quite arrived at the insight that in mathematics it might be *conceivable* to have an "existency" which is *totally* divorced from "constructibility"; and the fundamental theorem is an existence theorem not from choice but from necessity. There are no ready-made effective formulas, beyond the sieve of Eratosthenes, for actually constructing or finding the product (4) if N is some general integer; that is, for given N the number k of monomials in (4) is not only unspecified but incapable of being specified by a formula in the ordinary sense, and the various prime factors $p_1, \ldots, p_k$ with their exponents $n_1, \ldots, n_k$ are all unspecifiable and unpredictable in the same sense. But to form a product of many factors with such undetermined and unspecified data as this situation demands would have been much more than a Greek mathematician might have been expected to accomplish.[9]

In fact, the fundamental theorem of prime numbers, that is the assertion that any integer N can be (uniquely) represented as a product (4) of prime numbers, not only eluded Euclid but it continued to elude mathematicians for another twenty-one hundred years, and it eluded all the great mathematicians of the seventeenth and eighteenth centuries as well. The first time the theorem became generally known was in 1801 when it was presented in print by C. F. Gauss in §16 of his *Disquisitiones arithmeticae.* There is no substance to assertions that the fundamental theorem had been consciously known to mathematicians before Gauss, but that they had neglected to make the fact known. We think that the seventeenth, and even the eighteenth century was not yet ready for the peculiar kind of mathematical abstraction which the "fundamental theorem" involves; just as only the nineteenth century was comfortably prepared to conceptualize satisfactorily the notions of real number, limit, derivative, convergence, etc.

II. The Significance of Functions

5. A mathematical object of paramount significance to mathematics and science is the notion of "function." There are two main descriptions of the notion of function, one in terms of "correspondences" and another in terms of "relations," and they are both illuminating. But in reality the notion of function is not definable and any would-be definition is tautological.[10]

Functions are a distinguishing attribute of modern mathematics, perhaps the most profoundly distinguishing of all. Functions began to emerge in the Middle Ages, and one frequently associates the name of Nicole Oresme

[9] The gist of the above "explanation" for the absence of the "fundamental theorem" from Euclid's treatise was communicated orally by the present author to G. H. Hardy in 1933, as noted in the book of G. H. Hardy and E. M. Wright, *An Introduction to the Theory of Numbers,* 1938, Chap. XII, §12.5, p. 181 and a note to it on p. 188.

[10] Thus, H. Weyl in his *Philosophy of Mathematics and Natural Science,* 1949, says on page 8: "Nobody can explain what a *function* is, but. . . ."

(fourteenth century) with the first clear symptoms of their emergence. In its innermost structure Greek mathematics was a mathematics entirely without functions and without any orientation towards functions; and whatever rudiments of functions there may have been in later Greek mathematics could probably be traced to its merger with Babylonian mathematics in which something in the nature of functions was trying to break through.

It has been rightly emphasized by Oswald Spengler [11] that the presence of functions in mathematics signifies a certain all-around mobility and their absence a certain all-around stationarity, as evidenced by contrasts between Greek and modern civilizations in some of their essential aspects.

As far as physics is concerned, functions are certain mathematical expressions by which it is fitting to articulate states, events, and changes in nature; and mathematical symbols which emanate from contexts of physics are the letters and syllables of the "Language of Science" within which the mathematical functions of physics are not only mathematically coherent but also physically meaningful and productive. Mathematical functions in physics reveal and control causation, subordination, and other relations; they measure and appraise complexity and simplicity of events; their behavior expresses equilibrium or instability of a state of a system; and they assess generalizations, specializations, and individualizations.

The rise of mathematical functions during and after the Renaissance was intimately connected with the rise of analysis in mathematics. But eventually functions spread into all parts of mathematics and into many areas of rational thinking altogether. The Greeks had never "found" functions in a genuine sense, either in mathematics or in other parts of their system of knowledge. By outward appearance Greek mathematics was geometrical rather than analytical, and by inward structure it was representational rather than operational. Modern accounts of Greek assertions in science, especially of certain statements in the second half of Aristotle's *Physica,* sometimes express them in terms of present-day mathematical formulas. But it may be said that the role of mathematics in physics is not exhausted by merely letting mathematical schemata correspond to so-called physical realities. What is decisive is that suitable mathematical operations within the mathematical schemata shall be representative of meaningful connections and relations within the physical settings to which the mathematical schemata correspond; and in such mathematical operations functions are involved almost always.

To this kind of mathematical activity the Greeks were never able to advance, although some of them, like Democritus and Archimedes, may have come within a pulse beat of doing so. If appraised in retrospect, this incapacity of the Greeks may be said to have been one of their major intellectual failures and one for which no convincing explanation has ever been proposed. The usual rationalizations of the gradual decline of the Greek mathematics of Archimedes and Apollonius from political and socio-economical causes do not convincingly account for the fact that Greek

[11] *Decline of the West,* Vol. I, Chap. II, "The Meaning of Numbers."

mathematics even when in ascent and at the height of its development did not shape itself into an instrumentality for the creation and promotion of a mechanics and physics, as the mathematics of the seventeenth and eighteenth centuries did unceasingly.

In modern times a significant development which accompanied the mathematization of physics was the gradual emergence of the notion of a physical system and also a physical model. The Greeks never arrived at such a notion. The nearest to it in their thinking, especially in the thinking of Aristotle, was a certain notion of a " process-in-nature." The notion is very broad and equivocal and quite unsuited to being a background for mathematization. Aristotle had various terms for his " processes," such as motion (kinesis), locomotion, change, alteration, etc., and commentators of Aristotle, from the earliest onwards, have been endeavoring to find sharp distinctions between them. But it is frequently not easy to compare and delineate these terms as to their proper meaning and scope.[12] In most contexts these terms are unsuited for mathematization or any pre-mathematical formalization, even though the context sometimes suggests comparisons with so-called " functions of state " or " characteristic functions " of modern physics, especially thermodynamics.

The Stoics had a recognizable orientation towards making notions of physics specific by a formalization that would make them operationally tangible, but they were seeking to locate such a formalization in a " logic of linguistics " rather than in mathematics.[13] Even the Stoics never developed a trend towards a mathematically controlled physics in general, although Greek astronomy and optics were mathematical, and the mechanical achievements of Archimedes were mathematics at its sternest.

The Schoolmen of the Middle Ages (and perhaps even some late Hellenistical natural philosophers) did have an anticipatory trend towards a mathematization of physics. The Schoolmen were groping for a notion of function, both for mathematics and for physics, and they were trying to inject the notion of mathematical function into a physics which was still largely Aristotelian. Apart from this they were also pursuing a certain extra-mathematical linguistic " logification " of knowledge after the manner of the Stoics. But one must not exaggerate the sum total of their achievements. The limitations of the total achievement of the Middle Ages can perhaps be measured by the fact that the Schoolmen have not created a single treatise on mathematics or physics which would hold the attention of a non-specialist today as a work in its entirety.

The Renaissance did little for physics but much for mathematics. In mathematics it developed a " school syllabus " which in addition to Euclid's geometry embraced arithmetic, algebra, *Rechenkunst*, combinatorics, and eventually also trigonometry and logarithms. Trigonometry had been a

[12] The interpretations of classicists are summarized, for instance, in J. P. Anton, *Aristotle's Theory of Contrariety*, 1957, pp. 206-207. See also the treatise of Friedrich Solmsen, *Aristotle's System of the Physical World*, 1960.

[13] Max Pohlenz, " Die Stoa," *Geschichte einer Geistesbewegung*, second edition, Vol. I, 1959, pp. 22-55; Benson Mates, *Stoic Logic*, 1961; Antoinette Virieux-Reymond, *La Logique et l'epistémologie des Stoiciens*, 195–.

body of knowledge of considerable standing, but it had been a part of astronomy and geodesy. The Renaissance however began to study trigonometry as an area of mathematics in its own right, and in a certain vague sense it began to view the addition theorems for *sinx* and *cosx* as functional equations per se. Logarithms may have begun as an aid to trigonometry, but they in turn soon showed a tendency to become " independent." Altogether the general study of trigonometry and logarithms, as begun in the Renaissance, was a first phase in the study of the kind of analysis in which functions and manipulations of functions are the central issue. (Present-day reformers of training programs in mathematics ought to consider, and reconsider, that in eliminating trigonometry and logarithms they would abandon an area of mathematics on which an organic understanding of the concept of function has been seeded, grown, and harvested for the last five hundred years.) Apart from all this, the Renaissance also edited, printed, and propagated the works of Aristotle, Euclid, Archimedes, Apollonius, Pappus, and others, all of which were studied and exploited in the seventeenth century as never before and then recast into the style of analysis in due course.

The seventeenth and eighteenth centuries created modern physics by creating and giving first preference to a central division of physics which they themselves called Rational Mechanics and which comprised not only mechanics of finite systems and of rigid bodies but also mechanics of continua, whose main divisions are hydrodynamics, acoustics, and elasticity. Physics proper, that is the theories of light, heat, and electricity and magnetism were not forgotten or even slighted, but they were advanced at a gentler pace until such time in the nineteenth century when their turn for full attention would come. Thus Newton composed not only his formidable *Principia* but also an insinuating *Opticks*, at first not in Latin but in English, so that even poets might read it, as some did.[14]

The mechanics of the seventeenth and eighteenth centuries is virtually inseparable from mathematics. Beginning with the eighteenth century the mathematics with which mechanics was most closely intergrown was largely analysis of functions, and virtually all of this analysis was then made for mechanics or because of mechanics. With the beginning of the nineteenth century, mechanics became a model of mathematization for all of physics and gradually for other science as well. At the same time the analysis of functions began to widen and to pursue aims of its own, but nevertheless throughout the length of the nineteenth century a considerable part of it continued to be stimulated by physics directly.[15]

[14] It is worth noting that the *Opticks* even offers a thermo-radiative description of " black bodies " (the name is in Newton) in which the later definition of G. R. Kirchhoff and Max Planck is anticipated. In fact, " query 6 " in the Appendix to the *Opticks* reads as follows: " Do not black Bodies conceive heat more easily from Light than those of other Colours do, by reason that the Light falling on them is not reflected outwards, but enters the Bodies, and is often reflected and refracted within them, until it be stifled and lost? "

[15] A large amount of information on the interrelations between the growth of analysis and the growth of mechanics and physics is embodied in the great work of H. Burkhart, *Entwicklungen nach oscillirenden Functionen und Integration der Differentialgleidungen der*

6. According to Moritz Cantor[16] the word "function" was used as a mathematical term for the first time by G. Leibniz only in 1694, and even then not immediately in the present-day sense. It may be asserted however that an operational conception of function was in the thinking of mathematicians considerably earlier. An eminent geometer of the twentieth century, the late G. Castelnuovo, has pointed out[17] that as early as 1604, that is ninety years before the above given date, the mathematician Luca Valerio in a study on an Archimedian theme had *de facto* introduced the notion of a rather general class of (continuous) functions $f(x)$ on a finite closed interval $0 \leq x \leq \ell$. Valerio postulates, in words of his, that the function shall be monotonely decreasing from a positive value at $x = 0$ to the value 0 for $x = \ell$, and that nothing else shall be assumed. Castelnuovo remarks to this that Archimedes, in accord with other Greek geometers, "considers only curves whose construction or definition is already known (conics, spirals, etc.)." We concur with this judgment of Castelnuovo, but we must state that an elaboration of this judgment would require a comprehensive analysis. In fact, Archimedes himself gave a celebrated definition of what a "convex curve" is,[18] and this definition, if taken in isolation and by its formal wording, does not require constructibility or any other particularization. But, for instance, the very late commentator Eutocius (500 A. D.)[19] takes it for granted that Archimedes must have been thinking of curves which are put together of pieces of straight lines, circles, conics, etc., that is, of pieces of curves already known.

The *Principia* of Newton had an Archimedean *mise en scène*, and this kept point functions $y = f(x)$ out of sight. In the course of the eighteenth century, however, mechanics became totally analytical, and by the end of the century point functions, in one and several variables, were as ever-present and all-dominant as they are in all parts of physics today.

In many contexts of physics a function is expected to be given or constructed as an expansion (that is as an infinite series) of a certain kind. The seventeenth century immediately introduced expansions in power series, and they originated in mathematics directly but not yet in physics. But power series are the identifying "symbol" of analytic functions, and analytic functions occur everywhere, including physics. The eighteenth century introduced expansions in trigonometric functions and in related functions, specifically in Bessel functions.[20] These expansions were instigated mainly by requirements of mechanics of continua. They fall under the general notion of expansions in so-called eigenfunctions, and these are the leading instrumentality for the solution of boundary value problems in

mathematischen Physik, Jahresbericht der Deutschen Mathematikervereinigung 10_2, 1901-1908, 1800 pp.

16 *Geschichte der Mathematik*, Vol. III, second edition, 1901, pp. 215-216.

17 Guido Castelnuovo, *Le origini del calcolo infinitesimale nell'era moderna*, 1938, p. 11.

18 Archimedes, *On the Sphere and Cylinder I*, Postulate 2.

19 His commentary is included in Volume III of the *Works of Archimedes* by L. Heiberg.

20 See C. Truesdell, *The Rational Mechanics of Flexible or Elastic Bodies*, Introduction to *Leonhardi Euleri opera omnia*, Ser. 2, Vols. 10, 11, 1960; p. 159 and later passages.

mathematical physics. If we collect all leading boundary value and eigenvalue problems into one theory, then we may say that a large part of analysis of real variables falls under this theory indirectly, and that a large part of the total body of mathematical physics can be viewed to fall under it directly.

In the course of the nineteenth century and afterwards, the notion of an eigenfunction became very familiar to physicists; and when, after 1926, E. Schrödinger (and also L. de Broglie) was able to transliterate the then latest version of quantum mechanics from the idiom of matrices into the idiom of eigenfunctions, the comprehension of the new theory and its elaborations were greatly accelerated. This did not mean that the original Heisenberg idiom of matrices was neglected, let alone abandoned. To the contrary, the parallelisms between the two approaches deepened the understanding of each, and a third approach, via operators in Hilbert space, fused the two even more closely.

7. In the nineteenth century, most of the growing physics, in electricity and magnetism, in optics, and in heat conduction (as in the work of Fourier), was mathematically modeled on paradigms from mechanisms and predominantly from mechanics of continua. Therefore, in many parts of physics the mathematics was uniformly the same, not only in technique but also in the manner in which mathematical and physical conceptions were correlated with each other. However, there was one part of physics which did not conform to this general pattern, and this was the theory of thermodynamics. This became very manifest in the second half of the nineteenth century when the thermodynamics of E. Clapeyron and R. Clausius metamorphosed into a kinetic theory of matter which in its turn was mathematically linked to a novel kind of mechanics, namely the so-called statistical mechanics.

Now, this statistically oriented thermodynamics was a counterpart to and descendant of a somewhat earlier thermodynamics, which by no means has been superseded by it. This is the so-called "classical" or "axiomatic" thermodynamics, and it has the distinction that, in its "pure" form, it is not directly subordinate to the mechanics of the eighteenth century, mathematically or even conceptually. It does not obtain its functions by solving differential equations, as does mechanics, but it posits them directly, and these are the "equations of state" or "characteristic equations." Chief among these is the relation

$$pv = RT \tag{5}$$

for a perfect gas which, for $T = t + 273$, was notationally "standardized" by E. Clapeyron in 1834 [21] and, for T as absolute temperature, by Lord Kelvin later; but in its content, and by slightly different formulations, the formula had been well known before, inasmuch as it fused the law of Boyle and Mariotte (seventeenth century) with the law of Charles and Gay-

[21] In his "Sur la puissance motrice de la chaleur," *Journal de l'École polytechnique*, *14*: 153.

Lussac (1802), which, according to C. Truesdell, had likewise been known since the seventeenth century (Amontos, 1699). Relation (5) and similar relations do not involve coordinates of space and time, and there is no express reference to the three-dimensionality of the spatial volume v of unit mass in the formulas. In the textbook of Max Planck, *Vorlesungen über Thermodynamik,* no theory of mechanics is presupposed or developed in passing.

It is a remarkable fact that, in the nineteenth century, chemistry and biology were more attracted to thermodynamics than to any other large theory of physics. Aristotle was a biologist and, like many Greek natural philosophers, was greatly attracted to thermodynamics, much more so than to mechanics as the seventeenth century initiated it. This may help to explain why he did not arrive at the law of inertia of Galileo and Descartes, even though physics of locomotion was much in his thinking, and motion in general (kinesis) was the central datum of his biology and of his psychology too. The thought patterns of Aristotle in physics are reminiscent more of syllogisms in classical thermodynamics than of those in mechanics, and especially his notion of equilibrium in nature would fit much more a thermodynamics of irreversible events than a mechanics of reversible processes. Most of the criticism of Aristotle by Galileo, Descartes, and other "mechanists" of the seventeenth century would not stand up before the forum of classical thermodynamics, if it rendered judgment solely within its own jurisdiction.

III. The Significance of Real Numbers

8. It is generally asserted that modern physics (and science, even social science) is largely quantitative and that older physics was rather more qualitative. This is indeed so, but we wish to say that the distinction between the notions of quantity and quality is subject to some variations of interpretation, and that the determination of the actual meaning of quantitativeness in physics is itself a topic of physics or at any rate is involved in some major topics of it. Thus, in the theory of relativity, the problems of defining and determining the meaning of temporal simultaneity and of deciding simultaneity by so-called measurements do in a sense involve the over-all problem of the meaning of quantitativeness of physics; and so do, in another sense, in quantum theory, the problems of determining the meaning of observability and of determining the bounds of optimal observation of several magnitudes jointly (uncertainty relation).

We think that the Middle Ages tended to view the terms "quantity" and "quality" as representing contraries of antithetical status, but that in Greek philosophy this was not so or not yet so. In Aristotle, whenever the notions of quantity and quality appear in the same context, they usually are what Aristotle calls "categories." These two categories, qua categories, are indeed very different from each other, but we think that there is no presumption of contrariety between them. Furthermore, Aristotle usually has a table of categories before him which he sometimes enumerates. The table varies but it never consists of quantity and quality only; even in *Physica*, Book V, Chapters I and II, there is the additional third category of " place " left in the table after others have been discarded. Finally, in modern physics, reflections on the meaning of quantitativeness are usually directed only towards determining it by itself as such, and comparisons with other " categories " are made only for purposes of explanation and illumination. Last, in working physics, a chain of reasoning or the result of a chain of reasoning is termed " qualitative " if it is intended to be only heuristically suggestive, or " not-yet-quantitative " from whatever cause.

The demand of quantitativeness in physics seems to mean that every *specific* distinction, characterization, or determination of a state of a physical object, and the transmission of specific knowledge and information, must *ultimately* be expressible in terms of real numbers, either single numbers or groupings of numbers, whether such numbers be given " intensively " through the medium of formulas or " extensively " through the medium of tabulations, graphs, or charts. Especially, the value of a basic physical magnitude like mass, charge, or energy is given by stating a real number in addition to the " physical dimension " of the magnitude in question. It has been rightly emphasized by Pierre Duhem that the real numbers which specify such a physical object are usually intended to be the real numbers of arithmetic, which represent the results of some actual or potential " measurements," and not just real numbers which represent the natural ordering of the one-dimensional linear continuum.[22] Thus, if one characterizes the colors of visible light from the limit of red to the limit of violet by real numbers from 7.2×10^{-5} decreasingly to 4.0×10^{-5}, then these are usually quantitative numbers inasmuch as they represent wavelengths (in cm) of the theory of optics. But if one only wishes to represent the successiveness of the colors in the natural linear ordering, then one can suitably label them by real numbers which traverse any finite closed interval in many different ways. But in such a general labeling, the real numbers would only help to establish what medieval philosophers called the " intensity " of color qua " quality " (and not qua quantity). That any quality or condition may be expected to have a certain range of " intensity " of its own was in some sense already known to Aristotle.

The first phase of the rise of quantitativeness in physics was the development of the fully analytical setting for the entire body of rational mechanics in the course of the eighteenth century; and this development in physics had been preceded by and then became intergrown with a development in mathematics itself. The latter development was the arithmetization of geometry, that is the introduction of coordinate systems into Euclidean space, which, to an extent, had been inaugurated by Descartes, although put into practice considerably later.[23] By these twin developments the notion of physical position in space became fully quantitative, and the vectorial objects

[22] P. Duhem, *La Théorie physique, Son object et sa structure*, 1906, p. 179.

[23] For the slowness with which the actual arithmetization of geometry proceeded, see the account in C. B. Boyer, *History of Analytic Geometry*, 1956.

of the Newtonian system, namely velocity, acceleration, momentum, and force, became fully quantitative too. Euclidean coordinates were soon followed by curvilinear coordinates, in mathematics and in mechanics conjointly, and a pioneer architect of them was L. Euler. In his kinematics of rigid bodies he introduced coordinates for the sphere and for the group space of orthogonal transformations, and these coordinates became the prototype of coordinates for certain general group spaces and homogeneous spaces very much later. A notion, or prenotion, of spaces in more than three dimensions and of general coordinates in them is involved in the Lagrangean mechanics of finite systems with general constraints, and such a general theory could not have been articulated in the prequantitative Archimedean setting of the *Principia*, say; although it must be said that Newton, as also Huygens, does reason masterfully within his setting whenever a constraining condition happens to occur within a context.

Beginning with the nineteenth century the arithmetization of geometry initiated a systematic development of differential geometry and of topology of general manifolds, and we think that the insights which have been accumulated in these relatively recent doctrines of geometry are in their sum already comparable to those which have been accumulated in the non-arithmetized geometry of the Greeks since Thales. Also, a particular manifestation of the arithmetization of geometry was the establishment of a proper theory of real numbers in the nineteenth century. It had many features of the theory of proportions for incommensurables in Euclid's famed Book V. The basic difference in outcome is that while the Greek theory is intended to be a theory for a large class of magnitudes simultaneously, the modern theory constructed one particular magnitude, namely the real number, and envisages a mathematical isomorphism between this particular magnitude and the magnitudes of the large class. Mathematics also has many types of magnitudes, which are not at all isomorphic to real numbers, and in mathematics there is no such compulsion as in physics to represent adequately such magnitudes by groupings of real numbers.

9. If the various states of a physical system are presented by groupings of real numbers, then usually the elements which constitute a grouping cannot be adequately interpreted singly, each by itself, but must be kept together, either all of them inseparably or at least in certain subgroupings. The various elements of such a subgrouping constitute one indivisible mathematical object, and the elements by themselves are the (real-valued) components of it. The most familiar mathematical objects of this kind are vectors and tensors. They are defined at points of coordinate spaces. The values of the components depend on the particular coordinate system of the space, and there are familiar rules for transformations of components under changes of the coordinate system. Until not long ago it was generally felt that vectors and tensors are, by their essence, mathematical objects of a special kind and for special purposes. But contemporary mathematics tends to subsume the familiar vectors and tensors under a much more comprehensive notion of mathematical objects of several components in

general, and this tends to make vectors and tensors into mathematical objects of much less specialized standing.

We have already stated [1] that it was the Isaac Newton of the *Principia* who posited that, in his mechanics, velocity, acceleration, momentum, and force should be vectors, thus creating the notion of vector in the process. Newton was the true originator of this mathematical object, even though a textbook conceptualization of it came very much later, perhaps only in the twentieth century. We have also stated [1] that, in this sense, tensors of two indices also originated in mechanics, namely in the mechanics of continuous media. The mathematization of the latter theory was carried out in the first half of the nineteenth century largely by A. Cauchy, and the latter introduced *de facto* the mathematical objects which represent the physical pressure and stress tensor and the accompanying deformation tensor. The name " tensor " for these objects of Cauchy was coined by the crystallographer A. Voigt towards the end of the nineteenth century and came into general use after 1910.

Formal tensor analysis became generally known when it was chosen as the mathematical idiom of the general theory of relativity which came into being after 1913. But the foundation of tensor analysis was already laid in the last part of the nineteenth century (mainly by G. Ricci, but also by T. Levi-Civita, E. Beltrami, and others before them), and its motivation came from physics even then. In fact, on the one hand, the antecedents of the Ricci calculus were in the differential geometry of C. F. Gauss, B. Riemann, and E. B. Christoffel, which had strong links to the Lagrange-Hamilton mechanics of finite system. And, on the other hand, the avowed purpose of Ricci, and his predecessors, in shaping the new calculus was to be able to formulate the partial differential equations of general physics, especially those of elasticity and of electromagnetism, in general curvilinear coordinates of Euclidean space. Thus, the Ricci calculus was intended from the beginning to become a means for expressing mathematical invariance of physical laws, many years before A. Einstein had adapted it to such a purpose of his own.

Furthermore there was still a certain mathematical development which interposed itself between the work of Ricci and its application by Einstein, and this was the work of Minkowski in 1907–1908 on the adaptation of the theory of Maxwell from Euclidean space-time to " Minkowskian " space-time. Minkowski apparently did not know the work of Ricci, because he starts out from the ordinary calculus of vectors (and tensors) as it was then generally known from potential theory and electrodynamics. Minkowski adapted the notion of ordinary vectors and tensors in Euclidean space to his own four-dimensional space, and he then constructed *in vacuo* a tensor of order 2 which brought together mathematically the Maxwell stress tensor, the Poynting flux vector, and the Kelvin energy density. This was mathematically a decisive step, and although the later theory of Einstein was much more comprehensive, both physically and mathematically, yet a crucial first step had already been taken. Also, the mathematical achievement of Minkowski was very clear cut, and we think that in the theory of relativity only the construction of the relativistic electron by P. Dirac, which came twenty years later, compares with it in incisiveness.

10. Although physico-mathematical objects of more than one component have become indispensable for expressing the richness of physics and science in all its parts, the scalar objects of one component, that is, basic " magnitudes " of Greek vision, are still the leading entities. First among these is the notion of energy. To Aristotle it was one of the most inscrutable notions in all of natural philosophy.[1] The Rational Mechanics of the seventeenth and eighteenth centuries was not equipped to cope with it, even within the confines of its circumscribed aims. Only the physics of the nineteenth century was able to bring it under cognitive control. But it required all the resources of thermodynamics to accomplish this, and much of the resources of electricity and magnetism too. In the nineteenth century, energy became one of the most talked-about concepts in all science and also in philosophy, and industrial and social revolutions were waged in its name.

Finally we wish to point out that one of the most radical innovations in physics was the outcome of a problem regarding energy which had been quietly building up throughout all of the nineteenth century. The problem was the problem of radiation (which is a certain mode of transportation) of energy across " empty " space, and this problem cut across several divisions of physics, thermodynamics, optics, and finally also electromagnetism (Poynting flux vector) ; and the preoccupation with this problem led at the very end of the nineteenth century to the creation of quantum theory to which the future of physics belongs.

Temperature and entropy are an unusual pair of scalars, which seem to belong together mathematically. Temperature is at first consideration an everyday " observable," and yet physics required two hundred years to conceptualize it. Entropy is not an outward " observable " but to the contrary an " internal " construction, and yet it is as controversial today as it was on the day when it was conceived.

Even the concept of mass, which is the scalar of all scalars, was firmly set down only by Newton in his *Principia*, after irritating natural philosophy for over two thousand years. And the concept of mass continues to be very mystifying. Thus the assertion in physics that mechanical mass and gravitational mass are equal will never cease to baffle the " layman," and the mystique of the statement that mass and energy are one and the same, $E = mc^2$, will never cease to stir his imagination.

Finally we venture to suggest that Greek mathematics was both made and unmade by the efforts of the Greeks to conceptualize simple scalar magnitudes like length, area, and volume. The Greeks created the *eidos* of mathematics, and it is this which made the *Elements* of Euclid one of the greatest best-sellers of all times, in spite of the tediousness of its schoolmasterly style. But creatively the Greeks were more philosophers and poets than mathematicians, and they did not have in their creativeness the instinctive realization that in the actual creation of mathematical subject matter the complex comes before the simple, the recondite before the obvious, the hidden before the manifest. Greek mathematics discovered almost at birth some deep-seated problems of the *eidos* of mathematics, one of which was the problem of the true mathematical structure of scalar "magnitudes." Modern mathematics postponed this problem to the nineteenth century and was none the worse for it. But Greek mathematics compelled itself to face the problem fully from the beginning and exhausted the strength of its youth in mastering it prematurely. Archimedes could, within his setting, define the concept of a "limit" as nobody in modern mathematics before Cauchy, and Newton was totally unable to define rigorously instantaneous velocity and instantaneous acceleration to the end of his days. And yet Newton created analysis and mechanics, and Archimedes did not; and Newton was the first mathematical physicist, and Archimedes was not. We note that the honor of having been the first mathematical physicist ought perhaps to be divided between Newton and Huygens, if not even bestowed on Huygens exclusively. Huygens was indeed a physicist among the greatest, and in his work on evolutes in his "Horologium" he virtually shamed Apollonius of Perga by using Apollonius' own syllogisms, and *seemingly* no others, towards conceiving of a "local" curvature for a general convex curve; whereas Apollonius himself, in the fifth book of his *Conics*, came near the concept of curvature and went repeatedly around it but did not locate it. But it is a fact that Newton did compose a *Principia* for the ages, and Huygens did not, just as Charles Darwin did compose an *Origin of Species*, and H. R. Wallace did not, and such facts do count.

IV. THE SIGNIFICANCE OF COMPLEX NUMBERS

11. We think that it is legitimate to say that in a certain broad sense, the problem of the significance of the continuum of real numbers for physics has been a perennial major topic of Western philosophy (including medieval Arabic philosophy) from around 500 B.C. until the present. In Greek philosophy, aspects of this problem reflected themselves in views of Pythagorean circles that "all is number," in the renowned and (to philosophers) timeless paradoxes of Zeno of Elea, in the Platonic doctrine of great-and-small (perhaps), and in the multiform analyses of the meaning of Infinite, Space, Time, and Motion in Aristotle.[24] Through the permanency of Greek topics in later philosophy, and through other impulses, the problem of real numbers continued to recur in all stages of natural philosophy afterwards.

However, in addition to real numbers contemporary physics also employs complex numbers. These numbers have been known to physics for two hundred years, and in the nineteenth century their occurrence in physics has gradually widened to such an extent that a veritable "complexification" of physics has come about. However there seems to be no recognized topic

[24] Most of what is known about Greek views on the nature of mathematics comes from the *Physics* and *Metaphysics* of Aristotle.

or subtopic in philosophy whose subject matter is to evaluate the significance of complex numbers in mathematics, physics, and science in general. Even the far-ranging philosophies of mathematics of Bertrand Russell and A. N. Whitehead in the first half of the twentieth century have apparently not been concerned with this situation.[25] We do not intend to undertake such an evaluation ourselves, but we venture to make some observations on the manner in which the " complexification " of physics expresses itself.

Complex numbers are pairs of real numbers, and by taking real and imaginary parts, any set of relations in mathematics and physics which involves complex numbers can be replaced by an equivalent set in which only real numbers occur. However, complex numbers can be multiplied with each other, and it is a fact that the manner in which the real and imaginary parts of complex numbers are linked with each other by the rule of multiplication somehow endows the system of complex numbers with a range of power, effectiveness, and suppleness which in mathematics itself is both inexhaustible and inexplicable, and which has been affecting the course and development of physics deeply and even radically. It is true that a contemporary physicist usually accepts complex numbers as something comfortably familiar and marvels only at mathematical operators and groups, if at all. But we wish to say that the operators which occur in physics are virtually all intergrown with complex numbers, and that groups which occur in physics, if they are Lie groups, are also linked with complex numbers, in their mathematical background at any rate.

In the physics of the nineteenth century, and also in the theory of relativity, which by the texture and spirit of its mathematization was a terminal phase of that physics, complex numbers were being used ever more frequently, both in calculations and in derivations, and they were beginning to penetrate into thought patterns also. But basic conceptualizations and basic formulations continued to be presented and expressed in real variables only. However, quantum mechanics changed even this. The very basic equations

$$pq - qp = \frac{h}{2\pi i}$$

$$\frac{h}{2\pi i}\,\frac{\partial\psi}{\partial t} = -H\psi$$

of the new theory display the symbol $i = \sqrt{-1}$, which is indeed only a symbol, openly and directly. Also in these and other formulas of quantum theory the mathematical imagery is quite intense, and it suggests more credibly than in previous theories the possibility of attaining penetration into physics through symbolization in mathematics and of finding reality of nature through irreality of mathematics.

By separation into real and imaginary parts, one can of course formally

[25] See, however, a recent article by E. P. Wigner, " The Unreasonable Effectiveness of Mathematics in the Natural Sciences," *Communications on Pure and Applied Mathematics*, 1960, *13*: 1-14.

eliminate complex values from the above equations and also from the computation of the Born probability density out of the wave function. And one can certainly eliminate the imaginary from the de Broglie wave

$$\exp\left\{\frac{2\pi}{hi}(Et - p_x x - p_y y - p_z z)\right\}$$

by taking its real or imaginary part, say. But any attempt to do this in general, and not for a specific purpose on a special occasion, would run counter to a development whose stream is irresistible and whose course is indeflectable.

12. The analytical strength of complex numbers began to be recognized towards the end of the seventeenth century, and the principal analytical findings up to the middle of the eighteenth century culminated in the " magic " formula

$$e^{ix} = \cos x + i \sin x$$

which was clearly stated by Euler in his " Introductio in analysin infinitorum " in 1748. In this formula the letter x denotes arc length so that in particular

$$e^{i\pi} = -1, \quad e^{2i\pi} = 1,$$

and these relations bring together the three symbols e, i, π of very different provenance. In the second half of the eighteenth century complex numbers took a short but notable step on a path which, after a considerable halt, began to lead them to an area in which mathematics and physics were frequently meeting in the course of the nineteenth century. The step was made by d'Alembert.[26] In a work on hydrodynamics he introduced in 1752, and even earlier in 1746, the system of equations

$$\frac{\partial p}{\partial x} = \frac{\partial q}{\partial y}, \quad \frac{\partial p}{\partial y} = -\frac{\partial q}{\partial x},$$

which nowadays are named after A. Cauchy and B. Riemann of the nineteenth century. And d'Alembert stated in his style that a general solution of the system can be represented as

$$p = \phi(x+iy) + \phi(x-iy) + i\psi(x+iy) - i\psi(x-iy)$$
$$q = -i\phi(x+iy) + i\phi(x-iy) + \psi(x+iy) + \psi(x-iy)$$

where, in our parlance, $\phi(z)$ and $\psi(z)$ are analytic functions of the complex variable $z = x + iy$ and have real values on the x-axis. In 1755 Euler hailed the result of d'Alembert,[26] observed that the above system of equations is equivalent to the one equation

[26] See C. Truesdell, *Rational Fluid Mechanics, 1687-1765*, pp. LIV and LV, in *Leonhardi Euleri opera omnia*, Ser. 2, Vol. 12, 1954; also Paul Stäckel, *Integration durch imaginäres Gebiet*, Bibliotheca Mathematica Series III, Vol. 1, 1900, pp. 109-128, and *Beiträge zur Geschichte der Funktionentheorie im achtzehnten Jahrhundert*, the same series, Vol. 2, 1901, pp. 111-121.

$$\text{(6)} \qquad p + iq = 2\phi(x + iy) + 2i\psi(x - iy),$$

and gave proofs for all statements involved. We note that the assertion of Euler is valid, according to present-day standards, in the following version. Let f (z) be defined and analytic in a domain D of the z-plane, and let the domain be symmetric with regard to the x-axis. Being connected, the domain D must then also include points of the x-axis. Due to the symmetry of D, there exists a function $\overline{f(\bar{z})}$ in all of D and it is also analytic, and so are therefore the functions

$$\phi(z) = \frac{1}{4}\{f(z) + \overline{f(\bar{z})}\}, \quad \psi(z) = \frac{1}{4i}\{f(z) - \overline{f(\bar{z})}\}.$$

Now, $\phi(z)$ and $\psi(z)$ are real on the x-axis, and also

$$f(z) = \phi(z) + i\psi(z),$$

which is formula (6).[27]

In the nineteenth century, analytic functions of a complex variable spread into many areas of physics, and the twin topic of two-dimensional potential theory and conformal mapping received all its impulses in this way. But during the first hundred years after d'Alembert's inspired result the tempo of these developments was quite slow.

Another major development which began in the eighteenth century and also led to complexification of physics in the nineteenth century was the emergence of trigonometric expansions and related expansions in rising theories of vibrations. During the eighteenth century the basic functions in trigonometric expansions were the real functions cos nx, sin nx, and not the imaginary exponentials e^{imx}, and during the whole nineteenth century there were physicists who almost always used the real functions. However, the complex exponentials were known, and whenever they were demanded imperatively they were used. In the eighteenth century Euler stated the general theorem, which is rightly named after him, that a solution of a differential equation

$$\text{(7)} \qquad \sum_{m=0}^{n} a_m \frac{d^m f}{dx^m} = 0$$

with constant coefficients is a linear combination with constant coefficients of monomials

$$x^{\mu} e^{\lambda x},$$

where λ is any root, real or complex, of the polynomial

$$\sum_{0}^{n} a_m \lambda^m,$$

[27] This very simple proof of the assertion is due to Professor R. C. Gunning; our own proof was complicated.

and where $\mu = 0, 1, \ldots, r-1$, if r is the multiplicity of the root. Now Euler was led to the investigation of general equations (7) by generalizing the " eigenvalue " equations

$$\frac{d^2f}{dx^2} = af, \quad \frac{d^4f}{dx^4} = af$$

to which Daniel Bernoulli had been led in his work on vibrations in 1735.[20]

13. When solving differential equations which describe undulatory phenomena by the method of eigenfunctions, it is nearly always mathematically very suggestive to set up the mathematical solutions of the equations as complex-valued functions and then to take, as actual solutions of the physical problems, the real parts of the resulting functions only. This has become such an indispensable routine that it has almost developed a habit of thinking in terms of complex-valued physical objects directly. However, in the resulting complexification of the physics of undulatory phenomena, concurrent with this habituation from mere procedural appropriateness, there has also been a motivation from a genuine cause, and the originator of this motivation was A. Fresnel in his theory of total reflection (1823, published 1831).[28]

For propagation of light in adjoining transparent bodies[29] Fresnel had established that if α and β are angles of incidence and refraction, with

$$\sin \alpha = \mu \sin \beta, \tag{8}$$

then for light polarized in the plane of reflection the ratio of amplitudes of reflected and incident light

$$-\frac{\sin (\alpha - \beta)}{\sin (\alpha + \beta)}. \tag{9}$$

If however $\sin \alpha > \mu$, then the angle β as computed from (8) is complex valued. The ratio (9) is also complex valued but it has absolute value 1, and thus it " represents " a pure shift $e^{i\theta}$, where

$$\tan \tfrac{1}{2}\theta = \frac{\sqrt{\sin^2 \alpha - \mu^2}}{\mu \cos \alpha}. \tag{10}$$

Fresnel now interpreted this to mean,

> sinon par des raisonnements rigoureux, aux moins par les inductions les plus rationelles et les plus probables,

that there is total reflection, that the amplitudes of the two waves are the same, but that the waves differ in phase by the amount θ as given by (10). Fresnel also made a corresponding statement for light reflected at right angles, and both statements were completely confirmed by experiment.

28 *Oeuvres d'Augustin Fresnel*, Vol. I, 1866, pp. 767-799, especially p. 782.

29 Compare E. Whittaker, *A History of the Theories of Aether and Electricity*, Vol. I, 1951, p. 126.

We think that this was the first time that complex numbers or any other mathematical objects which are " nothing-but-symbols " were put into the center of an interpretative context of " reality," and it is an extraordinary fact that this interpretation, although the first of its kind, stood up so well to verification by experiment and to the later " maxwellization " of the entire theory. In very loose terms one can say that this was the first time in which " nature " was abstracted from " pure " mathematics, that is from a mathematics which had not been previously abstracted from nature itself.

Almost until the end of the nineteenth century efforts were made by some physicists to rationalize the conclusion of Fresnel more " convincingly " and even Henri Poincaré took part in these endeavors.[30] But others were more interested in exploiting the idea of Fresnel for more elaborate problems, that is for problems in which at least one of the two bodies is optically opaque. From 1837 on, J. MacCullagh and A. Cauchy turned their attention to the then current topic of metallic reflection,[31] and they offered a first theory of the phenomenon by mathematically introducing into formulas of Fresnel a refractive index which is a general complex number $\mu = \nu - i\kappa$. This assumption has maintained itself in later developments, and the real part ν is a " real " refractive index and the imaginary part κ is an index of absorption.

Sometimes even a working physicist pauses to wonder at the extent to which the symbolization by complex numbers is penetrating into the mathematical representation of schemata of physics at their depth. For instance, Arnold Sommerfeld (1868–1951) was wont to operate freely, from the beginning of his career, with complex-valued solutions of divers wave equations, but apparently in the expectation that in the end the real and imaginary parts of the solutions would have a physical interpretation separately. However when analyzing a Schroedinger equation for the Zeeman effect of the hydrogen atom, he found that a certain very " natural " solution, which fitted well the case of the so-called normal (that is non-relativistic) Zeeman effect, was such that its real and imaginary parts did not have a physical standing by themselves. He apparently attached importance to this fact. He stated it expressly in 1929 on page 141 of his *Wellenmechanischer Ergänzungsband*, which was a supplement to his leading work, *Atombau und Spektrallinien*; and he fully restated this fact in 1939, on page 104 of Volume 2 of the fifth edition of the entire work, although in the intervening ten years the " complexification " of physics had been progressing even faster than at any time before.

V. The Significance of Mathematical Space

14. At present, mathematics is growing and spreading very fast, the so-called applied mathematics is steadily increasing in volume, and in certain areas of physics some difficult modern mathematics is being tried out. And

[30] See his *Théorie mathematique de la lumière*, Vol. I, 1889, p. 336.

[31] Whittaker, *loc. cit.*, pp. 161-169.

yet one can say that during the last few decades no stirring mathematical innovations have been injected into physics, or even provoked by physics, and that, for instance, the far-reaching innovations in nuclear physics have not been reflected in depth in the simultaneous innovations in mathematics which have been equally far-reaching. Apparently the " predestined harmony " between mathematics and physics is itself a phenomenon of variable intensity, and if this is so, it would be most important that philosophy of science should study its variability and determine whether it has " hidden periodicities " and what they are.

We think that the philosophical problem of the relation between mathematics and physics is not receiving sufficient attention. This is due in part to the fact that this problem is a very difficult one, but it is also due in part to the fact that present-day philosophy of nature is still heeding the task which the seventeenth century has imposed on it.

The science of the seventeenth century gave a primacy to the theme of space and time which it inherited from late medieval natural philosophy, Western and Arabic. The Kantian philosophical system of the eighteenth century, and some philosophies preceding it, made this primacy within physics into a paramountcy within all of metaphysics, and this paramountcy has remained unchallenged. The nineteenth century acquiesced in it, and the theories of relativity and gravitation in the first decade of the twentieth century even enhanced it. But quantum theory, which is a profoundly new method in scientific thinking, may eventually change this. It is true that in quantum theory the new vistas on space have been startling and are as yet not fathomable, but the theory nevertheless bids fair to lower the notions of space and time from their Kant-Einstein heights of dominancy onto the general level of other notions of the evolving physics of particles, some of whose concepts and phenomena are equally challenging and sometimes inscrutable. It is true that the uni-directedness of the " arrow of time " in biology, cosmology, and even thermodynamics suggests problems which invite commanding attention, but it can be said that in many philosophical works which assert and elaborate on the paramountcy of space and time such concrete problems have not even been featured. Furthermore, in large parts of physics and other science only mathematical space and time are operationally of consequence, and we think that in this respect the notions of space and time have not been causing undue difficulties. To the contrary, the mathematization of space and time has been usually more advanced than that of other notions and phenomena of physics in the same epoch.

15. According to various remarks in Aristotle, a mathematization of the everyday statement that space is three-dimensional was consciously begun by Pythagoreans,[7] and Plato refers to this in the *Timaeus.* Aristotle tries, in *Categories,* 5a, 8-14, to give a " logically " worded definition that the place (*topos*) of an object is its " extension," and we think that a three-dimensionality of this extension is there implied. Also, large parts of Aristotle's *Physica,* and of other works, deal in one form or another with the

problem of mathematization of the statement that space of physics is " continuous," which problem had apparently monopolized the attention of leading Eleatics. But Aristotle does not make it clear whether Pythagoreans were already concerned with this problem, except that in *Metaphysica* 1083b, 7-19 he reproves Pythagoreans for holding the view that bodies are composed of numbers. Such a view, he says, or they already said, leads to the conclusion that there are spatial magnitudes which are indivisible, and this is highly repugnant to Aristotle. (In contemporary physics, many an elementary particle has a finite non-zero radius associated with it and is " indivisible " until and unless it is transmuted into or collides with another particle.)

It is important to realize that in the thinking of Plato and Aristotle the notions of space and time in their role of " scientific " concepts were objects of physics, and of physics only, and not at all of metaphysics; except that the natural philosophy of Plato was still in the " archaic " state in which physics proper was not separable from cosmology and cosmological genesis. Thus, Plato deals with space and time mainly in the *Timaeus,* and hardly in the ontological dialogues *Theaetetus* and *Sophistes.* And Aristotle deals with them fully in the *Physica* and *De Caelo,* and occasionally in " descriptive " science, but not in the *Metaphysica,* although the latter re-discusses other notions from the *Physica* on a " higher " level of physical primacy and fully discusses the nature of mathematics and its relations to physics and to other areas of knowledge.

Plato and Aristotle have two appellations for spatiality, *chora* (" space ") and *topos* (" place ") . We have already stated elsewhere [32] that in our view, in the diction of Aristotle, the term *chora* is usually the everyday word for space in general to designate the " room " occupied by an everyday object. *Topos,* however (especially in Book IV of *Physica*) , is the " scientific " word which designates a certain attribute of a single substance or body, or of a collection of bodies which constitute a " physical system " or rather a " process-in-nature " of Aristotelian complexion (see above, Section II, paragraph 5) . We think that in the *Timaeus* of Plato (51E-52D) the difference between the two appellations is lexicologically the same, except that the *Timaeus* does not envisage the space pertaining to an individual physical set-up in isolation, but only the cosmological space of the universe as a whole in its cosmological origination when yet unformed and not yet made physically specific. And we think that this kind of cosmological space is nearer to the *chora* of Aristotle than to his *topos,* which would explain why in the *Timaeus* the appellation *chora* prevails. Also, Aristotle's interpretative observation, which has caused controversies among commentators, that for Plato *chora* (" space ") and *hyle* (" matter ") are the same (*Physica* 209b, 10-11) is in accord with the fact that in present-day cosmological theories creation of space is frequently paralleled by creation of matter. It is true that further on in the *Timaeus* Plato " identifies " elementary bodies of physics with

[32] S. Bochner, " Aristotle's Notion of Place (*topos*) in Physics," *Actes of the Xth International Congress of the History of Science,* 1962, to appear.

figures of Euclidean space without introducing *hyle* at all. However, in contemporary physics, an intimate cosmological association of space with matter would also be compatible with a thoroughgoing " mathematization " of elementary particles of which matter is composed.

16. As we have stated above, the term *topos* already occurs formally in Aristotle's *Categories*. But there it is not or not yet a concept-in-depth of physics, and Aristotle gives only the mathematico-logical " dictionary definition," which must have been generally familiar, that the place of an object is its three-dimensional extension.

In the *Physica* however, after a lengthy reasoning, Aristotle arrives at the characterization (212a, 5-6) that if a physical system is enclosed by another, the *topos* of the system is " the innermost boundary of the enclosure "; and a few lines afterwards (212a, 20-21) he suggests an alternate version that *topos* is " the innermost *motionless* boundary of what contains." We will not restate our observations on Aristotle's views [32] but we will presently make another observation which will take the present-day role of *mathematical* space in physics into account.

The space of Newton's mechanics may have been a nearly metaphysical three-dimensional *absolute* datum, and the space of Kant was most fully metaphysically so. But it is a fact that the space of Euler, Lagrange, Jacobi, Hamilton, and other architects of the vast doctrine of mechanics in the eighteenth and nineteenth centuries was, in its operational aspects, that is at least during their " working hours," entirely physical and mathematical only, and that, in a certain novel aspect or version, it additionally became something entirely different from what it is to psychology and philosophy even today. Namely, in addition to the three-dimensional ordinary space of common human awareness, Rational Mechanics also gradually introduced the concept of a purely mathematical space which is multi-dimensional without any restriction on the dimension, which has a general differential geometric structure, and which serves as a substratum of mathematization for a general physical set-up, not only in mechanics but also in physics proper, and in fact in science anywhere.

Such a space is variously called the " space of generalized coordinates," or " space of parameters," or sometimes " phase space," and it has as many mathematical dimensions as there are physical " degrees of freedom " in the given set-up. The admissible generality of the differential geometric structure of such a space became greatly emphasized in the theory of relativity, but it had actually played a role before, as we have already stated above in Section III, paragraph 9. Also, some " models " of infinitely dimensional spaces already had been introduced in mechanics of continua before infinitely dimensional space became the " base space " of quantum theory, and thus an everyday occurrence in it.

Now, we venture to suggest that an anticipation of such a conception of space, albeit only a very dim one, may be imputed to Aristotle in his " essay " on *topos* in *Physica*, Book IV; whereas in Plato only the very transparent and obvious conception of space à la Euclid is discernible. And we think that it is Aristotle's half-awareness of this which makes him assert, and with some emphasis, that although Plato had made some correct and satisfactory statements about space, there is still considerably more that can be said about it. At any rate we think that if one does impute such an anticipation of Aristotle, then some additional understanding of Aristotle's approach to the problem of space may accrue. In fact, in the light of such an anticipation, Aristotle's statement that for a circumscribed system the *topos* is " the innermost boundary of the enclosure " would be his way of saying that any state of such a system is determined by a subspace in a comprehensive general space of parameters and that, if the subspace is a domain, an appropriate way of determining the domain is to exhibit its boundary. We note that Greek mathematics did not find it easy to separate terminologically, and to an extent even apperceptively, a domain from its boundary and was inclined to view the boundary as the prime attribute of a domain. Also, in Aristotle's famous case of a boat floating on a river (212b, 15-20), which gave rise to controversy, if one wishes to understand fully the varying physical effect which the water flowing past the boat has on its hull, it does not suffice to consider only a parameter space for the boat per se in isolation, but a parameter space for the system consisting of boat and river must be introduced. And this then leads to the variant characterization that *topos* is " the innermost *motionless* boundary of what contains."

17. After Plato and Aristotle, a certain measure of awareness that a mathematization of space is a prerequisite for " exact " science may have been in the minds of some natural philosophers from Aristotle to Galileo, but only the seventeenth century began to act in the matter.

We will now make a few remarks on the extent to which the three-dimensionality of space enters the mathematical laws of physics. The planetary laws of Kepler and Newton remain literally in force in an n-dimensional Euclidean space instead of in a three-dimension space, as long as a planet, or the sun and a planet, are each a geometrical mass point. However for $n > 3$ it is no longer true that under the gravitational force $1/r^2$ a homogeneous n-dimensional ball of non-zero radius acts as if its mass were concentrated in the center, and inside a homogeneous shell the gravitational force is no longer zero (*Principia*, Book I, Section XII). However the general equations of motion of Lagrange, Jacobi, and Hamilton in generalized coordinates for a finite system of mass points with any constraints make no reference at all to a three-dimensionality of a " true " underlying space.

In the mechanics of continua, in the general equations of motion of A. Cauchy and others after him, some notice is taken of the three-dimensionality of the underlying space. In Maxwell's electrodynamics the three-dimensionality was very seriously used in the fact that the electric force and the magnetic force are both vectors, the first translatory and the second rotatory. This fact persisted in the special theory of relativity; that is, Minkowski's version of the Maxwell equations for space-time in $n + 1$ dimensions requires $n = 3$ and cannot be readily adjusted to $n > 3$.

Finally, although quantum theory presupposes infinitely dimensional Hilbert spaces for its operators, nevertheless the three-dimensionality of a space of reality enters seriously, and perhaps even unavoidably, in the manner in which algebraic properties of the three-dimensional rotation group are expressive of the crucial phenomena of spin and angular momentum.

Within relativistic physics some attempts were made [33] to deduce the three-dimensionality of concrete space from physical laws. It would be even more interesting to do the same within quantum theory. The justification for such an undertaking is that our mathematics by itself, as an instrument of cognition, has no bias for the integer 3 or any other integer in its make-up.

We will deal with the significance of mathematical time on another occasion.

18. Since the seventeenth century the intellectual orientation has become such that problems about space and time command overriding attention, as a matter of course. And yet we think that for the comprehension of the universe it is just as important to know how the reality of physics can be articulated by the symbolism of mathematics as it is to know, beyond the shadow of a doubt, what space and time are. The Greeks had little mathematics and less physics and no symbols, and yet Plato and Aristotle gave much less weight to the problem of space and time than to the problem of the relation between mathematics and physics or, what was virtually the same to them, than to the problem of the relation between mathematics and knowledge in general. From Plato onward, Greek *mathesis* was "mathematics" and "general knowledge" both. Leibniz tried to activate this latent ambivalence of the meaning of *mathesis*, but, although he received a respectful hearing, not much ensued. The second half of the twentieth century is about to try again, and this time there seems to be a gathering mood for a comprehensive accord on the need of a *mathesis* of universality at the root.

[33] See, for instance, H. Weyl, *Space, Time, Matter* (trans. H. L. Brose), 1922, p. 217. The nature and "verification" of three-dimensionality of space was a recognized problem within the Theory of Knowledge in the nineteenth century. The gradual "dissolution" of the three-dimensionality of space in the pictorial arts in the course of the nineteenth century, which culminated in the achievements of Paul Cézanne, has apparently not yet been brought into parallel with generalizations of the concept of space in mathematical physics and in mathematics proper.

Revolutions in Physics and Crises in Mathematics

There is no simple, single formula for the course of revolutions in science.

Salomon Bochner

In this article I deal with two topics which, although separable, are closely connected with each other. The first and larger part of the article is concerned with the conception of a revolution in physics, as recently blue-printed in a provocative book by Thomas Kuhn (*1*). I make observations which are seemingly in conflict with those of Kuhn, but I really intend to amplify and qualify some of Kuhn's theses rather than to dissent from them, and my approach is somewhat different anyhow. After that, I make some observations on revolutions in physics as far as the underlying mathematics is concerned. And, finally, I make some remarks on so-called "foundation crises" in mathematics, which may be viewed as a kind of revolution, and especially on a major crisis of this kind which is presumed to have taken place in the 5th century B.C.

Kuhn, in his investigations into the nature of revolutions in science, analyzes both the inward ontological and epistemological nature of such revolutions and the psychological and behaviorist attitudes, resistances, and responses of practitioners of science, before, during, and after a revolution. Kuhn finds that revolutions in science are mostly internal revolutions, brought about by some scientists and then forced by the initiators on the scientific community at large. There is even an implied suggestion that, in the beginning, a revolutionary innovation may be both desired and resisted by the same group of scientists, ambivalently. Kuhn makes a point of emphasizing that most scientists all the time, and all scientists most of the time, prefer peace to revolution, normalcy to anomaly, and the preservation of their "paradigms" to changes of paradigms, a "paradigm," according to Kuhn, being more or less a sum of "universally recognized achievements that for a time provide model problems and solutions to a community of practitioners" (*1*, p. x.)

This finding is indeed meaningful, and as already noted by Gillispie (*2*), it is one that can be easily accepted. For my part, I found nothing singularly disturbing in the realization that among scientists, as in other groups of human beings, the revolutionaries of today are likely to be the conservatives of tomorrow; that paradigms are not readily

Reprinted from Science, August 2, 1963, Vol. 141, No. 3579, pages 408-411 abandoned or changed unless anomalies make it imperative; and that there may be diehards who will not give in even then. But if one *is* surprised and disturbed to find that resistance to innovation is widespread and even dominant among "professors" who are expected to be professionally pledged and conditioned to emphatically seek the truth and nothing but the truth, I think that there is no clear reason for singling out scientists from among scholars in general. Kuhn's diagnosis of innate conservatism does attach a certain stigma, and it is restrictive to the entire study to stigmatize scientists for something which philosophers and humanists also practice.

Perhaps we can see evidence of the humanists' concern to preserve a paradigm in a well-attested event (*3*) which occurred in the Neoplatonic school at Athens during the last 50 or 60 years before its dissolution by the Emperor Justinian in A.D. 529. At the time, leading circles in the school were opposed to the increase in Aristotelian features in the "official" world picture, the result of certain influences from within. A leading exponent of Aristotelian ideas was Marinus, born in Neapolis (the Hebrew Shechem) in Samaria, who eventually became head of the school, succeeding the much-adored "divine" Proclus (A.D. 411–484). Marinus, when still in the junior position of tutor, was in charge of Isidorus, a student. One day Marinus showed Isidorus a commentary on the platonic dialogue "Philebus," which he had just composed. But, on the authority of Damascius, biographer of Isidorus, the latter prevailed upon Marinus to destroy the commentary, on the strange grounds that their great master Proclus, then head of the school, had already composed a commentary on the "Philebus" to end all such commentaries. Present-day scholarship maintains convincingly that the true motivation for this request was apprehension lest the work of Marinus inevitably show a bias against the Neoplatonic "paradigm" even if Marinus made an effort to keep it out.

The author is Henry Burchard Fine professor of mathematics at Princeton University, Princeton, N.J.

Max Planck

One of Thomas Kuhn's star witnesses is none other than the physicist Max Planck. In *Scientific Autobiography*, written in 1937 (*4*), Planck remarked that "a new scientific truth does not triumph by convincing its opponents and making them see the light, but rather because its opponents eventually die, and a new generation grows up that is familiar with it." This harsh statement has the ring of authenticity, but it so happens that it hardly applies to Planck's own discovery. It is true that Planck's discovery was not hailed on the instant as the great breakthrough that it was, but it is equally true that Planck did not have to wait for anybody to die before a considerable measure of recognition was meted out to him.

Although Planck's first papers on quanta began to appear only in the latter half of 1900, as early as 1911 renowned physicists of the day convened at a Solvay Congress in Brussels with an agenda devoted entirely to the new theory, and with Planck as the honored speaker (*5*). Before that, in an encyclop dic article (*6*) dated May 1909 and entitled "Theorie der Strahlung," the radiation specialist W. Wien, although unsympathetic to Planck's theory, gave it exhaustive coverage. On the other hand, Planck himself, as late as 1922, in his Nobel address, almost tried to play down the originality of his achievement by pointing out how his intense preoccupation with the then relatively new Hertz vibrator (or "resonator") was the catalytic setting that eventually brought forth the quantum hypothesis (*7*). In fact, in the list of references to the written version of the address, Planck cites the paper of Heinrich Hertz in *Annalen der Physik* (*8*), in which Hertz discusses the vibrator which he had constructed for purposes of testing the Maxwell theory. With this vibrator Planck, in 1900, by a kind of "ideal" construction in the black body, devised his quantization. It is regrettable that, in after years, Planck and his biographers (especially von Laue) rarely restated this illuminating biographical fact, if they ever mentioned it again.

In general, I would say that between 1900 and 1925, in the quantum theory "revolution" which was brought about by Planck himself and then by Niels Bohr and others, the changeover from one paradigm to another was peaceful and evolutionary, without any of the characteristics of a revolution by *bouleversement*, and without manifestations of a forced normalization of an anomaly, as Kuhn envisages it. One might retort that, although this particular transition was a peaceful one, yet the emergence of the theory of relativity within the same period, 1900 to 1925, and in fact within the narrower period 1905 to about 1918, did indeed conform to the Kuhn pattern, and that the second half of the quantum theory revolution, which began soon after 1925 and in which Heisenberg, Schrödinger, Dirac, and others were the protagonists, did also. To this I would reply that what gave the emergence of relativity the character and status of a revolution was not its effect on physics proper but its effect on the notions of space and time, which, long before that, had become objects of paramount importance to philosophy in general. I have been trying to show in recent articles (*9*) that space and time did not become paramount in general philosophy until after the Renaissance, whereas in classical Greece, for instance, in spite of a strong trait of spatiality in the general imagery of rational thinking, space and time, as specific notions, were notions of physics and of physics only. If the concepts of space and time had not attained the preeminence in general philosophy which they had through developments that had occurred since 1600, relativity would have been much less of a revolution than it was.

Furthermore, the sudden outburst of interest in quantum theory and in Planck's constant came after the proclamation of the uncertainty principle by Heisenberg, in 1925, and here again the interest arose from the fact that Heisenberg, in his "popularization" of his principle, emphasized the involvement of the notion of ordinary space. The overall fact that the uncertainty principle applies to pairs of conjugate operators in general, if stated without emphasis on particular pairs which correspond to ordinary Newtonian coordinates of position and momentum, would hardly have caused philosophers the malaise they felt when the emphasis was placed on these coordinates.

Also, while the cosmological revolution in the 16th century did indeed replace one paradigm with another—namely, the geocentric with the heliocentric—it cannot be said that the theory of relativity replaced Newtonian space with other spaces, in the sense of making Newtonian space obsolete or antiquarian. Planetary and particle mechanics, vast stretches of "phenomenological" mechanics of continua, and much of statistical mechanics, of thermodynamics, and even of electrodynamics continues to be Newtonian; any serious attempt to de-Newtonize them would make most of mechanics, much of physics, and virtually all of engineering unrealistically complicated and would be widely resisted. And even within the heart of physics, which did indeed become genuinely relativistic, there is no single paradigm in control, such as a Kuhn revolution by *bouleversement* would have terminated in. In fact, the Lorentzian space of quantum field theory and the substratum space of most cosmological models now in vogue are totally contradictory.

Finally, I wish to point out that, to many physicists, the theories of relativity, whatever their éclat, were terminal phases of the era of Newton, Lagrange, Hamilton, and Maxwell rather than initial phases of a new era. But Planck's original quantum hypothesis, even if its advent was rather peaceful and even if its antecedents in 19th-century radiation theory were comparatively unspectacular, was apparently the prologue to a tremendous epic-still-to-come, of which only the introductory scenes have been playing thus far.

Kuhn's notion of a scientific revolution may subsume too many possibilities under a single formula. The formula apparently is meant to apply to the emergence of modern science in the 17th and 18th centuries, viewed as one giant revolution, in spite of its size and in spite of the fact it was much more the direct emergence of something entirely novel than the transformation, by revolution, of something old into something new. But the formula is seemingly also meant to apply to many particular events, which occurred in succession but are viewed separately, such as the many turns and twists and even vagaries in the presystematic phases of electricity and chemistry in the 17th and 18th centuries, especially the 18th century. And finally, as has already been noted somewhat apprehensively by Gillispie (*2*), the formula might end up by being applied, through a circular mode of identification, to any kind of changeover which has a visible and recognizable trait of originality and creativity associated with it. Now, the question of what constitutes originality and creativity in *Homo sapiens* is probably one of the most difficult problems in philosophy and philosophical sociology, and not much would be gained by reducing the problem of what is a revolution of knowledge to the problem of what constitutes creativity in man. Also, it is probably feasible to distinguish between latent and active phases within creativity. This would lead to a corresponding distinction, during nonrevolutionary periods in science, between genuine "normalcy" and "latency" of anomaly, and the evaluation of this distinction, within Kuhn's schema, might become very troublesome indeed. Furthermore, in physics there seem to be periods of concentrated creativity which are made up of rapid successions of many small but sharply defined discontinuities of achievement, and to such periods Kuhn's formula of a "normalization of an anomaly" can be applied only with difficulty.

Scheme of Mathematics Underlying Physics

A novel situation arises, in the case of physics, if we turn our attention away from physics proper and toward the scheme of mathematics that underlies it. This mathematics seems to have paradigms of its own, and they are more inward ones. These inward, structural paradigms seem to behave differently from the outward, purely physical ones, sometimes even disturbingly so. Thus, neither electricity nor magnetism had any mathematical paradigm at all until, toward the end of the 18th century, Cavendish and Coulomb formulated the so-called Coulomb law, and until, at the beginning of the 19th century, Poisson initiated magnetostatics and electrostatics by introducing a mathematical theory of potentials into the context. In this sense, in spite of Lavoisier, one must say that chemistry began in earnest only with the laws of Dalton, Avogadro, and Dulong and Petit all in the beginning of the 19th century. I do not mean to suggest that there were no disciplines of electricity, magnetism, and chemistry before the 19th century. But I do mean to suggest that, to somebody in the field of mathematics, the cataloging of many separate revolutions which these disciplines are supposed to have gone through in the 17th and 18th centuries is bewildering and unconvincing.

A similar observation could even be made about the course of "classical" Greek physics in its entirety. The developments from Thales to Aristotle are frequently presented as a "motivated" succession of revolutions in which an emergent would-be physical or cosmological system was knowingly and militantly put forward to supersede an earlier system. But the fact is that Archimedes, who was the only mathematical physicist Greece ever produced, seemingly refused to be involved in these crazy-quilt developments, and to his sober mind they probably appeared to be "irrational" and unmotivated.

On the other hand, in the case of mechanics, physicists and other scientists are wont to view the 17th and 18th centuries as one unit of development. And yet, mathematically *(10)*, Newton's *Principia* has the appearance and many of the attributes of the works of Archimedes and Apollonius, whereas Lagrange's *Mécanique Analytique* (1788) is not radically unlike textbooks of today, 20th-century modernisms notwithstanding. Also, the mathematicophysical subject matter of the various divisions of mechanics which took shape in the 18th century was not at all a mere explication of what had already been presented in the *Principia*, implicitly if not expressly, even though the effect of Newton's treatise on later developments was an overwhelming one. In fact, by "juggling" mathematical paradigms one could make out a case for the assertion that there was a much greater distance between the mechanics of Euler and Lagrange and the mechanics of Newton than there was, in the 19th century, between the electrodynamics of Maxwell and Hertz and the hydrodynamics of, say, Helmholtz.

Apparently the mathematization of a science affects the role and nature of revolutions that may and do occur in it. And since, on the other hand, most of science is tending toward mathematization, even determinedly so, one should guard against generalizing from the shape of premathematical revolutions to the shape of revolutions in general.

Mathematics itself also has its revolutions, and developments in the 20th century have led to the singling out among them of revolutions of a particular kind, which are termed "foundation crises." It has been asserted, with emphasis and even with a dash of sensationalism, that the classical Greek mathematics which is known from the works of Euclid, Archimedes, and Apollonius went through such a crisis in an early Pythagorean phase in the 5th century B.C. The cogency of this assertion may be questioned, and the assertion has indeed been contested. But the clamor for a retroactive crisis in antiquity has been such that temperate counterassertions have not been able to mute it. An "anomaly" in Greek mathematics did indeed emerge; it was the discovery that the square root of 2 is irrational, or rather that, in a square, the diagonal is incommensurable with the side *(11)*. Greek mathematics did certainly react to this discovery, with attentiveness and resourcefulness as is evidenced by the reference to the problem in the platonic dialogue "Theaetatus," and by the erection of a Greek theory of incommensurables, apparently attributable to Eudoxus, which is the subject matter of the 5th book of Euclid. However there is no indication in Greek doxography of a "crisis" or of a "mathematical scandal," except perhaps for the late-Pythagorean tradition that Hippasus of Metapontum, an unruly, early member of the Pythagorean sect, was violating rules of the sect by divulging to outsiders details about research-in-progress on incommensurables and other problems, and that he suffered divine retribution for his indiscretions *(12)*. One might perhaps also adduce the fact that, in Archimedes's work on "Sphere and Cylinder I," in the prefatory letter to Dositheus, the puzzling assertion that the mathematical procedure of Eudoxus is "most irrefragable" *(13)* indicates that even at that time there were some "philosophers" who were not satisfied with the manner in which Eudoxus resolved the crisis of incommensurables. Against these very slender items of support one has to note that there is no allusion to a mathematical crisis or "scandal" in any of the passages in Aristotle from which, with due caution, most of what is known about Pythagorean mathematics and principles of science has to be abstracted.

A systematic theorizing about foundation crises in mathematics was begun about 50 years ago in response to the challenging discovery, around 1900, that there are paradoxical situations in George Cantor's "naive" set theory and hence in mathematics as such. In fact, the first foundation crisis was identified, in substance rather than in name, in 20th-century mathematics itself, and past crises were then uncovered in the wake of this one. The "Greek crisis" theory was received very attentively and sympathetically all around, perhaps in remembrance of Zeno's puzzles, which to some philosophers are inexhaustibly provocative. But I should point out that the authors of a recent book (*14*) are trying to arouse interest in a third foundation crisis, and the "anomaly" which underlies this crisis is the one whose normalization consisted in the "rigorization" of analysis in the 19th century by Cauchy, Weierstrass, Cantor, and others. This last alleged crisis is the one least deserving of the name. Perhaps it was "anomalous" for the infinitesimal calculus to pile up achievement upon achievement in the 17th and 18th centuries, without being frustrated by inadequacies of mathematical rigor, and to become introspective as to its rigor only afterwards; if so, this was the healthiest and the most wonderful "anomaly" that could have occurred. Newton, the Bernoullis, Euler, d'Alembert, Lagrange, Laplace, and others made advances beyond anything one might have asked for.

To say that their achievements landed mathematics in a "crisis" is incongruous, unless one is prepared to aver that Thales and Pythagoras plunged rational thinking into a crisis-in-perpetuity, inasmuch as they introduced mathematics into rational thinking inseparably, and (as Plato and Aristotle felt) inasmuch as there is no prospect of constructing a logico-ontological foundation of mathematics that will be absolutely and unqualifiedly satisfactory to all, forever.

References and Notes

1. T. S. Kuhn, *The Structure of Scientific Revolutions* (Univ. of Chicago Press, Chicago, Ill., 1962).
2. C. C. Gillispie, *Science* **138**, 1251 (1962).
3. The main reference is to an article by Otmar Schissel in *Real Encyclopädie der klassischen Altertums-Wissenschaft*, Pauly-Wissowa, Ed. (1930), vol. 28, cols. 1759–67; the article is an authoritative interpretative digest from R. Asmus, *Das Leben des Philosophen Isidoros, von Damaskios aus Damaskus* (Philosophische Bibliothek, Leipzig, 1911), col. 125.
4. See either M.K.E.L. Planck, *Physikalische Abhandlungen und Vorträge [Collected Works]* (Vieweg, Brunswick, 1958), vol. 3, p. 389; or M.K.E.L. Planck, *Scientific Autobiography and Other Papers*, F. Gaynor, transl. (Philosophical Library, New York, 1949), pp. 33–34. Kuhn quotes the passage in *The Structure of Scientific Revolutions* (see *1*, p. 150).
5. P. Langevin and M. de Broglie, *La theorie du rayonnement et les quanta* (Gauthier-Villars, Paris, 1912); A. Eucken, *Die Theorie der Strahlung und der Quanten* (Knapp, Halle, 1914).
6. *Encyclopädie der mathematischen Wissenschaften* (1904–1926), vol. V3.
7. See M.K.E.L. Planck, *Collected Works* (Vieweg, Brunswick, 1958), vol. 3, p. 122.
8. H. Hertz, *Ann. Physik* **36**, 1 (1889).
9. S. Bochner, "The significance of some basic mathematical conceptions for physics," *Isis*, in press; "Aristotle's physics and today's physics," in preparation.
10. S. Bochner, *Am. Scientist* **50**, 294 (1962).
11. It has also been suggested that it was with respect to the regular pentagon, not the square, that the incommensurability of the diagonal and the side was first recognized [see K. von Fritz, *Ann. Math.* **146**, 242 (1945)].
12. For references relating to the "crisis," see Walter Burkert, *Weisheit und Wissenschaft, Studien zu Pythagoras, Philolaos und Platon* (Carl, Nurenberg, 1962), p. 432, and "Hippasos" in the index. For information about Hippasos in particular, see also M. T. Cardini, *Pitagorici, Testimonianze e Frammenti* ("La Nuova Italia" Publishers, Florence, 1958).
13. T. L. Heath, *The Works of Archimedes* (Cambridge Univ. Press, Cambridge, 1897), p. 2.
14. A. A. Fraenkel and Y. Bar-Hillel, *Foundations of Set Theory* (North-Holland, Amsterdam, 1958), pp. 14–15.

Princeton

Aristotle's Physics and Today's Physics

Salomon Bochner

THE PURPOSE of the present paper is to make some observations on the nature of the physics of the Greeks, especially of the physics of Aristotle, and the aim of the observations will be to point up similarities and dissimilarities, analogies and discrepancies, between the physics of antiquity and the physics of today. The aim is not to find out how much of Greek physics survives in ours, but to verify that to a certain extent the similarities in the modes of our general thinking, ancient and modern, also reflect themselves in parallelisms between the doctrines of physics which these modes of thinking have produced.

1. Physics and Mathematical Physics

My first statement is of a rather general nature. It is a counterpart to the familiar assertion that Greek physics, while being observational and perhaps also experiential, never became properly experimental. Now, the counterpart is this, that the Greeks, both of the Hellenic and Hellenistic eras, never developed, or even truly initiated, a kind of physics that would correspond to the so-called "theoretical" or "mathematical" physics of today. This was not at all due to the fact that Greek physics was not "ready" for it. In fact, modern physics, that is, the physics which began to develop after the Renaissance, started out from rather modest beginnings, but it showed *from the very beginning* a tendency towards becoming mathematical, and towards becoming so in a manner and to an extent that is a hallmark of its "modernism" and an instrumentality for its ever growing success.

The Greeks did have areas of knowledge in which science and mathematics interpenetrated. Their astronomy was clearly mathematical, from Eudoxus to Ptolemy. Also, already in the times of Aristotle there was, on his testimony, something called *optike* which was subordinate to geometry, and something called *harmonike* which was subordinate to arithmetic (and according to the *Posterior Analytics* $78^{b}37$ there was even a kind of *mechanike* that was subordinate to "stereometry").[1] Furthermore it follows from various known statements in Aristotle that Pythagoreans envisioned some kind of mathematization of physics,[2] although the extent and depth of their insights is not easy to appraise even when the allusions to them in Plato's *Timaeus* are added to express statements in Aristotle.[3] In a known passage in the *Physica* (II.2.$193^{b}22$-$194^{a}15$) Aristotle even seems to be wondering, for a fleeting moment, whether astronomy, *optike*, and *harmonike* can indeed be distinguished from mathematics proper. But in the same passage, and in passages in the *Metaphysics*, he has no such doubts about what he invariably and firmly calls *physike*, and which apparently corresponds to physics in general in our sense. He insists that mathematics and physics (that is, Greek mathematics and Greek physics) are distinct and separate; and that this separateness is not one that can be overcome. In this, as it turned out, he was right. This separateness was indeed never overcome by the Greeks, not even by Archimedes who could have done so, or by physicists and mathematicians after him who should have done so. Archimedes lived about a hundred years after Aristotle and by innate originality he was one of the greatest mathematicians ever. His best known achievements, a law on the balancing of the lever and a law on floating bodies, clearly pertain to mathematical physics and were the first of their kind. But they did not have the effect of initiating a mathematical physics at the time. Archimedes the mathematician was revered in antiquity, and Stoic and Epicurean philosophers who specialized in physics and cosmology must have been aware of him. However, there is nothing to indicate that any of these philosopher-physicists was capable of comprehending that Archimedes was a species of physicist, too, or that any of them was having an inkling or premonition, if ever so vaguely, that it would be this species of physics to which a later future would belong.[4]

[1] For references see Thomas Heath, *Mathematics in Aristotle* (Oxford, 1949), especially pp. 11-12.

[2] Heath, *loc. cit.* (see index under "Pythagoreans").

[3] See F. M. Cornford, *Plato's Cosmology* (New York, 1937).

[4] A. N. Whitehead, *The Concept of Nature* (Cambridge, 1930), p. 24.

2. Physics and Philosophy

The Loeb edition of Aristotle's *Physica* by Wicksteed and Cornford, which is a very good commentary on the contents of the treatise, begins its "General Introduction" (p. xv) with the following paragraph:

> The title 'Physica' is misleading, and the reader must expect to find little or nothing that it suggests in this treatise. 'Lectures on Nature,' the alternative title found in editions of the Greek text, is more enlightening. But 'Principles of Natural Philosophy' (as the term would have been understood in the eighteenth and earlier nineteenth centuries) would be better still.

I do not agree with this statement, and I find that it is generally less misleading to view the *Physica* as a book on physics, pure and simple, than as a book on philosophy in whatever sense. In my own endeavors to make sense out of Aristotle, that is, to find coherence and meaning in his problems and in whatever answers he finds for his problems, I found it revealing and rewarding to approach Aristotle's treatise as a general book on physics, literally and sometimes even ingenuously. Aristotle is frequently discursive, repetitive, and resumptive, and especially in the *Physica* he sometimes involves himself in equivocations, inconsistencies, and even contradictions which discourage readers, distress and alarm commentators, and encourage detractors gratuitously. Now, philosophy has always been very sensitive to inconsistencies, and it is probably best that it should be so. Physics since the Renaissance has been professing to be so too, but only when in a holiday mood of solemn introspection. But when in a workday mood of extrovert activity it has been much less attentive to the pitfalls of inconsistencies and sometimes it has even been outright oblivious of them. It so happens that in the last part of the nineteenth century some very articulate leading physicists, especially French ones (Pierre Duhem, also Henri Poincaré and others), were sternly positivistic, and were demanding that the philosophers' ban against inconsistencies must also be strictly applied to the edifice of physics as an indivisible totality and at all times. But in the twentieth century developments in various parts of physics have been precipitous and uncontrollably multiform, and strict formal consistency for physics as a whole is not enforceable. Physicists, almost eschatologically, profess to wish and pray for consistency in the future, but they are willing to forego it "temporarily," at any time, for the then present.

In the light of this, if I view the *Physica* and related treatises of Aristotle as works on physics, and not on philosophy, then many commentators' difficulties in these works lose their poignancy, in that they appear to be much less damaging to the value and meaningfulness of the whole than commentators fear. In fact, by allowing myself the license of drawing up appropriate parallelisms to situations from contemporary physics, some difficulties in Aristotle, without being removed thereby, can nevertheless be made to even enhance the value of Aristotle's physics instead of subtracting from it. An instance of such a "saving" interpretation will be given later in Section 17 to a notorious difficulty in Aristotle's theory of *topos*. But before this other matters will be dealt with.

3. Time in Physics

Aristotle has an essay on time in *Physica* IV, 10-14, and we will begin our comment with a quotation from the Loeb edition :

> As to Time, Aristotle enters no profound metaphysical speculation as to its essential nature. He is content with attempting to bring precision into the thoughts of the plain man, who may easily fall into confusions and contradictions when he tries to give himself an account of what he really means by it. (I, 378)

No part of this statement has any pertinence if one views the *Physica* as physics. Aristotle does not address himself to "the plain man" but to the physicist and he does not indulge in metaphysical speculation because time is to him a concept of physics, and most exclusively so. He does mention "time" in other treatises, but only sparingly, and in no way does he analyze it outside the *Physica* in any detail. It is important to keep always in mind that to Aristotle time and space were only objects of physics and not also of his First Philosophy, as they did gradually become in the course of the twenty-three centuries since.

As viewed from present-day physics one can say that Aristotle is struggling to establish or clarify the following facts, in thought patterns of his own.

(i) Time is a "determinant" of any motion or change in physics, so that any physical process runs its course in time ("Time is the number of motion").

(ii) Time in physics has to be represented by a certain mathematical schema, which makes it into a "*megethos*" (magnitude), and qua *megethos* time is mathematically a so-called open linear continuum, which is also unidirected ("arrow of time").

(iii) Although time is a unidirected open line, nevertheless the only way of measuring it is by cyclical, that is periodic, events.

(iv) The oneness of the universe is somehow connected with a oneness of time.

(v) There is no awareness of time without a perceiving soul.

We note that (iv) and (v), if viewed as problems, have been raised "independently" by physicists in the twentieth century. Furthermore, statement (iii) is a profound truth, and it originated in the *Timaeus*. We also remark that until the "discovery" of the pendulum in the seventeenth century no "reliable" cyclic event other than stellar motion was available, and that competent astronomers knew that only observed time was true time.

4. Physics and Cosmology

It is probably right to say that many of Aristotle's significant statements on time already occur in the *Timaeus*, expressly or implicitly, and that there is nothing in Aristotle to match Plato's stirring phrase that "Time is the moving image of eternity." We wish to observe, however, that Aristotle's statements on time in the *Physica* are physics and have pertinence in the physics of today, but that Plato's phrase would not have a specific meaning within physics or even physical cosmology of today. The *Timaeus* may be admired as a great book, but it is nonetheless a fact that nowadays for a compendium on physics and cosmology the mythologically cosmogonic setting of the *Timaeus* would be intolerably archaic and backward looking; whereas Aristotle's division of his work into a sober and matter-of-fact treatise on physics, which comes *first*, and a non-cosmogonical book on the heavens, which comes *afterwards*, would be in keeping with arrangements of compendia today. Aristotle may have even been the first to introduce such an arrangement knowingly. Furthermore, Aristotle's total dispensing with "myths" may be viewed as a turning towards "scientificality," no matter how deeply Platonists may draw inspiration from myths in their master's discourses.

The *Timaeus*, after the introductory conversation, has a solemn invocation of the gods, and Plato then descends downwards from a divine on-high to a cosmic ordering of the universe, and thence to the actual creation of it in detail by chance, necessity, and law. Aristotle, however, after the manner of a "regular" scientist ascends upwards. Such a scientist begins by dwelling fully in physics proper, after which he very likely will want to advance to problems of general cosmology, analytical or evolutionary. And after that he may even feel an urge to move on further to questions of ethics, morality, and theology.

In briefer words, to Plato physics was a particularization of cosmology, whereas to Aristotle cosmology was an extension of physics. This is a significant difference of orientation; it is neither "minor" nor only "technical" and it does not indicate a paucity of imaginativeness in Aristotle's whole make-up, as Platonists sometimes conclude. This difference of orientation affected Aristotle's approach to many problems, especially to the problem of space, to a significant extent, as we will still see. But we will next discuss Plato's principle of necessity and Aristotle's reaction to it.

5. Necessity and Chance

A penetrating insight into Plato's *Ananke*, against the background of nineteenth-century knowledgeability, is the high point in George Grote's interpretation of the *Timaeus* in his work on Plato. I like best the following single sentence:

> By necessity Plato means random, indeterminate, chaotic, preexistent spontaneity of movement or force: spontaneity ἡ πλανομένη αἰτία upon which Reason works by persuasion up to a certain point, prevailing upon it to submit to some degree of fixity and regularity.[5]

Cornford's *Plato's Cosmology* has an elaborate analysis of necessity which is a rationalization of Grote's finding,[6] and Cornford quotes not the above sentence but the following passage:

> We ought here to note the sense in which Plato uses the word Necessity. This word is now usually understood as denoting what is fixed, permanent, unalterable, knowable beforehand. In the Platonic *Timaeus* it means the very reverse:—the indeterminate, the inconstant, the anomalous, that

[5] *Plato and the Other Companions of Socrates*, 3d ed. (London, 1875), III, 266.
[6] Pp. 159-77. The quote is from Grote, p. 249.

> which can neither be understood nor predicted. It is Force, Movement, or Change, with the negative attribute of not being regular, or intelligible, or determined by any knowable antecedent or condition—Vis consili expers.

Cornford draws Book II of Aristotle's *Physica* into his analysis and he rightly concludes that underlying its Chapter 8 there is a notion of "Necessity" or rather of the expression "of Necessity" ἐξ ἀνάγκης which closely conforms to the interpretation in Grote. To this we wish to add the following observation. In the preceding Chapters 4, 5, 6 of Book II, Aristotle has manifold reflexions on the role of *Tyche* (chance) and *Automaton* (fortuitous) in the relations between man and man and between man and nature. If viewed in retrospect, one might say that in these reflexions Aristotle was groping for an extension of Plato's *Ananke* from the realm of the physical to the realm of the anthropical, that is, from the area of physical science to a would-be area of social science, that is, for an extension of the physical science of Huygens and Newton to the social science of William Petty and James Bernoulli. The *Ars conjectandi* of Bernoulli (published posthumously 1713; he died 1705), which inaugurated so-called laws of large numbers in probability and statistics, is very much concerned with Grote's subordination of the "random and indeterminate" to the persuasion of mathematical reason, with the aim of securing a measure of "fixity and regularity." It extends its concern mainly to problems of human and moral standing of the kind that preoccupy Aristotle in Chapters 4, 5, 6, and it is this which made Bernoulli's book a starting book for all of social science to come. It is worth noting that the corresponding subordination of the "random and indeterminate" in the realm of physical "spontaneity of movement or force" was begun in the nineteenth century with the work of Clark Maxwell and L. Boltzmann, and that in the twentieth century it entered even more deeply into novel theories of physics which seem to be reaching out even more persuasively for a subjugation to our peculiar reason of the universe as a whole.

In order to avoid misunderstanding, let it be stated that we do not assert that Plato's *Ananke* was the direct antecedent of some conceptualization in modern physical science, or that Aristotle's *Automaton* was the direct antecedent of some conceptualizations in modern social science. The fact is that in a certain sense Greek physics and even Greek mathematics had each run down its own course in late antiquity and that, in an important sense, modern physics and even modern mathematics began from a new beginning. But something important in our manner of rational thinking comes from the Greeks. They created it and now it is fully ours. Because of this there are many traits of similarity between Greek physics and our physics and they force themselves on our attention. An injudicious exploitation of such similarities can be misleading, but it is nevertheless not necessary to dismiss them from our attention. When prudently assessed such similarities can be quite informative, both about the Greeks and about ourselves.

6. *Chora* ("Space") and *Topos* ("Place")

In Plato, space (and time) occurs mainly in the *Timaeus*. Although it occurs in the dialectical part of the *Parmenides*, it is referred to but very sparingly in the ontological analyses of the *Theaetetus* and *Sophistes*. As for Aristotle, so also for Plato, space is primarily a notion of physics, or at most of physical cosmology, but for Plato this does not hold quite so firmly as for Aristotle. The reason is not that for Plato space is some kind of "philosophical" entity, as for Kant, say, but in the *Timaeus* physics has not yet come out entirely from the cosmological "Receptacle of its Becoming."[7] In fact, the very name *physike*, as also the names *mathematike*, *mechanike*, and even *poetike*, solidified only in Aristotle, and very quickly and durably so, and have hardly changed a shading of their meaning since. This in itself was an Aristotelian achievement which is rarely seen in its magnitude.

Plato and Aristotle have two appelations for spatiality, *χῶρα* and *τόπος*. *Chora* is frequently broader, and it is less specific and determined than *topos*. A locus in mathematics, that is, a figure which is determined by, or results from specific requirements, became *topos* and not *chora*. In the *Meteorologica*, when Aristotle wishes to single out a geographic district in a country, then usually *chora* stands for country and *topos* for district. Also, *chora* seems to be the room occupied by a general object, and both in *Timaeus* and *De Caelo*, if "occupying" is indicated by a verb from the root *ἔχειν*, then, apparently for idiomatic reasons, space is *chora* as a rule.

Apart from that, in the *Timaeus* space in its cosmogonic nascency (52A, 53D) is *chora* and it continues to be so designated while being

[7] About Archimedes as mathematical physicist, see also S. Bochner, "The Role of Mathematics in the Rise of Mechanics," *American Scientist*, 50 (1962), 294-311.

"shaken down" into shape (53A, 57C) and even after that for a while as long as it still may be empty = κενόν (58A). But when finally firmly "emplaced" and when it has become "gravitational" space, in the sense that some bodies in it are bound to move from certain positions into other ones, then it is *topos* (57C, 58C, 60C). In *De Caelo* Aristotle observes the same distinction, scrupulously so, but since his cosmology is less cosmogonic, the occurrence of *topos* prevails.

Aristotle also uses the name of *topos* for the conception of place which he creates in *Physica* IV, 1-5. It is an original creation of his, and in Chapter 2 he asserts with pride that it was not anticipated by anybody, not even Plato. Aristotle's conceptions of place would fit well into the physics of today, better than into the small amount of physics of his day he had to press them into. Viewed and interpreted in retrospect, what Aristotle was aiming at was a notion of spatiality that would serve as a certain setting for a general "physical system," thus being a properly conceived attribute or attributes of a substance or of a collection of bodies rather than mere "extension" in a naive sense. Aristotle's great advance over his predecessors manifested itself in a clearly discernible aim—which however was marred by a serious miss—of creating a space of physics, as a datum of physics, and of physics (and science) only, without being compelled to create first—with the aid of receptacles, matrices, bastard reasonings, and what not—an all-comprehensive space of cosmology, and then to obtain the "ordinary space" of physics and of "individual" physical events from the all-comprehensive space by a downward specialization only. What Aristotle was groping for was what is today a physicist's "space of the laboratory." If a physicist nowadays wishes to conceptualize the spatial setting of an indoor experimental design, then he ordinarily does not imitate the *Timaeus* and engage the services of a *demiurgos*, and have him create a vast universe, isolate from it his laboratory for immediate use, and discard all the rest of the universe as a left-over. Rather, he "designs" his spatial setting more or less to the extent, or to the degree of sophistication, to which he will employ it. And I think that it is this that Aristotle envisioned, farsightedly, though abortively.

What was limiting his vision was the sad fact that, when creating his notion of *topos*, he could only ask himself what is happening to the places occupied by wine and air when one pours wine from a jug into a goblet, instead of being able to ask himself what it means to say that the electric charge on a condenser accumulates spatially on or near the surface; or, instead of being able to ask the much more intriguing question whether the so-called "creation" and "annihilation" of elementary particles in contemporary physics is also accompanied by a creation or annihilation of individual *topoi* that are to be associated with them. Yet it is astonishing to see how much Aristotle was able to construct out of how little.

We stated above that his admirable aim was marred by a serious miss. Now this "miss" was his inability, in Book IV of the *Physica*, to make there a *total* and *complete* separation between (i) the definition of "space of the laboratory" and (ii) the notion of a cosmological space, which, by its very nature, must also be "gravitational" space, in the sense that it must also be a *natural* place for certain bodies to move towards, from some predetermined compulsion. This confusing feature was clearly recognized by Solmsen in his recent book.[8] In *Physica* IV, Aristotle's main characterization of the *topos* of a body (or of a physical system) is this, that it is "the inner boundary of what contains." This has nothing to do with gravitational properties at all, and Solmsen rightly observes that Aristotle was in an inarticulate fashion aware of this and actually hesitated to bring in the gravitational properties in the context of *Physica* IV; but he somehow brought them in nevertheless, though reluctantly. Solmsen also observes that the above characterization of *topos* is of no use to Aristotle anywhere else. We should point out perhaps that Aristotle does occasionally mention this characterization again, as for instance in *De Caelo* 326^{b}8, but he does indeed make no use of it there in actual fact.

Another confusing feature is a seemingly serious inconsistency between the characterization of *topos* given above and an alternate one which follows in Aristotle not many lines afterwards. We will deal with this inconsistency on its merits in a later section, but for the present we only wish to state that in contemporary physics the notion of space is beset with inconsistencies, and that no serious harm arises. Thus, (i) in the *ordinary* physics of our planetary system, space continues to be Newtonian; (ii) in the theory of single electrons or similar particles (that is, in the so-called quantum field theory), it is the space of special relativity; (iii) in the physics of our galaxy and beyond it becomes the space of general relativity; and (iv) the "statistical" space of quantum mechanics may be viewed as being different from, and thus inconsistent with any "non-statistical" space, Newtonian or relativistic.

[8] F. Solmsen, *Aristotle's System of the Physical World* (Cornell, 1960), p. 128, also p. 443.

Finally we are bound to take note of the fact that Aristotle also mentions *topos* and *chronos*, in his *Categories*, in Chapter 6 which is devoted to the category of "quantities." Whatever he says there about space is in the nature of logico-verbal statements and it suggests itself to view it as a "dictionary definition," perhaps. It pursues no particular aim, and one must not view it as intending to be a full-fledged alternate definition of space in rivalry with other definitions.[9] It would also be quite dangerous to attempt to date the *Categories* and *Physica* relative to each other on the strength of this discrepancy in the description of *topos* alone. Aristotle might have offered both descriptions "during the same semester" to different audiences, or in different contexts.

7. Space and Matter

In 209b11-13 Aristotle states flatly that Plato identifies his (cosmological) *chora* with *hyle* (matter), but in point of literal fact, Plato never does so. Many commentators, however, especially those in the nineteenth century and notably E. Zeller, agree with Aristotle that in a certain qualified sense in the *Timaeus* matter is space and space is matter. But others disagree. Taylor[10] for instance says to *Timaeus* 52b4—and Ross[11] is in complete accord with him—that matter really plays no part in Plato's cosmology at all, and that the "permanent implied in change" which to Aristotle is *hyle* is not thought of by Plato as "stuff" or "substrate" or "substrate of change." And Taylor concludes by saying that Aristotle "was probably unconscious that he was falsifying the theory of the *Timaeus* by forcing his own technical terminology into it."

To this we wish to observe that mathematicians, physicists, and scientists in general, when functioning *qua* scientists (and not *qua* "philosophers" perhaps) are incessantly perpetrating such "falsifications" of the most atrocious kind. Ordinarily they do not "quote" each other with a care which would be anywhere near to the exactitude demanded by philosophers *qua* philosophers, not even in contexts which have a polemical setting as does the one under consideration. Reformulations of someone else's results and interpretations

[9] For such views, see P. Duhem, *Le Système du Monde* (Paris, 1913), I, 197-200; also Max Jammer, *Concepts of Space* (Harvard, 1954), p. 15.

[10] A. E. Taylor, *A Commentary on Plato's Timaeus*, (London, 1928), p. 347.

[11] W. D. Ross, *Aristotle's Physics*, (Oxford, 1936), p. 566.

in one's own technical terminology, even without much warning or any warning at all, are almost taken for granted, and at any rate would not be branded as "falsifications," even when amended or corrected intentionally.

But even apart from the general attitude of laxity towards "falsifications" of this kind, there would be, to a physicist, a question in the present case whether there is any "actual" falsification at all. It is a fact that in present-day physics the word *matter* is used to designate both the familiar "old fashioned" matter of a very "tangible" kind as well as over-sophisticated assemblages of so-called elementary particles which are the very opposite to being "tangible" in this sense. Now, these latter particles are only purely mathematical constructions in a sense. They are much more so than are Plato's "elements" in *Timaeus*, 53C-55C, and all properties of such elementary particles are totally inseparable from rather complex mathematical theories which are their setting and without which they would be beyond any possibility of description. And yet the physicist calls them "matter," not only from habit or indifference to appelation, but because, in their own theory, such "intangible" particles are indeed the ultimate elements out of which "ordinary" matter is composed. And since Plato fashions his elements out of figures which are structurally part of his *chora* a young physicist of today would probably see nothing wrong or noteworthy in the finding that *chora* and *hyle* are "the same."

I would also like to make an observation which is only a side remark to the present immediate context. It is my view, which others may contest, that Plato's mathematical imagery in *Timaeus* 53C-55C is naive, and relative to us most "antiquarian," and that Aristotle's criticism of it in *De Caelo III*, far from being the "curse to science" that A. E. Taylor denounces it to be,[12] is to a large extent well taken.

It is true that Plato's general views on the nature of mathematics were on the whole perhaps more fit to subsume modern mathematics than were Aristotle's views. But these were only conscious overall views of a philosophical tenor, and the differences between the views of Plato and Aristotle did not affect the actual "operational" fitness for bringing about a mathematical physics, say.

[12] *Op. cit.*, p. 403.

8. Gravitational Space

It is a fact that Aristotle did not arrive at the law of inertia of Galileo and Descartes or at any laws of bodies falling under gravitation, even though physics of locomotion, both in the sublunary part of the universe and in astral reaches, was very much in his thinking, and a general notion of motion which was much more comprehensive than locomotion proper was a central datum of his "chemistry," biology, and also psychology. Galileo was all his life castigating Aristotle for his inadequacies in (*Galilean*) mechanics and, following his lead, commentators and interpreters are wont to take a most serious view of these defects and put great emphasis on them.

But I think that it distorts history to burden Aristotle so pitilessly with a responsibility for some wrong guesses in physics which even the very great Archimedes could not right, and to allow this one aspect of failure of his to weigh so heavily against whatever else he might have achieved positively.

It is a most remarkable and extraordinary fact that the development of science after 1600 began with the establishment of the so-called rational mechanics which held the stage of science in the seventeenth and eighteenth centuries, and in the nineteenth became the paradigm for all of physics and much of other sciences which began to properly organize themselves only then. And it is an equally remarkable fact that Greek science simply did not show a tendency towards a pattern of developments of this kind, but towards a pattern of developments of another kind in which no theory of mechanics was in the vanguard of advance. Aristotle, for instance, was a naturalist and as such he had the relative standing of a naturalist of the nineteenth century, say. And in the nineteenth century, naturalists and also chemists were greatly attracted to thermodynamics, more than to any other field of physics or mechanics proper. Also, in the nineteenth century, a then prevalent version of thermodynamics, the so-called "classical" thermodynamics, was virtually independent of gravitational mechanics altogether. All this reacted on biology, and, for instance, the great treatise on nineteenth-century biology by D'Arcy Wentworth Thompson, *Growth and Form* (1917 and 1942), mentions the law of inertia but sparingly. As stated in a recent paper,[13] the thought patterns of Aristotle in physics are more remi-

[13] S. Bochner, "The Significance of Some Basic Mathematical Conceptions for Physics," *Isis*, 54 (1963), 179-205.

niscent of syllogisms in classical thermodynamics than those of mechanics. Most of the criticism of Aristotle by Galileo, Descartes, and other "mechanists" of the seventeenth century would not stand up before the forum of classical thermodynamics, if it rendered judgment solely within its own jurisdiction.

Finally, we wish to say that Aristotle's explanation of motion by *antiperistasis* (215^a15; 267^a16, 18), for which even fierce partisans of Aristotle feel apologetic, is in classical thermodynamics quite meaningful in a sense. In fact, in the latter theory "a sliding wall or piston will move only if, and as long as, the pressures on the two sides of the wall or the piston *in its immediate vicinity* do not cancel out."[14]

We also wish to state that even "atomists" like Epicurus or even Democritus did not actually show any tendency for initiating a mechanics of the seventeenth or eighteenth century variety, say.

9. The Void (*kenon*)

Aristotle himself motivates his denial of the existence of a void largely from an opposition to Atomism, but it is also a fact that in classical thermodynamics a "void" is envisaged but rarely, if ever at all. In such a thermodynamics a physical system is made up of gases and liquids which are presumed to be contained in an overall enclosure, to which there is no "outside," as far as the system is concerned. The notion of such an overall enclosure does not allow for the presence, within it, of what one can call a "void." All that one could encounter would be a "vacuum" in some part of the system. And such a vacuum has to be created forcibly and artificially and is never "absolute." In the nineteenth century, all this was not deemed to be in disagreement with the then basic fact that all matter in chemistry and thermodynamics was assumed to have a molecular structure. In spite of the molecular structure, for the derivation and application of large parts of the theory, gases and liquids were somehow assumed to be distributed continuously, with no "interstices" present.

[14] S. Bochner, "Aristotle's Notion of Place (Topos) in Physics," *Acts of the Xth International Congress of the History of Science* (to appear).

10. Processes-in-nature

In present-day physics there are certain overall notions like "physical system," "physical state," "physical process," "physical model," etc., which severally and jointly are an important background for the mathematization of physics. The Greeks never arrived at a stage of having notions that would adequately correspond to these. The nearest to this nexus of notions in the thinking of Aristotle was a certain notion of process-in-nature, but, as the notion appeared, it was too loosely broad and equivocal for becoming itself a background for a mathematization of physics, even if a tendency for such a mathematization had been present. In a rather forced interpretation one can say that the nearest to an effort for a would-be mathematization of the notion of a process-in-nature was made in Book V of the *Physica*, especially in Chapter 2, but it should be immediately said that this effort was a very abortive one.

Aristotle has various names for a process-in-nature, all of which already occur in Plato and earlier. The leading names are: *kinesis* (motion), *metabole* (change), *genesis*; and others are "alteration," "locomotion," etc. Commentators have been trying to put these various names into a hierarchical ordering, and various arrangements have been suggested, some of which embody alternate strands of ordering in one.[15] But none of the proposed arrangements is really convincing, and I venture to say that none is really possible. The fact is that the above conglomeration of names for the general notion of *kinesis* is fluid by origin, and such fluid conglomerations of notions in physics do not ordinarily solidify into sharply and meaningfully separated concepts, unless perhaps specific and definite meanings have been assigned to them by deliberate conventions. Another conglomeration of notions, which also causes great difficulties of interpretation, are words which are introduced in *Physica* V, 3, in order to express properties of linear ordering of a one-dimensional continuum (together, apart, touching, intermediate, successive, contiguous, continuous).[16] In this case the basic difficulty is this, that the Greeks at that time were not ready, or not yet ready, to think through with great precision something which for modern mathe-

[15] Compare J. P. Anton, *Aristotle's Theory of Contrariety* (New York, 1957), pp. 206-7; H. H. Joachim, Aristotle, *The Nichomachean Ethics* (Oxford, 1950), pp. 269-275.

[16] Ross, *Aristotle's Physics*, p. 626.

matics, and in its setting, only the nineteenth century was able finally to do ; although in this case the Greek achievements which are embodied in the fifth book of Euclid's Elements do have a timeless validity, virtually.

With regard to the notion of *genesis* I wish to say that Solmsen over-dramatizes its course of development.[17] He asserts that it flourished under earlier Presocratics, was suddenly put under death sentence by Parmenides, was however resuscitated by Plato, and led a normal life under Aristotle. In physics, things somehow do not happen in this manner.

11. The Scope of Processes-in-nature

The notions of *kinesis* and *genesis* are not yet in Plato what they will be later in Aristotle. In Plato they are sometimes looser and broader, more indefinite and undifferentiated; or at other times, they are representative of a partial aspect or of a smaller phase of what they will be in Aristotle; or they are still enmeshed in Theaetetean logicality or Parmenidean existentiality, and are not yet fully textured in physicality as later in Aristotle. In Aristotle the allusions inherent in the notions are technically more "scientific" in our sense. If Greek "theoretical" science would have continued itself directly into our science without first aborting in its own phase, the continuation would have become Aristotelian and not Platonic, and one would hear fewer criticisms that Aristotle's enlargements of the Platonic notions of *kinesis* and *genesis*, say, were too small to be of great consequence or even that they had a trivializing effect due to Aristotle's tendency towards putting notions into tighter frameworks, such tighter frameworks around Platonic notions having the effect of narrowing the width of their far-sweeping visions, as Platonists sometimes fear. It should be observed, however, that such tightenings of frameworks around notions in physics and mathematics are frequently of great importance and seemingly small changes of nuance can be of considerable consequence.

In large parts of the *Physica* and in other Aristotelian contexts, *kinesis*, or one of the parallel notions, if representative of a general event, may represent what would be nowadays a physical, cosmological, chemical, or biological process; it may also represent one of many human activities, professional ones or, generally, intentional

[17] *Aristotle's System*, Ch. 4.

ones, as for instance the planning of the building of a house and the actual realization of the plans, even beginning with the inception of the intent and going through the necessary and natural stages of it up to and including the very completion of the task; or it may be the total process of painting a picture or carving a statue; or the process of acquiring a specific skill or a specific area of knowledge, or of educating oneself in general; or of carrying out a specific program of physical exercising for the sake of maintaining or improving one's health; or it may be the process of healing somebody or even oneself; etc.

However, one must not assume, after going through this catalogue of possibilities, that every conceivable "event" is a *kinesis* of some kind. In fact there are some peculiar limitations, on which Aristotle insists and about which he is fairly consistent. They are stated frequently in passages of his psychological works (*De anima*, *De sensu*, etc.), but the leading exposition of them is incorporated in his theory of pleasure in the *Nicomachean Ethics*, X, 1-5. An important and indispensable commentary on Aristotle's injection of the theme of *kinesis* into this context is in Joachim,[18] and further studies on this topic would be welcome. In the *Ethics*, what Aristotle wishes to exclude from being a "*kinesis*" is "pleasure." It apparently is not a "process," that is, something that leads up to and terminates in a "completion," but it is a "completion" itself, directly and immediately not from some particular contraction or atrophy of what precedes a completion but because there just is nothing at all that precedes. In a short sentence, one can say that to Aristotle pleasure is not a *kinesis* but an *energeia*, but there is something unsatisfactory about making the sentence quite so short. The important thing is that to Aristotle, the nature of pleasure is parallel to and even linked with the nature of sense perceptions, like seeing, hearing, etc., which are also, as he frequently and consistently maintains, not processes proper but only completions. In support of this parallelism Aristotle also adduces the fact that neither pleasure nor sense perceptions require a specific interval of time for their occurring, as any regular *kinesis* does. To this argumentation it ought to be observed that in the *Physica* the dissertation on time is given in Book IV and that in various analyses of processes-in-nature in the preceding three books time is never mentioned and hardly alluded to, although in later books of the *Physica* and in follow-up treatises time does occur.

[18] See note 15.

For a moment one might be tempted to say, when guided by our own standards of "morality," that Aristotle wished to keep "kinetistic determinism" out of the sphere of ethical responsibility, but it probably would not be easy to bring this into accord with Aristotle's attitudes towards ethics in general or with pre-Hellenistic attitudes in general. It is probably less of an improper modernism to say that in linking pleasure to sense perception Aristotle is aware of a certain "neurological" sameness in both, with which one would certainly agree today. What would probably be harder to agree to is Aristotle's view that the "neurological" events which occur in both are not *kineseis* proper but something else, that is, straightforward "completions." But one might agree with Aristotle to the extent that the human consequences and concomitants of neurological processes are something other than physical processes themselves, and not entirely reducible to such.

12. Opposites

An insistent feature of Aristotle's physics, which it is hardest to assimilate to present-day physics, at first sight at any rate, is the ever recurrent emphasis on, and reiteration of the general theme of opposites, contraries, and contrarieties. Aristotle asserts and implies that the theme, as an active theme, was common to Greek rational physics from its earliest. In this he is ostensibly right, but he is overstating the role of opposites in earlier stages of development, as H. Cherniss has been arguing convincingly.[19]

Some of the opposites like "hot and cold," "dry and moist," "rarified and condensed" are to us common and ordinary, even too ordinary and commonplace and pre-scientific to be the central theme of a scientific development. Others, like act and potency, which are apparently due to Aristotle, do indeed have an originality of their own, but one wonders what one might be able to actively achieve with them in the end. Finally, a pair of opposites like privation and completion seem to be quite remote from us, and are even repugnant to our thinking.

It is true that opposites and contraries are significant and even recondite attributes of the universe around us. But Aristotle, and perhaps the Greeks altogether, stuck to the theme of opposites too

[19] H. Cherniss, *Aristotle's Criticism of Presocratic Philosophy* (Baltimore, 1935).

doggedly and too unprofitably, or so it seems. In a retrospective evaluation, if such an evaluation be deemed admissible, one might be tempted to say that the overintense preoccupation of Greek physics with the theme of opposites was the greatest single impediment to a mathematization of their physics, that is, to a development and unfolding of their science as it "should" have come about instead of as it actually did come about—abortively, that is.

13. Motion of Motion (*kineseos kinesis*)

Nothing is seemingly more innocuous and ordinary than the opposites "hot and cold," "dry and moist," "rarified and condensed." And yet in the seventeenth and eighteenth centuries, while mathematicians and physicists were erecting their monumental edifice of "Rational Mechanics" with briskness and boldness, they were proceeding very slowly and with great circumspection in the area of thermodynamics, which is the area of physics to which these three pairs of opposites belong. The first truly "theoretical" results pertaining to these opposites were beginning to be envisioned only in the last decades of the eighteenth century, and only in the course of the nineteenth century were they actually codified.

Hot and cold are of course no genuine contrarieties, inasmuch as they can only be manifested by different positions on a thermometric scale, with various intermediary positions between the two. As viewed in retrospect, there is in Aristotle a noteworthy attempt, an abortive one, at arriving at a would-be mathematical clarification of the role and meaning of positions intermediary between opposite or contrary ones. The attempt may be read into the argumentation in Chapter 2 of Book V of the *Physica*. Somehow, to me, Aristotle's doings in this chapter are crying out for a mathematical representation of a *kinesis*, as there investigated, by a function $y = f(x)$, $a \leq x \leq b$, where the endpoints $x = a$ and $x = b$ correspond to the starting stage and completed stage respectively, and the values x between a and b to the stages of the *kinesis* in between. If one views his *kineseis* in Chapter 2 of Book V in this way, then all of Aristotle's statements about them are wrong, totally and disastrously so. In fact all his statements are negative statements, that is, warnings against what cannot or must not be done with the *kineseis* under consideration. As strictures on functions f(x) they would prohibit the very operations with them without which there would not be any physics at all today, not even on the most elementary level, that is. Aristotle's main contention (225^{b}16-226^{a}23) is this, that there cannot be a motion of motion, and he adduces for it no less than five proofs.[20] If applied to functions this would mean either that (1) there cannot be a function of a function; or that (2) there cannot be functions of *two* independent variables; or that (3) an "operator" like differentiation cannot be applied twice, and hence arbitrarily often, in succession. And the last alternative would mean that while one can form the notion of a "velocity" of a motion, which requires the formation of a first derivative, one cannot iterate this to form the derivative of the velocity, which is "acceleration."

Now, it is precisely such an incapacity to conceptualize the rate of change of the rate of change which separates Archimedes from Newton by a threshold which Archimedes never succeeded in leaping across.[21] Purely mathematically, there is nothing in Newton's *Principia* that was not familiar to Archimedes, except the notion of the rate of change of a velocity. And even here, only the notion was alien to Archimedes, and not the power for formalizing the notion mathematically, if by some reversal of history it would have come within his purview. In fact, purely mathematically Archimedes was much better equipped for dealing with it formally than was Newton, seeing that Newton did not manage to learn to define really rigorously the notions of velocity and acceleration to the very end of his days.

Aristotle's statement (225^{b}15-16): *οὐκ ἔστι κινήσεως κίνησις οὐδὲ γενήσεως γένησις, οὐδ' ὅλως μεταβολῆς μεταβολή* (there cannot be motion of motion or becoming of becoming or in general change of change) is a devastating self-indictment of Greek rational thinking at its root. It decrees that, in general, Greek thinking would not, because it could not, form a relation of relation, or a property of properties, or an aggregate of aggregates. I venture to say that the Greeks never quite reached the stage of forming an abstraction which requires more than one major step of abstracting; meaning that they rarely formed an abstraction from an abstraction, that is an abstraction vertically superimposed on a preceding one.

Archimedes did not have coordinate systems or mathematical functions. Nevertheless, in his book *On Spirals* he came quite close, in his own setting, to forming the derivative of a function

[20] See the translation by H. Carteron in the Collection Budé, *Physica*, II, 15-17.

[21] J. P. Anton, *loc. cit.*, Appendix 1 to Chapter 4, already stated that Aristotle's denial of change of change was in a sense reversed in the mathematical calculus of the seventeenth century, especially in Newton's *Principia.*

$$\frac{dy}{dx} = \frac{df}{dx} \qquad (1)$$

which is the mathematical pre-requirement for the "abstract" conceptualization of the notion of velocity. However, in order to advance to the concept of acceleration, one has to be able to form a second derivative. This requires that one form the derivative (1) at each and every point x, and then view the resulting mathematical object as a new function in x, and then apply the "abstract" process of differentiation to this new function again. It is this kind of iteration of logico-ontological abstractions to which Greek thinking was never able to penetrate to any noticeable extent.

Or, take the general (mathematical, or logical) notion of a "set" or "aggregate" as Georg Cantor introduced it (a "set" being a multitude of "elements" about which nothing is assumed known except that any two elements can be distinguished from each other by the separateness of their identities). Such a notion of "set" does not occur anywhere in Aristotle, and it was not congenial to his thinking, but it may nevertheless be conceded that this notion would have been within the easy grasp of the Aristotle of the *Prior Analytics*, say. However, in order to "operate" with his sets, Cantor *very quickly* forms, or constructs, or envisages, sets of sets, that is, sets whose elements are themselves sets; and most of his operations with sets apply to such. Now, it is this kind of "iteration" of the notion of sets of which only the barest rudiments occur in the *Prior Analytics*, or anywhere else in Aristotle or his successors, if they occur at all.

The most striking application of sets of sets is involved in Cantor's proposition that with any cardinal number whatsoever one can associate a second cardinal number which is genuinely larger than the first. Cantor takes an arbitrary set A, and he conceives of the set B whose elements are the subsets of A; and, by a reasoning whose originalty will never dim, Cantor then argues that the cardinal number of B is genuinely larger than the cardinal number of A. It is this kind of concept formation and reasoning which lay outside the known ontology of the Greeks. In this sense, the universes of Parmenides, Plato, Aristotle, and Archimedes were tightly closed, with "no exit."

Plato's dichotomy, although it embodied a sequel of steps, did not embody a sequel of abstractions. It was vertical; but it moved vertically downwards and not upwards, and it was a succession of logical particularizations and not of abstractive self-generalizations. Plato did have an abstractive process which moved upwards, namely the process of abstracting a (Platonic) Form or Idea from its underlying realizations. However, a Platonic Idea, when viewed as an abstraction, was a terminal one, and no further abstraction could be superimposed on it; or, to put it differently, once the Platonic Ideas were arrived at, there was nothing that Plato, or anyone else, could actually "do" with them. In *Metaphysica*, Book XIII, one of Aristotle's arguments against Ideas was the assertion that Plato had not provided how, out of the *Idea* of three-ness and the *Idea* of four-ness, to form (by an "abstract" mathematical operation with Ideas) the *Idea* of seven-ness. In this argumentation Aristotle was irrefutably right; but we have to add that Aristotle's own advice of not forming Ideas at all was a council of despair, and made matters very much worse.

We also wish to point out that the Greeks were never driven to conceptualizing, or rather to formalizing, the notion of "relation." In the twentieth century, this notion is a most primary datum, and no fruitful thought can be conceived without it. The Greek antecedent of our "relation" is their *pros ti*. The "Dictionary" in the *Metaphysica* has a long entry about it (1020^b25-21^b12), but the entry is not inspiring reading. The Greeks did not make their *pros ti* into a substantive "category" for "operational" rationalization, as it is with us indispensably today; nor even did they make it into some serious linkage within "logification" of knowledge, as some medieval Schoolmen at least attempted to do, although not successfully so.

14. Simple and Composite Changes

As we have already stated above, the *kineseis* of Book V of the *Physica* are presumed to be such as traverse various states which are intermediate between terminal ones. In contrast to this, Book I envisages various changes (*geneseis*) in which states intermediate between the terminal ones do not occur at all. Such changes are indeed very "simple" and in the Loeb edition the "argument" preceding Chapter 7 interprets the text to imply that "change, however complex, can always be resolved into a combination of antithetical movements."[22] The text of the *Physica* does not quite say so; but the interpretation is a plausible one, and we will adopt it for the moment for the sake of an observation to be based on it. In present-day physics the decomposition of a physical process into "simple"

[22] I, 69.

processes each of which is an indecomposable one-step event (any so-called emission or absorption, or alternately, creation or annihilation of elementary particles) is in the very center of the entire physical interpretative thinking.

Altogether I would say, though with some reservations, that two-valued truth functions of logic and the succession of yes-or-no signals which constitute a context of information in a contemporary computer would have been more in accord with Aristotle's outlook on physics in Book I of the *Physica* than would be the mathematical apparatus of point functions in Euler-Lagrange mechanics, say, although the latter would have fitted into the outlook of Book V, in a sense.

15. Aristotle's Theory of *Topos*

We have already stated in section 6 that Aristotle's *topos* in *Physica* IV is meant to be a "spatial" setting for his pre-notion, such as he had, of an individual physical system and that for an understanding of his purposes and achievements his half-hearted attempts, against his better judgment, at also injecting "gravitational" properties of natural place into the theory of Book IV ought to be ignored.

After a lengthy deduction from would-be axiomatic demands, Aristotle arrives at the characterization (212^a5-6) already stated above that if a physical system is enclosed by another, the *topos* of the system is "the innermost boundary of the enclosure." The context of the *Physica* also shows[23] that this concept of *topos* also implies a certain aspect of "voluminousness," but not a point-by-point "extension" which links boundary elements across the bulk of the system. This characterization of space obviously does not fit a mechanical system of finitely many mass points, because in such a system only point-by-point localization can be realized. However, it begins to fit a system of hydrodynamics or electrodynamics, say, because frequently, for such a system, its equilibrium or alterations of equilibrium are determinable from events located on the "innermost" part of the boundary layer. And it fits best a system of gaseous substances in classical thermodynamics, say, because the so-called characteristic equations of the theory involve spatiality only through its volume but through no other structural determination of it; in particular it makes no reference to points or figures

[23] Taylor, *loc. cit.*, p. 676.

in space, nor even to the three-dimensionality of it. We also note that the organization of a living cell in biology is largely determined by events across innermost parts of "boundaries" which in cell biology are called "membranes."[24]

After stating the above characterization of *topos*, Aristotle continues to analyze the phenomena which led to it and he seemingly involves himself in inconsistencies. He argues as follows. If a boat moves on a river which is itself in motion, then, by the above characterization, the *topos* of the boat ought to be the innermost boundary of the water which surrounds the boat. But since this water is changeable, Aristotle somehow does not like to permit it to define a *topos* of the boat; and Aristotle suggests that such a boat does not really have a *topos* in its own right, and that only the river as a whole has a *topos*. Altogether he then proposes the alternate characterization (212^a20-21) that "place is the innermost *motionless* boundary of what contains."

Some commentators, ancient and modern, find the two alternate characterizations of *topos* irreconcilably contradictory.[25] Some recent ones, however, find that the second characterization only superimposes a "Principle of Relativity" on the first, and that in the end there is no inconsistency at all.[26] In my own view the inconsistency can be better resolved, although not absolutely so, by another interpretation which I think is more "informal" and more suitable to Aristotle's approach than the Principle of Relativity is. Instead of exemplifying our explanation on Aristotle's boat we will take another case, which is equally "ordinary."

Suppose that in a workshop we observe an iron pipe as to its common properties. To begin with, its "place" would be the ordinary "extension" it displaces. If, however, the pipe happens to be rusted and we want to account for this too, still in an everyday manner,

[24] See, for instance, L. L. Langley, *Cell Functions* (New York, 1961), Chapter 3 and other passages.

[25] See Duhem, *loc. cit.* We note that Duhem was altogether very critical of any kind of inconsistency in physics. Thus his book: *Les Théories électriques de J. C. Maxwell;* Étude historique et critique (Paris, 1902), was written mainly with a view to exposing various inconsistencies in the sum total of the electromagnetic work of Maxwell.

[26] The most detailed interpretation of this kind was given by B. G. Kusnezow, "Die Lehre des Aristoteles von der relativen und absoluten Bewegung im Lichte der modernen Physik," published in the book *Sovietische Beiträge zur Geschichte der Naturwissenschaften*, Herausgegeben by Gerhard Harig, (Berlin: Deutscher Verlag der Wissenschaften, 1960), pp. 27-61.

then we have to consider that rusting is combination with oxygen; and assuming that the oxygen in the workshop circulates freely, then the oxygen of the entire workshop is responsible for it. Thus, even if the state of the iron pipe alone is at issue, for the examination of the rust the place of the object must be enlarged to extend over the whole shop, in a sense. Furthermore, if, say, some magnetization of the pipe has been observed and it cannot be explained by events and installations in the shop itself, then, for the sake of an explanation, the "true" spatial setting of our object must be even further enlarged.

Returning to Aristotle's boat, we might elaborate on his own argumentation as follows. If we stay in that boat below deck and are only concerned with events inside it, then the genuine place of the boat is determined by the interior boundary of the water flowing past it. If we become concerned with the effect of the water on the condition of the hull, then the entire river becomes a spatial setting, and if we even wish to consider the effect of precipitations and winds on the level and behavior of the water, which in its turn affects the hull of the boat, then the atmosphere above the river must also be added.

Actually, Aristotle himself says (209^a32-35, Oxford translation): "... you are now in the heavens because you are in the air and it is in the heavens; and you are in the air because you are on earth and similarly on the earth because you are in this place which contains no more than you."

The above interpretation is the elaboration of a previous one,[14] but, in another context, we have also given an alternate elaboration. It is of a more mathematical nature[13] and somewhat different from the present one, but it is compatible with it.

16. The Universe as One Physical System

Of the five chapters on the theme of *topos* in Book IV of the *Physica* the last one, that is Chapter 5, is inconclusive but intriguing. As we have repeatedly stated, we are adopting the interpretation that *topos* in the *Physica* is meant to be (non-gravitational) place pertaining to an individual physical system. In the *De Caelo*, however, *topos* is, quite differently, the gravitational space of the *Kosmos* as introduced in the *Timaeus*, more or less. Now in Chapter 5 Aristotle seems to be raising the following question. What happens if one views the whole universe, *τὸ πᾶν*, as one vast though "individual" physical system, that is, if one views the "whole" not as the *Kosmos* of the *Timaeus* but as an individual physical system which happens to be coextensive with the entire universe? This, incidentally, is a very legitimate "scientific" question. Aristotle puts it somewhat vaguely, but discernibly, and it does not belong to Parmenidean dialectic, let alone to some kind of "sophistry," but to physics proper. Aristotle asks whether this comprehensive physical system also has a *topos* of its own, or not.

Aristotle has no solution to this problem, which is indeed a vexing one, but he ponders it and throws out fragments of thought about it. He thinks that the situation is ambivalent, and that in some respects *to pan* does have a *topos* and in others not. Perhaps one ought to distinguish between being in place "actually" or only "potentially," he suggests. This is a routine suggestion of his, and he observes a propos of this that one might say that in a homoeomeric substance the constituent elements have mutual places potentially, whereas in a granular substance they have them actually. Or, he thinks, even if the whole is viewed as one physical system, it might still be pertinent to subdivide it appropriately into parts and then assign a *topos* to each part, although this cannot be done for the whole itself. Or, one might distinguish between something having *topos per se* or only *per accidens* and then decide that the whole has it only *per accidens*.

After offering such suggestions for a stretch (212^a31-212^b23), he suddenly runs out of possibilities, and in a closing statement (212^b23-30), which is also a summary, he avers that he has achieved all that he had set out to do.

In Book I of *De Caelo*, which continues the *Physica*, Aristotle expounds his doctrine that the world (*οὔϱανος*) is finite. He is having difficulties with distinguishing between finiteness of matter and finiteness of the geometric substratum, and with determining what finiteness of either actually means. However, this is the kind of difficulty with which contemporary cosmology, on its own level of sophistication, has also to contend, fully so. Also, a scientific cosmology does not necessarily convince by logical neatness. There is too much at stake to be persuaded by a smoothly flowing phraseology, verbal, logical, or even physical.

For reasons which present-day physics might not find it easy to endorse, during and after the Renaissance much criticism was heaped on Aristotle on account of the doctrine. But it is good to remember that Johannes Kepler, the pacemaker of mathematico-physical astronomy of today, was stoutly championing the cause of Aristotle;

that Isaac Newton never uttered an unkind word about Aristotle, whatever Galileo, Descartes, and even Leibniz[27] may have said; and that in the twentieth century the cosmological work of Albert Einstein has made "Finiteness of the Universe" a very fashionable doctrine again.

17. Act and Potency

By a somewhat forced analogy even Aristotle's pronouncement in *Physica* III, 1 that *"change is actualization of the potential"* or, more literally that *"the fulfillment of what is potentially, insofar as it exists potentially, is motion"* can be made to correspond vaguely to something important in modern physics, although in the seventeenth century Galileo, Descartes, and apparently also Leibniz[28] were bluntly asking of what possible use this pronouncement could be to anybody, ever.

To myself the pronouncement evokes the imagery of the fact that in a "natural" and "free" mechanical process potential energy is transformed into kinetic energy, and that the sum of the two is conserved. And if one permits this analogy, then one can also forgive Aristotle for having reiterated his pronouncement with a somewhat wearying insistence. In the nineteenth century the theorem on the transformation and conservation of energy was comprehended by scientists but very gradually, and only after the most wearying repetitions of its meaning and purpose by the sponsors of the theorem; whereas in the twentieth century general theorems of this kind are taught to a student in physics briskly and in short order, and then quickly applied to situations of gravest import.

It should also be stated that although Aristotle's denial of the possibility of an *actual* infinite is open to argumentation, his drawing up of a distinction, with emphasis, between potential and actual infinitudes is one of the most durable specific achievements of logico-mathematical thinking ever.

18. Conclusion

If one compares the physics of the Greeks with our theoretical physics of the last few centuries, and if one does so pragmatically

[27] A. Koyré, *From the Closed World to the Infinite Universe* (New York: Harper Torchbook, 1958), p. 58 ff.

[28] Léon Robin, *Aristote* (Paris, 1944), p. 300.

and without condescending sentimentality, one may come to the conclusion that Greek physics hardly made a start. Except for the two laws of Archimedes, namely, the law of the lever and the law of the floating body, which Archimedes and others viewed as mathematics proper, Greek physics did not state a single law which could be put into a present-day textbook in the manner originally stated. Greek physics did in no way bequeath a book which would compare to works by Euclid, Archimedes, or Apollonius in mathematics. And yet, Greek physics produced the system of Aristotle, which, although Aristotle was a naturalist rather than a physicist, held the stage of physics for almost 2000 years, and which, by its flashes of insight and uncanny anticipations, is evoking fascination even today. And scoffers of Aristotle ought to tell us, or tell themselves, where we would be today if none of his works had survived.

Aristotle's Notion of Place (topos) in Physics *

Salomon Bochner
Princeton University, New Jersey, U.S.A.

Since the 17th century a major part of physics and almost all of natural philosophy have been dominated by cognitive attitudes and thought patterns of mechanics in the style of Newton, Lagrange and Laplace. Concurrently, beginning with Galileo and Descartes, the role of Aristotle in physics and natural philosophy has been evaluated and interpreted predominantly against the background of these attitudes and thought patterns, not only by those who frankly compare the physics of Aristotle with modern physics, but unwittingly also by those who want to interpret Aristotle against the background of Greek philosophy in the main.

It is true that in the 19th century the glamorous theory of Maxwell, which fused electromagnetism with optics, was closely and deeply modelled on paradigms from space-time mechanics, and that in the 20th century the theories on relativity have been continuing the developments from Newton to Maxwell closely. It is also a fact that in the 19th century the very important theory of "classical" (that is, non-statistical) thermodynamics was not dependent on the then regnant mechanics at all. The 20th century quantum mechanics, which is the mechanics of the future, does not subordinate itself to the conception of space which the developments from Galileo to Einstein have been trying to impose universally; also, in the 19th century the large doctrines of chemistry and biology were leaning much more towards classical thermodynamics than towards any other area of physics. And since Aristotle was strong in biology and had an affinity for chemistry, it is neither equitable to Aristotle nor enlightening to philosophy to over-emphasize his failings in Galilean mechanics to which his thinking was not attuned.

We note that the textbook of Max Planck, *Treatise on Thermodynamics* (many editions) never introduces a space coordinate, and never alludes to the law of inertia of Galileo and Newton which has been aweing philosophy for centuries past. Introductory text-books on chemistry rarely mention the Law of Inertia, and the great treatise on

* This research was supported by the Air Force, Office of Scientific Research, under Contract AF 49 (638)-578, Princeton University.

19th-century biology by D'Arcy Wentworth Thompson. *Growth and Form* (1917 and 1942) mentions it but sparingly.

In classical thermodynamics a sliding wall or piston will move only if, and as long as, the pressures on the two sides of the wall or the piston *in its immediate vicinity* do not cancel out. Aristotle was apparently aware of this, and he apparently took it for granted that the "inertial" motion of a body comes about in the same manner. In this he was wrong, but his error was neither naive nor unforgivable.

Plato and Aristotle have two appellations for spatiality, *chora* (space) and *topos* (place). We think that to Aristotle *chora* is usually the everyday word to designate the room occupied by an everyday object; *topos*, however, especially in Book 4 of the *Physica* is the scientific word which designates a certain physical attribute of a substance, or of a collection of bodies which constitute a physical system. We think that in the *Timaeus* of Plato (51E-52D) the difference between the two appellations is basically the same, except that the *Timaeus* does not envisage the space pertaining to an individual physical set-up in isolation but only to the cosmological space of the universe as a whole. Also, the *Timaeus* introduces space in its cosmological origination, when yet unformed and not yet made physically specific, and we think that this is the main reason why in the *Timaeus* the appelation *chora* prevails. In this connection we venture to suggest that Aristotle may have been the first to separate physics from cosmology entirely; at any rate, the natural ordering of his books is such that the *Physica* comes first and the *De Caelo* afterwards.

The term *topos* already occurs formally in Aristotle's *Categories*, 5 *a*, 8-14, but only as to its ordinary logical meaning and not as a concept of physics. Accordingly, in the *Categories* Aristotle gives the dictionary definition, which must have been generally familiar, that the place of an object is its three-dimensional extension; the implication being that for ordinary logic and its applications this unsophisticated description of place ought to suffice. In the *Physica*, however, after a lengthy reasoning, Aristotle arrives at the characterization (212 *a*, 5-6) that if a physical system is enclosed by another the *topos* of the system is "the innermost boundary of the enclosure". The context of the *Physica* also shows [1] that this concept of *topos* also implies a certain aspect of "voluminousness", but not a point-by-point extension which links boundary elements across the bulk of the system. This characterization of space obviously does not fit a mechanical system of finitely many particles, because in such a system point-by-point localization is of the essence. However, it begins to fit a system of hydrodynamics or electrodynamics, say, because frequently, for such a system, its equilibrium or alterations of equilibrium are determinable from events located on the "innermost" part of the boundary layer. It fits best a system of gaseous substances in classical thermodynamics say, because the socalled characteristic equations of the theory involve spatiality only through volume but through no other determination of it. Thus the fundamental relation for a perfect gas.

$$pV = RT,$$

which articulates the laws of Boyle-Mariotte and of Gay-Lussac in one, involves the pressure p, the absolute temperature T, and a constant R, and the Volume V. But it makes no reference to points or figures in space, nor even to the three-dimensionality of it. Every thermometer is in a certain sense a one-dimensional device and it would be so even if our universe were seventeen-dimensional. We also note that the organization of a living cell in biology is largely determined by events across innermost parts of boundaries which in cell biology are called membranes. [2].

After stating the above characterization of *topos*, Aristotle continues to analyze the phenomena which led to it and he seemingly involves himself in inconsistencies. He argues as follows. If a boat moves on a river which is itself in motion, then by the above characterization the *topos* of the boat ought to be the innermost boundary of the water which surrounds the boat. But since this water is changeable, Aristotle somehow does not like to permit it to define a *topos* of the boat; and Aristotle suggests that such a boat does not really have a *topos* in its own right and that only the river as a whole has a *topos*. Altogether he then proposes the alternate characterization (212 *a*, 20-21) that that "place is the innermost *motionless* boundary of what contains".

Some commentators, ancient and modern, find the two alternate characterizations of *topos* irreconcilably contradictory [3]. Some recent ones, however, find that the second characterization only super-imposes a "Principle of Relativity" on the first, so that in the end there is no inconsistency at all [4]. Our own view falls between these two. We find that there are indeed inconsistencies among various statements on *topos* in Aristotle, but we think that this in itself does not invalidate his investigation as a whole. In fact the notion of space in contemporary physics is beset with inconsistencies, and no serious harm arises. Thus, (i) in the ordinary physics of our planetary system, space continues to be Newtonian, (ii) in the theory of single electrons or similar particles, it is the space of special relativity, (iii) in the physics of our galaxy and beyond, it becomes the space of general relativity, and (iv) the "statistical" space of quantum mechanics is inconsistent with any "non-statistical" space, Newtonian or relativistic.

In *Physica*, Book 4, Chapter 5, Aristotle tries to reason that in a certain qualified sense the universe in its entirety is "in place". In modern mathematical terms, he views the universe as a compact three-dimensional spherical disk; its open interior has then a *topos* which is determined by the two-dimensional outer spherical shell surrounding it. — Alternately he argues that the open interior is a union of closed simplexes; each of these simplexes is of course "in place" by his characterization, and so therefore is their union.

It is important to note that the characterization of *topos* by boundaries is not at all involved in *De Caelo* where the gravitational properties of cosmological space are studied in detail [5].

With regard to Aristotle's denial of "void" *(kenon)* in *Physica*, Book 4, Chapters 6-9, we wish to point out that, in a certain sense, classical thermodynamics also does not recognize the possibility of a void, at least not in permanency, unless perhaps, produced artifically and maintained forcibly. The entire reasoning of Aristotle in this context is in close accord with his overall outlooks on inorganic and organic science, and Aristotle's opposition to Democritean atomism is only a very secondary lateral aspect of it.

References and Notes

[1] See A. E. Taylor, *A commentary on Plato's Timaeus*, 1928, p. 676.
[2] See, for instance, L. L. Langley, *Cell Functions*, 1961, chap. III and other passages.
[3] W. D. Ross, *Aristotle's Physics*, 1936, p. 575-6: see also Pierre Duhem, *Le système du monde*, Vol. I, 1913, p. 197-200. We note that Duhem was altogether very critical of any kind of inconsistency in physics. Thus his book: *Les théories électriques de*

J. C. Maxwell, étude historique et critique, 1902, was written mainly with a view to exposing various inconsistencies in the sum total of the electromagnetic work of Maxwell.

[4] The most detailed interpretation of this kind was given by B. G. Kusnezow, *Die Lehredes Aristoteles von der relativen und absoluten Bewegung im Lichte der modernen Physik*, published in the book *Sovietische Beitrage zur Geschichte der Naturwissenschaften, herausgegeben* by Gerhard Harig, VEB, Deutscher Verlag der Wissenschaften, Berlin 1960, p. 27-61.

[5] This important observation ("the definition of place as given in Physics IV does not prove useful anywhere else") stems from Friedrich Solmsen, *Aristotle's system of the world*, Cornell University Press, 1960, p. 443).

Reprinted from JOURNAL OF THE HISTORY OF IDEAS, Vol. XXVI, No. 1, January–March, 1965, pages 3–24.
PRINTED IN U.S.A.

WHY MATHEMATICS GROWS

BY SALOMON BOCHNER

1. *What Is Mathematics?*

This question, if asked in earnest, has no answer. But if it be asked in a mood of lightness and of forbearance with equivocation the following answer might be given. "Mathematics is a form of poetry which transcends poetry in that it proclaims a truth; a form of reasoning which transcends reasoning in that it wants to bring about the truth it proclaims; a form of action, of ritual behavior, which does not find fulfilment in the act but must proclaim and elaborate a poetic form of truth."

This appealing sentence is taken from a book on "ancient man" in which however it is a pronouncement not on Mathematics but on Myth. That is, in the original, the sentence runs as follows. "Myth is a form of poetry which transcends poetry in that it proclaims a truth; a form of reasoning which transcends reasoning etc. etc."[1]

Plato had an inkling of this, both for Mathematics and for Myth, but no more than an inkling. He made many efforts to grasp the relation between poetry and truth, especially in book X of the *Republic,* but he was never successful. And, according to many statements in Aristotle's *Metaphysics*—which are wearying in their repetition—Plato never found a proper relation between mathematics and his (Platonic) Ideas, to the detriment of the theory of Ideas, no doubt. Aristotle himself was sometimes more successful, sometimes less. His *Poetics* is so masterful because it does not raise the question of the relation between poetry and truth, and only tries to decide what makes poetry effective. But Aristotle's view that mathematics can dispense with "Ideas" altogether is less than convincing.

In an over-all sense, the Greeks attempted to create two methodologies of "scientific" cognition. One was their Ontology and the other was their Mathematics; Aristotle's Logic was somewhere between the two, and, by Aristotle's own intent, at any rate, was meant to be ancillary only, in a general sense. Their Ontology was characterized by the Heraclitean nuance of "Logos" and the powerful Parmenidean "Ens" (*to on, to einai*). In Mathematics, which, as methodology, outstripped ontology, the name itself, for whatever reason, did not turn out to be as strong and affirmative as were, and are, Logos and Ens. But I think nevertheless that something of the originality of the Greek creation is reflected in the manner in which the

[1] See the book of Henri Frankfort and others, *The Intellectual Adventures of Ancient Man* (Chicago, 1946), 8; also in a Penguin edition, under the title: *Before Philosophy,* 16.

name occurs, and I venture to make some remarks about it.

The word "mathematics" is a Greek word, and, by origin, it means "something that has been learned or understood," or, perhaps, "acquired knowledge," or perhaps, "acquirable knowledge," that is "learnable knowledge." Fuller statements about the meaning of the word will be found in passages from late antiquity,[2] in the first leading history of mathematics, which is a work from the 18th century,[3] in the standard history of Greek mathematics,[4] and in work of 20th-century classicists.[5] The meanings of the word mathematics, as suggested above, seem to agree with the meanings of cognate roots in Sanscrit, and even the *Dictionnaire de la langue française* of E. Littré (1873) takes notice of this fact under "mathématique"; the *Oxford English Dictionary,* however, does not refer back to Sanscrit. The Greek lexicon of the 10th century A.D., which goes under the name of "Suidas," does not have the word "mathematics" at all, perhaps from "medieval darkness"; it only lists some words relating to "geometry" and "arithmetic."

The contraction of meaning of the word mathematics from general knowledge to mathematics proper is virtually completed in Aristotle (384–322 B.C.), but not yet in Plato (427–348 B.C.). There is a similar contraction in the case of "Poetry." It is also a Greek word, and, by origin, it means "something that has been done, manufactured, achieved"; but in this case the contraction of meaning is all but completed in Plato.

The word "mathematician" (*mathematikos*) occurs in Aristotle, and earlier. For Aristotle it does not always designate the "professional" of today, but in most cases it does. However, the definitive contraction of meaning, in Greek, came about very slowly, if ever. In the middle ages, for instance, the word mathematician, both in its Greek and Latin forms, commonly denoted an astrologer.

Statements in Aristotle in accumulation suggest that the technical use of the word mathematics, and of cognate words, originated in the thinking of Pythagoreans, but there is nothing in the documentation to suggest that there was a similar tendency in the thinking of natural philosophers from Ionia.[5a] Also late tradition attaches

[2] See Ivor Thomas, *Greek Mathematics* (Cambridge, Mass., 1939–42), I, 3.

[3] J. F. Montucla, *Histoire des Mathématiques* (Paris, second edition, 1799–1802, 4 volumes; reprinted 1960), I, 1–3.

[4] T. L. Heath, *A History of Greek Mathematics,* 2 vols. (Oxford, 1921), I, 1–25.

[5] For instance, B. Snell, *Die Ausdrücke für den Begriff des Wissens in der vorplatonischen Philosophie* (Berlin, 1924), 72ff.

[5a] *Ibid.*

achievements in "pure" mathematics only to Thales, but to none of the other Ionians. This may help to explain why there is so little made of them in Plato; Anaximander and Anaximenes are never mentioned in Plato, and Xenophanes only once, casually; Thales is somebody about whom to spread anecdotes; and Heraclitus is, by and large, the author of the *bon mot* that "one cannot step into the same river twice."[6]

I think that the memory of the original comprehensive meaning of mathematics never lapsed entirely, not in musings of philosophers at any rate, although it would not be easy to adduce documentation, meaningful for the Middle Ages and Renaissance. But it is fair to say that the attempt of G. W. Leibniz (1646–1716) to create a *mathesis universalis* was also an attempt to re-universalize mathematics.[7] The efforts of Leibniz were noted by others, even approvingly, but his calls to action were not heeded even then. In the 18th century, Kant's apriorization of geometry and arithmetic was a species of re–universalization of mathematics, and philosophers in the 19th century continued in this vein. In the 20th century a distinct trend to universalize mathematics has become discernible, in the second half of the century at any rate. It is much more than an attempt to resume where Leibniz left off, but it is not easy to say how the trend has come about. In large part, and outwardly, this trend manifests itself in a widespread disposition and desire to "apply" mathematics wherever and whenever possible. But it is also a fact that only a fraction of contemporary high-level research activity in mathematics is motivated and instigated directly by applications. Much of newly created mathematics has no overt bearing on technology, or on applied science, or even directly on theoretical science; and yet, it somehow seems inconceivable that such mathematics should not have some purpose and significance within the texture of our total knowledge. And then there is also the question, the peren-

[6] E. Zeller, "Plato's Mitteilungen über frühere und gleichzeitige Philosophen," *Archiv für Geschichte der Philosophie,* V (Berlin, 1892), 165–89. Also, reproduced in Zeller's *Collected Essays.*

[7] See, for instance, *Leibniz Selections,* edited by Philip P. Wiener (New York, 1951), XXVI–XXXII and 1–46. The name and conception of *mathesis universalis* had already been introduced in Rule IV (Rules for the direction of the mind) of Descartes, but not yet in a manner prerequisite for eventual "logicalization." But it would be difficult, for instance, to isolate, for quotation, an anticipation of this conception amidst the expostulations on the nature of mathematics in the *Opus Maius* of Roger Bacon (1214–1294).

nial one, which is asked sometimes shyly and sometimes challengingly, why it is that mathematics is so peculiarly "abstract," in a manner which is discouraging to many who try to learn it; and whether mathematics cannot be made more "intuitive," and more accessible to larger audiences, without forfeiting its substance, and its mission, whatever these might be.

Questions like these are not easily answered, at any rate not in one sentence, or even in one paragraph. But we will offer certain reflections and comments on the vicissitudes and the nature of mathematics, which may carry with them partial answers to questions like these.

2. *Euclid's Elements*

The complexities which beset the problem of the nature of mathematics are already encountered when one attempts to comprehend the extraordinary career, in the course of many centuries, of Euclid's famous treatise in 13 "books." The work was composed around 300 B.C., and almost nothing is known about the author.[8] The two little anecdotes about him are uninformative and banal. Their "philosophical" content is trivial when held against pronouncements on mathematics by Plato and Aristotle before him, and even by Herodotus long before him. In one anecdote, Euclid chides a student of his, by a gesture and not by a rational explanation, for wanting to know wherefore he should study a proposition which was apparently assigned to him as homework. In the second anecdote, Euclid states before the King of Egypt that there is no Royal Road to mathematics; even a mighty ruler cannot command that he be taught mathematics in Ten Easy Lessons.

Euclid's work has survived "intact," that is without any manifest stigmata of textual corruptions, as few large works from classical antiquity have, although it is not a book that stirs "human interest" as do Homer or Herodotus. It does not speak of human passions as do the all too few extant plays of Eschylus, Sophocles, and Euripides; and it does not challenge the intellect with the immediacy and directness of the works of Thucydides, Plato, and Aristotle, and also of the fragments of Parmenides. It is not even a book that has the direct scientific interest, or the same appeal to adultness, as Ptolemy's treatise on astronomy; and it is not at all a book of "technological" interest as are the books of the large compilation which is attributed to somebody named Heron. Also, if examined "objectively," Euclid's work ought to have been any educationist's nightmare. The work presumes to begin from a beginning; that is, it presupposes a certain

[8] See Reference 4, 354ff. Also, T. L. Heath, *The Thirteen Books of Euclid's Elements*, 3 vols. (Cambridge, 1926); especially I, 1–6.

level of readiness, but makes no other prerequisites. And yet, it never offers any "motivations," it has no illuminating "asides," it does not attempt to make anything "intuitive," and it avoids "applications" to a fault. It is so "humorless" in its mathematical purism, that, although it is a book about "Elements," it nevertheless does not unbend long enough in its singlemindedness to make the remark, if only very casually, that if a rectangle has a base of 3 inches and a height of 4 inches then it has an area of 12 square inches. Euclid's work never mentions the name of a person; it never makes a statement about, or even an (intended) allusion to genetic developments of mathematics; it makes no cross reference, except once; the exception being in proposition 2 of book 13, where the text refers to, and repeats the content of, the "first theorem of the tenth book" which, as it happens, is Euclid's "Substitute" for the later axiom of Archimedes. Euclid has a fixed pattern for the enunciation of a proposition, and he never deviates from it. In short it is almost impossible to refute an assertion that the Elements are the work of an unsufferable pedant and martinet.

On the face of it, in comparison to the *Elements,* the works of Archimedes, Apollonius, and Ptolemy are "pulsating with life." They address themselves not to students but to regular professionals, have a considerable research level, to begin with, are mathematically more difficult to approach, and their style is sometimes over-solemn. But their manner of presentation, although very controlled, is by no means rigid. There are motivations, explanations, justifications. There are historical passages, with names of past scholars, and those in Ptolemy's *Almagest* are of immeasurable value. The *Sand-Reckoner* of Archimedes is, throughout, addressed to King Gelon. The epistolary preambles in Archimedes and Apollonius have personal touches; they are informative, and sometimes uninhibited, or even chatty. Also, in contrast to Euclid, Archimedes is obviously indifferent to would-be philosophical distinctions between "axioms," "postulates," and "common notions," and he uses the terms as the fancy strikes him.[9]

Euclid's work became one of the all-time best-sellers. According to "objective" Pestalozzi criteria, it should have been spurned by students and "progressive" teachers in every generation. But it nevertheless survived intact all the turmoils, ravages, and illiteracies of the dissolving Roman Empire, of the early Dark Ages, of the Crusades, and of the plagues and famines of the later Middle Ages. And, after printing began, Euclid was printed in as many editions, and in as many languages, as perhaps no other book outside the Bible.

[9] See K. v. Fritz, "DIE ARCHAI in der griechischen Mathematik," *Archiv für Begriffsgeschichte,* Bonn, H. Bouvier, I (1955), 13–103.

Euclid lived through the Scientific Revolution of the 17th century unscathed, and even the discovery of Non-Euclidean geometries in the first half of the 19th century did not tarnish the "image" of Euclid, or of his *Elements*. American towns and cities continued to name streets and avenues after Euclid, much more so than after Plato or Aristotle. And Abraham Lincoln, in his campaign biography of 1860 written by himself and published under the name of John L. Scripps by the Chicago Press and Tribune, ventured to assert about himself that "he studied and nearly mastered the six books of Euclid since he was a member of Congress." [10] Lincoln's assertion that he had "nearly mastered" these books was one of the boldest and blandest campaign statements in the annals of American presidential elections, and folkloristic embellishments of this assertion were even less restrained.[11] And it is astonishing that in the America of 1860 a consummate grass-roots politician of the then Mid-Western Frontier should have thought that adding to a mixture of log cabin and railsplitting a six books worth of Euclid would make the mixture more palatable to an electorate across the country.

Even the 20th century has its problems with Euclid. The present day reformers, all over the world, of mathematical programs in primary and secondary schools have not been able to decide, with irrevocable finality, what to do about the unreformable Euclid in the midst of it all.

3. *Pre-Greek Mathematics*

Pre-Greek mathematics is mathematics of Egyptians and Babylonians, say between 2500 B.C. and 500 B.C. In subject matter, certain parts of Euclid's Elements are to be found in Pre-Greek mathematics and it is likely that the Greeks acquired them by adoption. The Greeks themselves acknowledged that their mathematics had received its first impulses from Egypt and Babylon, and their tradition may have understated the extent of the subject matter thus acquired. But the Greeks endowed their mathematics, whether acquired or created, with an intellectuality of its own, and they themselves were aware of this.

Babylonian mathematics was rather advanced, even more so than Egyptian. But it is difficult to ascertain, with our customary modes of rational discriminations, which are derivative from Greek thought processes, to what extent Pre-Greek mathematics arrived at a stage

[10] *The Collected Works of Abraham Lincoln* (New Brunswick, 1953), IV, 62.

[11] See Herndon's *Life of Lincoln* (New York, 1949), 248; Carl Sandburg, *Abraham Lincoln, The Prairie Years* (New York, 1926), I, 423–4; Emanuel Hertz, *Lincoln Talks* (New York, 1936), 18.

of being cultivated for its own sake in a theoretical separateness.[12] By documentation, it was, preponderantly, mathematics for a purpose and enmeshed with applications. Preponderantly, it was mathematics needed in astronomy; or, mathematics which occurs in computation of areas for civil purposes of taxation or inheritance, or of areas and volumes for military purposes of construction of moats and breastworks; or, it was mathematics needed for international commerce, or for computation of supplies for Royal Households, temples, or armies in the field.

It is difficult to make out how Pre-Greek mathematics was taught and internally transmitted; especially, to what extent the rules of mathematical procedure were inculcated by presentation of examples only, and to what extent they were transmitted by some kind of "motivations," "reasonings," or elucidating comments of a general nature. The Greeks created conceptions of "mathematics," "physics," "astronomy," "knowledge," "science," etc. Egyptians and Babylonians seemingly did not have them; but it is difficult to see, from our modes of thinking, how they could have advanced as far as they did without having had some kind of intellectual constructs which corresponded to such conceptions.

From lengthy passages in Plato one can form an impression of how it sounded when Athenians of Plato's time were discussing the nature of mathematics and the peculiar difficulties of learning it; and from other indirect documentation one might even hazard to project such impressions backwards into the 5th century B.C. There seem to be no documents of this kind for Babylonian mathematics, and for Egyptian mathematics there is barely one document, namely papyrus Anastasi I,[13] which may be exploited, perhaps, to this end. The document, whose actual purpose is not known with certainty, is cast in the form of a lengthy memorandum from one royal scribe to another, in which the writer upbraids his correspondent for his incompetence in the discharge of various high offices with which he is entrusted; he especially taunts him with his inability to submit a mathematically correct estimate in certain engineering necessities which an army under his command is in need of. We note that this mathematical passage has been thoroughly examined with regard to its mathematical technicalities,[14] but that an examination with regard

[12] For an ascertainment which is very sympathetic to Pre-Greek Mathematics see Kurt Vogel, *Vorgriechische Mathematik,* I Vorgeschichte und Ägypten; II Die Mathematik der Babylonier (Hanover, 1959).

[13] A. Erman, *The Literature of Ancient Egypt* (New York, 1927), 214–234; especially the mathematical passage on the last two pages.

[14] O. Neugebauer, *The Exact Sciences in Antiquity,* second edition (Providence, 1957), 79.

to "philosophy" of mathematics is apparently not feasible.

Mathematics, as an organized pursuit, began surprisingly early, and it is tempting to speculate how it came into being. The Greeks were the first to speculate, and their views, although only speculative, are of interest. Herodotus (480–425 B.C.) spoke only of geometry, and a general notion of Mathematics may not have been familiar to him. Being an anthropologist, he suggested (*Histories,* Book 2, chapter 109) that geometry came to Hellas from Egypt, where it had come into being because yearly inundations demanded that the size of landed property be frequently resurveyed for tax purposes. And in the same breath he adds that the Greeks learned the use of sundials and the division of day length into 12 parts from the Babylonians.

Plato (427–348 B.C.) is already concerned with various parts of mathematics, and in the course of the whimsical myth of Theuth and Thamus, in the dialogue *Phaedrus* 274 C, D, he says: [15]

> The story is that in the region of Naucratis in Egypt there dwelt one of the old gods of the country, the god to whom the bird Ibis is sacred, his own name being Theuth. He it was that invented number and calculation, geometry and astronomy, not to speak of draughts and dice, and above all writing.

Aristotle (384–322 B.C.) at last, speaks of mathematics in the full meaning of the concept, that is of mathematics for mathematics' sake, and he says in Chapter I of book I of the *Metaphysics,* 981b 24–26, that the mathematical "sciences," or "arts" (*mathematikai technai*) originated with Egyptians because their priestly class was afforded the leisure for such pursuits. As it happens, the Egyptian priests did not constitute a leisure class, but this does not affect the wisdom of Aristotle's observation. In the context of the chapter, the reference to the Egyptians is only meant to round out the general contention that there is a knowledge for the sake of knowledge; that pure mathematics is a prime example of it; and that the rise of this knowledge is not due to the quest for "consumer goods" and luxuries. This lofty Aristotelian view may be opposed, but it cannot be refuted.

4. *Greek Mathematics*

This mathematics culminated in the works of Euclid, Archimedes, and Apollonius, and its structure is confirmed in the astronomical treatise of Ptolemy. Somewhat different in tone, perhaps only because it is intended to be "applied" mathematics, is the work of the "theoretical engineer" Heron. Distinct in texture is the number theoretic work of Diophantus, and almost nothing is known about its evolutionary relation to the geometrically controlled work of the others.

[15] R. Hackforth, Plato's *Phaedrus* (Indianapolis, 1951), 156.

All these works played a great rôle in the outburst of science and new mathematics in the 16th and 17th centuries; and in the very formative 17th century they were studied and exploited as never before. And yet it is a hard fact, that within Hellenistic civilization itself, mathematical activity eventually ceased entirely, and so-called "exact" science and technology, in a proper and general sense, did not come into being at all. There are various "cultural" and "accidental" reasons which historians are adducing for this singular and definitive failure of Greek civilization, but somehow none of such reasons is really convincing. Greek astronomy was mathematical from the outset, and its mathematization made it highly successful; the Greeks originated a doctrine of optics, and its geometrization endowed it with possibilities of development; and, already beginning with earlier Pythagoreans, there was some degree of awareness abroad that a mathematization of concepts and relations would be an effective approach to the comprehension of nature. And yet, from whatever reasons, the Greeks never penetrated to the insight that an effective manner of pursuing mechanics, physics, and other science is to articulate qualitative attributes by quantitative magnitudes, and then to represent connecting rules and laws by mathematical formulas and relations, in order that the application of mathematical procedures to the formulas and relation may lead to further explications and developments.

The Greeks never arrived at such an insight; in spite of striking first steps in Archimedes they never developed a "theoretical" science in general; and eventually they even lost control over their mathematics, which they set out creating magnificently. Perhaps the sharpest illustration of the Greek "failure" can be obtained by a contemplation of the collection of works which goes under the name of Heron of Alexandria, and for our purposes we may assume, what some historians doubt, that there was one such person who was the author of the bulk of it. Now, the author of this collection was an eminent mathematical engineer, and it is remarkable that, like other great scientific engineers after him, he even wrote his own textbooks in mathematics, both on the elementary and on an advanced level. He was a great inventor of "trap doors", "magical devices," and general "wonders," but, behind all this, there was no proper scientific and technological urge, in our sense, that would drive this inventiveness into channels of utilization, industrialization, or other exploration; and Heron's great talent virtually came to naught. Somehow, there was no determined scientific or technological "motivation" within the totality of knowledge and effort; and this absence of a proper scientific motivation was so fateful that eventually even Greek mathematics withered away.

It appears that present day mathematics, and mathematically oriented science with it, grows and thrives in spite of, and perhaps even because of a certain fruitful self-contradictoriness in its pattern of unfolding. There is a quest, a dominant one, for a mathematics-for-its-own-sake which is motivated internally and almost self-sufficiently. Parallel to this there is a quest, a strong one, for a mathematics-for-purposes which is motivated mostly externally. Also, for the most part there is no forcible conjunction of the two quests, or even of the responses to these different quests. And yet, the total outcome is very gratifying, even though there is, on the whole, no forceful effort towards reconciling the divergences of the quests by direct confrontations.

The Greeks had an uneasy premonition of all this. But they were the original and proud discoverers of the meaning of "logicality," in and out of mathematics, and it would have been intolerable to them to countenance any kind of "self-contradictoriness" in connection with mathematics, not even in an external genetic aspect of its growth. When posted on their frontiers of knowledge, the Greeks were rigid observers of all rules of discipline and of what they deemed to be "logical" conduct, in spite of the fact, or because of the fact, that all these rules had been drawn up by themselves.

5. *Mathematics on the Rise*

During the so-called Scientific Revolution, which spanned the 16th, 17th and 18th centuries and centered in the 17th century, mathematics was going apace, as if anew and from novel sources. But, in spite of the novelty of inspiration, the working techniques of Greek mathematics were explored and exploited to the limit of their yield, as evidenced by the fact that Isaac Newton cast his forward thrusting *Principia* (1686) in the mold of much-tested but backwards directed Archimedism. For the mathematics of this era it would have been very difficult to say what of it was mathematics to a purpose and what was for its own sake. Most of it was, in fact, mathematics to a purpose, immediately or eventually. But we contend that contemporaries would not have been able to assess degrees of purposefulness, even if they could have been made to comprehend the cogency of such assessments. We shall adduce three arguments in support of this contention, one from algebra, another from geometry, and a third from mechanics.

(1) In mathematics, the 16th century was the century of the rise of so-called algebra, whatever that be. Not only was this algebra a characteristic of the century, but a certain feature of it, namely the "symbolization" inherent to it, became a profoundly distinguishing mark of all mathematics to follow. This distinction is so pronounced that one sometimes encounters the (superficial) judgment that if

only the Greeks had happened to "discover" this algebra, the course of history of mathematics would have run differently. At any rate, this feature of algebra has become an attribute of the essence of mathematics, of its foundations, and of its nature of abstractness on the uppermost level of "ideation" *à la* Plato. And yet, if one tries to uncover where this algebra came from, what its motivations were, and, above all, why it grew, then, apparently, the only explanation ever proposed is this, that as a twin phenomenon to "arithmetic" it was brought into being, preponderantly, by the Rise of Commerce in the late Middle Ages; so that the challenges, to which the rise of algebra was the response, were, predominantly, the very unlofty and utilitarian demands of counting houses of bankers and merchants in Lombardy, Northern Europe, and the Levant.[16] Even Oswald Spengler, whose views run counter to those of economic determinists, when attempting to demonstrate, in his essay on Mathematics,[17] that modern mathematics has "cultural" roots in medieval Gothic, is concerned more with mathematical functions than with Renaissance algebra. His attempted derivation is attuned much more to functions than to algebra, although, on mathematical grounds, the rise of functions and the rise of algebra are closely linked.

(2) The case of coordinate geometry runs almost in reverse. Nowadays, there is hardly anything that is more commonplace and utilitarian than a "pictorial" graph or chart, especially when drawn with a pencil on a sheet of prelatticed graph paper. Such graphs, apparently, began to be used only in the second half of the 18th century, and they apparently were an outgrowth of the employment of geometric coordinate systems in the first half of the 18th century, first in mathematics, and then, gradually, in rational mechanics, for which they became the indispensable setting. Coordinate systems, in their turn, became the hallmark of the "new" geometry that had been inaugurated by (Fermat and) Descartes in the first half of the 17th century, so much so, that the name of "Cartesius" has been firmly affixed to them, although coordinate systems only *begin* to occur in the work of Descartes himself.[18] Now, it may be stated that there was nothing "practical" or "utilitarian" in Descartes' own conception of his geometry, nor in the conception of his contemporaries or immediate successors. Descartes motivated his innovation on abstractly theoretical, and even philosophical grounds; and to his contempor-

[16] Wallace K. Ferguson, *The Renaissance* (New York, 1940); the first systematic presentation of this explanation is perhaps due to Werner Sombart; see his book *The Quintessence of Capitalism* (London, 1915), Chapter VIII (the art of calculation).

[17] Oswald Spengler, *The Decline of the West* (New York, 1932), I, Chap. II, "The meaning of numbers."

[18] C. B. Boyer, *History of Analytic Geometry* (New York, 1956).

aries, the theory appeared to be so esoteric and recondite that regular commentaries on it were composed.[19] Mathematically, the most immediate purpose of Descartes was to initiate an appropriate technique for what eventually became known as "algebraic geometry." It is very remarkable that Isaac Newton, who comprehended the new geometry as deeply as anybody else, Descartes included, applied it in fact only to a study in algebra geometry.[20] In the *Principia* there is no sign of an anticipation that pre-introduced coordinate systems would become the ineluctable setting in all of mechanics, a century hence.

(3) Finally, our argument from mechanics is as follows. For the 18th century and in the first decades of the 19th century the amount of mathematics which was created for, and introduced into mechanics, theoretical and applied, was enormous, to the discomfort of many a "practical" scientist of the day, no doubt. For instance, in engineering mechanics nothing has a more "practical" sound than words like "stress," "strain," "pressure," "torque," etc. And yet, it is a hard fact, that the present-day systematic theory of these concepts was created in the 1820's by A. Cauchy, who, by all odds, was a mathematician first and foremost, and a very "pure" mathematician at that. Daniel Bernoulli, in the 18th century, was a physicist, very recognizably so. But virtually all other leading architects of the various parts of mechanics of discrete and continuous media were just as recognizably mathematicians, and among the most eminent since Pythagoras too: James and John Bernoulli, d'Alembert, Euler, Lagrange, Poisson, Laplace, Gauss, Cauchy, Jacobi, Hamilton, etc. Correspondingly, most of their theorizing emanated from "pure thinking," with very circumscribed entanglement in direct planned experimentation. And yet, the mechanics that has evolved, and the mathematics that was created in the process, are for the ages.

6. *Prefabricated Mathematics*

Not all mathematics from the 17th and 18th centuries, even when done by mathematicians who excelled in mechanics, was originally mathematics for a purpose; and some of it has remained mathematics for its own sake until today. Then with the beginning of the 19th century, Mathematics became wholly independent of mechanics and then physics, and its intellectual self-sufficiency began to be a philosophical verity. One began to hear the statement that, in the schema of our knowledge, Mathematics ought to be counted not among the Natural Sciences (*Naturwissenschaften*) but among the Humanities

[19] *Ibid.*

[20] *Enumeration of Lines of the Third Order*. Translated by C. R. M. Talbot (London, 1861).

(*Geisteswissenschaften*). Also, the commonplace dictum that "mathematics is the language of science", which even influences college curricula, is a partial corroboration of this statement inasmuch as it brackets mathematics with language. And yet, although the old intimacy between mathematics and mechanics was seemingly extenuated, another kind of relationship between mathematics and theoretical physics, in lieu of the old one, was beginning to develop. It was a rapport built more on parallelisms of pursuits rather than on identities of aims. The fruitfulness of the rapport manifests itself in the phenomenon that, from time to time, theoretical physics is able to seize upon a ready-made piece of mathematics, which had not been familiar to it before, and to turn it to instant use, as if it had been prefabricated for just the occasion. Mostly, such prefabricated mathematics is, by origin, mathematics for its own sake, and yet its yield, when it is invoked, is usually richer than that of so-called "applied mathematics," that is of mathematics which is mathematics for a purpose by intent and design.

A known illustration of all this, the first in this century, occurs in the Theory of Relativity. The Special Theory of Relativity did not pre-require mathematics that was unfamiliar to physicists; even Minkowski's placement of existing formulas from physics into a framework of four-dimensionality did not come about outside the mathematics already familiar from the theories of Maxwell, Lorentz, and Poincaré. But when Albert Einstein was pondering the formulation of the General Theory, he was lacking a mathematical technique that would lead him towards it. But then a mathematician directed his attention to recent work in tensor analysis by G. Ricci and also T. Levi-Civita.[21] As it happens, this "analysis" had arisen in a context from theoretical mechanics. However the subject matter had been a mathematically esoteric one; physicists in general had had no provocation to become familiar with it, and it was mathematics for its own sake by general tenor. Einstein, however, found it to be a mathematical device most providentially prefabricated for his purposes, and he adopted it with gusto. This analysis then entered the very texture of the theory. It imparted to physics as a whole a certain procedural flexibility, and, at the same time, it directed the attention of mathematicians to a significant theory, as yet little noticed, in their own midst. There developed a collaboration between physicists and mathematicians, which, in this specialized area, at any rate, has remained unbroken; and mathematicians began earnestly to cultivate the novel "differential calculus" for its own sake. The calculus was steadily gaining in significance and eventually it became important to Topology. Furthermore, the theory of relativity, as cultivated

[21] Philip Frank, *Einstein, His Life and Times* (New York, 1953), 103.

by physicists, began to be interested in areas of topology in which the tensor calculus domiciled itself, so that a new, and even more sophisticated contact between relativity and differential geometry has come about.[22]

In quantum physics, which is the core of physical science, it happened also that a setting was fashioned out of prefabricated mathematics, not immediately but rather soon. The original disparate versions, of W. Heisenberg, and of E. Schrödinger, were soon merged, by Schrödinger, into one, and the union of the two was mathematically consummated, undivorceably and fruitfully, in the precincts of so-called Hilbert space. The notions, in quantum physics, of "physical state," and of "physical observable," are vague, disquieting, and never-to-be-assimilated. But their mathematical counterparts, namely the notions of "vector in Hilbert space" and of "linear operator in Hilbert space" are determined, reassuring, and assimilable. D. Hilbert had constructed his theory about 20 years previously. It was a magnificent theory, but mathematicians did not quite know whither to carry it further. But, since entering physics, the theory of operators has been studied unremittingly, by mathematicians and physicists alike. Also, the preoccupation with it has made the concept of an operator from a tool in physics to a reality in nature, and it has raised the mathematization of physics to new levels. There is hardly a purely mathematical statement in Hilbert space which some physicist would not, by a transliteration, interpret as an event, or as a property of an event in nature.

Furthermore, quantum physics appropriated for itself large parts of group theory, so much so, that physicists began to compose their own mathematical monographs about it, or at least to give their own accounts of what is relevant to physics.[23]

A very different case of intervention of prefabricated mathematics occurred in the theory of shock waves. A knowledge of the theory became urgent with the advent of atomic fission in the 1940's, and it was found that the best extant study about this subject was a full length book that had been written, in 1903, by the eminent mathematician J. Hadamard.[24] The book in its turn was based on a much earlier memoir by the great 19th-century mathematician, B. Riemann,[25] and on subsequent work by others. In 1903 Hadamard

[22] J. A. Wheeler, *Geometrodynamics,* Italian Physical Society (New York, 1962).

[23] One of the first of this kind was the book of E. P. Wigner, 1931, whose translation in New York, 1959, is entitled: *Group Theory and Its Application to the Quantum Mechanics of Atomic Spectra.*

[24] J. Hadamard, *Leçons sur la propagation des ondes* (Paris, 1903).

[25] B. Riemann, *Über die Fortpflanzung ebener Luftwellen von endlicher Schwingungsweite,* in Gesammelte Mathematische Werke (Leipzig, 1892), 156.

was at the height of his career, and he elected to write a book which became topical only 40 years afterwards.

Finally we may point out that contemporary studies in network and information theory, mechanical and human, in their beginnings had to fall back on the work of George Boole (1815–1864) a century ago, and that the work of Boole had been done with no such specific applications in prospect.

7. *Relation Between Mathematics and Science*

It is not easy to explain how it comes about that, every so often, it is possible to apply, directly and potently, a mathematical development of yesterday to an actuality in science of today; and the problem of explaining this possibility has been, in one formulation or another, a central problem in the Theory of Knowledge since Plato. Before Plato, the problem occurs in the poem of Parmenides. But Parmenides was an ontologist, the first and unsurpassedly greatest ontologist ever, and the problem appears as an ontological dilemma, without any reference to mathematics. His famous disciple Zeno of Elea began to see the dilemma as a problem affecting mathematics. However, it appears from Aristotle's report about Zeno, in his *Physics* and from other evidence about him, that Zeno envisaged the problem in certain particular "puzzling" sub-problems, and not in one comprehensive tableau.

To judge by Aristotle's Metaphysics, Book I, Chapters 5–8, Pythagoreans were the first to raise the problem, in pre-articulate versions, of the validity and significance of mathematical knowledge for other knowledge, especially for scientific knowledge. Plato inherited their problem, and articulated it better. And, as can be seen from the dialogue *Meno,* at some time, Plato widened the problem by merging with it the Socratic problem of establishing a social and economic doctrine by rationalization from logicality. All this contributed to the paramountcy of Greek philosophy and to the perdurance of Greek philosophemes; but the problem of the relation between mathematics and science remains.

The difference between mathematics and science is that they have different subject matters. Therefore, cognitively, mathematics is not part of science, or subordinate to science as a whole. A scientist in his working mood may please himself to view mathematics as nothing other than a "handmaiden of science"; but, philosophically, this is a shallow statement. Any natural science deals with the "external world," whatever that be. This even applies to the psychology of the human mind, say. In a scientifically conducted psychological introspection or self-inspection, the internal object of the inspection becomes "externalized," by the procedure itself. Strictly speaking, however, mathematics deals only with objects of its own imagery, which are internally conceived, internally created, and inwardly structured, even though they are tangible, substantive, selfsame, and somehow communicable and shareable.

When one is concerned only with the effect of mathematics on science, the farthest one can go, cognitively, in subordinating mathematics to science is suggested by the following simile. If science is viewed as an industrial establishment, then mathematics is an associated power plant which feeds a certain kind of indispensable energy into the establishment. The counterparts to mathematicians would be the designers, maintenance men, and administrators of the power plant. Of these, the majority need be interested only in the requirements of the power plant itself; and solely a minority need be aware of the actual workings of the establishment itself, let alone be expert in its activities. This simile is unfavorable to mathematics in that it does not account for the great power of creativity which resides within its compass. We wish to say, however, that there also is danger in overglorifying mathematics and attempting to subordinate all of science to it, naïvely and forcibly. Pythagoreans, and even Plato, after having discovered, or created, the metaphysical status of mathematics, were prone to endow mathematics with a somewhat lop-sided priority in the realm of cognition; and genuine scientists like Archimedes and Ptolemy, had to brace themselves against such exaggerations. In some respects, it is almost safer for mathematics to remain the handmaiden of science than to be forcibly enthroned as the dictatorial mistress of it; and yet, in our very presence, that is in the second half of the 20th century, there seem to be trends developing to bring about just that.

8. *The Abstractness of Mathematics*

There is a statement abroad that mathematics is abstract, and the statement is frequently made with emotion; either approvingly or disapprovingly; or critically with the implication that the abstractness is carried too far and ought to be kept within limits. Apart from emotional value judgments, if the statement itself is right, then objectively, it can only mean that in mathematics abstractness has a quality of its own which is peculiar to it. We shall make a brief examination whether such is indeed the case.

In general, to think "abstractly" means to exercise a certain mental faculty of intellectual selectivity, to bring about suitable selections, and to concentrate one's intellectual attention on what has been selected. Now any rational thinking has to be selective and concentrative, and thus, in a sense, all rational thinking is abstract thinking. The cursory reading of an ordinary telephone directory, or of a very plain listing of historical or geographical names and facts, is perhaps not yet an "abstract" process, not discernibly yet, at any rate; but the reading of any other book is. A cookbook is abstract, and so are the Yellow Pages in a telephone directory. Any academic textbook, on whatever level, and in whatever subject, must be abstract in a more than perfunctory sense, or it is useless. This general value of abstraction does not yet single out mathematics to any extent.

One might say that what makes mathematical abstraction different is the formation of many difficult conceptions whose definition and scope must be meticulously adhered to. But the same applies very much to jurisprudence, say. Also, for instance, contemporary literary criticism has become severely conceptualized, so much so, that a non-professional reader frequently may find it easier to understand a literary work directly, even a recondite one, than to grasp a piece of literary criticism about it. And, furthermore, in sociology, for instance, conceptualizations, and even incomprehensibilities, seem to grow more profusely than anywhere else.

At this stage one may be tempted to say that if stringent conceptualization is not yet characteristic of mathematical thinking, then its linking and pairing with symbolization is. Now, the use of symbols has certainly a powerful place in mathematics, but, as one can quickly see, if the use of symbols is to be made into a significant part of the characterization of mathematical thinking, considerable clarifications have to be proffered. Chemical theories of valency and bond are steeped in a highly conventionalized symbolism from which they cannot be separated. But it is not a symbolism which is interlocked with mathematization; in fact, in fully theoretical works on quantum chemistry this symbolism is displayed less emphatically than in regular manuals of chemistry.

Furthermore, as is well known, the classical Greek mathematics of Euclid, Archimedes, Apollonius and others, does not operate with symbols at all; not even in "disguise," as did Cardan in the 16th century. This is even viewed as a defect of Greek mathematics, and as one which has causally brought about the decay of Greek mathematics in the end. And yet, it would not be proper to say that, in general, the Greeks did not have an affinity to thinking in symbols, mathematical or other. This is simply not true; in a certain qualified sense the Greeks even had a marked affinity to thinking in symbols. For instance, by all evidence, it was the Greeks who were the first to use the letters of the alphabet as "symbols" for numerals. The alphabet itself was not of their own creation, this they took from the Phoenicians. However, the Phoenicians themselves never used their alphabet for numerical purposes, but had separate signs for numbers; whereas the Greek installation of the alphabetical system of numerals is presumed to reach back into the 8th century B.C.[26]

Also, Aristotle in his *Prior Analytic* employs letters as symbols for logical data, and he is not in the least self-conscious about it; neither Aristotle himself nor anybody else in antiquity saw anything unusual or original in this. Modern commentators, however, seem much impressed by this. Paul Tannery[27] notes that Aristotle uses letters as symbolic representations of objects of thought, and Tannery finds that the Greeks were only a step away from the algebraic algorithm of Viète. He is distressed over the fact that Pythagoreans did not take the same one step as did Viète, but, instead, abandoned themselves to sterile reveries. J. Jørgensen,[28] echoes Tannery when saying that "in his treatment of propositions and syllogisms, Aristotle uses letters as symbols for the terms, thus laying the first foundation of all subsequent symbolic logic and algebra." These two statements are not correct. The Greek employment of letters in lieu of numerals, propositions, and syllogisms, falls indeed within the general area of mathematical abstraction, and does indeed testify to a measure of Greek capacity for it; but it is not, or not yet, the kind of abstraction which constitutes the essence of the mathematics of today. The difference between the two levels of abstraction, ancient and modern, is a very considerable one, and not much is gained by viewing it as a difference of a single cognitive step only, which the Greeks somehow failed to take, as if by accident. As viewed in retrospect, the Greeks, for all their cleverness, were not able, or not yet able, to make abstractions that were more than idealizations from immediate actuality; their abstractions were rarely more than one-step abstractions, and as such they remained within the purview of what is called "intuitive" in an obvious and direct sense. They did not make second-level abstractions from abstractions, or, alternately, abstractions from intellectually conceived possibilities and potentialities, and such abstractions of higher order as were made remained rudimentary and operationally unproductive. In fact, our notion of "abstract" remained alien to the classic Hellenic and Hellenistic idiom of philosophy, and the word itself cannot be translated into classical Greek with idiomatic ease.

The Greeks clearly knew that a geometrical line is the idealization of a physical wire whose non-zero diameter is allowed, ideally, to shrink to zero. Archimedes was more or less in possession of the notion of a mass-point, knowing that it is an "ideal" limit of a small ball of variable radius but fixed mass whose radius has shrunk to

[26] See Heath, Reference 4, 32–33.

[27] *Mémoires Scientifiques* (Paris, 1920), IV, 199.

[28] *A Treatise on Formal Logic,* I (New York, 1931), 47.

zero. The Greeks abstracted the notion of a "general" triangle from the set of all individual triangles, and similarly the notion of a general quadrangle, pentagon, etc. They even went further, and they created the notion of a general polygon, the number of sides of the polygon being unspecified, and even variable.

The latter notion was already an abstraction from an abstraction, but it was a relatively easy one. It would have been much more notable if the Greeks, by suitable abstractions, had created the conception of a space of more than 3 dimensions, say of 4 dimensions, or, even more comprehensively, of any (unspecified) number of dimensions. But this they most decidedly never did, or ever remotely foreshadowed. Now, the statement just made about the Greeks might be objected to on the ground that the Greeks could not have possibly envisaged a multidimensional space for the following reason. The introduction of space of several variables was begun by Lagrange in his *Mécanique Analytique* in the second half of the 18th century, and it was strongly dependent on the introduction of coordinate systems into ordinary space, and such coordinate systems gradually evolved out of the *Géométrie* of Descartes (1637). The Greeks however, for whatever reasons, never introduced coordinate systems. Thus, one might object, it does not make sense to reproach the Greeks for not having introduced multidimensional geometry, if they happened not to be in possession of a device which is an indispensable prerequisite for it. To this potential objection I would like to make two rejoinders.

Firstly, from our retrospect, it is incomprehensible and "inexcusable" that the Greeks did not indeed introduce coordinate systems into their space. What is much worse, and much more inexcusable, they were hardly able to introduce a coordinate system on the (one-dimensional) time-axis, that is, in other words, to introduce a common calendar for general use. The citizens' chronology of the Greeks of the classical period is not a testimonial to their greatness.[29] For instance, Thucydides was a very eminent historian, and before him there had been a galaxy of scientist philosophers, Thales, Pythagoras, Anaximander, Anaximenes, Heraclitus, Xenophanes, and Parmenides, and others, whose names glitter from the pages of history books. But between them they did not generate enough spontaneity for the construction of a usable calendar, so that there would be no need for Thucydides to have recourse to his notorious "summer-and-winter" calculus of dates, than which there is nothing more embarrassing in the entire extant corpus of Greek writing.

But secondly, the Greek failure to conceptualize the notion of

[29] See the article "Aera" by T. W. Kubitschek in Pauly-Wissoway, *Real Encyclopädie*, I, part 1 (Stuttgart, 1894), beginning with column 606.

space reaches back into the early stages of the notion, in which the absence of coordinate systems would not yet be of consequence. Pythagoreans, and certainly Plato in the dialogue *Timaeus,* had a clear and mathematically satisfactory awareness that the space of events, that is our so-called Euclidean space, has three dimensions, that a surface is a figure having two dimensions, that a line is one-dimensional, and that a single point is, in our parlance, zero-dimensional. Furthermore, the bulk of Euclid's geometry, namely Books I–X of the *Elements,* takes place in a two-dimensional space, and Euclid is much more at ease in this space than in the three-dimensional space of Books XI–XIII. And yet, neither Euclid, nor even Archimedes, clearly conceptualized this two-dimensional space of "ordinary" geometry into a special entity in its own right and of its own standing. It somehow always remained, conceptually, a planar figure (*epipedon*) of relative existence only, that is one which was somehow immersed in the three-dimensional space of totality. And the three-dimensional total space itself was not even conceptualized, as a mathematical datum, to the stage at which it would receive an independent mathematical name of its own.

May I also observe that Greek natural philosophy in general was under the sway of a certain feature of anthropocentric sphericity, and that the failure of Greek mathematics to install coordinate systems can be linked to this feature. Lucretius, when arguing the unboundedness of the world in *De rerum natura,* (bk. I, lines 951–983), envisages the physical universe not, for instance, as a cube whose side is unbounded, but as a sphere whose radius is unbounded. In the center of the Lucretian sphere is a perceiving human mind, and Lucretius states expressly that, due to the infinitude of the radius, it is possible to shift the center of the sphere from any point in space to any other point. In the case of an unbounded cube the possibility of freely shifting its center suggests Cartesian coordinates, but not so in the case of an unbounded sphere. At best, the sphere would suggest so-called polar coordinates, but the calculus of such coordinates is indeed more than Archimedes and Apollonius could have mastered mathematically, at the time, that is. But all this does not "explain" why Greek mathematics, in its strength, did not somehow somewhere crack the confines of this sphericity and open a pathway towards something new and fresh, for itself, and for Greek knowledge in general.

The efficacy, in mathematics, of abstractions from possibility, and conjointly with this, of abstractions from abstractions, reveals itself most manifestly in so-called algebra, if, in the present context, we understand by algebra the topic of calculatory operations with symbols. For instance, if we take ordinary rational numbers, that is quotients of integers, as an actuality, then the general real numbers, which are limits of rational numbers, may be viewed as abstractions from *actuality,* or, alternately, as one step abstractions from reality. This given, the concept of a complex number $a + bi$, $i = \sqrt{-1}$, in which a, b, are real numbers, is an abstraction from *possibility,* inasmuch as it is *possible* to extend to such objects the basic arithmetical operations of addition, subtraction, multiplication, and division, in a suitable manner. Also, even if one assumes that complex numbers are only a one-step abstraction from real numbers, then, in order to reach them from rational numbers, it would take two successive abstractions at any rate. Next, if one proceeds to quaternions, one has to make at least one further abstraction, and the additional abstraction involved is certainly a very exacting one, inasmuch as the law of commutativity for multiplication no longer holds. General Cayley numbers require a further abstraction, and so on.

The Greeks rarely ventured beyond abstractions from actuality, in mathematics, in natural philosophy, and in other philosophy; and we wish to state, with emphasis, that the absence of algebraic symbolism from Greek mathematics was a symptom of a general Greek limitation, and not the cause of it. And all Platonists notwithstanding, this gross deficiency in Greek "operational" thinking makes Plato's "idealist" epistemology entirely unsuited to deal with the torrent of abstractions from possibility, which is the hallmark of our time. It is illusory to presume that Platonic idealism can cope, interpretatively, with quantum field theory of today, even if Heisenberg himself should say so.

At times, Plato or Aristotle would arrive at the very edge of a "higher" abstraction, but without then further advancing into it. In Plato's early dialogue Charmides, which deals with *sophrosyne* (= self-control, or so), on page 166C-E, Socrates elicits from Critias the description that *sophrosyne* is an *episteme* (= knowledge) of other *epistemai,* including its own self; and Socrates insists on adding, as a further element, a species of negation-of-its-own-self (= ignorance). Mathematically, such a *sophrosyne* would be a set whose elements are: (1) itself, (2) a bizarre anti-itself, and (3) other suchlike elements.[30] Viewed schematically, this was a daring construction. But nothing followed in its wake, and the construction remained solitary, dialectical, and woefully sterile. Also, on page 169A, Plato even coined the momentous conception of knowledge of knowledge (*episteme epistemes*), which, set theoretically, would be a set consisting

[30] For information and references see: T. G. Tuckey, *Plato's Charmides* (Cambridge University Press, 1951); and Klaus Oehler, *Die Lehre vom noetischen und dianoetischen Denken, bei Platon und Aristotles,* Zetemata Monographs, no. 29 (München, 1962), 123ff.

of one element only, namely of "itself." Perhaps in imitation of this,[31] Aristotle introduced, in Metaphysics, Book 12, Chapter 9, as an attribute of God, the breathtaking notion of "thinking of thinking" (*noesis noeseos*) which, as a conceptualization, captivated commentators from Theophrastus to Aquinas.[32] This conception did enter into the fabric of our thinking, penetratingly so, but non-mathematically; and none of the commentators themselves dared to break the Aristotelian seal of transmundanity in which the conception had been handed down to them.

Modern mathematics, that is mathematics of the 16th century and after, undertook abstractions from possibility from the outset. At first, these abstractions were not overtly very venturesome, but they have become so ever more. In the 20th century abstractions of this kind have become very commonplace, and there is much clamor for introducing them into school curricula, under the cloak of a so-called New Mathematics. Sometimes, abstractions from possibility are performed to excess; but the basic privilege for making them cannot be curtailed.

There is a growing demand for abstractions from possibility in ever wider areas of organized knowledge, especially in natural science and social science; and in physics such abstractions, which were initiated in Newton's *Principia,* are beginning to abound almost as much as in mathematics. Mathematics continues to be the main repository for abstraction from possibility, old and new. It is this which makes many sciences seek affiliation with mathematics; they want abstractions of this kind, much more than the mathematical techniques themselves.[33]

Princeton University.

[31] For Aristotle's indebtedness to Plato in this case, see Paul Shorey, "Aristotle's *De Anima,*" *American Journal of Philology,* XXII (1901), 149–164, esp. 160–164; also, Oehler, *loc. cit.*, note 30, pp. 201ff.

[32] O. Hamelin, *La théorie de l'intellect d'après Aristote et ses commentateurs* (Paris, 1953); also, J. Tricot, *Aristote, La Métaphysique,* Nouvelle edition (Paris, 1953), Vol. 2, Commentary to Chapters 7 and 9 of Book 12, especially pages 672–674 and 701–704. For axiological observations see W. J. Oates, *Aristotle and the Problem of Value* (Princeton, 1963), 114, 235–236.

[33] This research was supported by the U. S. Air Force Office of Scientific Research.

Plato and the One-ness of Knowledge

SALOMON BOCHNER

Plato of Athens (427–347 B.C.) was the first, and most authentic, universalist of cognition and knowing. He was heir to Parmenides, Heraclitus, and Pythagoreans, one-time pupil of Cratylus, and full-time pupil of Socrates; and there would have been no Plato without them. But we, in turn, have inherited our intellectuality, directly or indirectly, from Plato; and without this we would not have the ability to discern what there was for Plato to inherit or learn from whomsoever he had done so.

Plato recognized, almost from the first, that there is an important structure of one-ness and sameness in the diverseness of our knowing, and that this one-ness unifies and links all our knowledge: the knowledge of the external world with the knowledge of our internal self; the knowledge of morals with the knowledge of mathematics; the knowledge of the individual with the knowledge of society; the knowledge about the schoolchild with the knowledge about the adult; the knowledge about a profession with the knowledge about poetry; the knowledge of what is right with the knowledge of what is holy; the knowledge about man with the knowledge about God.

Plato maintains, tirelessly, that the true texture of this one-ness is our rationality, whatever that be. There is only one way to knowing things, namely the rational way, and this way is neither selective nor restrictive; if it leads to knowing any one thing, it must lead to knowing every other thing. That is, there is only one single ratio, take it, or leave it.

But Plato also saw the dilemma—and felt the anguish of the insight—that this very same rationality, however much it must be used to explain

---

Salomon Bochner received his doctorate at the University of Berlin, studied at the University of Copenhagen, Oxford, and Cambridge. In 1933 he joined Princeton's mathematics department, now occupies the Henry Burchard Fine Chair in Mathematics. A versatile mathematician, insatiable reader, and prolific writer, especially concerned with analysis and topology, he has also a strong interest in the history and philosophy of science, the theme of his latest book, *The Role of Mathematics in the Rise of Science*, from which this article is excerpted with permission.

This paper was reset using $\mathcal{A}_\mathcal{M}\mathcal{S}$-TEX, the American Mathematical Society's TEX macro system.

anything else, is entirely unable to explain its own self; rationality is not self-rationalizing, and the provenance of rationality is cloaked in an irrationality of its own. Thus, although even the knowing about God can only be arrived at rationally, yet, reversely, any rational knowing at all, about anything there is, must somehow flow from a fountainhead of divinely inspired Supreme Intelligence; and Plato gave to this Intelligence an aspect of human-ness which he termed the Supreme Good.

Plato knew—not very articulately, but sufficiently—that this dilemma also applies to, and is at its starkest in, mathematics. I think that by knowing this Plato had a certain superiority over mathematicians of his time; but there is no evidence for assuming that in matters of technical competence Plato was superior to, or even equal to, "professional" mathematicians in the Academy.

I think that philosophemes of Parmenides and Heraclitus made Plato also realize that it is difficult to separate the problem of Cognition from the problem of Re-cognition, and these two problems from the dilemma above. Plato somehow attempted to solve these problems and the dilemma, all in one, for better or worse, by the introduction of Ideas which are eternal, and of Souls which are immortal. The details of the solutions are inexhaustibly provocative, and the problems are as fresh, and as insoluble, today as on the day when Plato was first inspired to pose them.

Plato had his peculiarities. From boyhood he dreamt of becoming not only a philosopher but also a "king," that is, a politician and statesman of consequence. From whatever reasons he never succeeded; probably because he simply did not have an aptitude for the game of politics. And in order to relieve his frustrations he wove into the *Republic* fantastic proposals which cannot but baffle the reader; and even in his old age, in the *Epistles*, he was still insisting that the world of politics had been in a conspiracy against him, and that he had been right and everybody else wrong.

Plato did not stop berating sophists. Whatever their faults, they did constitute the educational system of Greece, and Plato should have known that every educational system always has shortcomings, and always needs reforming. And it is more effective to work hard at proposals for remedies than to harp on abuses.

Fortunately, Plato polished his written works more than any philosopher before or after him; and it is incongruous to have him reiterate that oral teaching is the primary need, and that putting things in writing is of secondary import only.

Finally, Plato's insistence on the theme of Eros and pederasty is misleading, and the presumed prevalence of the theme in Greek life and civilization is a near-invention of Plato's; there is almost nothing about the theme in Herodotus and Thucydides, in the extant plays of the tragedians, in Aristotle or other philosophers, or in the orators. Also, for instance in Homer, Achilles and Patroclus are comrades-in-arms and nothing else; and *Honi soit qui mal y pense.*

The Size of the Universe in Greek Thought

S. BOCHNER

ABSTRACT. The cosmological achievements of the Renaissance can be put under two headings: —1) A group of mathematically oriented scientists, consisting mainly of Cusanus, Copernicus, Galileo, Kepler, and Newton, gradually and for all time to come, divested the traditional Greek model of the Universe of all its impeding [sic] and frustrating features of centrality, circularity, and sphericality. — 2) Another group, consisting mainly, but not exclusively, of philosophers, with Giordano Bruno in the vanguard, strove to divest the traditional Greek model of all features of finiteness of size, and to make the Universe as wide-open and as all-infinite as possible. Now, the cosmological theories of the 20th century, both the relativistic and the telescopic ones, have restricted and even repudiadle [sic] the assertions of the second group, and, in our view, have vindicated and enhanced Greek presumptions about the size of the universe. The Greeks maintained, very judiciously, that the universe has an «indefinite» size, or that it is «complete» by size, and we are attempting to elucidate the meaning of these two possibilities in the light of present-day knowledge.

1. INTRODUCTION. — IS THE UNIVERSE INFINITE OR FINITE?

Before proceeding with any analysis of this question we might be expected to define, with some precision, what, for our context, the attributes of Infinite and Finite are meant to be; but this we cannot and will not do. The purpose of the present study is to force the above question on Greek cosmology, and to try to elicit from it what kind of answer or answers to the question were implied, or contained in the development of Greek thought. However, Greek cosmologies were not built on anterior concepts of Finite, Infinite, and suchlike, but, on the contrary, were an important setting for the development of such notions, and it would therefore be contrary to our purpose to start out with fixed conceptions on our part. Rather, we are assuming that we are given citizens' conceptions of Finite and Infinite, common to a general

This article was typeset by the American Mathematical Society. The original version appears in Scientia **103** (1968), 511–531.

stratum of intellectuals, and we are asking again: is the universe infinite or finite?

The 19th century took it generally for granted that the universe is infinite, even as all-infinite and wide-open as Giordano Bruno (1548–1600) had depicted it [**1**], although as seen from our retrospect, there were some adumbrations of other developments yet to come [**2**]. A decisive change occurred early in the 20th century, with a publication, in 1917, of a memoir by Albert Einstein [**3**]. He bluntly suggested, as a scientific speculation, that the universe might be not only finite, but even closed, that is topologically compact, which is the most restrictive mode of finiteness there is; and ever since, our whole world, both of reality and of thought, has not been the same again, literally so. There are nowadays many cosmological models under scrutiny, some finite, and some not, and, for instance, the models with «continual creation » (Bondi, Gold, and Hoyle) are rather removed from ordinary finiteness, but even the latter universes are not as untrammeled in their unboundedness as the public's universe between 1700 and 1900 A. D. used to be.

The following can be taken as a measure of the change in out-look. In England, in the 17th and 18th centuries, the Platonist Henry More (1614–1687) and the scholar Richard Bentley (1662–1742), among others, insisted on the infinity of the universe virtually as a profession of Anglican faith [**5**]; but, in the 20th century, in 1931, the bishop of Birmingham, the mathematician, natural philosopher and church scholar Ernest William Barnes (1874–1953), stated: «It is fairly certain that our space is finite, though unbounded. Infinite space is simply a scandal to human thought ... the alternatives are incredible» [**6**].

On the whole, in our contemporary thought the size of the universe is somewhere between finite and infinite, and the theme of the present paper is the fact that, from naiveté or precociousness, such was also the case in Greek thought. Especially, Aristotle's universe was not quite as finite as usually presented, and the universe of the first atomists Leucippus and Democritus was certainly not as infinite as sometimes asserted. As to its extent, the Greek universe was, for the most part, «indefinite», but in Parmenides its structure is «complete», and it is the main aim of the present study to suggest that this notion of Complete lends itself to a certain measure of mathematization, even if the notion of Indefinite probably does not.

Yet, the notion of Indefinite, however unrigorous, runs through the entire course of modern cosmology too. René Descartes (1596–1650), the modern paragon of Right Reason and Clear Thinking, called his universe, which he termed *extension*, indefinite, and no amount of badgering by Henry More made him change this into infinite [**7**]. Also, Nicolas Copernicus (1473–1543) before Descartes, and Isaac Newton (1642–1727) after Descartes, both called the size of universe «immense», which was their version of «indefinite», and nothing would make them change either [**8**]. Finally, presentday assessments of «the Depth of the Expanding Universe» are couched in similar terms. Even if one considers the possibility that ever more powerful telescopes would reveal even more galaxies, yet the red-shift in the waves coming from the distant galaxies — whether it be explained on the theory of Edwin P. Hubble (1889–1953) or on the theory of E. A. Milne (1896–1950) — reduces the «effective» presence of distant galaxies relative to us, the more so the more remote the galaxy; thus, in a peculiar sense there arises a «rim of the universe» at which the galaxies «fade into nothingness» **[9]**, which is a manner of saying that the size of the universe is indefinite rather than infinite.

Before turning to our topic itself, I wish to state, expressly and with emphasis, that we are not in the lest suggesting that, in cosmology, the 20th century resumed where Classical Antiquity left off. Far from it. The Greeks, most clumsily, clogged their models of the universe with various features of centrality, circularity, and sphericality, which were restrictive, frustrating, and exasperating; and it required protracted exertions by Cusanus, Copernicus, Galileo, Kepler, and Newton, to clear these impediments away **[10]**. We are only suggesting that in the question of the overall size of the universe the Greeks had remarkable anticipations, and that they arrived at insights which are somehow perdurable.

2. The Infinite and the Finite

The common Greek word for «The Infinite» is *to apeiron* (τὸ ἄπειρον; *a* = non, *peras* = bound, limit) **[11]**, and it has many uses, in Science and Letters, and Poetry and Prose **[12]**. In its meaning — which has to be elicited from the context — *apeiron* ranges from: infinite, illimited, unbounded, through the exceedingly important byways of: indefinite, undetermined, undefinable, down to: immense, vast, huge, etc. **[13]**.

Apeiron became a word of Western stature at the very beginning of Greek Rational Philosophy, when it was made into a keyword of natural philosophy by Anaximander of Miletus (610–545 B.C.), [sic] a younger contemporary of Thales, of whom Diogenes Laertius ventured to say, that it was Anaximander — and thus not Thales — with whom (Greek) speculative philosophy actually began **[14]**.

Long before that, *apeiron* occurs, directly or by a near cognate, a few times in Homer, and the *Iliad* virtually opens and closes with it. In fact, in chant I of the *Iliad*, in line 13, the priest Chryses is said to carry to the ships of the Achaeans, a *huge* ranson (απερείσ αποινα; apereisi' apoina) for the freeing of his daughter. And in chant 24, line 776 expresses that it was a *countless* throng (δῆμος απειρων; *demos apeiron*), before whom Helen spoke for the last time, wailingly.

On the other hand, long after Anaximander, on the threshold of the second phase of Greek civilization, nemely [sic] in the beginning of Hellenism (325–300 [sic] B.C.), *apeiron* occurs in one of the most celebrated sentences from Greek mathematics, namely in Euclid's axiom on parallels, thus:

«That, if a straight line falling on two straight lines makes the interior angles on the same side less than two right angles, the two straight lines, if produced *indefinitely* (επ απειρον; ep' *apeiron*), meet on that side on which are the angles less than two right angles» **[15]**.

There was such a diversity of shadings of the meaning of *apeiron* that no overall antonym to it could be formed. If *apeiron* means *infinite* in a quantitative sense, then its usual antonym, with the meaning of *finite*, is *peperasmenon* (πεπερασμένον) **[16]**. But, unlike *apeiron*, *peperasmenon* never acquired a literary cachet. It was not used by poets or poeticaly [sic] oriented Presocratic philosophers, and even Pythagoreans were not attracted to it. Aristotle reports on a Pythagorean table of 10 pairs of opposites, which is headed by the pair finite-infinite. The word used for infinite is of course *apeiron*; however, the word for finite is simply the noun *peras* (bound; πέραζ) **[17]**, and Plato imitates this Pythagorean usage in a lengthy passage in the dialogue *Philebus* (23C–26D) **[18]**.

An interesting expression for *finite* occurs in Parmenides (6th century B.C.). Actually, as we will try to show later on, the Parmenidean notion of *finite* is nearer to our conception of *complete* than to our true conception of *finite*, but for the present this need not be taken into account. Parmenides constructs, out of his own imagery, an extra-ordinary universe, a kind of sphere (σφαῖρα; sphaira) **[19]**, in which cosmological and ontological features strangely blend. This sphere is in some sense *finite*, and Parmenides expresses this by saying, twice, that it is *en peirasi* (εν πείρασι) **[20]**, which can be translated either by: *in* bounds **[21]**, or by *within* bounds (or limits) **[22]**. The second translation has the advantage, if an advantage it be, that in presentday mathematics it could pass as an equivalent for the technical term *finite*; although, in its own contexts in Parmenides, the word *peiras* exudes a vigor and mystique of poetry **[23]**.

3. The Indefinite

Biblical poetry glories in the Divine Creation of the Universe, but there seem to be no allusions to the infinite or indefinite size of cosmic dimensions. But in the works of Homer and Hesiod, although they are much less theocratic, there are formulaic expressions about the vastness of Earth and Sea, and the word *apeiron* and equivalents occur **[24]**. The oldest Greek expressions of vastness refer to horizontal extensions, and statements of an indefinite extent of what is vertically above or below the horizon were slower in emerging. There is in Hesiod an allusion to the depth of Tartarus but not a statement that it is indefinitely great **[25]**.

However, from the 6th century B.C. there are preserved two clear poetic expressions about the indefinite (*apeiron*) extent of what is underneath the surface of the Earth, one in a fragment attributed, somewhat uncertainly, to the great lyrical poet Pindarus, and the other in a fragment of Xenophanes, a poet of Theology and Natural Philosophy.

The first fragment (#130 Bergk), a dirge (*threnos*), says this about rivers of Hell:

«From the other side sluggish streams of darksome night belch forth a boundless (*apeiron*) gloom» [26].

The fragment of Xenophanes (B28, Diels-Kranz), in an excellent translation by Guthrie, is as follows:

At our feet
We see this upper limit of the Earth
coterminus with air, but underneath
it stretches *without limit* (es apeiron) [27].

The *apeiron* in this fragment clearly refers only to what is under the surface and it is not justified [28] to apply it also to what is above the horizon. It is less clear whether *apeiron* in this fragment means indefinite or infinite. Commentators have been debating the point for the last 100 years, and most of them opt for indefinite [29]. For my part I suspect that Xenophanes was somewhat fuzzy in logical thinking, did not have in his mind a clear distinction between infinite and indefinite, and would not have grasped what the debating of our commentators is about.

About 200 years after Xenophanes had made this pronouncement about the Depth of the Underneath, Aristotle was apparently reminded of it, when in *De Caelo* (294a 11–294b 30) he collected views of predecessors on how and why the Earth is at rest in the universe, as it seemingly is. Aristotle avers that some of the solutions of the problem offered are worse than no solution, and he outright assails Xenophanes — of whom he altogether did not have a very high opinion [30] — for saying that the Earth «pushed its roots to infinity» (*ep' apeiron*), thus suggesting that it extends downwards indefinitely. And Aristotle buttresses his indictment of Xenophanes by quoting, verbatim, a passage from Empedocles which — without mentioning names — sharply rebukes all those who, ignorantly but mouthfully, assert that the deeps of the Earth and the ample ether are unbounded (*apeiron*) [31].

We have recited this because, very long after Aristotle, in the 6th century A.D. in the twilight time between Latest Antiquity and Earliest Dark Ages, Simplicius, the great Aristotle commentator, has made to all this the following brilliant observation (Xenophanes, fragment A47, Diels-Kranz) which could be lifted from a chapter on relativity of motion in a presentday [sic] book on Principles of Mechanics.

«I have never had an occasion to read the pertinent lines of Xenophanes, and thus I do not know whether in saying that the underneath of the earth is infinite (*apeiron*) he understood that it [that is, the earth] is immobile, or that the space, and the ether under the earth, is infinite so that the earth appears to be at rest even though it is unceasingly moving downwards».

This is followed in Simplicius by the following mild criticism of Aristotle and Empedocles:

«For, neither does Aristotle clarify the matter, nor do the verses of Empedocles analyze it clearly; in fact, «deeps of the earth» could also signify those «stretches» into which it descends».

The observation on the relativity of motion is good Science, the critcism however is doubtful History of Science. There is nothing to suggest that the original text of Xenophanes, even in full, would have enlightemed [sic] Simplicius on his alternative, or in fact whether Xenophanes was intellectually equipped to comprehend its meaning. Scientific pronouncements in the 6th century B.C., however «rational», were still aphoristic and oracular, rather than systematic and analytical, and only in the course of the 5th century B.C. would the query of Simplicius have become generally comprehensible.

Aristotle was already half-aware of this complication incident to historical developments [**32**], and, on the whole, when judging a philosopher's formal rigor of reasoning, Aristotle tended to be the less exacting, the earlier the activity of the philosopher. Thus, although Aristotle was already fairly critical of Xenophanes, he was still quite understanding towards Anaximander, who had lived about half a century earlier than Xenophanes, and Aristotle was probing the «true» meaning of Anaximander's *apeiron* with restraint, patience and respect.

It is a known fact that Aristotle is obviously reluctant to mention Anaximander by name. In the lengthy essay on *apeiron* in Physica, Book 3, chapters 4–8, Anaximander is mentioned only once, in 203b 14, and even there not in the midst of one of the basic statements about the conception of the Infinite, and of its modes of existence and occurrence, but only within a near-liturgical declaration that the Infinite is divinely immortal and everlasting. This reluctance can be explained by the assumption that Aristotle, even with the book of Anaximander before his eyes, was not able to make out, in depth, what Anaximander's own *apeiron* really was, or was intended to be — as a datum of science, that is, — and that therefore Aristotle had to establish on his own what *apeiron* really was, «objectively», so to speak.

Aristotle's difficulties can be understood if one considers that the book of Anaximander was already over 200 years old when Aristotle was writing his own essay on *apeiron*: that it was the first book on the Science of Nature [**33**] perhaps a compendium of various sciences with a chapter on *apeiron* prefixed [**34**]; and that, by a statement in Simplicius — which may have originated with Theophrastus, the disciple and successor to Aristotle — Anaximander, in his writing, was using metaphors which were not easy to interpret [**35**]. Also, it happens frequently in pioneering works of science that the first author's statements about a novel conception of his, or about a novel interpretation of a previously occurring conception, are voluble, ambiguous, over-determined, and inconsistent. There are indications that Amaximander's *apeiron* was something «Undetermined», and that it combined various aspects of indeterminacy which were somehow merging into each other. In

Anaximander's vision of *apeiron* there were apparently involved, many-in-one, a «macroscopic» indefiniteness of size; a «microscopic» indefiniteness of divisibility; and an indefiniteness of the physical nature of a peculiar generative substance, an *arché*, which was neither space nor matter, yet perdured in an indefinite eternity, was responsible for growth and decay, and embodied the divine attributes of the imperishable and the immortal.

In science, such accumulations of traits of indefiniteness in one overall conception are not at all a mark of primitiveness or of hoary antiquity, and they are likely to come about whenever observation, or experimentation, has outrun ratiocination. By any account of the physics of Descartes [**36**], his *matrière subtile*, especially when «fused» with his *extension*, has all the attributes of an *apeiron* as enumerated. Especially, as already stated in the Introduction, Descartes firmly maintained that the size of his *extension* is indefinite and not infinite; and we are going to adduce another example of the proximity of basic tenets of Descartes to Greek antecedents. The Greeks never identified their space of Nature, which was *chora* or *topos* (= place) [**37**], with their space of mathematics; and for the latter space they even did not have a conceptual name, not anywhere in Euclid, Archimedes, Apollonius, and other sources. Descartes does not have a conceptual name for the space of his *La Géometrie*, calling it, very maturely, *espace qui a trois dimensions* [**38**], but he apparently does not «identify» it with his *extension*, which, from what I can see, is his space of physics that corresponds to the Greek *chora-topos*.

A more recent *apeiron*, which, in a sense, is descendent from the Cartesian «subtle matter», is the Aether of the 19th century electrodynamics; it has retained, in modified form, the principal traits of Cartesian indefiniteness, as can be inferred from systematic works about it [**39**].

Finally, the 20th century has created an entirely novel *apeiron*, namely the Elementary Particles, and all features of indefiniteness of the original *apeiron* of Anaximander are present, and yet startlingly different. The macroscopic indefiniteness of size has become the law of Max Born that for large assemblages of particles in space the density of occurrence is a probability and not a certainty; the «microscopic» indefiniteness is the so-called indeterminacy law of Werner Heisenberg that for two or more quantitative attributes of a particle there are theoretically fixed lower bounds for the optimal degrees of precision with which the quantities can be observed conjointly; the indefiniteness of the physical nature of the *arché* has become the law of Louis de Broglie that every elementary particle has two possible realizations, a corpuscular and an undulatory, depending on the setting; they are «imperishable» and «immortal», in the sense that «annihilation» and «creation» of particles is nothing other than a transmutation of one kind into another; and lastly the particles are paired in opposites of «Particle» and «Anti-Particle», and presentday explanations for this are, in truth, not more convincing than Pythagorean explanations of the even-odd, one-many, square-oblong, etc. [**40**].

Returning to the Greeks we observe that then as today physics was propelled by visionaries and by workers, and if Anaximander was a visionary then his younger compatriot Anaximenes was a worker, one of the model-physicist-workers of all times. If «le style, c'est l'homme», then a penetrating characterization of Anaximenes was given by Diogenes Laertius when saying of Anaximenes, eight centuries after his time: «He writes simply and unaffectedly in the Ionic dialect» [**41**]. Also, Diogenes usually tries, if he can, to report some enlivening anecdotes about his «eminent philosophers», but Anaximenes is one of those about whom he knows none.

Anaximenes, being a solid physicist of his era, was much concerned with the occurrence of various states of aggregation in nature, and he taught that they arise by suitable physico-chemical transformations of a single generative substance, which he called *air* (*ἀήρ*; *aer*), whatever it really was. Apparently the universe as a whole was for Anaximenes one large physical system, the largest there is, and he was reluctant to make a statement about his air or his universe, that it was *apeiron*, that is, indefinite or infinite [**42**]; except that in the only surviving fragment of his (Bl, Diels-Kranz), which is not very reliable, Anaximenes says: «Just as our soul, which is air, holds us together, so *pneuma* (*πνεῦμα*; breath?) and air surround the whole cosmos»; and this is usually taken to imply that, for Anaximenes, the cosmos is immense or *indefinitely large*, or so, because the following intriguing parallelism.

According to Aristotle, a fairly early Pythagorean speculation in cosmology links *pneuma* with *apeiron*, together with *Void* (*κενόν*; kenon) and *Number* (*ἀριθμός*; arithmos), and Aristotle puts it thus (Physica, Book 4, chapter 6; 213 b22):

«The Pythagoreans too, hold that Void exists and that *pneuma* and voir enter from the *apeiron* (the Unlimited ?) into the Heaven itself, which as it were, inhales; the void distinguishes the nature of things, being a kind of separating and distinguishing factor between terms in series. This happens primarily in the case of numbers; for the void distinguishes their nature» [**43**].

This Pythagorean suggestion that there is an extra-cosmic vastness outside and beyond the cosmos proper, and that, in some sense, it is an *apeiron*, has been made from time to time since; and the Pythagorean assertion — made in terms of «inhaling» — that this vastness interacts with «substances» has an uncannily modern ring. Almost any presentday theory of the Expanding Universe somehow assumes that the remotest galaxies at the rim of the Universe, when «advancing», are also creating in front of them a «Void» into which they advance, and that this Void becomes a medium for the gravitational, electromagnetic, and perhaps also hydromagnetic interactions of the galaxies among themsvelves [sic] [**44**].

4. Innumerable worlds

Presocratic philosophers, and also later ones, were wont [sic] to suggest that there are, or might be, innumerably many worlds, where «world» stands for *ouranos* (ουρανόζ; Heaven) or *cosmos* (κόσμοζ) [**45**], and we wish to point out that these suggestions did by no means automatically imply that the *total* universe (τό πᾶν; the *All*) has to be outright infinite.

In some cases in which the Ancients speak of innumerable worlds it is, form [sic] our contemporary attitude, not even appropriate to do so at all, because the multiplicity of the worlds was meant to be only a successive one. Thus, in the cosmology of Empedocles [**46**], and also in the system of the Stoics [**47**], there apparently was a single self-same universe which however was cyclically decaying and regenerating in a periodic rhythm. The cosmological system of Heraclitus [**48**], if he had [**49**], may have also been of this kind; and the cosmological views of some Pythagoreans may have also been so oriented [**50**]. It is my personal judgment that Greek speculations about cosmic cycles were intellectually far beneath Greek scientific achievements in general, and that they were somehow so prescientifically vague that it is not even appropriate to inquire whether the universes in question were finite, or infinite, or what else. In a vague sense, though, these universes were bounded, and to this we observe that 20th century cosmology has mathematically envisaged cyclic universes which are even closed and strictly periodic; and by relaxing the strictness of periodically one another could even allow for thermodynamic irreversibility too [**51**].

While the purely cyclic universes were still mythologically tainted, there was a deceptively similar type of a Presocratic universe which is scientifically very appealing. According to Aristotle, there was an important cosmological hypothesis, professed by some, that there are «innumerable *cosmoi*, some in process of becoming and others in process of perishing», and that «these processes of becoming and perishing of necessity involve motion» [**52**]. Aristotle does not associate names of originators with this hypothesis, perhaps because it was, or was becoming rather widespread [**53**] and also had not had a clearly specifiable first proponent. But it seems certain that the chief expositors and propagators of this hypothesis had been Leucippus and Democritus (5th century B.C.), the first and most imaginative atomists ever [**54**], although some versions of the hypothesis may have been conceived earlier, perhaps already by Anaximander [**55**].

It is doubtful whether Leucippus and Democritus had it in their thinking that there are infinitely many worlds which co-exist simultaneously. The few mentions of innumerable worlds in testimonials about them do not suggest this [**56**], but they suggest that, in the thinking of the first atomists, the accumulation of *cosmoi* (= worlds) in their universe behave like an accumulation of stars in a galaxy of today, or better yet, like an accumulation of galaxies in a cluster of galaxies of today. In fact, their *cosmoi* were formed out of atoms by some kind of «gravitational clustering» and they were dissolved again by some kind of «gravitational break-up», with vortices (*vortex*= δινη, *dine*) of inexplicable origin [57] playing a part in either process. The number of such *cosmoi* is invariably stated as *apeiron*. If we translate this by *immense* then this fits most evolutionary theories of today, and the translation: *indefinite* would fit the «continual creation» theories of today. But it is neither necessary nor appropriate to translate it by: *innumerable*, in the sense of *outright infinite*, as usually done.

Finally, if Anaximander, who lived considerably earlier than the atomists, did indeed say that there is a multiplicity of worlds, it would seem hopeless to try to decide today whether he meant temporal succession or spatial plurality. Anaximander may not have felt an intellectual urge for absolutely distinguishing between successive and synchronous plurality, and at the same time he may have vaguely felt that, as in contemporary relativistic continual creation cosmologies, such distinctions cannot even be made absolutely. From our retrospect, an early Greek philosopher could very well have been «primitive» and «advance» at the same time.

5. The Sphere of Parmenides

Parmenides of Elea envisaged a twofold universe; an extraordinary sphere which was the embodiment of Pure Being, and which we have already introduced in Section II above, and a «stellar system» which was the world of experiential reality. We will concern ourselves only with the sphere of Pure Being.

Parmenides was neither a mathematician nor a physicist, nor even a natural philosopher, but only an ontologist, that is, a species of philosopher which he himself created, and the Parmenidean sphere has, for all its ontological splendor, the irreality of a glass universe. Yet, when searching for an ontoglical attribute that would express the size of this universe Parmenides adumbrated a general mathematical conception of mathematics of today. From our retrospect, Parmenides does not say that by size his sphere is infinite, or indefinite, or that it is finite, as a first impression might suggest; but what he says is that by structure it is *complete* (τετελεσμένον; *tetelesmenon*), and he says it twice, once «positively», and once «negatively».

The positive statement is (Fragment B8, 42–45):

«But since there is a furthest limit, it is in every direction *complete* (*tetelesmenon*); like the body of a well-rounded sphere, from the middle everywhere of equal strength»,

and the negative is (Fragment B8, 30–33).

«For powerful Necessity holds it in the bounds of the limit, which encircles Being, because it is *not* right for Being to be *incomplete*, for it is not in need» **[58]**.

In presentday mathematics there are two separate but related notions of completeness, one for a vectorspace [sic] with perhaps topological or algebraic structures on it, and another for a topological space with an additional structure superimposed, say, for a manifold with a Rienammian metric on it. Since our sphere is a manifold, we are imputing to Parmenides that he adumbrated the second case.

Now, let us call a combination of a manifold with a structure a universe and then view such a universe as *complete* if it is non-continuable; meaning that the manifold cannot be extended, *under the preservation of the given structure*, into a larger manifold of the same dimension. To put it differently but equivalently, in this view a universe is complete, if there is no other equi-dimensional manifold, with the same kind of structure superimposed, of which the given manifold is a proper part.

By this definition, the most assuredly complete universe arises if, in the terminology of presentday topology, the underlying manifold is compact (or «closed»), so that as a combinatorial complex it has boundary zero, because, for such a manifold, the universe is complete, in our sense, for whatever structure may be superimposed on the manifold additionally. 20th century cosmology is somehow aware of this, and it is this which makes cosmologists revert to closed universes ever so often.

But many, and even most universes under investigation «open», that is not compact, and we are going to verify, by exhibiting a model, that a universe may be complete and yet open, even «maximally» open in the following sense. We take it for granted that a single Euclidean cell is «maximally» open, and that its combinatorial boundary is topologically «maximal»; and we are going to exhibit a universe of the following kind: (i) the underlying manifold is a topological cell, (ii) there is a structure superimposed which makes the resulting universe complete, and yet (iii) the structure is such that, in a very meaningful way, every point of the topological boundary of the manifold is relative to the structure, a «furthest limit» (πεῖραζ πύματον = *peiras pymaton*) as demanded in the first Parmenidean passage quoted above.

The manifold of our universe will be a part of 3-dimensional Cartesian coordinate space, and the superimposed structure will be a Riemannian metric. We denote the three coordinates by (x, y, z). We are introducing the distance from the origin

$$r = (x^2 + y^2 + z^2)^{\frac{1}{2}}, \tag{1}$$

so that $0 \leq r < \infty$, and for $r \neq 0$ we are introducing the direction cosines

$$\alpha = \frac{x}{r}, \qquad \beta = \frac{y}{r}, \qquad \gamma = \frac{z}{r}, \tag{2}$$

for which

$$\alpha^2 + \beta^2 + \gamma^2 = 1. \tag{3}$$

We chose for our manifold the interior of a ball of some finite radius r_0, that is the pointset given by

(4) $$x^2 + y^2 + z^2 < r_0^2,$$

or, equivalently, by

(5) $$0 \leq r < r_0.$$

It is topologically a cell, and the boundary of the cell is the 2-dimensional spherical surface

(6) $$x^2 + y^2 + z^2 = r_0^2.$$

Thus, the boundary consists of the points

(7) $$x = r_0\alpha, \quad y = r_0\beta, \quad z = r_0\gamma$$

for all possible triplets of number (α, β, γ) satisfying (3). Our Riemannian metric in (4) shall be given by a line element

(8) $$ds^2 = h(r)^2(dx^2 + dy^2 + dz^2),$$

in which the factor $h(r)$ is defined, continuous and differentiable on the half-open interval (5), and is positive there,

(9) $$h(r) > 0, \quad 0 \leq r > r_0. \text{ [sic]}$$

Since $h(r)$ depends only on the quantity (1), this metric is invariant with respect to rotations around the origin, and thus our universe is indeed «from the middle everywhere of equal strength», as stipulated in the Parmedidean passage above. Consequently any radius vector of the universe, that is any pointset

(10) $$x = r\alpha, \quad y = r\beta, \quad z = r\gamma; \quad 0 \leq r < r_0$$

for fixed (α, β, γ) is a geodesic line. It terminates in the boundarypoint [sic] (7), which thus becomes a «furthest limit» as Parmenides requires it. The geodesic distance of a point (x, y, z) within the ball from the center of the ball is

(11) $$\int_0^r h(t)\, dt,$$

and if we let this point move towards its «furthest limit», this distance becomes

(12) $$\int_0^{r_0} h(t)\, dt,$$

whether the value of this integral is finite or infinite.

If the value is infinite, our universe is complete. Contrariwise, if it were not complete, our ball would be part of a larger manifold with an extension of the given Riemannian metric. Therefore, at least one of the points (7) would become an interior point of the larger manifold, and its geodesic distance from the center of the ball would again be the integral (12), which would have to be finite, contrary to the fact that it is infinite.

In particular, the integral (12) is infinite if

$$h(r) = (r_0^2 - r^2)^{-1} \text{ [sic]}$$

In this case our universe is the Cayley-Klein model of the non-Euclidean hyperbolic geometry of Bolyai-Lobachevsky **[59]**, and thus we see that this universe is complete.

Its «dual», namely the non-Euclidean elliptic geometry of Riemann-Beltrami, is also complete, on the very strong grounds that the underlying manifold is closed, that is, topologically compact.

This leaves us with the Euclidean case, in which the ball (4) is endowed with the metric

$$ds^2 = dx^2 + dy^2 + dz^2, \tag{13}$$

which is subsumed under (8) for

$$h(r) = 1.$$

In this case the integral (12) has the finite value

$$\int_0^{r_0} l \cdot dt = r_0, \tag{14}$$

and our ball (4) is indeed not complete. In fact every domain of the Euclidean space which contains the ball is a continuation of it. Furthermore, if in (14) we let r_0 tend to ∞, then the length becomes infinite, and this can be interpreted to mean that the total Euclidean space, if endowed with the metric (13), is a universe which is likewise complete. Also, even in the case every radius vector (10) points to a «furthest limit» at infinity which is located in the direction (α, β, γ); and these furthest limits taken together can be viewed as a spherical surface at infinity which encloses the universe as in Parmenides **[60]**.

As evident from a lengthy passage in Aristotles's essay on the infinite (206b 33–207b 32), in which Parmenides is mentioned by name, Aristotle was aware of the fact that the universe of Parmenides was intended to be complete, and he gives the following very good brief characterization of completeness:

«When there is nothing left over and beyond, there is something *complete*, that is a *whole*, and a *whole* (for example, a whole man or a whole box) is, by definition, what has nothing wanting» **[61]**.

However this good insight is marred by Aristotle's misconception that completeness is not compatible with infinity, and his views on the relation between infinity and completeness can be obtained from the following three sentences taken together:

«That is infinite, of which it is always possible, in regard to quantity, to take a part outside what has already been taken (207a 7–8). [sic]

«The dignity and all-embracing quality of the whole is only ascribed to the infinite because of a superficial confusion of the infinite with the whole (207a 20–22). [sic]

«As infinite, the infinite does not contain, but is contained; as infinite it is therefore also unknowable, for matter is formless. Evidently then, the infinite is to be treated as an *aspect* [μόριον; morion] rather than as a whole (207a 26–28)».

From whatever reasons, Aristotle was opposed to (outright) infinitude, especially in physics and cosmology. Therefore Aristotle was pleased with Parmenides for not displaying infinity in his universe, and he was angry with Melissus, a follower of Parmenides, for inserting into the universe of Parmenides some feature of Infinitude of his own **[62]**. Sometimes Aristotle speaks quite contemptuously of Melissus, but in the present context he puts it relatively mildly thus:

«The view of Parmenides that the world being equal in every direction from the middle is limited, is to be preferred to the view of Melissus that the whole is infinite (207a 17–19)».

If Aristotle was not right in assuming that completeness and infinity are incompatible, then, 250 years later, the great Latin poet Lucretius (Titus Lucretius Carus; 1st century B.C.) was not right in asserting (*De natura rerum*, Book 1, 958–987) that our universe cannot be represented by a ball of finite radius r_0, as in the hyperbolic geometry, but only by a ball of infinite radius, as in the Euclidean geometry. Lucretius uses a «flying-dart» argument against the finiteness of r_0 and the following lines are frequently quoted:

«Suppose now that all space were created finite, if one were to run to the end, to its furthest coasts, and throw a flying dart, would you have it that this dart, hurled with might and main, goes on whither it is sped and flies after, or do you think that something can check or bar its way» **[63]**.

Thus, Lucretius takes it for granted, that if, by chance, the underlying manifold of our universe can be represented by a ball of finite radius r_0 in a Cartesian space, then the Cartesian exterior of the ball, that is the pointset given by

$$x^2 + y^2 + z^2 \geq r_0^2, \tag{15}$$

somehow intellectually clings to our universes and could not be structurally broken away form it.

In defense of Lucretius, and of many other scientists and poets after him, it can be said that only within the last 100 years or so did mathematics establish, clearly and firmly, that it is possible to view a ball (4), even for finite r_0, as a «space» entirely by itself, self-sufficiently and self-containedly, without any intellectual compulsion to associate the pointset (15) as its «exterior» with it. A mathematical anticipation of this relatively recent insight can be traced back to Nicolas of Cusa (15th century A.D.), but not beyond, except that physical and cosmological adumbrations of it can be detected in Aristotle, as we will still see in Section 7.

But before turning to this we will make a further observation on completeness of the universe, even if it does bear on the problem of the size of the universe, which is the avowed topic of the present paper.

6. Stability and Stationarity

The main ontological conclusion of Parmenides, in its physical aspect, can be decomposed into two successive steps. Firstly, since the universe is structurally complete it is physically stable, and it thus persists in a certain uniform sameness, eternally, even non-temporally; and secondly, since the universe is stable, it is inert and immobile, and everything is also ungenerated and imperishable.

The second half of the second step, namely that everything and anything is ungenerated and imperishable was, by itself, not too disturbing. Thus, the atomists Leucippus and Democritus, to whom matter was atomistically discrete, and Anaxagoras, to whom matter was molecularly granular [**64**] could interpret it to mean that generation and corruption came about simply by association and dissociation of immutable particles, provided the particles, could move about. And in the outlook of Aristotle to whom matter was distributed continuously by granulation [**65**], all «genuine» generation or corruption of matter is nothing but a transformation of one kind of substance into another [**66**]; just as in our contemporary physics of Elementary Particles, the so-called «creation» and «annihilation» of particles is nothing but the transformation of particles from one kind into those of another kind.

But the first half of the second step, namely the assertion that the universe, being stable, is totally immobile, is unreal and unacceptable. There are nowadays two procedures for avoiding this assertion, and both were discovered by the Greeks.

The first procedure simply denies the first step in Parmenides that the universe is unalterably stable; it assumes that there are events which violate stability, and it is usually assumed that the manifestation of such an event is the formation of a vortex. The Greeks knew this [**67**]; even Heraclitus and Empedocles, in their very vague ways, assumed that there were some kind of vortical instabilities occurring periodically. But, much more maturely, Anaxagoras and also the atomists, in their several ways, assumed that a vortical unstability [sic] occurred once, and only once, and brought about the creation of the universe; in modern times there have been many such assumptions, beginning with the Kant-Laplace Hypothesis in the second half of the 18th century, and any such assumption must concede the working of an extra-rational Power, that is, of a *Nous*, as Anaxagoras called it.

A second procedure was inaugurated by Aristotle (Physica, Book 1, chapter 2). Aristotle does accept the first step of Parmenides that our universe is unchangeably stable, but, very scientifically, he equates ontological stability with physical stationarity; and in physics stationarity is not at all identical with immobility, far from it. Thus, in presentday acoustics and hydrodynamics, a stationary process may admit a general periodic motion, physically similar to the circular motion which Aristotle posits for his universe in Book 8 of the *Physica* and in *De Caelo*. Furthermore, in presentday cosmology, even steady-state models which admit «continual creation» are stationary in a manner that makes them subsumable under the stability of Parmenides, in a sense.

7. Place and Matter

In Aristotle's *Physica*, the essay on Infinity (Book 3, ch. 4–8) is followed by an essay on Place (*topos*, *τόποζ*; Book 4, ch. 1–5), which we have previously analyzed from our own approach **[68]**. For any proper part of the universe Aristotle defines its place as «the inner boundary of what contains it», meaning that the place of any proper part of the universe is determined by the internal boundary of what surrounds it in the universe. In the last chapter of the essay, that is in chapter 5, Aristotle tries, by various means, to assign a place to the universe in its entirety, and we are adding the following observation to what we have said about this before. Aristotle tries, however inarticulately, to express the fact that the universe, although as finite as a ball (4) of section 5 above, does *not* have an exterior, like (15), but nevertheless has a boundary, like (6), which is internally defined, and that it is this boundary which determines a space for it.

Understandably, Hellenistic and medieval expositors of Aristotle had almost insuperable difficulties with section 5, when they tried to fully and unequivocally elucidate it. Thomas of Aquinas in his own Commentary on the *Physica* (in Lecture 7 on Book 4) **[69]** reports, critically, on efforts by Alexander of Aphrodisias (260 A.D.), Themistius (360 A.D.), Avicenna (ibn Sina; 980–1037), Avempace (ibn-as-Saigh ibn-Bajji; –1138), and Averroes (ibn Rushd; 1126–1198). The nearest to our knowledge of today is the opinion of Alexander that «the ultimate sphere in no way is in place» (ultima sphaera nullo modo est in loco) **[70]** which we can interpret to imply that the intrinsic boundary of the universe — that is our spherical shell (6) — is not itself a part of the universe, even if it determines it.

In *De Caelo* Aristotle deals with the physics proper of the universe, and in Book I, ch. 5–7, he asserts and derives that, in a suitable sense, the diameter of the universe is finite. He deduces this mainly from his a-priori Hypothesis that the total mass of the universe finite, and, even more convincingly to himself, from his a-priori Hypothesis that the total angular momentum of the universe is finite. These two a-priori Hypothesis are highly respectable, and have many adherents among leading physicists and cosmologists of today; but they are compatible with a non-finite spatial extent of the universe if the possibility of an absolute vacuum is admitted. Aristotle however was a thermodynamicist [**71**], and in thermodynamics there is no absolute vacuum as a part of a physical system. More specifically, in *De Caelo* it is taken for granted that there is body (soma; σῶμα) everywhere, and, in our terms, it is apparently assumed that it has a minimal positive density; because Aristotle reasons that if the diameter of the universe were unbounded then so would be its total mass, which he finds incongruous, and moreover, due to the diurnal rotation of the globe, the total rotational momentum of the universe would be infinite, which Aristotle finds even more absurd [**72**].

Aristotle is less persuasive in his reasoning (*De Caelo*, Book 1, ch. 8–9) that there is only one «world», meaning in our actuality, only one solar system, or, perhaps, only one galaxy. What he actually tries to prove is that the gravitational force field of the universe cannot have more than one major singularity. Now, Aristotle admits both «heaviness» and «lightness», meaning that his gravitational constant between two different bodies may be positive or negative, and he tries to persuade himself that there can be no gravitational clustering into more than one major unit. But such clustering can occur in our present-day models of the universe, under the validity of the Newtonian force field, and it can probably occur in more complicated gravitational force fields too.

An intriguing remark on multiple worlds occurs in Aristotle's *Physica*, within his essay on Time. Aristotle mentions (218a 30–69), when listing predecessors who had spoken out on the nature of time, that, according to some, Time is the movement of the universe, and, according to others, even the celestial sphere itself. Aristotle observes to this that on such views, if there were more than one world, each of them endowed with its own standard of time, there might result various standards of time, being in effect simultaneously. In this conclusion Aristotle was nearly right. It is true that in contemporary cosmology the various time standards in different galaxies can be compared with each other by intercommunicating light signals, but the intervening time lags by millions of light years, the red shifts, and the bendings of gravitation do keep the various time standards separated from each other to a considerable extent.

Aristotle's rigid mandate that there must be matter everywhere in space, was undoubtedly not good for the development of physics in the long era between Antiquity and Late Renaissance. Nevertheless, the nucleus of the mandate, namely the presumption that in physics there cannot be space without any physical data at all, was greatly vindicated in the 20th century by the insight that every bit of space in physics is some kind of field, either a gravitational field generated by masses as in Relativity, or an electrodynamic or other de Broglie field as in Quantum theory, or both. However, in presentday physics the field which is dual to particles may expand and become indefinitely large even if the supply of corpuscles is finite and finite; that is, in Aristotelian terms, the total amount of *soma* may be finite, and yet the diameter of the universe not.

We will terminate on a daring statement. We dare to state that this duality between matter and field — which is an opposition and a correlation in one — was foreshadowed, peculiarly and quaintly, yet sharply recognizably, almost 2500 years ago, in the Atomism of Leucippus and Democritus; and we are suggesting that it is the adumbration of this duality which is the central achievement of Greek Theoretical Physics proper, and not just the introduction of a plurality of ultimate constituents of matter, in which Empedocles and even Anaxagoras also participated. Aristotle was not a partisan of Atomism, but he was aware of its originality, and especially in *Metaphysic*, Book 1, ch. 4, 985b 5, he states it very clearly, almost emphatically, that the Atomists' opposition between full and void — when transliterated into Parmenidean terms — becomes a polarity between Being and Not-Being; and, what does singular honor to the Atomists, and to Aristotle who reports on them, Aristotle clearly states that this Not-Being also exists, just as much as does the Being [**73**]. Diogenes Laertius was certainly right in saying that Democritus had been the Prince of Greek physics proper [**74**].

Notes

[**1**] This is apparently still taken for granted in Alexandre Koyré (1892–1964), *From the closed world to the infinite universe*, first printing by the Johns Hopkins Press, 1957. The report on Bruno is in chapter II, and in other places (see index there under Bruno).

[**2**] J. D. North. -*The measure of the universe, a history of modern cosmology*, Clarendon Press, Oxford, 1963, chapter 5, beginning on p. 70. This is a comprehensive work of its kind. A smaller book is Hermann Bondi, Cosmology, Cambridge University Press, 2nd edition, 1960.

[**3**] North, (see note 1), p. 72.

[**4**] North, (see note 2), 198–217, and passim (see index there under steady-state cosmology).

[**5**] Koyré (see note 1), chapters V, VI, VII; see also my forthcoming book: S. Bochner, *Eclosion and Synthesis*; *the Outburst of Knowledge after the Enlightnement*, [sic] chapters 12 and 14.

[**6**] North, (note 2), p. 134; also pages 271, 374.

[**7**] Bochner, (note 5).

[**8**] ibid.

[**9**] James A. Coleman. -*Modern Theories of the Universe*, Signet Science Library, 1963, p. 63, see also other passages.

[**10**] Bochner, (note 5).

[**11**] For another possible etymology see Charles H. Kahn, *Anaximander and the Origins of Greek Cosmology*, Columbia University Press, 1960, p. 231. Many of our statements about apeiron will closely resemble those made in the work of Kahn yet be different in intent, and we will not attempt to elaborate on the underlying differences of approach.

[**12**] There are several cognates of *apeiron* which are closely related to it in language and meaning; they do not occur frequently and, for our specialized purposes, we will not separate them from *apeiron* itself.

[**13**] When intending to express that something is infinite rather than indefinite the Greeks would usually employ an expression consisting of more than one word;

but when wanting to say that something is indefinite rather infinite they might use the adjective *aoriston* (ἀόριστον), for instance.

[14] Diogenes Laertius. -*Lives of eminent philosophers*, Book I, chapter 13; Leob edition, vol. I, pp. 14–15.

[15] Thomas L. Heath -*The Thirteen Books of Euclid's Elements*, Vol. I, p. 202. We will not deal with the problem whether the *ep'apeiron* of the axiom on parallels should be translated by *indefinitely*, as done in English, or by *in infinitum*, as done in Latin, French, Italian, German, etc.

[16] The antonym to *aoriston* (note 13) is *orismenon* (ὡρισμένον).

[17] Aristotle. -*Metaphysica*, 986a 24; see also, for instance, G. S. Kirk – J. E. Raven, *The Presocratic Philosophers*, Cambridge, 1957, p. 238. The most recent comprehensive account of Presocratic Philosophy is in W. K. Guthrie, *A History of Greek Philosophy*, Cambridge University Press, vol. I, 1962, vol. II, 1965. In volume I, p. 245, Guthrie lists the 10 pairs of opposites thus: limit-unlimited, odd-even, one-plurality, right-left, male-female, at rest-moving, straight-crooked, light-darkness, good-bad, square-oblong.

[18] See, for instance, K. Hackforth, *Plato's examination of Pleasure* (*The Philebus*), Cambridge University Press, 1945, pp. 37–49.

[19] The word *sphaira*, in the meaning of *ball* («to play at ball») occurs already in the Odyssee, Chant VII, line 100.

[20] Fragment B8 (Diels-Kranz), lines 26 and 49.

[21] So in Diels-Kranz, *Die Fragmente der Vorsokratiker*, and in many commentators.

[22] So in Kirk and Raven, (note 17 above), p. 276. We observe that in the fragment B8 of Parmenides, in line 26, the expression in full is *en peirasi desmon* (ἐν πειρασι δεσμῶν), which means: within limits of fetters.

[23] Guthrie (note 17), vol. II, p. 37, says that «In Parmenides the word *peiras* falls, again and again, on our ears with the force of a blow. [sic]

[24] To this and the following see Rodolfo Mondolfo, *L'infinito nel pensiero dell' Antichità Classica*, «La Nuova Italia», Firenze, 1956; especially «Parte Quarta». See also Kahn, (note 11). Mondolfo quotes a remark of Robert Rivaud that Chaos and Heavens are never called *apeiron*.

[25] Mondolfo, (preceeding note), p. 275–278.

[26] We are rendering the translation in the Loeb Edition of *The Odes of Pindar*, edited by John Sandys, Heinemann, London, 1927, p. 591. About the authenticity see Louis Richard Farnell, *The Works of Pindar, The Commentary*, Chicago University Press, 1932, p. 434. The French translation by Aimé Puech, *Pindare*, vol. IV, Fragments (Collection Budé), 1923, p. 196, is as follows: «De là sortent, en vomissant des ténèbres infinies (*apeiron*), les fleuves, au cours lent de la nuit obscure».

[27] Guthrie, (note 17), Vol. I, page 394. The French translation in Paul Tannery, *Pour l'histoire de la science hellène*, 2nd edition, Paris, 1930, p. 148, is as follows: Nous voyons sous nos pieds cette limite de la Terre, en haut, du côté de l'éther, mais le bas s'en va à l'infini.

[28] As done in Mondolfo (note 24), p. 275.

[29] See Mario Untersteiner. -*Senofane, Testimonianze e frammenti*, «La Nuova Italia», Firenze, 1956, p. CLIV–CLVII.

[30] S. Bochner. -*The Role of Mathematics in the Rise of Science*, Princeton University Press, 1960, p. 370.

[31] Compare Untersteiner (note 29), pp. 86–90 and other passages. See also Guthrie, (note 17), Vol. I, p. 394, footnote 2.

[32] In *Metaphysica*, 995a 30, Aristotle calls such a complication a «knot»

($\delta\varepsilon\sigma\mu\acute{o}\zeta$; *desmos*) in the development. Compare Kahn, **[11]**, p. 18, last two lines.

[33] See Kahn, **[11]**, p. 6.

[34] Guthrie, loc. cit., Vol. I, p. 29.

[35] Kirk and Raven, **[17]** p. 107.

[36] For instance, Mary B. Hesse, *Forces and Fields*, Nelson and Sons, London, 1961, 102–108.

[37] Bochner, **[30]**, chapter 4.

[38] La Géometrie, p. 368 (p. 146 in *The Geometry of René Descartes*, translated by David Eugene Smith and Marcia L. Latham, Chicago, Open Court Publishing Co., 1925).

[39] Sir Joseph Larmor (1857–1942), article «Aether» in *Encyclopedia Britannica*, 11th edition; also his extensive book: *Aether and Matter; a development of the dynamical relations of the aether to material systems on the basis of the atomic constitution of matter, including a discussion of the influence of the earth's motion on optical phenomena, being an Adams prize essay in the University of Cambridge*, Cambridge University Press, 1900. Also, Edmund Whittaker, *A history of the theories of Aether and Electricity*, Vol. I, and also Vol. II.

[40] Bochner, **[5]**, chapter 9.

[41] Diogenes Laertius, Book 2, chapter 3; Anaximenes, fragment A13, Diels-Kranz.

[42] See the excellent footnote in Antonio Maddalena, *Ionici, Testimonianze e frammenti*, Firenze, 1963, p. 180.

[43] Kirk and Raven, p. 252; Guthrie, Vol. I, p. 277.

[44] In addition to North and Bondi, **[2]**, see also G. J. Whitrow, *The structure [sic] and Evolution of the Universe*, Hutchinson, London, 1950; G. Gamov, *The Creation of the Universe*, Viking Press, New York, 1952; Jagjit Singh, *Great Ideas and Theories of Modern Cosmology*, Constable, London, 1961.

[45] About the meaning of *ouranos* and *cosmos* see the enlightening article of F. M. Cornford, Innumerable worlds in Presocratic philosophers, *The Classical Quarterly*, vol. 28, 1934, 1–16.

[46] Kirk and Raven, pp. 326–7; Gutherie, Vol. II, 167–183.

[47] E. Zeller, *Die Philosophie der Griechen*, 5th edition, 1923 (reprinted 1963), Part III, 152–160; there is an English translation of an earlier edition by O. J. Reichler, under the title: *The Stoics, Epicureans, and Sceptics*, London 1880, pp. 165–169.

[48] About Heraclitus there is a very extensive account in E. Zeller - R. Mondolfo, *La filosofia dei Greci*, Parte prima, Vol. IV (Eraclito), Firenze, 1961; see especially pp. 218–269.

[49] In my personal judgement Heraclitus was an aphorist, and not a systematic philosopher, let alone a scientist, and he did not have a cosmological system. See Bochner, **[30]**, pp. 332–335.

[50] Guthrie, Vol. I, p. 282.

[51] North, **[2]**, pp. 132–33.

[52] *Physica*, Book 8, chapter 1, 25b 18–21; our translation is adapted from the Oxford and Loeb renditions.

[53] For a critique of late reports about this hypothesis, see Cornford, **[45]**.

[54] Kirk and Raven, 409–414; Guthrie, Vol. II, 404–413.

[55] For a full account of various opinions relating to Anaximander see Guthrie, Vol. I, 106–114.

[56] For references to such mentions see Kathleen Freeman, *The Presocratic Philosophers*, 3rd edition, Oxford 1953, 302-3; virtually all passages there quoted are translated in Milton C. Nahm, *Selections from early Greek Philosophy*, 3rd edition,

New York, p. 100 and following pages.

[57] In presentday cosmology there are also vortices of inexplicable origin. See *The Structure and Evolution of Galaxies*, Solvay Institute, 13th Physics Conference, 1965, Interscience Publishers, p. 20.

[58] *Not Incomplete* is the translation of *ουκ ατελεύτητον* ; (ouk ateleuteton); «in need» = incomplete. Our translations are from Leonardo Taran, *Parmenides*, Princeton University Press, 1965, p. 86.

[59] R. Bonola, *La geometria non-euclidea; esposizione storico-crilica del suo sviluppo*, Bologna, 1906.

[60] Guthrie, Vol. 2, p. 38.

[61] Physica 207a 9–11. In this and the following several excerpts we are quoting from Richard Hope, *Aristole's Physics*, University of Nebraska Press, Lincoln, 1961, pp. 54–55.

[62] Presentday commentators do not find it easy to identify these features of infinitude. See Kirk and Raven, 299–302; Guthrie, Vol. 2, 206–209.

[63] Book I, 969–974. We are quoting from Cyril Bailey, *On the nature of things*, Oxford, Clarendon Press, p. 59.

[64] About Anaxagoras see Daniel E. Gershenson and Daniel A. Greenber, *Anaxagoras and the Birth of Physics*, Blaisdell, New York, 1964.

[65] About the granular aspect of matter in Aristotle, see Andrew G. van Melsen, *From Atomos to Atom*, Harper Torchbook, 1960, 38–41.

[66] W. D. Ross, *Aristotle, a complete exposition of his works and thought*, Meridian Books, 1959, pp. 101–2.

[67] Mondolfo, [**24** above], 309–326, also 383–389.

[68] Bochner, (note 37).

[69] See the English translation by Richard J. Blackwell, Richard J. Spath and W. Edmund Thirlkel, Yale University Press, 1963, 114 ff.

[70] For similar brief Latin sentences about the views of the other four commentators see Hope, [**61**], p. 86, footnote.

[71] Bochner, [**30**], pp. 158–9.

[72] Descartes also did not tolerate a vacuum, but spread his «subtle matter» all over his *extension*; however was only half an Aristotelian and so, as we have already stated in section III above, he only maintained that the size of the universe is indefinite, and not that it is finite. In the 19th century, the universe was filled with electromagnetic aether which was a descendent from the «subtle matter;» but the intellectual distance from Aristotle was such that there were apparently no doubts about the size of the universe on account of that.

[73] Kirk and Raven, 404–5.

[75] Bochner, [**30**], 319–20.

PRINCETON, PRINCETON UNIVERSITY, DEPARTMENT OF MATHEMATICS

THE RISE OF FUNCTIONS*

by Salomon Bochner

Summary. The heart of analysis is the concept of function, and functions "belong" to analysis, even if, nowadays, they occur everywhere and anywhere, in and out of mathematics, in thought, cognition, even perception.

Functions came into being in "modern" mathematics, that is, in mathematics since the Renaissance. By a rough division into centuries, the 17th and 18th centuries made various preparations, the 19th century created functions of one variable, real and complex, and the 20th century has turned to functions in several variables, real and complex. In the realm of complex variables, the 20th century has been largely working on themes set by the 19th century, some themes beguiling and pleasant, others harsh and unpleasant. Automorphic functions in several variables are intriguing, beguiling, and pleasant; singularities of functions or of manifolds of several variables are harsh, forbidding, and most unpleasant. Which of the two topics will last longer, and perdure into the 21st century is another question; probably the unpleasant one.

For both real and complex variables, the 19th century molded the general concept of function and also created large classes of special functions, but there was a difference between the cases of real and complex variables. In the case of real variables the molding of the general concept of functions and the creation of special classes of functions proceeded independently from each other and were performed by different authors, even though both activities received their motivations from mechanics and physics. In the case of complex variables, however, the molding of the concept of function and the creation of special classes of functions were proceeding in close intimacy and interaction, with Bernhard Riemann being the chief architect of the dual enterprise.

It is this which made the 19th century into an era of analytic functions *par excellence*. Also, in memory of this, the word "Funktionentheorie" in the title of German books used to indicate, and still indicates, the theme

* Supported in part by a grant from the United States Air Force Office of Scientific Research.

of functions of *complex* variables, and of no other; it being understood that functions of a *real* variable will *expressly* identify themselves as such. Thus, C. Carathéodory, a leading 20th century master of the theory of functions of a real variable, published in 1918 a large "Theorie der Funktionen einer *reellen* Veränderlichen." But, as recently as in 1950, a treatise of his on complex variables appeared under the title "Funktionentheorie," *tout court*. It is true that the treatise was put through publication posthumously. But this title was undoubtedly so found in the author's papers, and would have most likely become Carathéodory's own, had he lived.

The 19th century also created an in-between topic, namely the topic of the "Poisson summation formula" so-called (this appellation is a 20th century coinage), which seems to fall between real and complex variables, combining the two with each other and with the analytic theory of arithmetical forms and of algebraic numbers. Among other things, the Poisson summation formula generates, by way of "theta relations," a very remarkable class of special functions. These are the so-called zeta functions of number theory, algebraic geometry, and the theory of automorphic functions. When viewed by themselves, zeta functions appear to be rather isolated objects of analysis, but the Poisson summation formula as a substantive background links them with analysis at large.

In a very broad sense, the Poisson summation formula is the key to all and any "dualities" and "reciprocities" in mathematics, and hence also in mathematical physics. Dirichlet injected the formula into number theory, for all time to come, when he demonstrated that, by using the formula, it becomes "child's play" to fully derive the reciprocity law for Gaussian sums, over which Gauss had labored long and hard. The formula is also the natural setup for dealing with the remainder term in so-called lattice point problems for euclidean space. Finally, and most gloriously, Erich Hecke used the formula, and only this formula, for the derivation of the Riemann-Hecke functional equation for zeta functions over algebraic number fields.

Regrettably, there is no book as yet dealing with the derivation of various known zeta functions by means of the Poisson summation formula.

Antiquity. The Greeks, mathematicians and others, did not have the concept of a (mathematical) function in their thinking. It is not possible to discern in the body of Greek mathematics something that could be interpreted to be an adumbration of the notion of a function $y = f(x)$ as we know it today, or, at least, as it is discernible in the mathematics of the 16th and 17th centuries, say.

The Greeks did, of course, have some familiarity with categories of cognition such as "correspondence," "dependence," "mapping," even "binary relation," which enter into our present-day notion of a function. But the mere occurrence of such categories does not yet make for the presence of functions. Even the occurrence of assertions or conclusions which can be readily translated or transliterated into functional relations is not yet enough. It is still necessary, and this is decisive, that something also be "done" with those functions, or with the trains of reasoning corresponding to them, that is, that some kind of "mathematical" operations be performed with, or on, these functions, or with the trains of reasoning that "stand in" for functions. Now, it is this kind of "operational" activity, or only attitude, which it is difficult to discern in the realm of Greek thought, mathematical or other.

Thus, although the Greeks — and, in fact, Aristotle single-handedly — created the syllogistic aspects of our modes of formal deduction, they never broke through to a satisfactory conception of "relation," binary or *n*-nary, reflexive, invertive, or correlative. Book 5 of Aristotle's *Metaphysics* is a dictionary of some basic philosophical terms, and among others it has a lengthy entry on "relation" (*pros ti*). But the content of the entry is so embarrassingly ordinary that philosophers and logicians in general are not aware of it, and only "all-inclusive" commentaries of Aristotle take note of it. And of an *algebra* of relations, as begun by Leibniz (1646–1716) and insisted upon by the American mathematician-philosopher Charles Sanders Peirce (1839–1914), there is hardly a trace among the Greeks. Furthermore, historians of logic have been recently asserting, even heatedly, that the Greeks remained creative in the field of logic even after Aristotle, and that, specifically, some principles of our "modern" propositional modes of implication were already discovered by some of the Stoics [*1*, pp. 6–8]. But, here again, of propositional *functions*, that is of propositional schemes that involve "all" or "any," there is no trace [*2*, p. 32].

Even the great Archimedes does not have functions in his thinking, meaningfully that is. Isaac Newton's treatise on mechanics [*3*] is ostensibly composed in the style of Archimedes, that is, in terms of curves and geometric paths, all without coordinates. Yet Newton's treatise is, by its internal direction, function-oriented, whereas the work of Archimedes is not. For instance, Newton views the tangent to a planetary orbit at a point as the limiting position of a secant through this fixed point and a variable neighboring point of the orbit, meaning that he performs the operation of differentiation on "hidden" coordinate functions. Archimedes however adheres to the euclidean definition that a tangent to a curve is a straight line which *globally* intersects the curve at one point only, and he pretends to observe this definition even in his essay on (archimedean) spirals $r = c\theta$ (polar coordinates). Archimedes is aware of the fact that any straight line in the plane of the spiral intersects it in more than one point, and he apparently observes the euclidean requirement only *half*-globally, for a half-coil of the curve. But there is no tendency in the essay to make the requirement a properly local one.

Furthermore, Archimedes' law of the lever is a "conservation law" for the rotational momentum

$$f(p,l) \equiv p \cdot l, \tag{1}$$

(l = length of the arm, p = the suspended weight), meaning that

$$p_1 \cdot l_1 = p_2 \cdot l_2. \tag{2}$$

However, Archimedes could not envisage "operationally" a "function" like (1). Thus, he was unable to conceptualize the physical datum of rotational momentum, and he had to express the equality (2) in the euclidean (that is, Greek) manner as a proportion

$$p_1 : p_2 = l_2 : l_1.$$

This explanation of the intellectual limitation of Archimedes is seemingly different from, yet in fact very cognate with, our previous explanation that Archimedes was unable to conceptualize a product like $p \cdot l$ as a ring operation within the semi-ring of positive real numbers [*4*, pp. 181 ff.].

We note that this ring operation was first introduced by Descartes at the head of his *La Géométrie* (1637), and that the first formal definition of a mechanical momentum was given by Newton in his treatise. Newton introduced not the *rotational* momentum but the *translational* momentum. He called it "quantity of motion" and characterized it as the product of mass and velocity [*4*, p. 1], or rather as a bilinear functional on the cartesian product of mass and velocity.

Middle Ages. In the Middle Ages there were some stirrings of the kind of analysis in which functions are domiciled, and the most function-oriented medieval mathematician was Nicole Oresme (1323–1382). He devised a version of graphing for which he is renowned, and he also envisaged exponentiation a^r (for fractional exponents $r = p/q$) [*5*, pp. 288–295]. These two achievements, when taken together, certainly suggest functions of the kind that occur in later developments. Oswald Spengler in his *Decline of the West* (1922–1924) — whatever the shortcomings of the work as a whole —

rightly insists that with Oresme's anticipation of functions something very new was added to the classical Greek mathematics of Euclid, Archimedes, and Apollonius.

Since the second half of the 19th century, Oresme's standing as a harbinger of analysis in general and of the concept of function in particular has been steadily on the rise. But, on hard tangible evidence, it seems impossible to assay what Oresme's effect on subsequent developments of analysis actually was. His works were known in the 15th and 16th centuries [*6*, p. 88]. But he, or his mathematical works, are apparently never mentioned by name in the decisive 17th or 18th centuries [*6*, p. 165], and may have been unfamiliar to them. It can only be recorded that in the 19th century the Great Rehabilitation of the Middle Ages, which had become a state of mind, somehow remembered Oresme and began to restore his achievements one by one.

Renaissance. It may be said that in the 16th and 17th centuries almost anything mathematics achieved stimulated the eventual emergence of functions. Thus, the rise of formulaic algebraic expressions undoubtedly contributed to the rise of functions which can be given by such expressions. Also, the intensive preoccupation with logarithms could not but lead to the introduction of the pair of functions $\{\log x, e^x\}$ and to the realization that these functions are inverses of each other. Finally, trigonometrists may have sensed that the addition theorem

$$\sin(x + y) = \sin x \cos y + \cos x \sin y$$

is a "functional equation" by which to define $\sin x$ and $\cos x$ for angles greater than 360°; this suggested itself to me when reading the great work of von Braunmühl [*7*], although I would not be able to adduce a specific reference.

It can be said even more affirmatively that the analysts of the 17th century, certainly beginning with Descartes and Fermat, always had functions in their thinking, even though they spoke of "curves," as they had to. (The term "function," a dictionary word since the 16th century, began to be used as a mathematical term only in 1694, in a publication of Leibniz.) Thus, Fermat certainly dealt with functions in his famous paper on maxima and minima, which he composed sometime between 1629 and 1638 (and in which the word "analyst" occurs several times). He considers a general "parabola"

$$y = P(x),$$

in which $P(x)$ is a polynomial of any degree, and he asserts that its maxima and minima occur among those points for which

$$\frac{dP(x)}{dx} = 0. \tag{3}$$

Now, Fermat does not actually form a derivative by a limiting process, and he does not express his condition as we have just done; but he applies an operational procedure which ought to delight an algebraist of today. He replaces the variable x by $x + E$, forms the expansion

$$P(x + E) = P(x) + P_1(x) \cdot E + \cdots + P_n(x) \cdot E^n,$$

and asserts that the maxima and minima are among those points for which

$$P_1(x) = 0, \tag{4}$$

[*8*, pp. 183 ff.]; [*5*, p. 382]. The algebraic purity of the procedure is commendable, but there is a price on it. With this derivation of his criterion (4), Fermat cannot properly prove his assertion, and he knows it. And so he exorcizes the ghost of (the "algebraist") Diophantus (ca. 250 A.D.) to stand mathematical surety for him that his assertion is all right. The editors of the collected works of Fermat were apparently puzzled by this invocation of the shades of Diophantus, and in a terse footnote they seem to make Diophantus into an analyst for the nonce. Moritz Cantor, however, observes perspicaciously that to Fermat "even infinitesimal considerations were emanations of number theoretic conceptionalizations" [*9*, p. 858]. Yet as late as 1934 an editor of a German translation of Fermat's essay most gratuitously remarks that "Diophantus employs the word 'approximation procedure' (Arithmetica, v. 14 and 17) in a sense different from Fermat's" [*10*, p. 44].

Continuity. Leibniz was apparently the first knowingly to associate with the notion of function the attribute of continuity. This was a meaningful "first," and we are going to make some remarks on the meaning of it. It must be stated however that, in the main, Leibniz reflected on this association philosophically rather than mathematically, so that working mathematicians probably did not become aware of this association, directly, that is. Indirectly they may have indeed been influenced by it, but it would be difficult to establish this. Specifically, it would not be easy to trace back Cauchy's preoccupation with the phenomenon of continuity of functions in working mathematics to Leibniz' reflections, over a century before, on continuity of functions in natural and other philosophy.

We must return to the Greeks for a proper beginning. Greek rationality was aware of continuity from the first [*11*], and the Greeks had a standard word for "continuous" (*synechés*), which, literally, can best be rendered by "holding together," or "interlocking." The word occurs already in Homer in both a spatial and a temporal sense, and in its temporal sense, that is, when referring to the flow of time, its meaning is already semi-figurative, foreshadowing the connotations of meaning of today.

Centuries later the word appears, even profusely, as a technical term in Aristotle's *Physics*, in the kind of technical meaning which it might have in a scientific or philosophical context of today. The meaning of *synechés* in Aristotle might not be exactly the same as the meaning of *continuous* is today, but in a cursory reading of the *Physics* the translation of *synechés* by *continuous* is good enough.

As against this, there is the remarkable fact, which cannot be sufficiently stressed or overinterpreted, that Greek mathematics proper, that is, the Greek mathematics as it is known from the works of Euclid, Archimedes, Apollonius, Heron, Ptolemy, etc., never, but never, states, asserts, suggests, or negates that something in mathematics is *synechés* in a technical meaning of the term, nor does it ever take recourse to an obvious verbal equivalent of it.

In the *Physics*, however, *synechés*, when used technically, occurs in a manner which would be recognizably mathematical nowadays. When occurring there, it is intended to describe, in rather involved thought patterns of Aristotle, the essential mathematical feature of the linear continuum $(-\infty, \infty)$ of today, namely its "completeness" in the sense of Dedekind and Cantor. Aristotle has great difficulties separating denseness from completeness, but even professional mathematicians in the 17th and 18th centuries might have had such difficulties too.

This is all the continuity that Aristotle is aware of. He never mentions "topological" continuity, that is, continuity of a function or of a "mapping" of any kind, except that he is aware of the fact (which he labors most repetitively) that in a *uniform* motion $x = ct$ the continuity structures of the spatial continuum $\{x\}$ and the temporal continuum $\{t\}$ are isomorphic.

The absence of topological continuity from Western thought lasted very long. In fact, topological continuity is discernible for the first time only in Leibniz, not in straight working mathematics, but in many expostulations of something which Leibniz called a Law of Continuity (*lex continui*). This Law was not really a hypothesis or principle of the metaphysics of Leibniz, but rather a leitmotif of it. Among other things the Law asserted, or only implied, that the data and features of the universe are all continuous, whether asserted individually, in mutual correlation, or in functional dependence. Thus, anybody so disposed may detect in Leibniz insights of the following kind: the rudiments of a conception of space as a Hausdorff neighborhood space, together with the corresponding definition of continuous (function or) mapping in terms of neighborhoods; the rudiments of the fact that for a system of differential and other functional equations the solution usually depends continuously on initial conditions and other parameters; the rudiments of the hypothesis that the mathematical laws of physics are constant in space and time, or at most vary continuously; the rudiments of the law that biological species evolve continuously; etc.

After Leibniz, among philosophers and mathematician-philosophers, the greatest proponent of a universal law of continuity was C. S. Peirce, whom we have already mentioned above. Like some other 19th century philosophers before him (J. F. Herbarth, G. T. Fechner), he spoke not of continuity but of *synechism*, and he did not acknowledge any indebtedness to Leibniz. Peirce was familiar with the work of Georg Cantor and with the methodology of working mathematics, and through the length of his philosophical career he endeavored to find a conception and principle of continuity that would apply to mathematics and ontology both. In this he utterly failed, as he was bound to, because no ontology worthy of its name is a mere "extension" of mathematics, and because in mathematics continuity may vary with the context and purpose, even freely so, whereas in ontology proper this freedom is greatly curtailed if it is present at all.

Piecewise Analytic Functions. In the 18th century, working mathematicians of the stature of Euler, d'Alembert, and Lagrange were trying to find out what functions are or ought to be, and how and when they are "given," and mathematicians even began to classify functions, somewhat ingenuously at times. Somehow their findings were uncertain, ambiguous, and inconclusive, so much so that even historical accounts of them do not quite agree with each other, or, at any rate, try to be as circumspect as they can. There is a reason for this. In the 18th century "there was a near-perfect, richly yielding, fusion of mathematics and mechanics" [*4*, p. 7], so that a mathematical function was not only an object of mathematics, but, by equal priority, also an object of mechanics, and thus had to satisfy needs and expectations of both in equal measure. For instance, it seems that in the thinking of Lagrange an analytic function was, in equal parts, a function likely to occur in mathematical analysis and a function likely to occur in a typical situation of his *Mécanique analytique*. Now a rather simple situation arises if one throws a ball against a wall from which it bounces back. The coordinates of the ball, as functions of time, cease to be analytic at the time point of impact but are analytic in the adjoining time intervals. Thus, Lagrange would have had difficulty in firmly deciding whether an analytic function of t has to be indeed analytic throughout, or may be only piecewise analytic in finitely many adjoining intervals [*4*, pp. 287–288].

At first glance, the formation of a functional object by putting together pieces of analytic or other "well-defined" functions may appear to be, mathematically, a makeshift operation, an ingenuous one. But the 19th century learned to respect, explore, and exploit such formations; and in the 20th century there would hardly be any topology if it were not for simplicial and related decompositions and approximations.

Trigonometric Series. Amidst all its uncertainties about the nature of a function, the 18th century somehow managed to make the capital discovery — which, in a way, has been unmatched since — that functions of a very "general" class can be represented in the form

$$f(x) = \tfrac{1}{2} a_0 - \sum_{n=1}^{\infty} (a_n \cos nx + b_n \sin nx). \tag{5}$$

In the early 19th century, Fourier greatly emphasized what had been known before, that for a given $f(x)$ the corresponding "Fourier coefficients" in the expansion (5) usually have the values

$$a_n = \frac{1}{\pi} \int_{-\pi}^{\pi} f(x) \cos nx dx, \qquad b_n = \frac{1}{\pi} \int_{-\pi}^{\pi} f(x) \sin nx dx, \tag{6}$$

and he also greatly emphasized that "any" function $f(x)$ has a representation (5), even if the function is *absolument arbitraire*.

As it turned out, this *absolument arbitraire* was a great "challenge" (à la Toynbee), and, in a sense, the creation of the theory of functions of a real variable was a "response" to this challenge.

Firstly, Dirichlet made the following contributions (1829–1837):

(i) He gave his famed "definition" of a truly "arbitrary" numerical function $y = f(x)$, as a "general" correspondence from x to y.

(ii) He introduced — perhaps for the first time — a *specific* class of functions of a real variable to a *specific* purpose. It was the class of piecewise monotone functions; and Dirichlet established the fact that for such a function the Fourier series converges at all points.

(After the rise of set theory, towards the end of the 19th century, these functions of Dirichlet "engendered" functions of bounded variation and also rectifiable curves.)

Secondly, Riemann made the following contributions (1854, published 1867):

(iii) He was the first to create a *precise* class of *integrable* functions, so as to be able to define the Fourier coefficients (6). Furthermore, in presenting his criterion for (Riemann) integrability, he prominently used, probably for the first time, the notion of "a necessary and sufficient" condition, literally so.

(iv) He sharply distinguished between a *trigonometric* series

$$\tag{7} \tfrac{1}{2}a_0 + \sum_{n=1}^{\infty} (a_n \cos nx + b_n \sin nx)$$

and a *Fourier* series. For the latter the coefficients $\{a_n\, b_n\}$ are given by the formulas (6) by means of some function $f(x)$, but in the first case no such formulas are assumed at all.

(v) For a Fourier series he created the concept of "localization" of convergence (and he also conceived the Riemann-Lebesgue lemma), thus creating the concept of a "local" property for mathematics at large.

(vi) For any trigonometric series, with

$$a_n, b_n \to 0,$$

he introduced the sum function

$$F(x) = \tfrac{1}{4}a_0x^2 - \sum_{n=1}^{\infty} \frac{a_n \cos nx + b_n \sin nx}{n^2}$$

and treated it as a present-day Schwartz distribution of level 2. That is, he introduced "testing" functions" $\phi(x)$ and "defined"

$$\int \phi(x)\, \frac{d^2F(x)}{dx}$$

by

$$\int \phi''(x)F(x)dx.$$

(I venture to remark that, long before Schwartz, such "distributions" were introduced by myself as "generalized" Fourier transforms; see [*12*, Ch. VI].)

Thirdly, and very decisively, Georg Cantor, after studying closely the work of Riemann, added the following proposition:

(vii) If the trigonometric series (7) is convergent, and to the limit value 0, at *all* points of the interval $-\pi \leqq x \leqq \pi$, then the series is identically 0, meaning that $a_n = b_n = 0,\ n = 0, 1, 2, 3, \ldots$

This last proposition was a very technical statement from working mathematics having no unusual features at all and yet it had the following momentous consequences. After proving the proposition, Cantor asked himself whether, in the hypothesis, the convergence to 0 has to indeed be known for all points x of the interval, or whether there might not be in the interval an "exceptional" set E for the points of which nothing is stipulated. Cantor found, successively, that the following sets are exceptional: (a set consisting of) a single point; a finite number of points; a set having a single accumulation point (and Cantor defines an accumulation point for the occasion); a set having finitely many accumulation points; a set E whose set of accumulation points E' has a single accumulation point or finitely many accumulation points; etc.

This led Cantor to the conception of a transfinite ordinal number, and thence to the conception and theory of pointsets and also of general aggregates; and the world of thought, of any thought, has not been the same since.

Orthogonal Systems. The trigonometric representation (5) of $f(x)$ is an expansion of $f(x)$ into an orthogonal system, and it is remarkable that mathematicians did not realize this feature of trigonometric functions for a century or longer. The 19th century discovered many other complete orthogonal systems among various families of special functions, functions of Bessel, Lamé, Lagrange, Laguerre, Hermite, Jacobi, Heine, Gegenbauer, and others. It was discovered for each such system separately that general functions can be expanded in terms of it, and it was even known, after a fashion, that every system was a complete set of eigenfunctions associated with an elliptic differential equation. But somehow in the 19th century these separate facts were not properly linked up; the accumulated knowledge was broad and eclectic rather than compact and systematic and did not contribute to the general theory of functions of a real variable. Riemann, for instance, took no notice of the special functions of this kind (except for hypergeometric functions, which, however, were holomorphic functions for him), and there does not seem to be a single "Riemann formula" about them.

But after the turn of the century the study of orthogonality suddenly became a serious mathematical occupation. Its achievements were spearheaded by the Riesz-Fischer theorem, which was a great triumph of the newly conceived Lebesgue integral, and above all by Hilbert's spectral decomposition of a bounded self-adjoint operator in Hilbert space. This in turn led to L_p-spaces, Banach spaces, and functional analysis.

Early functional analysis in the 19th century distinguished between "functions," "functionals," "operators", etc., as if they were entirely heterogeneous objects [*13*, passim]. But the 20th gradually realized that they are all functions in a broad sense, if a function $y = f(x)$ is conceived to be a mapping from a general set $X : \{x\}$ into a general set $Y : \{y\}$, with no relations between the sets X, Y stipulated. Logicians maintain that some such broad definition of a function had already been anticipated by the logician Gottlob Frege in the 1870's. Maybe so. But even if this was so, it is most unlikely that working mathematicians were aware of it, or influenced by it. Rather, they arrived at their insight by their own "hard" way.

Riemann Surfaces. In the realm of complex variables, the pioneering achievement of Bernhard Riemann (1826–1866) was to characterize certain classes of functions, which are initially given in the complex plane in terms of certain expressions, by overall properties of the functions on suitable compact Riemann surfaces, or, as later developments explicated it, on the universal covering surfaces of the compact ones. A leading case, in which the universal covering surface need not be envisaged, is the following.

Consider an equation

$$P(w, z) = 0,$$

in which $P(w, z)$ is an irreducible polynomial of the two variables w and z, where z varies over the Gaussian sphere S. The solution $w = w(z)$ of the equation is an n-valued algebraic function (n is the degree of the polynomial in w), and with it we can form the class of functions

$$F_1 = \{R(w(z), z)\},$$

for all possible rational functions $R(w, z)$. Now, Riemann characterizes the class F_1 as follows. He forms the n-sheeted Riemann surface T over S, on which $w(z)$ is properly defined and memorphic, and he considers the class F_2 of all those functions which are defined and meromorphic on T. Then,

$$F_2 \equiv F_1.$$

It is easy to see that $F_1 \subset F_2$, and the burden of the assertion is that also $F_2 \subset F_1$; for a present-day proof see [*14*, p. 155], and note that this proof does not proceed by mere "talking" and "cerebration" but also by a recourse to the Lagrange interpolation formula. Also, if for some element $t(z)$ in F_2 and some z_0 in S over which T is not ramified, the n functional elements of $t(z)$ over z_0 are different from each other, then $w(z)$ is also a rational function of $\{t(z), z\}$.

On one occasion Felix Klein gave an evaluation of Riemann's originality which mixed much admiration with a dose of incomprehension [*15*, pp. 118–119]. We translate it thus:

> Riemann invariably investigates functions thus: at the head of the investigation he places their defining properties, and from these properties he deduced everything else, especially the formulas holding for the functions.
>
> This procedure of Riemann appears difficult only as long as it does not lean on any concrete knowledge. As soon as the latter takes place [that is, the leaning], it appears most peculiarly simple and transparent.
>
> We could also express this in the following way: Riemann's procedure is scientifically excellent, but pedagogically unusable [*unbrauchbar*]. One must not begin with it, but bring it at the end. A first example of this Riemannian treatment is his theory of abelian integrals; he defines them as functions which by a closed circumambulation on the Riemann surface vary only by an additive constant. A second topic which Riemann treats in this way is linear differential equations . . . .
>
> According to Riemann, in the theory of linear differential equations, one considers simultaneously n functions $y_1, \ldots, y_n$, which, on the Riemann surface, experience linear transformations after encircling certain "points of ramification" as also closed circumambulations.

Klein's criticism of Riemann, even if limited to "pedagogy," was not prophetically inspired because in the second half of the 20th century, a hundred years after Riemann's death, his "procedure" is flourishing with a vengeance. There is an expanding mathematical "industry" which for any compact complex manifold — algebraic or not, in one or several variables — conceives and examines all sorts of classes of objects, scalar or tensorial, tangential or fibral, holomorphic or meromorphic, "periodic" or automorphic; and generally each of the classes has some kind of *finite* basis, additive, multiplicative, or algebraic. Also, from time to time the name of Riemann injects itself into the context.

Analytic Continuation. Riemann showed little interest in "arbitrary" analytic functions, that is, in functions not "generated" in some algebraic manner from the complex variable z. He knew, even for arbitrary functions, that their analytic continuation is unique, but he did not make much of the fact. Thus, Riemann never formulated the statement, of which he was undoubtedly aware, that a holomorphic (or meromorphic) function in a domain D of the complex plane — for instance, in a disc — gives rise by analytic continuation along paths in the complex plane to a *unique maximal* Riemann (covering) surface $\tilde{D}$ over the complex plane into which $f(z)$ can be analytically continued (and similarly for the Gaussian sphere instead of the complex plane). We will call this analytic continuation of $f(z)$ from D into $\tilde{D}$ its "Weierstrass continuation" and denote it on $\tilde{D}$ by $\tilde{f}(z)$.

As we have already noted before in [*16*], there seem to be some misunderstandings as to the meaning and extent of the feature of *uniqueness* with which the Weierstrass continuation is endowed. The fact is that this uniqueness is much less "absolute" than sometimes vaguely taken for granted, and we are going to explain what we mean by this. Our present explanation will be somewhat different from and more detailed than the one given in [*16*].

Let D_1 be a bounded simply connected domain in the z-plane, and D_2 such a domain in the w-plane; let

$$w = \phi(z) \tag{8}$$

be a one-one holomorphic mapping from D_1 to D_2; let $f_2(w)$ be a holomorphic function on D_2, and let $f_1(z) \equiv f_2(\phi(z))$ be its preimage on D_1. If we form the two ensembles

$$\{D_1, f_1\}, \quad \{D_2, f_2\} \tag{9}$$

then (8) is also a holomorphic transformation of the first ensemble into the second, in an obvious sense. We now form the Weierstrass continuation of each of the functions f_1, f_2. This gives rise to two "larger" ensembles

$$\{\tilde{D}_1, \tilde{f}_1\}, \quad \{\tilde{D}_2, \tilde{f}_2\}, \tag{10}$$

and, contrary to what one might vaguely expect, these two larger ensembles need no longer be holomorphic images of each other, either by the mapping (8) itself, or by any other mapping. One can easily construct counterexamples, and we now choose the following.

Let D_1 be the disc

$$D_1 : |z| < 1. \tag{11}$$

Let D_2 be the interior of a Jordan curve B_2 no arc of which is real-analytic, and let

$$f_2(w) \equiv w.$$

Then

$$f_1(z) \equiv \phi(z).$$

By known properties of conformal mapping, the function $\phi(z)$ has no analytic continuation at all, so that

$$\{\tilde{D}_1, \tilde{f}_1\} \equiv \{D_1, f_1\}. \tag{12}$$

But $f_2(w) = w$ can be continued analytically from D_2 into all of $\mathbf{C}^1(w)$, so that

$$\{\tilde{D}_2, \tilde{f}_2\} \equiv \{\mathbf{C}^1(w), \tilde{f}_2\}. \tag{13}$$

Now there is no conformal mapping from the disc (11) into all of $\mathbf{C}^1$, and thus the two ensembles (10), as given by (12) and (13), cannot be homeomorphic.

The conclusion to be drawn from this counterexample is this: that the Weierstrass continuation process does not at all apply to an "abstract" ensemble

$$\{D, f\}, \tag{14}$$

in which D is an "abstract" Riemann surface and f a holomorphic function on it. But if one does start out with an abstract ensemble (14), then the process does become applicable if there is given an additional Riemann surface S (which in the "classical" case of Weierstrass is the complex plane or the Gaussian sphere, but which, in fact, can be quite general) and a holomorphic *unbranched* mapping of D into S. In fact, if we denote this mapping by g, then the process applies not to the ensemble (14) but to the greater ensemble

$$\{D, f, g), \tag{15}$$

and the assertion is that this greater ensemble has a continuation

$$\{\tilde{D}, \tilde{f}, \tilde{g}\} \tag{16}$$

which is both *maximal* and *unique*. Also, the symbol f in (15) need not represent a single holomorphic function, but may represent an assemblage of holomorphic functions, or, in fact, of scalar or tensorial holomorphic objects, and may also subsume mappings into some complex spaces, which need not have anything to do with the fixed space S. Also, the scalar or tensorial objects need not be strictly holomorphic, but they may also be meromorphic, provided they are so both in (15) and (16).

All this follows readily by adapting the reasoning in Chapter 1 of Weyl's book [*14*]. It also applies to the case of several complex variables, if D, S are equidimensional and they and their extensions are assumed to be arcwise connected. It should be noted that for $n \geqq 2$ general complex spaces like D, S need not be separable, and if they are indeed not separable then uncountably many sheets of $\tilde{D}$ may be lying over the same points of S. If however D and S happen to be separable, then $\tilde{D}$ will be so too, and there will be only countably many sheets of $\tilde{D}$ spread over S.

We have noted before that the object f in (15) may subsume general holomorphic mappings into unspecified complex spaces. This suggests that it ought to be possible to dispense with the "special" unbranched mapping g into an equidimensional space S and to put some "general" restriction in its place. Now, we have found in [*16*] that it suffices instead to demand that the totality of objects $\{f\}$ separates points on D in the following sense.

We assume that in any holomorphic coordinate system around any point of D, every object f is characterized by a set of components each of which is a holomorphic function in the given coordinates. Now, take any two points P, P' on D, different or not, a coordinate neighborhood $N : \{z\}$ of P, and a coordinate neighborhood $N' : \{z'\}$ of P', such that there exists a one-one holomorphic mapping $z' = \phi(z)$ from N to N'. Take all functional elements $\{f(z')\}$ in N', and form their transforms $\{g(z)\} \equiv \{f(\phi(z))\}$ in N. Now, our hypothesis is that whenever the totality of elements $\{g(z)\}$ is the same, object by object, as the totality of elements $\{f(z)\}$, then the two points P, P' are identical, the coordinate systems $\{z\}$, $\{z'\}$ are identical, and $\phi(z)$ is the identity mapping.

With this definition we proved in [*16*] the following proposition.

It follows from mere analyticity of the data that any ensemble $\{D, f\}$ *whatsoever always has maximal extensions* $\{\tilde{D}, \tilde{f}\}$*, but there are usually many such. If however the ensemble* $\{D, f\}$ *has the separation property just described, and if we consider extensions* $\{\tilde{D}, \tilde{f}\}$ *with the same separation property, then the maximal extension is unique.*

For complex dimension $n \geqq 2$ an interesting case of non uniqueness can be exhibited by use of the Hopf blow-up as follows. Let V^n be a compact complex manifold, say algebraic, and let P^0 be a point on it. Let D be the manifold $V^n - P^0$ (that is, V^n minus the one point P^0), and let f represent all meromorphic functions on V^n which are non singular at P^0. The "natural" maximal extension is $\tilde{D} = V^n$, and the resulting $\{\tilde{D}, \tilde{f}\}$ is indeed maximal because $\tilde{D}$, being compact, is non continuable. However, instead of adding merely the point P^0 we can also add, by performing a Hopf blow-up, a projective space P^{n-1} of $n - 1$ complex dimensions. The resulting complex manifold

$$\tilde{D}' = \{V^n - P^0\} \cup P^{n-1}$$

is again compact, and there is a corresponding maximal extension

$$\{\tilde{D}', \tilde{f}'\},$$

in which $\tilde{f}'(P)$ arises from $f(P)$ by assigning to all points P of P^{n-1} the constant value $f(P^0)$. In this new maximal extension, $\{\tilde{f}'\}$ no longer separates points.

We remark that a quotient of holomorphic functions at P_0 can be similarly extended from a neighborhood of P^0 to a neighborhood of P^{n-1} by extending numerator and denominator separately, so that it was not even necessary to exclude from the class $\{f\}$ meromorphic functions on V^n that are non regular at P^0.

If now we couple this remark with the observation that the Hopf blow-up of an algebraic variety is again algebraic [*17*], we arrive at the following insight which ought to have a sobering effect on any devotee of functional analysis of our times. *A compact complex manifold V^n, of complex dimension $n \geqq 2$, even when algebraic, cannot be "completely characterized" by the assemblage of (holomorphic and) meromorphic functions on it, even if to the scalar functions all possible tensorial functions (that is, tangential vector bundles) be added. For $n \geqq 3$, even non tangential holomorphic vector bundles may be added* (see [*18*, pp. 192–195]).

A Parting Thought. The statement just made and its rationale exemplify a developing trend that bids fair to take over and prevail in geometrically oriented analysis for decades to come. During nearly a hundred years, since after 1870, geometrically controlled analysis was searching for and striving to articulate "harmonies," "symmetries," "homogeneities"; and among cognoscenti, the credal inspiration for this mathematical state of mind drew from something called the Erlanger Program, whatever that was. Mighty achievements ensued: the theories of Lie groups, Lie algebras, symmetric spaces, even, in part, of automorphic functions among such. And yet, all along, something new and different was burgeoning, something that tried to overcome, or at least to make itself independent of, the "retrogressive" seeking of bigger and better symmetries and homogeneities. All truly exciting achievements in analytical and differential topology of recent years, beginning with the pioneering efforts of Marston Morse decades ago, have been of this novel kind, and they seem to be a truer fulfillment of the general aspirations of our century than that which had preceded. And there is wisdom to such aspirations. For instance, in the realm of several complex variables, nothing is less accessible to present-day analysis than a compact complex manifold that is simply connected and does not have a single complex automorphism acting upon it. Yet a "random" compact complex manifold is probably of this kind, and it is crying out for something to be done about it.

In the second half of the 20th century, our universe of thought, feeling, perception, and physical and cosmological reality somehow refuses to be placidly "symmetrical," and if it is symmetrical, then only in a crude surface approximation with innumerably many "local" deformations (like Hopf blow-up) deeply affecting, if not totally destroying, whatever "pleasant" consequences symmetries might entail. Such are facts of our life, mathematical and other, and the oncoming generations of mathematicians will simply have to cope with them.

REFERENCES

[1] BOCHENSKI, I. M., A History of Formal Logic, trans. and ed. by Ivo Thomas, University of Notre Dame Press, Notre Dame, Indiana (1961).

[2] MATES, BENSON, Stoic Logic, California University Press, Berkeley, Calif. (1953).

[3] NEWTON, ISAAC, Mathematical Principles of Natural Philosophy, ed. by Florian Cajori, California University Press, Berkeley, Calif. (1962).

[4] BOCHNER, SALOMON, The Role of Mathematics in the Rise of Science. Princeton University Press, Princeton, N. J. (1966).

[5] BOYER, CARL B., A History of Mathematics, John Wiley & Sons, New York (1968).

[6] ———, The History of the Calculus, Dover, N. Y. (1949).

[7] BRAUNMÜHL, A. VON, Vorlesungen über Geschichte der Trigonometrie, 2 vol., Teubner, Leipzig (1900–1903).

[8] OEUVRES DE FERMAT, ed. by Paul Tannery and Charles Henry, Vol. I, Gauthiers-Villars, Paris (1912).

[9] CANTOR, MORITZ, Geschichte der Mathematik, Vol. II, Teubner, Leipzig (1913).

[10] FERMAT, PIERRE DE, Abhandlungen über Maxima und Minima (1962), Trans. and annot. by Max Miller (Collection: Ostwald's Klassiker), Akademische Verlagsgesellschaft, Leipzig (1934).

[11] BOCHNER, SALOMON, Continuity and Discontinuity in Nature and Knowledge, Dictionary of the History of Ideas, Scribner, New York (to appear).

[12] ———, Vorlesungen über Fouriersche Integrale, Akademische Verlagsgesellschaft, Leipzig (1932).

[13] PINCHERLE, S., Funktional Operationen und Gleichungen, Encyklopädie der Mathematischen Wissenschaften, II A 11, Vol. 2.1.2, Teubner, Leipzig (1904–1916).

[14] WEYL, HERMANN, The Concept of a Riemann Surface, transl. by G. R. MacLane, Addison-Wesley, Reading, Mass. (1964).

[15] Klein, F., Vorlesungen über die Hypergeometrische Funktion, Springer, Berlin (1933).

[16] Bochner, S., Intrinsic analytic continuation and envelopes of holomorphy, Proc. Nat. Acad. Sci. U.S.A. **53** (1965), 904–907.

[17] Calabi, Eugenio, and Maxwell Rosenlicht, Complex analytic manifolds without countable base, Proc. Amer. Math. Soc. **4** (1953), 335–340.

[18] Trautmann, Günther, Ein Kontinuitätssatz für die Fortsetzung kohärenter analytischer Garben, Arch. Math. **18** (1967), 188–196.

Rice University

MATHEMATICAL REFLECTIONS

SALOMON BOCHNER

Part I. Purely Mathematical Americana

My first awareness of American mathematics came soon after my enrolling as a student at the University of Berlin, in the late fall of 1918. In the early part of 1919 there began to appear Leonard Dickson's three-volume work *History of the Theory of Numbers* (1919-23), the scholarship of which made a deep impression. For instance, it was a rumor among students that the proof-reading of the galleys took Dickson two whole years.

It is a curious feature of this work, which I discovered only much later, that it apparently nowhere states, — certainly not with due emphasis, — the "fundamental theorem" on prime numbers, namely that every natural number is a product of primes, and uniquely so. The report does have statements, in abundance, which correlate properties of a natural number to properties of its prime number factors, but it is shy of stating — emphatically at any rate — that every natural number is a product of prime numbers and uniquely so. This reluctance to present the fundamental theorem in an otherwise very conscientious historical account may be due to the fact that, while this theorem is sometimes vaguely attributed to Euclid, it had apparently not been stated expressly before 1801, when Gauss featured it in his *Disquisitiones arithmeticae*, with some emphasis, and rather early in the work. The nearest that Euclid himself came to this theorem was his proposition (vii, 24): "if two numbers be prime to any number, their product also will be prime to the same."

G. H. Hardy and E. M. Wright, in their *An Introduction to the Theory of Numbers*, first edition 1938, make the following statement, on page 182, which, as they acknowledge on page 188, was triggered by remarks made to G. H. Hardy by myself, and I well remember that the remarks were made, sometime in 1933, in Hardy's large living room in Trinity College, in Cambridge, England:

> "It might seem strange at first that Euclid, having gone so far, could not prove the fundamental theorem itself; but this view would rest on a misconception. Euclid had no formal calculus of multiplication and exponentiation, and it would have been most difficult for him even to state the theorem. He had not even a *term* for the product of more than three factors. The omission of the fundamental theorem is in no way casual or accidental; Euclid knew very well that the theory of numbers turned upon his algorithm, and drew from it all the return he could." (Hardy and Wright, *loc. cit.*).

This is a 20th century insight, whereas Dickson's book is still rooted in Victorianism, and the difference is a telling one. Altogether, the work of Hardy and Wright, however technical, is a rich source book for the history of mathematical ideas.

Of course, what had discomfited Dickson was not just the fact that the fundamental theorem does not occur in Euclid, but, undoubtedly much more so, that none of the great number-theorists in the 17th and 18th centuries, like Fermat, Wallis, Euler, Lagrange, etc., had chanced upon it either. This may have been a general malaise among Victorian number theorists from which Dickson was unable to free himself. In fact, some textbooks in the 19th century did present and analyze the fundamental theorem in a forthright and satisfactory manner. But even when they expressly credited the fundamental theorem to Gauss, they did so in a very circumspect manner, as if to leave for themselves an escape tunnel in case somebody should suddenly discover that the theorem had been explicitly known in the 18th century after all.

But their circumspection was not called for. It was a very natural development of mathematical ideas that the fundamental theorem reached the stage of explicit articulation in the 19th century only. In my book *The Role of Mathematics in the Rise of Science*, (1966) on p. 16, I put it thus:

> "Arithmeticians have been showing embarrassment over the fact that the express formulation of the [fundamental] theorem came so late, and they have been trying to "pre-date" it. But there is no substance to assertions that the fundamental theorem had been consciously known to mathematicians before Gauss, but that they had neglected to make the fact known. We think that the 17th, and even the 18th, century was not yet ready for the peculiar kind of mathematical abstraction which the "fundamental theorem" involves; just as only the 19th century was comfortably prepared to conceptualize satisfactorily the notions of real number, limit, derivative, convergence, etc."

I am stating this here because, while Leonard Dickson was not yet receptive to this kind of explanation, a Charles Sanders Peirce, who was even almost a generation older than Dickson, already had been. I could not off-hand quote a reference in C. S. Peirce to this kind of comprehension, mainly because Peirce never wrote any kind of comprehensive treatise himself and because his purely mathematical writings have not yet been published at all. But they are being edited, by Carolyn Eisele, in 4 or 5 volumes, and some day I hope to trace this sort of thinking to Peirce, who was very advanced in many things of this kind.

Altogether, something about Dickson's *History* makes it a very "old-fashioned" number theory, and if one wants to have a brief concise summary of this old-fashioned number theory, then one may take to hand an article, from around 1928, by E. Bessel-Hagen, in the so-called Pascal's *Repertorium der höheren Mathematik*. But if one wants a good whiff of modernity, of genuine modernity, not just of surface "radicalism," then one may take to hand the 1923 book on number theory by E. Hecke (*Vorlesungen über die Theorie der algebraischen Zahlen*), which, I think, has never been translated. One of the first statements in the book is a formulation and proof of the then "abstract" theorem that a finite abelian group is a product of cyclic groups. Before Hecke's book, this beginner's theorem was only presented in the midst of applications, usually in the theory of the cyclotomic equation for primitive roots of unity. Also, around 1923, there was still something "ancillary" about "abstract" theorems of this kind. Thus, among my fellow students in Berlin there was some brow knitting over the fact that Heinz Prüfer, who was several years older, dared to initiate his academic career in mathematics, for which he was obviously heading, with a doctoral thesis in 1921, — entitled *Unendliche Abelsche Gruppen von Elementen endlicher Ordnung* — on nothing other than the structure of abelian groups, albeit his groups were infinite ones. Some of us students took it even for granted — I don't know why — that the great algebraist Issai Schur, who was Prüfer's thesis advisor, was rather lukewarm towards this work.

Returning to Dickson it must be said that he himself was also quite "advanced" in some specialized directions. Thus, Issai Schur in a middling advanced classroom course, referred emphatically to his book: *Linear groups with an exposition of the Galois field theory*, which appeared, in English, in the prestigious Teubner Collection in 1901, as *the* source book for Galois Fields, i.e., finite fields. Around that time only very few books of this Collection were in a language other that German. However, several years after my student days van der Waerden's *Algebra* came out, in which the Galois Field theory occupies only several pages altogether; and the book of Dickson was soon eclipsed and forgotten.

Another book by an American, of which I became aware very early, was the *Elementary Principles in Statistical Mechanics* by Josiah Willard Gibbs, which appeared in 1902, and was almost instantly translated into German by somebody who later became a big name, namely by Ernst Zermelo, creator of the *Axiom of Choice*. And Zermelo performed the translation very admiringly even though Gibbs did not even quote Zermelo for an important theoretical objection — the so-called *Wiederkehreinwand* — to the assertion that the entropy is steadily increasing. This objection, raised by Zermelo in 1896, argues that "the usual kinetic model of a completely and permanently isolated gas behaves quasiperiodically" in time, and thus should not be endowed with a quantity, like the entropy, whose growth in time is in no wise periodic. An earlier objection, the so-called *Umkehreinwand* raised by Joseph Loschmidt already in 1876, is the "crude" argument that a mechanical system, and thus also any physical process described by a mechanical system, is reversible with regard to time, whereas the growth of the entropy is not. For references and details see Paul and Tatiana Ehrenfest, *The Conceptual Foundations of the Statistical Approach in Mechanics* (1912), Cornell University Press 1959, especially pages 14 and 15.

I do not remember whether I ran up against the name of Gibbs in a lecture course or by myself; but I do remember most distinctly that it was in a lecture course, again of Issai Schur, that I was introduced to the name of Joseph Henry Maclagan Wedderburn, which Schur wrote out, fully and solemnly, on the blackboard, as that of the "true" expert on "*Schiefkörper*," that is "non-commutative fields." Schur also mentioned that Wedderburn was at Princeton, in the U.S.A., and this was the first time that I was introduced to the name of the place with which I would become associated for so long.

I do not think that Schur at the time introduced the word "hypercomplex system," or some such expression, and I am most certain that he did not speak of AN algebra, as a mathematical object with certain properties. Algebra was a part of mathematics, namely the part dealing with symbols, and AN algebra was, around 1920, still totally unfamiliar to a student on the Continent, like myself. In fact, AN algebra, as a term and object, is in my present-day view, the most outstanding 19th century "native" American contribution to present-day mathematics, and I must briefly report on how I found out about it. I heard the term for the first time sometime in 1925, in Oxford, at a small seminar of G. H. Hardy, in which there were, among others, myself, Besicovitch, Oppenheim, perhaps Titchmarsh and Echo Dolores Pepper, a young American girl on a fellowship. She came from Chicago, and she gave at this seminar a lecture on algebras, which left all of us totally uncomprehending what it was all about. She spoke in a well-prepared, well-articulated self-confident manner, but none of us had remotedly heard before the terms she used, and we were lost.

Still, to my own astonishment and education, only ten years later, in 1935, there appeared a German book by a young algebraist — no more than 28 years of age — one Max Deuring, with the title: *Algebren*, (*Algebras*). The word *Algebren* in German sounded to me harsh and unappealing, and yet when I turned over the pages in the book, the references were to mathematicians all over the world, with American mathematicians perhaps not even predominating.

This did not still in me the quest for knowing when the concept of AN algebra had really come into being. However, this quest was dormant in me for many years, and only about 10 years ago did I make an earnest attempt to answer the quest. I had to go back surprisingly far, until I finally landed in Benjamin Peirce's famous essay, *Linear associative algebra*, first published, quite informally, in 1870-71, based on lectures given in 1866–70 to the National Academy of Sciences, and then published formally in volume 4 of the American Journal in 1882.

I now wish to make some external observations on this article. It not only speaks of *an* algebra, but it also uses the plural "algebras," and this probably for the very first time anywhere in the English language. It also created the terms *idempotent* and *nilpotent* in our present-day sense. It tried to introduce a *nilfactor* which was used sporadically, but this is nowadays a *zero divisor*. Correspondingly, it also tried to introduce an *idemfactor*, to indicate a solution of $ab = a$ for $b \neq 1$, but this has apparently not caught on at all.

Furthermore, and very importantly, the edition of the essay in the *American Journal* is annotated by the son of Benjamin Peirce, Charles Sanders Peirce, by far the greater intellect of the two. C. S. Peirce created American pragmatism, which I do not understand at all, and which is, objectively speaking, not easy to define or characterize. For the latest about "Pragmatism," see the article under this title, by Philip P.

Wiener, in the *Dictionary of the History of Ideas*, 1973. In a different direction, C. S. Peirce created the algebra of relations, or algebra of "relatives," as he called it. This is probably his most specific achievement of a mathematical turn, and probably his most durable. But he also tried to create something specifically philosophical of a mathematical hue, which, if successful, would have become his greatest feat. In this he failed, even decisively so, but to this I want to devote the entire second part of this paper, because it was his most deeply mathematical attempt, although in a very non-technical, purely philosophical sense. In continuing, I want to report on three Americana, which I have chanced upon, one from the 20th century, and two from the 19th century.

The one from the 20th century is a very peculiar one, one that is hardly realized. Carathéodory published three major works: firstly his large book on real variables which appeared in 1918; afterwards, a book on the Calculus of Variations which appeared in the early 1930's, and finally a two-volumed work on functions of one complex variables (*Funktionentheorie*) which was put through the press posthumously in 1950. Actually Carathéodory was planning the book on complex variables ahead of the book on the calculus of variations, but what was holding up the book on complex variables was a widespread presumption among analysts in the 1920's that any book on complex variables, if to be complete, and to deserve the name, has to contain a complete and rigorous proof, without any prerequisites, for the theorem of C. Jordan that a simple closed curve in the plane decomposes the plane into two disjoint parts, an interior and an exterior. The analysts in the 1920's knew full well that this was a "topological" theorem, which already in 1911 had been extended by L. E. J. Brouwer from a simple closed curve in two dimensions to an $(n-1)$-dimensional topological sphere in n dimensions. Also, in 1923, K. Kerekjartó published an extensive work on two dimensional topology in which the Jordan curve theorem, and all other two dimensional prerequisites of complex function theory, as then needed, were treated in great rigorous detail. But, for all that, the complex variable analysts of the time were still vying with each other in the quest for producing, for the Theorem of Jordan, a proof to end all such proofs, and Carathéodory was experimenting for many years with a proof of his own.

What finally did persuade Carathéodory, however reluctantly, to dispense with such a proof of his own, was the following. About 1930 he made a grand tour of the United States, in which he was received almost regally in all great places on the East Coast, on the West Coast, and in the Mid-West. Now, at a stopover at Princeton, he met James Alexander, who had recently not only given a very elegant topological proof of Jordan's theorem for $(n-1)$-dimensional spheres in E_n, but had even succeeded in giving a meaningful extension to lower dimensional spheres in Euclidean E_n, something European mathematicians did not have in their vision at all. This finally gave Carathéodory the excuse for abandoning efforts of his own, and the published version of Carathéodory's work accepts two-dimensional topology as known.

Another American mathematical event in the 20th century, also a topological one, that had a sobering effect on Carathéodory was the great work of Marston Morse around 1928. Until then, Carathéodory mastered everything and anything ever produced in the Calculus of Variations, but I think the Morse Theory was something he no longer understood. I remember an evening, sometime in 1931 or 1932, in the home of Carathéodory, with S. Lefschetz, who was passing through Munich, also present, at which a student of Carathéodory tried to present the main paper of Morse. It was obvious to me that nobody present understood much. I was looking into the face of Lefschetz trying to gauge how much he understood at least, but he was very good at not giving himself away.

But now, for two very separate Americana from the 19th century. For the first I must return to my earliest student days. One of the first mathematics books I ever bought, perhaps literally THE first one, was Hermann Weyl's *Die Idee der Riemannschen Fläche*, (1913), or, as the translation has it, *The Concept of a Riemann Surface*. I studied this book for years on end — on and off, of course —, and many a piece of work of mine originated this way. But there was one thing in it that I have never been able to assimilate, to my own way of thinking, and this is the so-called Riemann-Roch theorem. On page 137 it reads as follows:

On a surface of genus p, if d is a divisor of order m ($m = m_1 + \cdots + m_r$; $d = p_1^{m_1} \cdots p_r^{m_r}$, $m_q \gtreqless 0$), *if B is the number of linearly independent* (in the complex sense) *differentials which are multiples of d; and if A is the number of linearly independent meromorphic functions which are multiples of* $1/d$, *then*

$$A + (p - 1) = B + m.$$

This is undoubtedly an aspect of a very comprehensive duality theorem, which lies athwart all of mathematics. Schematically, as I see it, the most important typical aspects of the comprehensive duality theorem are the following ones. If both sides of the duality are continuous sums, then the duality is a functional equation of the Riemann-Hecke type for zeta functions in number theory, and the functional equation is somehow a key to the Riemann hypothesis. If in the duality, one side is a continuous sum and the other is discrete, then in the simplest classical case this is the relation between a periodic function and its Fourier series, or, almost equivalently, it is the Cauchy residue formula; and in a very modern recondite version it seems to be the general Selberg trace formula. (The inversion formula for Fourier integrals, in which the sums on both sides are continuous, is a "limiting" case of this, and does *not* belong to the Riemann-Hecke functional equation.) The role of this formula is clear to me, I think. It is, in classical physics, the duality between an electromagnetic field on the one hand, and electric and magnetic "discrete" charges on the other hand. And in quantum physics it is the much more trenchant Janus-like de Broglie duality between the undulatory and corpuscular aspects of the same elementary particle valid simultaneously.

But in the Riemann-Roch theorem both sides are discrete sums, and both finite, and the relation to the other aspects of the duality theorem is not easy to state. (The reciprocity formula for Gaussian sums, in which both sums are finite, does not belong here, but is a degenerate case of a Jacobi modular relation for theta functions which underlies the Riemann-Hecke functional equation; see my paper: Remarks on Gaussian sums and Tauberian theorems, *Journal of the Indian Mathematical Society*, vol. 15 (1951), 97–104.) Undoubtedly the Riemann-Roch duality is linked to the duality between gradience and co-gradience, and homology and cohomology, both of which have first been stated by Poncelet in projective geometry and by Grassmann in his nascent tensor calculus; but these links are somehow too obvious, and not too enlightening to me. Now, in my search for an understanding of the Riemann-Roch theorem, I have been looking at the references in the footnotes of Weyl's book. There, on pages 136 and 137 (of the English edition) there is a reference to one E. Ritter who in 1894 was one of the first who, except for one relatively minor flaw, gave a proof of the Riemann-Roch theorem even for so-called fractional divisors. And later on in Weyl's book, in a footnote to page 147, this E. Ritter is even credited with formulating an extension of the entire Riemann-Roch theorem from system of differentials and functions that are univalent on the closed Riemann surfaces to such differentials and functions that are defined on the covering space of the Riemann surface, but are reproduced by the elements of the Poincaré monodromy group by a fixed multiplicative character of the group. And Weyl adds that in a later paper in 1896 (Math. Annalen, vol. 47), E. Ritter even further generalizes this to solutions of certain types of differential equations on the Riemann surface.

This sounded to me like the work of a very good mathematician, and I wondered why I had not heard from Ritter otherwise. So, one day I decided to look up the last quoted paper, and I had a surprise. The very last sentence of the paper announces a continuation, but there is a sad footnote by the editor attached, stating that, alas, there will be no such continuation. The author, Dr. Ernst Ritter, had accepted a call to Cornell University at Ithaca, but on the way to Ithaca, in New York, he succumbed to a typhus attack, on September 23, 1895. Also, the footnote announced a detailed obituary in the *Jahresbericht der Deutschen Mathematiker Vereinigung*.

This obituary was done by F. Klein, and is, oddly enough, dated September 25, 1895. It made me doubly unhappy for Cornell, and the U.S.A. in general, for having missed out on such a gifted mathematician. Ritter had been born in January 9, 1867, and thus was only 28½ years old when he died. He had become a *Privatdozent* at Göttingen only two terms before receiving the call from Cornell. It was meant to be a big, purely scientific position by which Cornell wanted to become aligned with other mathematically forward-striding great American universities. Klein added that at sailing time Ritter did not feel well, but having a robust constitution ignored it. Immediately on arrival he was taken to the Government hospital at Ellis Island, where he passed away, surrounded with all the care which friendly American mathematicians could show him. I have been unhappy for American mathematics since.

Our last item of Americana is about a mathematician at the other end of the 19th century, at the beginning of it. This one *did* arrive in this country, he *was* active here, very much so, in fact, he was even hyperactive in a sense, and he even managed to carve out for himself a niche, an extremely tiny one, but a niche in the Hall of Immortal Mathematical Fame, so that he was something of a pioneer of American mathematics, beyond question; and yet it is extremely difficult to assess what it is that American mathematics really owes to him, durably, that is.

His name was Robert Adrain, and the best and most accessible source of information about him is an article by the somewhat redoubtable Harvard mathematician Julian Lowell Coolidge, entitled "Robert Adrain, and the beginning of American Mathematics," in the *American Mathematical Monthly*, volume 33, 1926, p. 61–67, in which earlier biographical information is listed. Of interest is also a subsequent brief appraisal, in the *Dictionary of American Biography*, volume 1, 109–110, by the able historian of mathematics, David Eugene Smith, who taught at Columbia University.

Adrain was born on September 30, 1775, at Carrickfergus, Ireland, came to the U.S. around 1800, and lived here until his death on August 10, 1843, on his farm in New Brunswick, New Jersey. He must have been a tempestuous person, because he was frequently changing jobs, not always for better ones. He was a master in the Academy of Princeton, New Jersey; then principal in similar schools in York, Pennsylvania, and in Reading, Pennsylvania; then Professor at Rutgers College (then called Queen's College) in New Brunswick, New Jersey; then Professor at Columbia College, New York City; then again at Rutgers; then at the University of Pennsylvania where he even became vice-provost; then, teacher at the Columbia College grammar school, yes grammar school, after which he retired, or was retired to his farm at New Brunswick. Far the most part he published only middling-interesting mathematics, mainly through books and in a short-lived journal of his own, called the *Analyst, or Mathematical Companion* (1808); but even this middling-interesting activity produced, in a journal called the *Mathematical Correspondent* (founded 1804 by George Baron), a curve called by Adrain the *catenaria volvens*, which half a century later was rediscovered, quite independently and very systematically, by the leading algebraic geometer Alfred Clebsch.

But Adrain's actual claim to fame was the presentation of two versions of what the 19th century called "the exponential law of (accidental) errors (of observation)," one of which even appeared in print a year before the publication of the version of Gauss, by whom the law was subsequently known and after whom it is usually named. The 19th century produced a large number of versions of the exponential law of errors (see, Emanuel Czuber, *Theorie der Beobachtungsfehler*, Leipzig 1891), and Adrain's versions are perhaps among the least satisfactory ones (see the article of Coolidge). But Adrain did achieve a measure of firstness in the matter, although adumbrations of the law reach back into the 18th century, to the work of De Moivre (see also I. Todhunter, *A history of the mathematical theory of probability from the time of Pascal to that of Laplace*, 1865); so that D. E. Smith may by forgiven for the extravagant statement: "If, instead of being a self-made mathematician, he had come under the influence of men like Laplace, Legendre, and Gauss, [Smith might have here added the name of Poisson], he (Adrain) might have been a great leader. As chance had it, he was in a mathematical desert."

It is rash to assert, as D. E. Smith seems to do, that Adrain received no mathematical stimuli from the "ecological" setting in this country. The fact is that throughout the 19th and into the first decades of the 20th century there was in this country a *broad* interest in, and cultivation of, the kind of mathematical probability and statistics into which the 19th century law of errors naturally falls. The best evidence for this is the fact that even the above-named gentleman-mathematician J. L. Coolidge wrote a book on this subject. By mathematical avocation, Coolidge was a geometer, as evidenced by a string of textbooks, all published by the Clarendon Press, Oxford, and they were as follows: (1) *The elements of Non-Euclidean geometry*, 1909; (2) *A treatise on the circle and the sphere*, 1916; (3) *The geometry of the complex domain*, 1924; (4) *Algebraic plane curves*, 1931; (5) *A history of geometrical methods*, 1940; and (6) *A history of the conic sections and quadratic surfaces*, 1945. And he even ended his book-writing career with, — also for the Clarendon Press — (7) *The mathematics of great amateurs*, 1949. Now, amidst all such books of a rather uniform trend, he also had a very different book in 1925, likewise at the Clarendon Press, under the title: *An introduction to mathematical probability*, which also had chapters on "errors of observation," "errors in many variables," "indirect observations," "the statistical theory of gases," and, *note*, on "the principles of Life Insurance." And in the introduction the author states that "the present work is based upon the lectures which I have delivered, usually in alternate years, at Harvard University." This book, however hodge-podgey it may appear to be to a mathematician of today, was still very much in the style of 1925, and, in fact, the book of Coolidge was quickly translated into German, and I myself so read it.

But it is a curious fact that only a few years afterwards, this kind of book simply went out of style; and it is not the fault of J. L. Coolidge, if very soon after the appearance of his book, the work, original and translation, began to collect dust on the shelf which it has not shaken off since. Suddenly, around 1930, different conceptions of the cognitive nature and of the role of probability and of the role of statistics, and of their mutual relations came into being, and a different kind of textbook became necessary. All of a sudden, "exponential law of errors," which in the 19th century statistical activity was a household word, a state of mind, as it were, almost disappeared from textbooks and their subject indexes; and what — to a discerning antiquarian — was left of it were simply some versions of the so-called law of large numbers, more-or-less, and of some other asymptotic laws, perhaps. And it should be added that very soon after the change-over entirely different American crews joined in the "New Directions," both in probability and in statistics, and have been on top of developments since.

In some peculiar manner, the American folks-genius must have had some kind of awareness in the late 1920's that an important phase of mathematical statistics had come to an end, because an American professional statistician even wrote a valedictory summary on this entire period. It was called *Studies in the History of Statistical Method, with Special Reference to Certain Educational Problems* (1931), and the author was Helen M. Walker, then Assistant Professor of Education, Teachers College, Columbia University. It is also a remarkable feature of this book, — and a comment on the still very nineteen-centuryish American outlook of it on Life and Education — that the educational aspect in the title of the book is taken very seriously indeed in the body of the book. Thus Chapter VII has the title: *Statistics as a subject of Instruction in American Universities*, and the sections of the chapter are as follows: (1) Purpose and method of the chapter, (2) Instruction in Statistics in 1890, (3) First College Courses in Statistics, (4) Pioneer Courses in Various Departments, (5) Outline of the Development of the Teaching of Educational Statistics in America, (6) The present situation, and, MARK!!, (7) Remarks by Florence Nightingale concerning Instruction in Statistics.

To come back to Adrain, I would like to repeat that it is not at all certain that by being in this country he was lost in a mathematical desert. It is even more hazardous to speculate that by staying in Europe, and in the right places in the vicinity of the right people, he might have come under the influence of men like Laplace, Legendre and Gauss, and thus become a great leader. Adrain's true activity falls into the period between 1790 and, say, 1830, a so-called *Age of Upheaval* in the terminology of the *New Cambridge Modern History* (I myself called it, for my purposes, *The Age of Eclosion*, and extended it from 1776–1825); and this era is, on record, one of the most difficult and recalcitrant to analyze and comprehend with regard to intellectual developments and spontaneities, and with regard to the influence of intellectuals upon each other. By what is known, but not comprehended about this era, it is equally valid to speculate — or at least could not be gainsaid — that it was precisely the relative isolation in the United States which brought out whatever originality there was in Adrain, and that, if he had stayed in Europe, whatever originality he was endowed with might have been nipped in the bud altogether, or smothered and choked off entirely. Because, if it is true that in the United States this era was in some respect a desert, then in Europe it was, in the same respect, a suffocating jungle, and it would be rash to say that Adrain would have fared better in a jungle than in an alleged desert.

What was reminiscent of a jungle in Europe at that time was an overwhelming outburst and uncontrollable proliferation of all kinds of knowledge, on all kinds of levels, all over the landscape. Leaving aside at *first* (and I mean at *first*) mathematics, and mechanics in all its parts (mechanics of point systems, and of continuous and of hydrodynamic and elastic media), it can be asserted that a whole spectrum of sciences, natural social, and even humanistic, came then into being; and if the sciences did not all actually come into being only then, it can nevertheless be stated that they then organized themselves in such a way, as if this were their true beginning, finally.

During that period there was the true beginning of physics *proper* (of thermodynamics, electricity, magnetism, even of optics, ancient as this discipline was), of chemistry (beginning with Dalton and Avogadro), of geology, biology (Lamarck), psychology (Herbart); of systematic sociology (Auguste Comte); even of economics, certainly of mathematical economics, or, at least of economics built on models; of linguistics; of the so-called "higher criticism" in all kinds of ancient literature, the Old Testament and Homer among others; of various specialized histories, like history of astronomy, of medicine, of Roman law, etc. (For all this, compare my book *Eclosion and Synthesis*, 1970.)

Even in mathematics the developments during that period were much more radical, and much more difficult to account for rationally than commonly realized, even if, externally, the developments did not appear to be as tumultuous as in other areas of knowledge, scientific or other. During this era mathematics and mechanics separated from each other; the conception of an existence theorem, and of a uniqueness theorem, and of a necessary and sufficient condition, was finally born, and only then. One need only compare the spirit of, and approach to analysis in Lagrange and Cauchy, — who after all were in a common French tradition — to see how radically analysis was reoriented. Lagrange and Cauchy simply do not speak the same mathematical language, whatever the similarity of vocabulary and grammar. For instance, to single out only one point, no matter how much Lagrange may assert and insist that a function is for him an "abstract" mathematical object, in his thought patterns it somehow is residually a mechanical orbit or perhaps a physical function of state; whereas in Cauchy, orbits and forces and pressures are always functions, as they are for us today. The history of this transition in mathematics must of course be pursued, to the limit of possibility, as a problem of the genetic unfolding of mathematics as mathematics. But it is also a fact that this transition in mathematics has parallels to transitions in many other areas of academic pursuits; and in order to round out the comprehension of the transition in mathematics such other transitions must be taken into account too.

The leading transition mathematicians were Laplace, Gauss, and Poisson. The provenance of Laplace is easy to see; he was in a continuing French tradition, from Lagrange to Cauchy. Poisson, although also a French mathematician, is much harder to understand and to place. And it is a curious fact that he is the only French mathematician of consequence whose *Oeuvres Complètes* have never been edited, and about whom there is no decent scientific biography. And Gauss, finally, is totally inexplicable as to his "ethnic" origin. His only mathematical "forbear" would be Leibniz. But it is very farfetched to attach him to Leibniz, and it is simply impossible to say where Gauss came from, mathematically, that is. He was suddenly there, just as some of the greatest prophets of the Old Testament were suddenly there. For instance, the first such great prophet in the Old Testament was Elijah, and the Bible introduces him suddenly, without any warning, as "Elijah the Tishbite, who was of the inhabitants of Gilead." That is all; no father's or grandfather's name, no name of a clan, nothing.

Also, our assertion that Gauss, for all his greatness, was the figure of a transition period, is borne out by the fact that while he already felt the mathematical need for giving a rigorous proof, and in fact more than only one proof, for the fact that every polynomial with complex coefficients has a zero, he was not yet in an intellectual frame to define the convergence of a sequence, or of a series of numbers, satisfactorily, that is. This was done only by Cauchy, who had the unusual distinction of being the first to give a reasonably satisfactory definition of continuity as it occurs in mathematics.

Part II. Charles Sanders Peirce

The statement just made about Cauchy is my cue for presenting my last and most significant item of Americana. In a very peculiar sense, which I shall try briefly to sketch, the conception of continuity was in the center of a dominant theme of 19th century intellectuality, in all its compartments, in and out of mathematics. And it is a remarkable fact that Charles Sanders Peirce was most peculiarly preoccupied with this theme, overtly and covertly, directly and by contrast. It was an intellectual obsession with him. At any rate, it is possible to view much, or even most, of his philosophical work, in all its complexity, as variations, however diverse, on the theme of continuity. If his work is now so viewed, it is admissible to say, that the preoccupation with this all pervading 19th century theme of continuity made Peirce into the greatest philosopher America has produced thus far, but that it also un-made him, as evidenced by the fact that Peirce was never able to compose a key treatise through which to express himself and in which to summarize himself, as philosophers are wont to do. His was a real American Tragedy. A great philosopher, and an even greater failure.

This harsh judgement was not originally mine. I encountered it in the book of Murray G. Murphey, *The Development of Peirce's Philosophy*, Harvard University Press, 1961, but found it very appropriate and just, however harsh it be. But Murphey did not substantiate it sufficiently, and I undertook to rationalize it from my own approach in the article: "Continuity and Discontinuity in Nature and Knowledge," which has appeared in the *Dictionary of the History of Ideas*, 1973.

Before briefly expounding my own view on the matter, I would first like to quote the decisive statement in Murphey, which is the last paragraph in his book. This is its text.

> "As one reads through the thousands of pages of manuscript which are all that remain from Peirce's life's labor, one cannot escape the feeling that these are the ruins of a once great structure. Every paragraph and every doctrine seem to be fragmentary parts of some larger whole. As Morris Cohen has remarked: 'In his [Peirce's] early papers, in the *Journal of Speculative Philosophy*, and in his later papers, in the *Monist*, we get indeed glimpses of a vast philosophic system on which he was working with an unusual wealth of material and apparatus.' (Charles S. Peirce and a Tentative Bibliography of His Published Writings,

Journal of Philosophy, Psychology and Scientific Methods, XIII, 727 December 21, 1916.) But this is an illusion — Peirce's illusion: the grand design was never fulfilled. The reason is that Peirce was never able to find a way to utilize the continuum concept effectively. The magnificent synthesis which the theory of continuity seemed to promise, somehow always eluded him, and the shining vision of the great system always remained a castle in the air."

We ought to observe though, that Murphey was in no wise implacable in his criticism of C. S. Peirce. Thus, after the book, Murphey also contributed the article on Peirce to *The Encyclopedia of Philosophy*, and there this kind of criticism does not occur at all.

But now we are turning to our own analysis, in which we will repeat much from our above-mentioned article in the *Dictionary of the History of Ideas*.

Continuity is a perplexing concept. The mere name of it poses a puzzle. Our common familiar name of it comes from the latin verb *con-tinere* which means "to hang or hold together," and forms of the verb *continere* are attested since the 3rd century B.C., from which the earliest large-scale Latin texts have come down. But long before that, a Greek equivalent of the entire word-complex had been in use. In fact, the Greek word for our "abstract" noun "continuity," as standardized by Aristotle, is the adverbial form *to syn-echés* (*τὸ συνεχές*), and the cognate verb *syn-echéin* is likewise translated in dictionaries by "to hang or hold together." To a layman this would suggest that the two words *con-tinere* and *syn-echéin* ought to be close linguistic relatives, especially since the prepositions *con* and *syn* have such similar sounds. But, surprisingly, the works on Indo-European linguistics which have been consulted in my layman's way, do not, in general, "identify" *con* and *syn*, and, in particular, they do not assert that the Greek and Latin stem words for *continuity* had a common root in Sanscrit.

Now, in Greek, this *synechés* has a venerable ancestry, meaning that it already occurs in Homer, than which there are no earlier texts. In fact, both the verb *synechéin* and the adverb *synechés* occur in Homer, and it is most remarkable that already in Homer they occur on different levels of abstraction, and that this difference has not worn off entirely ever since. The verb occurs on a lower level of abstraction. It occurs in *Iliad*, 4, 133, in the expression "the golden clasps of the belt were *held together*," in which its meaning is quite concrete. But the adverb, which is used twice, is both times used in a semi-abstract meaning, namely in the meaning *continually* (in time). The first occurrence is in *Iliad* 12,15–25, thus: "Zeus made it rain *continually*"; and the second occurrence is in *Odyssey* 9, 74, where Ulysses, in the narration of an adventure, uses it thus: "There for two nights and two days we lay *continually*." This peculiar intellectual tendency to have a heightened awareness of the abstractness and intricacy of continuity when dealing with continuity of Time has been with Philosophy, in all its compartments, ever since. For instance, a good deal of the philosophy of Henri Bergson can be so envisaged. But, of importance to us, is a section in Peirce,

6.86, entitled Time, in which the statements are rather intricate and even obscure, even by standards of the writings of C. S. Peirce, in which obscurity is a recurrent phenomenon.

Returning to Homer, to the passage in the *Odyssey* in which *synechés* means *continually* (in time), it is remarkable that the adverb *synechés* is there reinforced by the adverb *aiei* (*αιεί*) which means *always, ever, eternally*. The *Odyssey* thus adumbrated a tripartite bond between continuity, time, eternity; and this bond — by the inclusion of eternity — has been variously contemplated and exalted in theologically oriented (general) philosophy ever since. This bond is nowhere stronger than in the Old Testament, which, in a sense, is a vast dissertation on the Lord's *continuous* (because unceasing) concern for his "Chosen People," but the extant canon of the Old Testament does not have at all a word whose functions would correspond to those of *synechés*. It is further remarkable that both this bond and the intellectual preference for temporal continuity are verified by the order in which the cognates of our English word *continuity* have come into use. According to the entries in the *Oxford English Dictionary*, our word *continual* (in time) was the first to emerge. It occurs as early as 1340 A.D., in the phrase: "great exercise of body and continual travail of the spirit," in one of the so-called *English Prose Treatises* of the hermit Richard Rolle of Hampole (1290–1349). But for all other cognates of *continuity* the same dictionary quotes only from Chaucer, who wrote about half a century after Hampole, or from later sources, even much later ones. Thus, according to this dictionary, our adjective *continuous* gained currency in the 17th century only.

In mathematics, continuity is *nowadays* defined precisely and unequivocally, but only *nowadays*, since Cauchy, that is. There is no reason for being overly surprised at the fact that continuity has become a precise mathematical concept so relatively recently only, and one must not judge, as sometimes done, that mathematicians before that were "confused," or in some sense "defective" in their cognition. Continuity is nowadays as vital a mathematical conception as can be, but, for all that, it is not at all a mathematical conception to begin with, very far from it. This cannot be sufficiently emphasized. It is one of the greatest triumphs of mathematics that it has somehow managed to turn continuity into a genuinely precise and sharply defined concept as no other academic area of cognition has, although to begin with, continuity is not a peculiarly mathematical category of cognition at all. Rather, to begin with, it is a sprawling all-pervading notion, occurring, in all precincts of perception and intellection, ideation and abstraction; and outside of mathematics, it is ambiguously conceived and loosely applied, and it merges and fuses with neighboring concepts, like uniformity, steadiness, constancy, etc., all of which have, in mathematical contexts of today, definitions of their own.

That continuity has many meanings, or, at any rate, many shadings of meaning, can be recognized by the fact that it has many antonyms. The leading antonym to "continuous" seems to be not "discontinuous" but "discrete." But others are: saltatory, sudden, intermittent, indivisible, atomic, particulate, even: monadic.

A Monad however, being a kind of synonym to Unity and One-ness, may suggest both continuity and discreteness, self-dually. The monad of Leibniz, as presented in his *Monadology*, is apparently of such a kind; that is, it also suggests continuity, even if it is an irreducible ultimate unit, not only of physical structure but also of consciousness, cognition, and metaphysical coherence. Charles Sanders Peirce has a much more sophisticated version of duality between continuity and discontinuity, or at least an anticipation of such a duality, a version of duality which reached into the 20th century and even became a hallmark of it — but, of this, later on.

In one matter, though, Leibniz' insight may have been deeper than Peirce's. It concerns the relation between "function" and "continuity." A young mathematician nowadays is taught, even in high school, that the concept of a "function" $y = f(x)$ is a primary datum, whereas the property of "continuity" of a function is a secondary datum. A function, he is taught, is, to begin with, an "arbitrary" association of some y with each x, whereas continuity is a restrictive mode of association, which can only be brought about if the spaces of the objects x and of the objects y carry "topological structures," so called. Now, in the actual genesis of the two conceptions in the mathematics since 1600 there was no such hierarchical subordination of "continuity" under "function." It was almost the other way around; it was the urge to come to grips mathematically with the concept of continuity that was greatly responsible for the gradual emergence of the conception of a mathematical function too. Now, from my reading in Leibniz and in Peirce, such as it has been, I have the impression that Leibniz somehow knew better than Peirce what, cognitively, the true hierarchical relation between continuity and function really is.

Returning now to continuity *per se* we note that a most challenging problem in the History of Ideas and Knowledge, which has not yet been studied at all, are the manner and stages by which the sharp mathematical strand of continuity of today has arisen and evolved within the sprawling, weaving, undulating texture of continuity in the broad spread of cognition and knowledge in our total mental awareness. Also, most surprisingly, the study of Greek Rationality does not help at all. On the contrary, the Greek situation makes the problem ever so more perplexing, from our retrospect at any rate; the brutal fact being, that Greek mathematics, as known to us from the many documents that have come down to us, has absolutely never and in no wise come to grips with the conception of continuity at all, in a plain, honest, direct confrontation, that is.

In extant Greek mathematical texts, the word *synechés* occurs extremely rarely, and never, but never, in a technical sense; nor does any kind of recognizable express circumlocution appear in a technical sence. From knowing *all* extant Greek mathematical texts, but only those, and no philosophical texts, it would be impossible to surmise, that *synechés* was a potent philosophical concept that appeared in very meaningful philosophical contexts, meaningful and intricate ones. The most articulated such occurrence is in Aristotle's *Physics*, especially in his analysis of Zeno's puzzles (the flying arrow, Achilles [and the tortoise], the midpoint, the race-course).

But already well over a century before, *synechés* occurs in the great ontological poem of Parmenides; and the following two passages in which it occurs must be read slowly and thoughtfully, if the meaningfulness of the occurrence is to be savored. (The passages are lines 3–6 and 22–26 from fragment 8 in Diels-Krantz; all our translations of Parmenides are those of Leonardo Tarán.)

> "Being is ungenerated and imperishable, whole, unique, immovable and complete. It was not once [upon a time only], nor will it be [only in the future], since it is now altogether, one, continuous (*synechés*)."

> "Nor is it divisible, since it is all alike. Nor is there somewhat more here and somewhat less there that could prevent it from holding together (*synechéin*); but all is full of Being. Therefore it is all continuous (*synechés*), for being is in contact with Being."

Now, at first hearing this may sound like just another philosophical gibberish. Try however to replace here "Being," by "(Dedekind-Cantor) linear continuum," be it one-dimensional, two-dimensional, three-dimensional, etc., and then the passages immediately sound much more meaningful. The second passage especially has a thrilling ring, thus!

> "Nor is it divisible, since it is all alike. Nor is there somewhat more here and somewhat less there, that could prevent it from holding together. But all is full of continuum. Therefore it is all continuous (*synechés*), for continuum is in contact with continuum."

After this take the first passage, but in the sense of natural philosophy, and in a suitable paraphrase, thus:

> "Continuum is ungenerated, and imperishable, whole, unique, immovable and complete. It was not just sometimes in the past, nor will it be only in the future, since it is now altogether, one continuous (*synechés*)."

Before continuing with the subject matter proper, we wish to comment briefly on a famous passage in Parmenides in which he suggests that the universe has the shape of a homogeneous sphere: (lines 42–49 of fragment 8). The passage becomes rather luminous if one not only replaces "Being" by "(Cantorian) continuum," but also "non-Being" by "a gap (as in a Cantor ternary set)" in the following text.

> "But since there is a furthest limit, it is in every direction complete (*tetelesmenon*); like the body of a well-rounded sphere, from the middle everywhere of equal strength; for it need not be somewhat more here and somewhat less there, for neither is there non-Being to prevent it from reaching its like, nor is there Being so that it could be more than Being here and less than Being there, since it is all inviolable; for from every point it is equal to itself, staying uniformly in the limits."

That — from our retrospect — it is justified to impute to Parmenides a measure of preoccupation with the nature of the linear continuum is corroborated by developments engendered by Parmenides, not so much in the work of Melissus of Samos, as in the famed work of Zeno of Elea, both of whom were prominent successor-disciples of Parmenides. Melissus introduced some features of infinitude into the universe of Parmenides, and thus contributed, however slightly, to its eventual mathematization. And Zeno's puzzles contributed much more so, however slowly and even circuitously.

The work of Zeno, as known today, had two major aspects, which for our purposes are separable, however much a philosophical analysis in depth may find them overlapping, or even coextensive. One aspect was the general problem of One-and-Many, with which we need not concern ourselves as such; and another aspect was the more special, and more renowned, problem of the "puzzles" or difficulties relating to motion, with which any study of the notion of continuity in its metaphysical entirety must concern itself, in one form or another. The manner in which we are concerning ourselves with the puzzles is a purely negative one. We are not going to examine the puzzles at all on their own Zenonian standing and merit, or on the many philosophical expostulations and critiques of them, but we are pointing out, emphatically, the following closely interlocking facts.

(1) Whatever, nowadays, the philosophical interest in the puzzles continues to be, our *working* science and working mathematics is totally unaffected by them. There is not a trace of a hint of the puzzles in Newton's *Principia*, or in Lagrange's *Mécanique Analytique*, and a physicist of today is not less of a physicist if he does not know anything about the puzzles at all. In fact, there is not a trace of a hint in all of Greek scientific documentation that Archimedes was in the least interested in the puzzles, or that he had ever heard of them at all. I am not asserting or even suggesting that Archimedes never heard of them, but I am only pointing out that in the extant corpus of Greek writing there is not a trace of a hint of an allusion to his having known anything about them, and I do maintain that it is very pertinent and very important so to point out.

(2) The puzzles are known only because Aristotle reports on them, in Book VI of the Physics, and his purpose in reporting them is not "to save them for posterity," but to refute them. Now, from our retrospect, Aristotle's refutation consists, in his own thought patterns, in the following interconnecting attempts. (a) He attempts to establish the prerequisite continuity properties of the (Cantor-Dedekind) linear continuum. (b) He attempts to represent a *uniform* motion on the line by a linear function $x = ct$, and this is especially difficult for him, because the Greeks simply had no kind of analytical anticipation of a function in our sense. (c) And he finally attempts to argue that in this would-be linear function, the mathematical structural paradigm of the two quantities x and t is the same, namely that of the linear continuum so that, indeed, as we know it today, the puzzles simply do not arise.

(3) Now, a most remarkable commentary on these interlocking facts is this, that

these heroic attempts of Aristotle to establish such very basic and indispensable verities of analysis are not reflected in the least in the mathematics of the Greek professionals at all. While Aristotle is "experimenting" not only with the notion of "continuity" (*synechés*), but also with neighboring notions of the linear continuum, like: "together," "continuous," "overlapping," "everywhere dense," etc., professional mathematics is deathly silent about any such notions. One is forced to conclude that Greek professional mathematics was simply too cowardly to face the conception of continuity head-on, front-face, in a direct confrontation, as Aristotle, and perhaps also some other philosophers, had so urgently been prodding it to do. And the consequences of this "cowardice" were immeasurable. If nothing else, it took the Evolution of Ideas almost 2000 years fully and unequivocally to recognize that the puzzles of Zeno are, if properly viewed, problems of mathematics. At any rate, it appears that the first such recognition can be documented by a book of 1647 only, namely by the *Opus geometricum quadratura circuli et sectionum coni* of Gregory St. Vincent. Before that, the puzzles belonged to *Physics*, more-or-less, as they had so belonged in Aristotle, whatever Physics may have been understood to be, in Antiquity and in the Middle Ages.

Of course, it would be wrong, even preposterous, to say, that Greek mathematicians did not do anything about continuity at all. Of course they did, and even some of their most glorious achievements pertained to that part of our present-day analysis in which continuity is domiciled. Such achievements were: their theory of proportions, as presented in Book V of Euclid's *Elements*, and their theory of exhaustion that pervades the work of Archimedes. But, as far as continuity is concerned, these "substitute" achievements, when viewed coldly and unsentimentally, were maneuvers of evasions, instead of battles of confrontation; and Greek mathematics, and mathematics long after the Greeks, paid a high price for putting up with such evasions, the price being some unbelievably crude and obvious gaps in syllogistic reasoning — gaps which only the 19th century identified and filled.

Such a crude gap occurs in the very first theorem of Eculid's *Elements*. In it Euclid tacitly assumes that if a circle is given in the plane, and if a point inside the circle and a point outside the circle are connected by an arc (or a segment, or any simple curve), then the arc must meet the circumference of the cirle in at least one point. This gap is a gap in the knowledge of the structure of the linear continuum. And since Greek mathematics never faced up, directly, to the problem of the structure of the continuum — as the work of Aristotle should have urged it to do — it managed not to become aware of the gap at all. Not only did Archimedes accept this kind of gap unquestioningly, but even the great Gauss still did, when, in one of the proofs of the theorem that a polynomial with complex coefficients has a complex root, he takes it for granted that a polynomial with real coefficients of odd degree meets the real axis somewhere. And it must be stated again unsentimentally, that this gap in Gauss is of the same size and shape as the gap in the first theorem in Euclid, 21 centuries before. Only after Gauss, even soon after Gauss, did analysis finally deal with the problem frontally, and even so it took mathematics the best part of the 19th century to master the problem.

About a century before Cauchy and others began to make the conception of continuity mathematically rigorous and operationally productive, Leibniz began to make it philosophically tangible, articulate, and useful. Thus he, finally, overcame the Aristotelian (meta-) physical contrariety between motion (*kínesis*) and rest (*eremía*) by pointing out, in terms of today, that rest is the limiting case of motion when velocities reduce to the value 0. But this was only a particular statement of Leibniz, which, in his view, was only a particular instance of a general Law of Continuity (*lex continui*) which runs through his entire metaphysics and science. Leibniz, from prudence or incapacity, did not present this law — important as it was to his thinking — in a systematically thought-out study of its own, but he frequently reverted to it in divers contexts, presenting some of its aspect each time. Perhaps the most representative single statement of the Law of Continuity is the maxim of cognition that "when the essential determinations of one being approximate those of another, as a consequence, all the properties of the former should also gradually approximate those of the latter." (*Leibniz selection*, edited by Philip P. Wiener, 1959, p. 187; see also pertinent passages in Hermann Weyl, *Philosophy of Mathematics and Natural Science*.)

This leading statement may be interpreted to assert that functions, and functional dependencies in nature are usually continuous. We know nowadays that this is an over-assertion, since chemical processes and nuclear processes can be discontinuous, even violently so. And Leibniz' emphasis on continuity was to an extent a part of his general attitude of optimism — which is a *leitmotif* of the 18th century althogether — that the world is rational, progressive, self-ameliorating, and perhaps even the best of all possible worlds. Nevertheless, the *lex continui* was an important event indeed, in that it revealed an aspect of (mathematical) continuity that had been hidden from view until then. Whatever insight into continuity Aristotle and other ancient and medieval philosophers may have had, this insight always referred only to the structure of the linear continuum, that is, to the topological structure associable with linear ordering; whereas Leibniz discovered, even if in a blurred way, the possibility of continuity of a *function*, that is, of a mapping from one topological space into another, and he even advanced towards a conception of a general function in the process. Professional mathematics and philosophy of mathematics pride themselves on the insight — and they spent the entire length of the 19th century on arriving at the insight — that a function, whatever that be, need not be continuous. But, as already stated before, it is a most significant fact that, in relatively recent Western Thought, the conceptions of function and of continuity have evolved simultaneously and in close intellectual interpenetration with and in dependence on each other, however much they were later disengaged from each other, in fact, or perhaps only in intent. The lasting residual dependence of these two conceptions upon each other comes about in the following way. It is all very well to say that a function is an "arbitrary" correspondence between two sets of objects, the domain and the range. It can be argued that however "arbitrary" a correspondence may appear to be, it does follow some kind of pattern nonetheless, and if one philosophically extends and generalizes the concept of continuity to subsume the presence of "any" kind of pattern, regularity, orderliness, etc., then, of course, most functions in Thought and Cognition are continuous indeed. The possibility of such an extension has been recognized and attempted by Charles Sanders Peirce, and it was this attempt in which he failed, or rather in which he never succeeded. Of course, his attempt was not quite as simplistic as our first statement, just made, would suggest, and we will try to describe, to an extent, how un-simplistic it was.

Returning to Leibniz, we have to say that it is not easy to assess the direct effect, if any, of the Law of Continuity on the growth of professional mathematics and physics or even on philosophy itself. There seems to be no manifest reference to the *lex continui* in Immanuel Kant's analysis of his *a priori* Space and Time, however much continuity of one kind or another is involved in their structure and mystique. Nor is there a manifest reference to the *lex continui* in the unusually slow unfolding of the conception of continuity in the working mathematics of the 18th and 19th centuries.

Of course, the "Law" of Leibniz may have been burrowing deep inside the texture of our intellectual history, thus affecting the course of Knowledge and Cognition. But the only outright corroboration for this is the fact that about 150 years after Leibniz a *lex continui* came into being in the thinking of C. S. Peirce and permeated it to a considerable extent. Except that Peirce called it not *lex continui* or *Law of Continuity*, or something similarly synonymous, but *synechism*, thus going back to the Greek form of the word continuity instead of the Latin. And may I state in parentheses that it was this "accidental," or perhaps not so accidental, trait of nomenclature that attracted my attention to Peirce, the general philosopher, at first. Peirce obviously did not want to admit, to himself or to others, his indebtedness to Leibniz in this matter, and he has been very successful in this respect, inasmuch as nobody states it outright that *synechism* is a *lex continui* by another name, however much Peirce may intend to differ from Leibniz in particular or general attitudes. It is altogether remarkable that in the vast collection of Peirce's papers the name of Leibniz appears extremely rarely, although Peirce apparently had Leibniz very much on his brain (see Max H. Fisch, Peirce and Leibniz, *Journal of the History of Ideas*, 33 (1972), 485–496). And, finally, Peirce absolutely never refers to the fact, which must have been known to him, that the verbal switch, in philosophical jargon, from "continuity" to "*synechés*" had been made earlier in the 19th century by the philosophers Johann Friedrich Herbart (1776–1841) and Gustav Theodor Fechner (1801–87), the first famous for psychology and pedagogy, and the second for the Webner-Fechner Law of quantitative psychology (intensity of sensation varies as the logarithm of the stimulus).

It must be stated though that Herbart uses *synechés* only in a *de*scriptive manner, while Peirce uses it very *pre*scriptively, whereas Fechner uses it somewhere in-between. Thus Herbart has in his *Metaphysics* a long section on what he terms "Synechology," in which he *de*scribes, from the approach of a peculiarly Herbartian compound of psychology and realism, various basic features of continuity that are inherent to space, time, and matter; Peirce however maintains very *pre*scriptively, that continuity — of a Peircean hallmark, of course — pervades the decisive categories of cognition. Whereas Fechner, in a position somewhere in between, — which cannot be illuminated because Fechner is there impenetrably obscure — enlarges on the "synechological outlook versus the monadological outlook" (*Die Tagesansicht gegenüber der Nachtansicht*, 2nd edition, 1904, p. 204). Also, when Herbart and Fechner turned to the Greek root word *synechés* for the formation of "synechology" then this was much less of a linguistic affectation than when Peirce coined his "synechism." Because Herbart and Fechner wanted the adjective "synechological," in the sense of "pertaining to, or descriptive of continuity," which cannot be readily formed from "continere" whereas Peirce's coinage of "synechism," to take the place of Law of Continuity, had no such justification.

Now, Peirce's failure to establish his Law of Continuity as he envisaged and wanted it, had the following overall reason, stated schematically. Peirce was aiming at a conception of continuity that would be metaphysically at least as all-pervasive as Leibniz had envisioned it, be cognitively even vastly more general than the notion of Leibniz, and yet be as rigorous, unequivocal and precise as the mathematics and mathematical logic of Peirce's era could make it. But with this ambition Peirce was striving after an impossibility. Carrying out this program would have amounted to fusing mathematics with metaphysics and general cognition in their entirety, which, however, seems impossible. Mathematics is not as large as are philosophy and cognition in their full scope. To formulate this precisely, "if a conception from general philosophy has been made mathematically rigorous, then it can wear the vestments of mathematical rigor to advantage only when moving about in areas of mathematics proper, or, at best, in border areas which mathematics is in the process of penetrating, but certainly not when moving about in areas which are well outside of mathematics' sphere of influence" (*Dictionary of the History of Ideas*, vol. 1, p. 502), and outside of the philosophy of mathematics also.

It is not clear from the statements of Peirce, and it may have never become clear to himself, whether synechism is intended to be present effectively outside of areas of philosophy of mathematics, or whether, conversely, the philosophy of mathematics is meant to extend into every precinct of metaphysics in which the presence of synechism is detectable. Peirce was one of the first of a species of philosophers who by trend, intent, or circumstances had been blurring the several demarcations between mathematics, mathematical logic, philosophy of mathematics, and general philosophy. Outside of Peirce, very prominent representatives of this new species of philosopher were A. N. Whitehead and Bertrand Russell; and it is worth observing that they frequently mused that it ought to be possible to trespass on philosophy proper with conceptions from mathematics, and they even made attempts in this direction. But they were prudent enough, especially Whitehead, not to become entangled in difficulties into which Peirce was stepping only too boldly. Philosophers like Whitehead and Russell were saved from the kind of impasse into which Peirce maneuvered himself, because they disciplined themselves to write up their findings in full-scale texts and monographs from time to time. Whatever did not fit into a monograph or become a monograph, simply had to be dropped or at least de-emphasized. Peirce — from whatever reason, perhaps from reasons of personal hardships — did not compose a single basic philosophical treatise of this kind, and it is probably idle to muse, whether he did not write a treatise because he was unable to think through synechism, or whether, conversely, he was not forced into making an intellectual decision with regard to synechism because he did not compose a systematic treatise about it.

After leveling these criticisms against Peirce, it must be stated that although his efforts to "synechize" our Universe, in its external and internal manifestations, did not succeed — and in fact, from our retrospect, were bound to fail — he did go about this self-appointed task in a very knowledgeable and interesting way. His outlooks and presumptions were much less "naive" than those of Leibniz; not only were they attuned to some of the scientific and other cognitive insights-in-depth of his late Victorian times, but they even foreshadowed some of the radical intellectual change-overs of the early decades of the 20th century too.

First of all, Peirce knew as sharply as anybody before him, that one cannot have continuity all by itself, only and exclusively, without any kind of discontinuity also present, however much the discontinuity be subordinate to continuity, existentially. This had been known long before Peirce. For instance, in the thinking of Leibniz, his *lex continui*, however universally valid, was inseparable from his *monadology*, which posited that anything there is anywhere, in nature and life, physics and metaphysics, logic and being, reality or mysticism, thought and impressions, *per*ception and *con*ception, is somehow built up of some ultimate units, which Leibniz called *monads*, and which, by their One-ness, somehow represented continuity and discreteness, both.

A very crude, low-leveled, but effective way of summarizing the monadology of Leibniz is to point out that even a physicist who abhors mechanics of discrete mass systems, but delights in mechanics of continuous media — and professes a scientific creed that only a mechanics of continuous media is metaphysically and ontologically meaningful — nevertheless knows that, in everyday life, all matter is somehow granulated, that there could be no chemistry without the assumption, however fictitious it be, that there are molecules, and that, however beautiful, perfect and meaningful Maxwell's purely field theoretic system of electrodynamics may be, it somehow is impossible not to perceive, at least in fictitious thought, that something like electrons are conceivable too. Now, this kind of physicist is not just a strawman of ours, there had indeed been such a one, long before Leibniz, and it was none other than Aristotle. Aristotle was strongly opposed to the atomism of Leucippus and Democritus, and for very meaningful scientific reasons of his; but he understood atomism masterfully, and greatly respected it as an intellectual achievement. Furthermore, a relatively recent study of the history of "philosophical" atomism by Andrew G. van Melsen, in his illuminating book *From Atomos to Atom* (1952), has shown that, nevertheless, Aristotle had a theory of smallest particles, which the middle ages called *minima*, and that actually all along, after Aristotle, there had been, however ambivalently, two kinds of theories, or rather two kinds of irreducible elements, one being *atoms* and the other being *minima*.

Now, Peirce has also something corresponding to Leibniz' monadology as a companion to the *lex continui;* but, going far beyond that, Peirce has something which, incipiently, corresponds to the 20th century de Broglie duality between (continuous) waves and (discrete) corpuscles; and he furthermore has an insight into something entirely different that goes far beyond what Leibniz perceived, namely an insight into the fact that *evolution*, in its post-Darwinian explication, straddles and fuses continuity and discreteness both. Now, for a few condensed details about this.

Peirce's counterpart to the monadological element in the *lex continui* of Leibniz is imbedded in Peirce's recondite doctrine of a basic tripartite category of cognition, the parts of which Peirce denotes, in hierarchical ascent, simply by Firstness, Secondness, and Thirdness. For our purposes the role of this entire category of cognition can be exemplified by the following occurrence, which, however simplistic in outward appearance, is actually representative of Peirce's thought pattern, in its outward scheme, at any rate. Suppose that a working mathematican has to consider a certain general "space," that is, a general Cantorian point-set with some kind of structure, whether the "structure" be topological, algebraic, or relational. Now, in Peirce's scheme of cognition the steps by which the mathematician becomes fruitfully cognizant of this set are as follows. He first of all becomes aware of the set in a first tentative over-all perception, and this amounts to the logical intervention of the category of Firstness. Then, he "verifies" in his mathematical mind, that it is indeed a Cantor set, consisting monadologically of "separable" elements — so that from Peirce's approach, a feature of discontinuity is involved in all existence. But, finally, and decisively, the mathematician becomes aware of, and verifies the cogency of the actual structure of the space, that is of the manner by which the various elements of it are combined and linked with each other; and this he does by the exercise of the Category of Thirdness, which is the decisive one. However simplistic this above example perhaps is, it does legitimately explain how, from this entire approach, both Firstness and Thirdness involve continuity (Firstness does so tentatively and Thirdness decisively) and Secondness contributes a feature of discontinuity, simply by envisaging existence in its most elemental form. A trenchant statement of the synechological link between Firstness and Thirdness and of the generality of the

Peirce notion of Continuity is contained in the following passage from a small but helpful book about Peirce.

> "Thirdness is mediation, generality, order, interpretation, meaning, purpose. The third is the medium or bond which connects the absolute first and last, and brings them into relationship. Every process involves Continuity, and continuity represents Thirdness to perfection."

(See Eugene Freeman, *The categories of Charles S. Peirce*, 1931, p. 19.)

Next, Peirce's anticipation of the 20th century duality principle of quantum theory appears in the following insights and assertions of his.

The 19th century took it for granted that not only a mechanical system but also any other physical system is subject to a law, usually called determinism, that initial conditions, especially initial temporal conditions, *when precisely known*, fully determine the entire future course of events and developments within the system. This "law" was expressly stated by Laplace, and is attributed to him, but it actually was already contained in the *lex continui* more-or-less. But, seemingly inconsistently with this, the 19th century also arrived at the insight that it is possible to arrive at conclusions that are well-determined, even if initial conditions are not known precisely but only "statistically," that is, probabilistically; and the 19th century erected the mighty kinetic theory of gases and of other matter in this way. In this theory, position, direction, and velocity of molecules are assumed to be known by probabilistic averages only, however much, by deterministic hypothesis, they might be precisely determinable at every point. Furthermore, the 19th century took it for granted that this statistical approach was directed by "intellectual" and "practical" convenience, as it were, namely by the difficulties of truly precise observations, and by the mathematical inconvenience of dealing rigorously with a mechanical system of very many particles.

Now, Peirce liked the statistical approach but not its prevalent rationale, and he radically modified the approach by introducing the concept of "absolute chance," by which he meant the existence of a genuine indeterminacy as opposed to an indeterminacy arising merely from ignorance. Thus he says (6.46): "Try to verify any law of nature, and you will find that the more precise your observations, the more certain they will be to show irregular departures from the law. We are accustomed to ascribe these, and I do not say wrongly, to errors of observation; yet we cannot usually account for such errors in any antecedently probable way. Trace their causes far enough and you will be forced to admit they are always due to arbitrary determination, or chance." Peirce even introduced a "law" which maintains the existence of "absolute chance." and he called it "Tychism," from the Greek word *Tyche*, which was the name of the goddess of chance.

A very systematic and influential version of Tychism was created — or perhaps one should say that Peirce's own Tychism was vindicated — in the 1920's, around

1926, in quantum theory, and it came into being in several stages. In 1925 Heisenberg introduced his so-called law of uncertainty. Soon afterwards, Born and Heisenberg were led by this to the statistical interpretation of the Schrödinger Quantum theoretic wave function, which became known about the same time. And this statistical interpretation soon led to a deepening of the understanding of the de Broglie duality between wave and corpuscle which had been proclaimed by him in 1924, even before Heisenberg and Schrödinger made their discoveries.

It is no discredit to Peirce to say that he did not specifically anticipate any of these wonderful achievements of quantum theory. But this is a fitting occasion for "scolding" him, more sternly than ever, for not having collected his philosophemes into a systematic treatise. Doing so might have led him to confront and compare Synechism with Tychism in such a way as to make him a harbinger of the pronouncement of de Broglie that wave and corpuscle, that is continuity and discreteness, are theoretically and phenomenologically compatible and inseparable. What is even worse, although Peirce knew Aristotle extremely well, he did not comment on the observation of Aristotle that there are cases of genuine chance (*tyche*) which are disturbingly "un-random." For example, if someone goes "by chance (*tyche*)" to the market-place where he unexpectedly meets someone he has been wishing to meet, then this is, after all, a consequence of his "determined" decision to go to the market (*Physica* 196 a 4), and not of some "random" desire to meet the person whom he did meet.

And yet, Peirce had still one more insight to his credit. This one was in a way the profoundest of all, but it was also the least worked-out of all. Biological Evolution, Darwinian, or even Lamarckian, has very much the appearance of a phenomenon of continuity, and its emergence and development in the 19th century is, undoubtedly, one more corroboration, and a very powerful one, of the overall fact that the 19th century was very much continuity-oriented, that is, overwhelmingly *synechistic*, as Peirce might have put it. And yet the Late 19th century became very restive over the fact, that however synechistic a specific evolutionary change-over may appear to be, it still does represent steps of discontinuity, however small ones, and that it was not even able to account to itself how these steps of discontinuity really come about. In the early 20th century the uneasiness was compounded by the fact that *Mendelism* and *mutationism* entered upon the scene of genetics, and were, at first sight, even at cross-purposes with natural selection. The biologists finally resolved these difficulties, and peace now reigns on this particular battlefield.

Now, it is most remarkable that Peirce had had some kind of inkling of the intricacy of this biological situation, and that he even introduced a special law by which "intellectually" to control the situation. He should have called this law "evolutionism" or by some such name, but instead he injected "Love" into the context, and called it the Law of Evolutionary Love. This "Love" was not even sexual love, as would be quite befitting a context of Evolution of Species, but christological Love, the Love of the Gospel of John, which in Greek is called *agapé*. And so, Peirce renamed the Law of Evolutionary Love *Agapism*, and as a name for some kind of semi-continuity this is one of the oddest coinages conceivable.

Altogether, Peirce created two triads. One was Firstness, Secondness, Thirdness — and the second was Synechism, Tychism, Agapism, and there seems to be some parallelism between these two triads. But, having never composed a treatise, Peirce cannot be called to account to explain how these two triads correspond to each other, and let alone how they also connect with the philosophical doctrine of Pragmatism of which C. S. Peirce was an architect. There is a rising flood of studies on the problems and puzzles which are the legacy of Peirce's philosophical efforts, and we want to terminate with referring to one relatively brief article that perhaps comes closest to justifiying our own critique of Peirce. It is Charles Hartshorne, "Charles Peirce's One contribution to philosophy and his most serious mistake", in Edward C. Moore and Richard S. Robin (editors), *Studies in the philosophy of Charles Sanders Peirce, Second Series*, Amherst 1964, pages 454–474.

Mathematics is the oldest organized knowledge there is. Yet "what makes mathematics so effective when it enters science is a mystery of mysteries" (*The Role of Mathematics in the Rise of Science*, p. v); and any philosopher bent on solving the mystery once for all, as C. S. Peirce tried, may come to grief.

(A comprehensive collection of Peirce's writings appears in *The Collected Papers of Charles Sanders Peirce*, Vols. I-VI, edited by Charles Hartshorne and Paul Weiss, Cambridge, Mass., 1931–1935.)

This is a paper which was presented at the Conference on History of American Mathematics, sponsored by Texas Tech University, May 1973. The article will also appear in "Graduate Studies of Texas Tech University" at a later date. The Association is grateful for permission to use material from the article "Continuity and Discontinuity in Nature and Knowledge," *Dictionary of the History of Ideas*, copyright 1973, by Charles Scribner's Sons.

DEPARTMENT OF MATHEMATICS, RICE UNIVERSITY, HOUSTON, TEXAS 77001.

Vistas in Astronomy, Vol. 19, Pt. 2, pp. 133–161. Pergamon Press. Printed in Great Britain

MATHEMATICAL BACKGROUND SPACE IN ASTRONOMY AND COSMOLOGY

SALOMON BOCHNER

Department of Mathematics, Rice University, Houston, Texas

CONTENTS

I. Introduction 133
II. Statement of our problem 135
III. Nicholas of Cusa (1401–64) 137
IV. The *machina mundi* of Copernicus 139
V. Epicycles and twentieth-century Analysis 142
VI. Kepler's Laws The stormy rise of Analysis 145
VII. Space in Newton 149
VIII. Infinity in cosmology 152
IX. "Immensity" of the Universe in Newton and Copernicus 154
Appendix. Maximal universes 156
Notes. 158

I. INTRODUCTION

It is as difficult as it is provocative to explain how it came about that soon after the Middle Ages, Astronomy (and Cosmology)—which had been an established "exact" science, with solid achievements to its credit—somehow began from a new beginning, the new developments starting initially with Cusanus, "officially" with Copernicus, systematically with Kepler, and irrevocably with Newton. An explanation for the emergence of these "New Directions" in Astronomy can be sought from different approaches, and it is the purpose of the present paper to suggest one more approach, perhaps a novel one.

It is the thesis of the present paper that what set off the new developments in Astronomy from the preceding ones, was the infusion and even irruption of a radically new mathematics, namely of mathematical Analysis, into the texture and structure of Astronomy, leading to new outlooks and outcomes in the process.

We must emphasize, however, and ask the reader to keep constantly in mind, that in our interpretation of the mathematical developments themselves, the rise of Analysis in the sixteenth and seventeenth centuries was by no means articulated mainly by the rise of the infinitesimal calculus, as frequently taken for granted. Rather, it was articulated, even more pertinently by the emergence, covertly yet potently, of certain mathematical objects and constructs that were beyond the cognitive reach of Archimedes, Appolonius, Ptolemy, and of their intellectual dependents in Late Antiquity and the Middle Ages. The principal such objects are: mathematical background space, real numbers, and mathematical functions, and we will make statements about each. These three objects are hard to separate from each other at their roots. Yet the role of real numbers and of Mathematical functions becomes fully manifest in mechanics of physical systems in general, rather than of planetary systems in particular; whereas in Astronomy proper, as developments appear to us, it is mathematical background space that is all-present and all-dominant, however covertly at times. Whence the title of this paper.

It is the temper of the twentieth century to think in terms of scientific revolutions, whatever that be, and a so-called Copernican Revolution looms large among such. From our approach, if there indeed was a Renaissance "Revolution" in Astronomy, it took place most tangibly and radically in the underlying mathematics, which changed over from a finitely limited fatigued and exhausted Greek mathematics to an infinitely unbounded, ever buoyant and inexhaustible "modern" mathematics. The total change-over, in all aspects, lasted several centuries, but somehow the name of Copernicus became prominently attached to it. Such being the case, we will begin with Copernicus, and give a certain priority to matters relating to him, although it was the work of Kepler in which the new Astronomy became glaringly and dazzlingly "new" indeed. Another reason for emphasizing Copernicus none the less is the following. The present article is a rather expanded version of a contribution by this author to a recent Copernicus Symposium on "The Copernican Achievement",[1] and it has been both convenient and appropriate to retain the original subject matter in skeleton.

Actually, in our interpretation, the leading protagonist by far will be Kepler and not Copernicus. Yet the attention given to Copernicus will allow us to highlight two contrasting features in the development of astronomy and cosmology which are central in our account.

One of the features is the fact that the thinking of Copernicus was dominated by the conception, or rather imagery, of a *machina munda*, which, however important it be to the comprehension of Copernicus, simply and totally disappears soon after him, beginning with Kepler at any rate (Sections IV and VI).

The second feature is the distinction between *World* (or *Universe*) and *Heavens*, that is between *mundus* and *caelum*, that is between *to pan* (or *to holon*, or *kosmos*) and *ouranos*. Antiquity and Middle Ages tended to blur the distinction. But Copernicus put it and kept it in focus; and this distinction has become ever more justified in subsequent developments (Section IX).

According to the *Thesaurus linguae 'latinae*,[2] the expression *machina mundi* occurs for the first time in the first century B.C. in *De Rerum Natura* of Lucretius (book V, line 96) and afterwards a number of times in Later Antiquity; but it would not be easy to determine what, for instance, the poetic vision of Lucretius, enchanting as it be, really saw in it, tangibly that is. In the Middle Ages the expression is encountered several times in the thirteenth century, namely in two separate books entitled "The Sphere", one by Robert Grosseteste, and the other by Iohannes de Sacrobosco, and, of course, in commentaries on these two, for instance, still in the thirteenth century, in the commentary of Robertus Anglicus on the work of Sacrobosco.[3] But it still is not easy to say what it is really meant to be.

But then in the first half of the fifteenth century it occurs in the work of Nicolas of Cusa in a context which, to me, is suggestive of an important meaning, even if the meaning is not operationally articulated (Section III). Finally, I find it easiest to explain certain features in Copernicus by identifying the presence of a *machina mundi* in his thought patterns, even if the expression is never mentioned. Copernicus also never mentions the name of Cusanus, whose work he must have known.

That, in general discourse, the distinction between World and Heavens tended to blur was already noted by Plato, in the fourth century B.C., who put it thus: "Now the whole Heaven, or Cosmos, or if there is any other name which it specially prefers, by that let us call it,—so, be its name what it may, . . ."[4] Four centuries later, in the first century A.D., Pliny the Elder opens book II with the following line:

Mundum et hoc—quocumque nomine alio caelum appelare libuit

The world and this—whatever other name men have chosen to designate the sky . . .[5]

Yet further on in book II, namely in II, 102, Pliny seems to imply a distinction, making Heavens less comprehensive than World.[6]

Furthermore, back in the fourth century B.C., Aristotle, in his work on the Heavens, states, in 278 b 11, that *ouranos* can have various meanings,[7] whereas several Stoics—more-or-less contemporary with Pliny—performed similar dissections of meaning for *kosmos*.[8]

And yet considerably later, around A.D., 400 Macrobius in his *Commentary on Scipio's Dream*, XVII, 5, in a context in which he clearly distinguishes *mundus* from *caelum*, also pointedly states that *mundus* corresponds to *to pan*, spelling the latter in the original Greek.[9]

In sum, there had always been some kind of distinction between *mundus* and *caelum*, however latent at times. Copernicus put the distinction into focus, and it was a good thing to do.

II. STATEMENT OF OUR PROBLEM

It is not easy to gain an insight into the intellectual development of Copernicus, and in particular it is not easy to determine when, and by what stages the *Commentariolus*, which was Copernicus' informal "first version" of this theory in manuscript, came into being. In a somewhat vague way, it had been generally assumed that the *Commentariolus* belonged to Copernicus' middle years, but in 1962 J. R. Ravetz suggested, on rather speculative evidence, that Copernicus composed it even before leaving Cracow for Italy in 1496.[10] This was opposed with rather brief statements by E. Rosen,[11] and, quite recently, with an impressive array of arguments by Curtis A. Wilson.[12] The decisive points at issue seem to be whether certain specific astronomical data which seem to underlie the reasoning or rationale in the *Commentariolus* were already in existence before Copernicus left for Italy, or at any rate would have been readily available to him even then.

Now important and meaningful as the resolution of such issues may be, the problem of the true intellectual origin of the *Commentariolus*, that is, of Copernicus' true intellectual motivations, is not answered hereby. Copernicus may well have started developing his heliocentric insights quite early in his academic career, long before he had the "library facilities" for testing and verifying them, or the skill of articulation for expressing them. Thus, his dissatisfaction with the Ptolemaic assertion that the huge sphere of stars performs daily rotations need not have come about at all by the knowledge of astronomical detail from specific astronomical works. It may have been simply the expression of a metaphysical malaise, which was brought about by certain nascent general developments in the relation between mathematical space and cosmic universe, which were destined to profoundly alter astronomy and cosmology in due course. These novel relations between Space and Universe are already manifest in Cusanus, almost a century before Copernicus, except that in Cusanus they are involved in theologically motivated cosmology rather than in operationally articulated astronomy.

We will describe these developments as we see them, and we will find that various features in Copernicus are a reflection of them, some more, some less. We will even take note of Copernicus' preference for uniform motion on or of circles and his aversion for equants, even if in these cases the link with our general developments will be precariously artificial because blandly anachronistic. But very natural and pertinent will be our "explanation" of the following pair of occurrences in book I of *De Revolutionibus*, one in chapter 6 and the other in chapter 8. In chapter 6, Copernicus declares, with a certain solemnity, that the Heavens (*caelum*) is "immense", but he shows no disposition to amplify what "immensity" actually means. In chapter 8 he then touches upon the parallel question of the size of the World (*mundus*), whether it is finite or infinite. He states, somewhat rapidly, that he is not going to become involved in the question, but he does add an amplification, in a sense, namely, that this is the kind of question that should be left to *physiologoi* (philosophers of nature?) to argue about. These two statements seem oddly assorted, but our approach will harmonize them nicely (see Section IX).

In our attempts to elucidate the role of space in astronomy, which is the actual subject matter of the present article, we will have to refer to previous findings of ours on the role of space in science and knowledge in general. And we begin by citing from a recent article on "Space"[13] in general, that although the Greeks did have space, they had it in much fewer areas of Cognition and Knowledge than we have it today, and that only after the Middle Ages did a noticeable penetration of the notion of Space into other and newer areas of perception begin. Thus, the Greeks did have space in physics and cosmology, and perhaps also theology, but they did not have space of logical, ontological, or psychological perception, which only the Renaissance and the Scientific Revolution began to discern.

After suchlike preliminaries the article on "Space" continues verbatim as follows.

> More conspicuous, and almost fate-sealing, was the absence from Greek thought of a general conception of space for geometry and geometrically oriented analysis. Greek mathematics did not conceive an overall space to serve as a "background space" for geometrical figures and loci. There is no such background space for the configurations and constructs in the mathematical works of Euclid, Archimedes, or Appollonius, or even in the astronomical work *Almagest* of Ptolemy. When Ptolemy designs a path of a celestial body, it lies in the astronomical universe of Ptolemy; but as a geometrical object of mathematical design and purpose, it does not lie anywhere. In Archimedes, who, in some respects was second only to Isaac Newton, the mathematical constructs were placed in some kind of metaphysical "Nowhere" from which there was "No Exit" into a mathematical "Future".
>
> Such a background space was rather slow in coming. Thus, Nicholas Copernicus did not have it yet. He was an innovator in astronomical interpretation, and not in mathematical operation. His mathematics was still largely Ptolemaic, and only a bare outline of a *mathematical* background space is discernible in his *De revolutionibus orbium coelestium*. Nevertheless, already a century before Copernicus, Nicholas of Cusa, churchman, theologian, mystic, and gifted mathematician, in book II of his leading work *Of Learned Ignorance* (*De docta ignorantia*), adumbrated an overall space of mathematics by way of an overall mathematical framework for the space of the universe. But the leading statements of the metaphysics of Cusa were enveloped in theology and mysticism and not very comprehensible to his contemporaries and to others after him.

These two paragraphs were written several years ago and I have since come to a somewhat different assessment of Copernicus and his intellectual dependence on what had preceded him. In the case of Ptolemy, whatever his greatness, I find, on reconsideration, that he indeed did not have a mathematical background space anywhere in his thinking. The nested system of spheres in Ptolemy's *Planetary Hypotheses*,[14] even though there were specific numerical distances between the spheres, was but the paradigm of a *machina mundi*, which, as we will still see (Section IV), was something very different from a space-filling universe in our sense. This *machina mundi* lived on very long, and was still quite dominant in the thinking of Cusanus and Copernicus. Now, beginning with Cusanus the first outlines of a background space are breaking out here and there, and my original impression was that they are fainter in Copernicus than in Cusanus. But on reconsideration I find that these outlines are not fainter in Copernicus than in Cusanus, but only less discernible, and they are so because in the *De Revolutionibus*—other than in the *Docta Ignorantia*—these outlines are immersed in operational astronomy without yet being of operational consequence themselves.

After Copernicus, in the second half of the sixteenth century, many kinds of space suddenly come to the fore. Mathematics initiated the space of descriptive and projective geometry, and the representational arts the space of standard (rectilinear) perspective;[15] and there were many future-oriented statements on space in the works of philosophizing scientists like Bernardino Telesio (1509–88), Franciscus Patrituis (= Patrizi, 1529–97), Giordano Bruno (1548–1600), Tommaso Campanella (1567–1639),[16] and others. Professional astronomers themselves were seemingly lagging behind, but only in the course of the sixteenth century itself; whereas in the early seventeenth century Kepler burst in on the scene, parading his New Astronomy all over the scene, and dazzlingly revealing mathematical background space of astronomy for all time to come. Mathematical background space reveals itself, and shines through, clearly, unmistakably, and ineradicably, in Kepler's twin statement that the orbit of Mars and of each other planet lies in a plane, and, in fact, in a separate plane for each planet, and that in this plane the radius vector issuing from the sun sweeps out equal areas in equal times. That these two statements belong together is tacitly but powerfully corroborated in Newton's *Principia*, in which they constitute the very first theorem of book I thus: "The areas which revolving bodies describe by radii drawn to an immovable center of force do lie in the same immovable planes, and are proportional to the times in which they are described."[17] Newton could not have tendered a more powerful endorsement to Kepler than by starting off the *Principia* in this way. Also, as if in preparation for this theorem, in a famous scholium which precedes this theorem and is separated from it only by some purely mathematical lemmas, Newton very expressly states, in words of his own, *for the first time* in the evolution of exact science but for all times to come, that there can be no such science without expressly positing a background space (and time) to begin with.[18] Also, in operational truth this space is first and foremost and most indispensably a mathematical space, which simply must be posited if there is to be a mathematically controlled science at all. Newton knew this perfectly well in the purely scientific compartment of his intellectual heart, even if he termed it *absolute* space, thus attributing to it physico–metaphysical features beyond the call of operational necessity. Nevertheless, many philosophers and scientists insisted on analyzing these extra-operational features of absolute space, but nothing came of such attempts. The analysis proved fruitless and obfuscating even if it was conducted with passion by many among the great, with Leibniz and Einstein among them.[19] Except that the great pragmatic mathematicians of the eighteenth century showed admirable judgement in refusing to be drawn into this debate. We quote from Max Jammer's book on space.

> It is interesting to note how little the actual progress of the science of mechanics was affected by general considerations concerning the nature of absolute space. Among the great French writers on mechanics, Lagrange, Laplace and Poisson, none of them was much interested in the problem of absolute space. They all accepted the idea as a working hypothesis without worrying about its theoretical justification. In reading the introductions to their works, one discovers that they felt that science could very well dispense with general considerations about absolute space.[20]

Also Diderot and D'Alembert, in the article "Espace" of their *Encyclopédie*, say with regard to the existence of absolute space: "...cette question obscure est inutile à la Géometrie & à la Physique."[21]

To this I would like to add for my part that in physics and cosmology a mathematical background space must be given initially under all and any circumstances, even in any cosmogeonic theory in which the creation of a physically meaningful intrinsic space is presumed or intended to be part of the total creation process, and a vital part too. For instance, H. Alfvén in a recent theory of the creation of a magagalaxy out of pre-given ambiplasma makes the following statement: "The starting point in our model is an ambiplasma, which fills the huge sphere. Its density is uniform throughout."[22] Thus, before the actual cosmological space has come into being there was somehow somewhere a "sphere" filled with ambiplasma, and it was obviously a Euclidean sphere. The Euclidean structure of this primordial sphere is implied in the requirement that there shall be everywhere a point density of the ambiplasma and that it be everywhere numerically the same. Of course, it would be compatible with the above assumption to have not a Euclidean space but any kind of metric space in which spheres and point-densities can be defined. But such a metric space would still have to be posited and specified to begin with.

A very incisive situation prevails in any present-day cosmology that is based on the General Theory of Relativity. A universe is then a space–time object, with various physical data on it, the Riemann–Lorentz metric, which represents gravitation, being one of the data. Underlying it is a four-dimensional differentiable manifold, which corresponds to a background space in our sense, and all the other data correspond to the physical structure and content of the Universe. Whatever boundary-value problems occur involve the physical data on the given differentiable background space and not the differentiable structure of the background space itself which is intended and presumed to be given in a maximal extent to begin with.[23] Albert Einstein himself would have probably felt uncomfortable at a sharp separation between background space and physical space, and so would have probably even H. P. Robertson, who was quite pleased with his slogan that "geometry is a part of physics"; but the present-day generation of professionals in the Theory of Relativity takes the separation in its stride.

See also the Appendix.

III. NICOLAS OF CUSA (1401–64)

Nicolas of Cusa, or Cusanus had a mathematical turn of mind, and he had some technical mathematical achievements to his credit. He was the first "modern" analyst, inasmuch as contours of a background space are discernible in his vision of a cosmological universe as perceived by him in the last three chapters of book 2 of his famed *Docta Ignorantia*.[24] There Cusanus makes intriguing pronouncements on the Universe, referring to a *machina mundi* at times. These pronouncements are not mathematically operational. But they are more than merely philosophical, and they call for the introduction of a mathematical background space in which to become the more meaningful. They are somehow future-oriented, pointing towards later mathematical developments, even much later ones. A fine study of the cosmological outlook of Cusanus is contained in a general work of Koyré,[25] which presents a sequence of excerpts from Cusanus with an illuminating commentary by which to link them up.

Cusanus made two major statements about the mathematical shape of the Universe, and they are interlocking yet separable. At first they sound irrationally mystical, but they have proven themselves to be rationally scientific.

(A) The world is neither finite nor infinite, or, what is virtually the same, it is both finite and infinite.

(B) The *machina mundi* has its center everywhere [and thus nowhere], and its circumference is nowhere.

Statement (A) can be interpreted to mean that the world has features by which it is finite and others by which it is not. Whether Cusanus himself so interpreted this statement or not, the fact is that the statement tries to resolve a problem of great complexity which has been vexing Western Cosmology from Presocratics till today. Very few scientists have made the Universe unequivocally finite. Even Aristotle has statements in his *Physica*—in the essay on Infinitude in book 3, chapters 4–9—which are not quite in accord with his contention in *De caelo* that the Universe is finite. In the twentieth century Arthur S. Eddington, at the height of his career, boldly asserted that the Universe consists of a finite number of elementary particles only and he even dared to suggest the numerical value N of this cosmological number of his[26]; but it is not at all clear whether N was meant to subsume all conceivable elementary particles or protons and electrons only.

On the other hand, the *non*-scientist Giordano Bruno made the Universe unequivocally infinite, by positing that it consists of infinitely many "worlds" that are distributed uniformly all over a Euclidean space. But very few genuine scientists after him made it quite so unequivocally infinite too. In fact, in 1931 the redoubtable churchman Ernest William Barnes, bishop of Birmingham, declared: "It is fairly certain that our space is finite though unbounded. Infinite space is simply a scandal to human thought.... The alternatives are incredible...." When quoting this on a previous occasion I ventured to add that "beginning with Newton, scientists who seemingly acquiesced in infinitude did so not only silently, but also restlessly and even sullenly".[27]

Statement (B) of Cusanus was not new by paradigm, only by interpretation, inasmuch as there had been a medieval aphorism that God is nowhere and everywhere.[28] Perhaps because of this "mystical" background, statement (B) did not find ready *entrée* into "modern" science, but it was vindicated eventually, even with a flourish. It was vindicated spectacularly when Einstein suggested a closed universe of the mathematical type of a sphere; it fits statement (B) to a tee. It was this kind of space to which Bishop Barnes referred as infinite but unbounded.

Of course, ordinary Euclidean space fits (B) too, even ideally, but only mathematically. To make the fit cosmologically right too, the universe has to be a Giordano Bruno universe, more or less. However, such a universe is not only astrophysically unstable, and thus—as so beautifully and irrefutably argued by Kepler[29]—cosmologically very unlikely, but it is also most emphatically infinite, without any feature of finiteness by which to conform to statement (A) as well.

Still, from our retrospect, there does remain a possibility for constructing a universe that, in a sense, satisfies statements (A) and (B) both, and is open "on all sides" nonetheless. It suffices to take our "ordinary space", divest it of its ordinary Euclidean structure with Gauss–Riemann curvature 0, and invest it instead with a hyperbolic structure with constant negative curvature, after the procedure of Bolyai and Lobachevsky. By known constructions of A. Caley and F. Klein, the Bolyai–Lobachevsky geometry in any dimension n can be realized by introducing a suitable non-Euclidean metric in the open unit ball

$$x_1^2 + \cdots + x_n^2 < 1$$

of Euclidean E_n. In this metric the unit ball is homogeneous relative to a transitive group of motions, so that there can be no "absolute" center, and it is also infinite in this metric.

But in the metric of the ambient Euclidean space its diameter is of course finite.

It would, of course, be very anachronistic to impute to Cusanus an adumbration of a space, or universe of the specific Bolyai–Lobachevsky structure. However, it is my view that a search for a universe whose size is "in between" finite and infinite is as old as the earliest speculations of the Presocratics.[30] In particular, a very interesting "in between" was envisioned by Parmenides (born *ca.* 540 B.C.) for his intriguing ontological universe that he conceived being structured of Pure Being (*to eínai*), Pure Thought (*to noeín*), and also Ontological Truth (*Alétheia*). He asserted that it is *tetelesmenon*, a Greek word meaning "completed", or simply "complete". Aristotle understood this to be a near-equivalent of "finite", and generations of commentators, even very recent ones, have so understood it too. And this interpretation has been encouraged by the fact that there is a nuance, but only a nuance, of "ordinary" finiteness in Parmenides too. But for my part I find, and have argued before, that his "completeness" was an anticipation of the same spatial quality which made the Cusanus universe both finite and infinite, as stated above for the Cayley–Klein model of hyperbolic space, say. It is also a peculiar fact that in present-day topology there is a property called "completeness", literally so, which is such that both a closed space à la Einstein and an open space à la Bolyai–Lobachevsky are fully endowed with it.

The reader may have wondered why we drew Parmenides into our "comparative" study at all, seeing that his universe had the "outlandish" features of consisting of Being, Knowledge, and Truth, and thus can hardly qualify for a meaningful confrontation with space and universe of today. Obviously, one would think his universe was something that only somebody in Antiquity could have dreamt up, and it seems incongruous to interpret and analyze this dream in the context of modern theories. To suchlike objections we wish to observe that each of them is perhaps valid, except for the one that a universe composed of such outlandish ingredients could have been dreamt up in antiquity only. Because it was dreamt up all over again, in our present twentieth century too, and to the applause of an admiring following too.[31] But there is one very striking difference between the two constructions, the ancient and the modern, and it is a difference which very much corroborates the main thesis of the present paper. Namely, this twentieth-century universe is expressly accompanied by something corresponding to our mathematical background space, whereas the universe of Parmenides, being a Greek construction, does not have such a companion at all.

The modern construction I am referring to is the "logical space" (*logischer Raum*) of the linguo-philosopher Ludwig Wittgenstein in his famed *Tractatus*.[32] This logical space is some kind of aggregate or congeries of logical entities like "facts", "atomic facts", "states of affairs", "propositions", etc. Some commentators of the *Tractatus* ascribe to this space some specific structural features, but even these features are not very geometrical in a traditional sense. Furthermore, the ontological traits of the world of *Tractatus* could be taken straight out of the universe of Parmenides—except that the Truth in Parmenides, although already "two-valued", was still ontological rather than logical. And yet there is a difference between the two. Namely, the *Tractatus* introduces not only the logical space but also a "world" or "universe" (*Welt*), and the ontological traits occurring adhere to this world. And the logical space is in a sense but a background space to the World, although in a peculiar Wittgensteinian way the logical space "seems to be more primary than the world, inasmuch as the constituents of the world are only some kind of 'pictorial' representation of the constituents of the logical space". Now, it is this presence of a background space to the world which makes the Wittgensteinian fancy so acceptable and even luring to many twentieth-century analytical philosophers of various stripes, but makes the Parmenides universe, which does not have a background space, so strangely esoteric and antiquarian. Yet the Parmenidean space is logically much more "tidy" than the Wittgensteinian, and as regards the size of the Universe the view of Parmenides was decidedly the more interesting one. The "completeness" of the universe of Parmenides was a subtle trait of it. Wittgenstein, however, asserted quite unnecessarily that his space is "infinite". He does not make any use of this infinitude; and he made the assertion of it so casually that it obviously was only the standard philosophers assertion since Giordano Bruno, and nothing more.

IV. THE *machina mundi* OF COPERNICUS

Copernicus introduced two simultaneous motions of the Earth, the yearly translation around the Sun and the daily spinning around its axis; and this has become standard since. However, Copernicus' operational procedure of forming the resultant of these two motions was still traditionally medieval, and radically different from how it is formed today.

Nowadays the two motions are vectorially independent in the underlying mathematical three-dimensional space, say in the absolute space of Newton. The superposition, or addition of these two motions in this space is commutative, that is either of these two motions can come first. In fact, in suitable coordinates (x,y,z), the location of a point of the Earth at time t is represented by

$$\begin{aligned} x(t) &= r \sin\theta \cos(\varphi + nt) + \xi(t), \\ y(t) &= r \sin\theta \sin(\varphi + nt) + \eta(t), \\ z(t) &= r \cos\theta \qquad\qquad\quad + \zeta(t), \end{aligned} \tag{1}$$

where r is the distance of the point of the Earth from the center of the Earth, (θ,φ) are its latitude and longitude, n is a constant depending on the normalization of time, and $\{\xi(t),\eta(t),\zeta(t)\}$ are the coordinates of the center point of the Earth at time t. The formulas (1) were introduced in the eighteenth century, mainly by Euler and Lagrange. But their substance was created by Isaac Newton in the seventeenth century, even though he chose to sheath them in an Archimedian garb. And, before Newton, the substance of the formulas (1) was perceived and exploited by Johannes Kepler, except that he was aware of them only for the planetary system, whereas Newton created them for all of mechanics, celestial and terrestial both.

Copernicus, however, did not have a universe, in the sense of a physical system in a space, but only a "space-less" *machina mundi*, which was, vaguely, an arrangement of wheels, gears, and rods. The movements of the parts of the *machina* were "centrally" controlled, coordinated, and hierarchically ordered. Nothing was "loose", or "free-floating" as if in a "space", and no two submovements in the *machina* were conceived as being "independent" and "commutative".

A first sequence of this entire conception was that every movement had to be composed of a finite number of circular and retilinear motions, because they were the only ones that a *machina* could produce. This entire outlook may have also been the "true" reason why philosophers since Plato and Aristotle had philosophized that motions on circles (and straight lines) were the aesthetically basic ones. They simply made a virtue of a necessity. In the case of Copernicus, C. C. Gillispie, in a fine article about A. Koyré in which he also interprets Koyré's view of Copernicus, comes to the conclusion that "he [that is Copernicus] was archaic in his addiction to the Platonic aesthetics of circularity, making it into a cosmic kinematics".[33] This may have been so. But Copernicus also had an operational dependence on a *machina*; and it would probably be futile to try to decide which influence was the stronger one, the addiction to circularity or the dependence on a *machina*.

Secondly, the image of a *machina* explains why, still in Copernicus, it is the *orbis* on which a celestial body is located which moves and not the body itself; the body itself being firmly affixed to a spot on the *orbis*. Otherwise, the body would be "on the loose" on its *orbis*, and beyond any central control, which was not in accord with the image of a *machina*.

Thirdly, and most importantly to our present context, the two motions of the Earth, the yearly and the daily, could not be commutatively independent, but, as in every conjunction of motions, they were hierarchically ordered.

Now, in the present conjunction, Copernicus undoubtedly deemed the motion along the ecliptic to come first, and the daily rotation to be subordinate to it. Such being the case, the Earth "ought" to be firmly in place in the moving *orbis* of the ecliptic. In particular, the axis of daily rotation "ought" to be firmly affixed to the plane of the ecliptic, and thus form a fixed angle with the polar axis of the ecliptic. However, the axis of the Earth does not form such a fixed angle, but remains parallel to itself in "absolute" space, and, in Copernicus' vision, this can come about only by an extra motion of the Earth, a third one, which is designed to bring this about. In this motion, "The axis of the Earth describes the surface of a cone in a year, moving in an opposite direction to that of the Earth's center, that is from east to west".[34]

This is Copernicus' Motion of Declination, which, in its time, provoked much head shaking. Kepler, who, piece by piece, discarded all paraphernalia of a *machina*, brushed the Third Motion aside with the curt statement that it was "something nominal rather than factual".[35] And Kepler was right, because, from our retrospect, the Third Motion was nothing but a piece of clumsiness, a testimonial to Copernicus' incapacity to add up two motions vectorially. But it is also a testimonial to his greatness as an astronomer that, once he had installed this piece of clumsiness, he was nevertheless able to extract some very crucial astronomical yield from it.[36] He made the period of the Third Motion slightly less than a year, and in this way he explained the precession of the equinoxes, which was now at last correctly explained as a slow motion of the axis of the Earth.[37]

It must be stated, however, that by operational paradigm, this third motion was not original with Copernicus, but was Ptolemaic too. It was an imitation of something that Ptolemy had done in connection with the epicycles of the outer planets, in order to account for the fact, astonishing to him, that the planes of these epicycles were parallel to the ecliptic; which, from our retrospect, was as it should be, because "the epicycle of an outer planet was nothing but the earth's annual orbit around the sun transferred to the planet in question".[38] The interpretation of the Universe as a *machina*, which shines through in Ptolemy's *Planetary Hypotheses*, would have demanded a different location of the epicycle,[39] and Ptolemy accounted for the difference by "the introduction of a small auxiliary circle, . . . which revolved in the zodiacal period of the planet".[40]

That Copernicus was still largely Ptolemy-oriented comes through, almost inadvertently, in the following sentence in Dreyer's work: "In his sixth book, the shortest of all, Copernicus deals with the latitudes of the planets, and this is the part of his work in which he keeps closest to Ptolemy".[41]

In sharp contrast to this, it was precisely the problem of latitudes that was the jumping-off problem for Kepler, from which he soared into the vastness of his New Astronomy. And that in the matter of latitude Copernicus was still clinging to Ptolemy is closely connected with the fact that, in a very crucial respect, his transition from a geocentric to a heliocentric outlook was hardly of any operational consequence at all. Namely, unlike Kepler after him, Copernicus neither expressly states nor shows an awareness of the two-part fact that, due to the heliocentric hypothesis, the orbit of each planet is in a plane and that these planes are various planes in a common (three-dimensional) overall space. Copernicus does speak of a plane (*planus*) of the ecliptic in chapter 1 of book 6 of *De Revolutionibus*, but this *planus* is a direct translation of *epipedon* in chapters 2 and 3 of book 13 of the *Almagest*. And if Copernicus also speaks of the line of intersection of the planetary orbits with the plane of the ecliptic then he apparently refers only to the line of nodes,[42] without introducing the plane of the planetary orbit to.

Derek Price, in a critique of Copernicus, speaks, none the less of a "Copernican invention of a planetary *system*".[43] I think that he concedes to him too much. Copernicus only modified and refurbished an age-old *machina mundi*. The glory of having created a planetary system belongs to Keplar, with some "assistance" from Tycho Brahe, perhaps.

And yet in one context Copernicus does explain a certain movement in a plane as a resultant of two motions which he adds vectorially although it can be produced, in a related manner, by two motions which are linked hierarchically. The movement is the so-called "simple harmonic motion", $x = a \cos t$, $-\infty < t < \infty$, and its first known construction out of circular motion occurs in the work of the thirteenth-century astronomer Nasir ad-Din at-Tusi (1201–74), director of the observatory of Maragha. It was a hierarchical composition of two circular motions, which Dryer describes thus: "If there are two circles in one plane, one touching the other internally and of a diameter equal to *half* that of the other, and if the greater one rotates, and a point moves along the circumference of the smaller one in the opposite direction with *twice* the velocity and starting from the point of contact, then that point will move along a diameter of the greater circle."[44]

Copernicus, without naming at-Tusi, presents a very similar construction of the simple harmonic motion on a line, but his mechanism is better described as follows. Put the harmonic motion into the (x,y)-plane, thus

$$x = a\cos t \qquad -\infty < t < \infty,$$
$$y = 0, \tag{2}$$

and introduce the complex variable

$$z = x + iy.$$

Then

$$z = a\cos t = \frac{a}{2}e^{it} + \frac{a}{2}e^{-it},$$

which is the sum of the two vectors

$$v_1 = \left(\frac{a}{2}\cos t, \frac{a}{2}\sin t\right), \; v_2 = \left(\frac{a}{2}\cos t, -\frac{a}{2}\sin t\right),$$

and the resulting movement is as follows. There are two circles of *equal* radius, moving with the *same* velocity in opposite direction, and the second having its center fixed on the first. Then any fixed point of the second moves "harmonically" on a straight line.

As a famous cancelled passage in the manuscript of Copernicus shows, he knew even more. Consider the movement (2) as the degenerate case, $b = 0$, of the more general movement

$$x = a\cos t, \qquad y = b\sin t,$$

whose orbit is the ellipse

$$\frac{x^2}{a^2} + \frac{y^2}{b^2} = 1.$$

But now

$$z = x + iy = a\cos t + ib\sin t = \frac{a+b}{2}e^{it} + \frac{a-b}{2}e^{-it}$$

and the ellipse arises in the above construction if the two circles have different radii, everything else being the same.[45]

V. EPICYCLES AND TWENTIETH-CENTURY ANALYSIS

In the next chapter we will turn to Kepler who emancipated astronomy from all those component motions on cycles, epicycles and straight lines. But before doing so we will make some observations on these pre-Kepler motions in the light of mathematics of today. This is justified since the ancient and medieval astronomers well knew, in thought patterns of theirs, that these component motions were only "hypothetical"; the purpose of the "hypotheses" being to calculate the position of the planets, account for their motion, and exhibit their harmony,[46] and generally to "save (= explain, rationalize) the phenomena" in the operationally simplest manner. Chapter 2 of book 13 of the *Almagest* has particularly beautiful philosophemes about this. Our account will be mainly analytical, not historical; but the reader need not shy from drawing historical conclusion by himself. Much too little is known about mathematical thought processes to draw a firm line between anachronisms and adumbrations.

Take any periodic motion which we consider not only in three dimensional space but in an n-dimensional Euclidean space $E_n\{x^1,\ldots, x^n\}$ for any n. It is given by functions

$$x(t) = \{x^1(t), \ldots, x^n(t)\} \tag{1}$$

all periodic in t, with the same period, which we assume to be 2π. If the functions have, say, continuous first derivatives, then we have Fourier expansions

$$x^p(t) = \sum_0^\infty \left(a_k^p \cos kt + b_k^p \sin kt \right), \tag{2}$$

and our vector (1) can be written as a sum of vectors

$$x(t) = \sum_{k=0}^\infty \xi_k(t) + \sum_{k=1}^\infty \eta_k(t)$$

where, for fixed k,

$$\xi_k(t) = \{\xi_k^1(t), \ldots, \xi_k^n(t)\}, \tag{3}$$
$$\eta_k(t) = \{\eta_k^1(t), \ldots, \eta_k^n(t)\} \tag{4}$$

such that

$$\xi_k^p(t) = a_k^p \cos kt, \tag{5}$$
$$\eta_k^p(t) = b_k^p \sin kt. \tag{6}$$

The movement (5) is a simple harmonic motion on the line in E^n which extends from the point $(-a_1^p, -a_2^p, \ldots, -a_n^p)$ to the point $(a_1^p, \ldots, a_n^p)$; and the same holds for (6) because $\sin kt = \cos(kt - \pi/2)$. Thus we obtain the following conclusion.

THEOREM 1. *Any periodic motion* (1) *with period* 2π, *in* E^n, *for any* n, *with continuously differentiable components* $x^p(t)$, *say, can be represented, as a convergent, countable vectorial sum of simple harmonic motions, and uniquely so.*

Now, if for every one of the harmonic motions, we place the line of motion into *some* two-dimensional plane in E^n, and apply the construction of at-Tusi and Copernicus, we obtain the following result.

THEOREM 2. *The periodic motion* (1) *can also be represented as a countable vectorial sum of circular motions with constant angular velocities, each in a certain two-dimensional plane. The periods of the motions are integer multiples* $2\pi k$, *but no general statement can be made about the mutual position of the various two dimensional planes.*

Remark (important). The vectorial representation of Theorem 2 is in no-wise unique, and I do not know of any method by which to single out a preferred one among them. Thus the epicyclic arrangement does indeed *only* "save the phenomena", as the Ancients maintained, without endowing the participating cycles with a physical substance or meaning individually.

It would be hardly possible to impute to Copernicus a measure of adumbration of Theorem 2. He did not have enough of a background space in which to perform a vectorially superimposed addition of an unbounded number of various circular motions in various planes without "hooking" up the various component motions in a serial succession as in a *machina*. Newton had a superb insight into vectorial additional of physical data in space,[47] yet even he does not yet show a readiness— certainly not a spontaneous one— for the kind of "Fourier synthesis" that is embodied in Theorem 2.

But quite apart from this it is questionable whether our Theorem 2 is *astronomically* as meaningful as might appear at first glance. Nowadays, in the case of a planetary motion in E^2, the Fourier decomposition (2) is embodied in the pair of formulas

$$\cos u = -\frac{e}{2} + 2\sum_1^\infty \frac{1}{s} J_s'(se) \cos sl, \quad \sin u = \frac{2}{e}\sum_1^\infty \frac{1}{s} J_s(se) \sin sl, \tag{7}$$

in which u is the eccentric anomaly, e the eccentricity, $J_s(t)$ is the Bessel function, and l is the mean anomaly which is a normalization of the time variable to make the orbital period 2π.[48] Now, formulas (7), whatever their importance, are apparently not used for an epicyclic decomposition of the elliptical orbit as our Theorem 2 would suggest. It is different in the case of acoustics in which a suchlike interpretation of the mathematically self-same Fourier decomposition does occur. It states, for instance, that the "noise" produced by a vibrating string is a vectorial sum of infinitely many simple tones whose intensities converge to zero even though their frequencies tend to infinity.

In celestial mechanics the analog to this would be that the observed approximation to a periodic motion can be successively improved by successively adding (epi-) cyclic corrections in which the radii of the circles tend, on the whole, to 0, but the frequencies are ever higher, meaning that the periods on the added circles are ever shorter. In celestial mechanics, however, improved observations and theory suggest the addition not only of "nutations", with *short* periods which correspond to "overtones" in acoustics, but also of "hidden periodicities", like the Hipparchian precession, with *long* periods, even very long ones, to which there are no obvious basic analogs in acoustics.

Now, there is a class of functions, the so-called almost periodic functions of Harald Bohr,[49] each of which has, in complex form, a (generalized) Fourier expansion of the form

$$x(t) \sim \sum_{0}^{\infty} a_n e^{i\lambda_n t}, \tag{8}$$

in which the countably many exponents $\lambda_1, \lambda_2, \ldots$, can be any real numbers whatsoever, with no arithmetical restrictions at all. We must emphasize that these functions are a creation of the twentieth century, and adumbrations of them go back at the farthest to Lagrange and not farther back. Still, since we are on the subject, we will formulate a "true" mathematical theorem, which in a sense is a partial extension of Theorem 2 to objects of the type (8). But we ought to state, emphatically, that we are going to formulate hypotheses which will be so restrictive as to make any application or verification of the resulting theorem quite illusory.

Hypotheses. We consider an "idealized" planetary system, that is, in E^3, a system of n moving masspoints

$$X: x_p(t) = (x_p^1(t), x_p^2(t), x_p^3(t)), \qquad p = 1, \ldots, n, \tag{9}$$

moving according to Newton's law of motion, and subject to Newton's law of gravitation. The given solution (9) exists in the entire time-line $-\infty < t < \infty$, and all components $x_p^j(t)$ are bounded for all times and have continuous first and second derivatives. Any two of the mass points (9) never collide, and in fact, they stay at a "safe" distance away from each other, that is there exists a constant $\alpha > 0$, such that

$$|x_p(t) - x_q(t)| \geq \alpha \tag{10}$$

for all t, and all $p \neq q$. This is a basic condition of "regularity". But we need much more, namely a condition of "separation", which is most restrictive, and we formulate it as follows.

Let R be the closure of the range of the "regular" orbit (9) in E^{3n}, and consider all other "regular" solutions

$$Y: \{y_p(t)\}, \quad Z: \{z_p(t)\}, \quad \text{etc.}, \tag{11}$$

whose range is *contained* in R. It even suffices to consider only such solution, for which the lower bound α in (10) is the same as for (9). Now the condition of "separation" demands that there exists a $\delta > 0$ such that any two such solutions (11) which are not identical be spaced apart by δ, that is

$$|Y(t) - Z(t)| \equiv \sum_{p=1}^{n} |y_p(t) - z_p(t)| \geq \delta, \ -\infty < t < \infty.$$

THEOREM 3. *If a solution* (9) *of the n-body problem satisfies all these hypotheses, then its components* $x_p^j(t)$ *are Bohr almost periodic functions.*[50]

Now, it is a central point of the theory of almost periodic functions that each such function (8) can be approximated to, uniformly in all of $-\infty < t < \infty$, by *finite* exponential polynomials of the form (8), and thus we can conclude as follows.

THEOREM 4. *If Theorem 3 applies, the motion of any celestial unit occurring can be approximated to, uniformly in all of* $-\infty < t < \infty$, *by suitable finite sums of (epi-)cycles, with a constant angular velocity each, the angular velocities occurring in the same solution being totally unrestrained as to mutual size.*

Operationally and cognitively, the mathematical analysis involved in Theorems 3 and 4 was beyond the horizon of Copernicus, still beyond the reach of Kepler, and not yet familiar to Newton, although perhaps not beyond his ken. Nevertheless, Copernicus' reluctance to allow equants can be beautifully rationalized within the settings of these two theorems.

If, as we have done, we interpret the series (8) as a "vectorial" sum of uniform circular motions then the admission of equants would replace (8) by a sum

$$x(t) \sim \sum_{0}^{\infty} c_n e^{i\varphi_n(t)}, \tag{12}$$

in which $\varphi_n(t)$ is no longer a linear multiple $\lambda_n t$, but a more general function. But a "general" series (12) is radically different from a series (8) and radically more difficult to manage. In the case of a series (8) if the function $x(t)$ is presumed to be given, the coefficients a_n can be obtained by forming the limit

$$\lim_{T\to\infty} \frac{1}{2T}\int_{T}^{T} x(t)\,e^{-i\lambda t}\,dt.$$

It exists for every λ, has the value a_n for $\lambda = \lambda_n$, and the value 0 for other λ's. For general series (12) all such "simple" statements are lost, or become atrociously complicated.

VI. KEPLER'S LAWS. THE STORMY RISE OF ANALYSIS

Kepler may not have been much better equipped than Copernicus to anticipate an analytical formula like formula (8) of Section V, let alone to operate with it. But by some spontaneity of developments, Kepler somehow acquired, or generated within himself, the perception of the kind of mathematical background space that a proper physical interpretation of such a formula prerequires. This perception of a background space was part of a new kind of thought patterns which sprang up in Kepler's cognition. Within these new thought patterns Kepler was able to discover and rationalize—by means of the then nascent Mathematical Analysis, more-or-less—his very basic second law of planetary motion, and later on also his third law, $T^2 = cd^3$.

Kepler is also credited with highly innovatory achievements in mathematics proper, especially in his *Nova stereometria doliorum*, which starts from the technological problem of determining volumetrically the best proportions of a wine case. Also in his *Astronomia nova* he computes a formula amounting to

$$\int_0^t \sin t dt = 1 - \cos t.$$ [51]

But these achievements were not of the kind to be named after Kepler, let alone to immortalize him. Things like these "were in the air". If Kepler had not done them others would have undoubtedly done them within a short time. But his assertion that a planetary orbit lies in a plane and that the radius vector sweeps out equal areas in equal times is something that only a Kepler could have conceived. Nobody would dare to suggest that if Kepler had not conceived it, some near-contemporary of his would have "surely" done so instead. And his third law was, as an achievement, even more "individualistic" than that.

What keeps historians of mathematics from viewing and emphasizing the second law, say, as a creative achievement in mathematics proper—namely as an important step in the creation of Analysis—is the fact that the syllogistic reasoning of Kepler in his would-be "derivation" of this proposition is hopelessly wrong. It is very wrong indeed. No casuistic intellectual contortions by Kepler "apologists" will change this; and it is in fact an insult to Kepler to so "apologize" for him. It is almost pathetic to contemplate that such an eminent scholar as the Kepler editor Max Caspar, in his monumental translation of the *Astronomia nova*, declares that assertions of the faultiness of Kepler's reasoning are "misleading" (*irreführend*).[52] But they are not in the least "misleading", not to me at any rate. They have led me, gradually but forcibly, to the recognition that Kepler's mathematical mind was working on two levels, an old, traditional, Greek, level and a new, modern, analytical level, and that the *modus operandi* of the second level was very different from that of the first, in Kepler's time, and for a long time to come, for almost two centuries in fact. The conception and attitude of "mathematical rigor", that is, of syllogistic consistency, and deductive cogency was, in truth, alien and even repugnant to this nascent Analysis, in which everything was new, stormy, growing, bursting forth. Analysis was then in the making, genuinely so, and at a very early stage of development too, at which "anything goes", as long as the outcome is right; and in the case of Kepler it was uncannily right indeed.

In his rational–irrational way, Kepler may have known this himself, because when, on the height of his intellectual powers, he introduced his third law $T^2 = cd^3$, he did not even go through the pretence of offering a syllogistic demonstration for it.[53] In a way there had been nobody like Kepler since Pythagoras, with such a knack for arriving at momentous conclusions by false or fictitious syllogisms. And this is not surprising, since Analysis was as new with Kepler as Greek mathematics had been with Pythagoras.

It is noteworthy—and it corroborates our thesis of the two levels of mathematics in Kepler—that Kepler is substantially, if crudely, correct in his volumetric work; undoubtedly because this work, although by outcome it strongly pointed towards the incipient infinitesimal calculus, was somehow still done in the body, shape and even spirit of the classical Greek mathematics, which, after being in the doldrums for over a thousand years, was trying strenuously to make a comeback. This Greek mathematics did still play then a part, even a greater one in the seventeenth century than in the sixteenth century, however not in a comeback, but in a finale, even a grand finale. Some aspects of the finale were highly dramatic, others almost comical.

For instance, I find it comical to read in Descartes, in the introduction to his *La Géometrie* (1637)—which was the building material for the forthcoming coordinate geometry and as radically future-oriented as any creation of the seventeenth century—that all that he, Descartes, wants to achieve with this work is to solve or improve on the solution of certain problems with which the Late Hellenistic mathematician Pappus had been preoccupied. Now, Pappus was a very gifted mathematician of his time, and had he lived in an era of mathematical creativity rather than sterility, he might have become even a great mathematician too. Nevertheless, the difference in level between Pappus and Descartes was enormous, and yet it is somehow laughable to me for Descartes to imagine that all that motivated him deep inside him was the drive to produce some geometrical devices that could deal more efficiently with some problems that were the day-by-day preoccupation of Pappus.

In contrast to this, a highly dramatic aspect of this finale manifests itself in the structure and *mise-en-scène* of Newton's *Principia*. The work was future-oriented for the ages, and Newton undoubtedly knew this, but he nevertheless articulated it in the Archimedian idiom of an epoch long gone-by. He may have done so from his sheer mastery of the idiom in which he hardly had a peer. This apparently was the view of the great historian of science William Whewell who, unforgettably, put it thus.

> The ponderous instrument of synthesis [meaning Greek mathematics], so effective in his hands [that is Newton's hands], has never since been grasped by one who could use it for such purposes; and we gaze at it with admiring curiousity, as on some gigantic implement of war, which stands idle among the memorials of ancient days, and makes us wonder what manner of man he was who could wield as a weapon what we can hardly lift as a burden.[54]

But no such formal mastery can be attributed to Newton in the case of analysis. Far from it. He was already much better in handling analysis proper than had been Kepler, but he was still very far from perfect; as we are going to verify by two major occurrences in the *Principia*.

We have already stated above that Newton's very first concern in the *Principia* was to put Kepler's second law on as sure a syllogistic footing as he, Newton, could make it. Not only did Newton make this law the very first Proposition 1, Theorem 1, of book I of the *Principia*, but he also clarified its extent, by stating and establishing it for any kind of central force, and by enunciating very clearly that the orbit of any body moving under such a force lies in a (two-dimensional) plane, and that, in this plane, the radius vector sweeps out equal areas in equal times. The level of clarity in the enunciation in the *Principia* is distinctly higher than it is even in Kepler's *Epitome*, where it is already quite satisfactory indeed. And yet, it must be stated—although Newton commentators seem to be shy of stating it—that Newton's own proof of this Proposition, which of course must be done with Analysis, is far from perfect. For my part, at any rate, I am not embarrassed to admit that I have never been able to understand it in Newton's own reasoning, which is simply not intelligible to me, but could convince myself of its correctness only by taking recourse to coordinates and algebraic symbols, as textbooks on mechanics have been doing it since Lagrange, say.

Even less in accord with our present-day expectations of syllogistic style and mathematical rigor is, in the *Principia*, Newton's definition, or rather description of a derivative of a function. He termed it an "ultimate ratio" and phrased its description thus:

> For those ultimate ratios with which quantities vanish are not truly the ratios of ultimate quantities, but limits towards which the ratios of quantities, decreasing without a limit, do always converge: and to which they approach nearer than by any given difference, but never go beyond, nor in effect attain to, until the quantities have diminished *in infinitum*.[55]

It is, of course, always possible to "glamorize" such a verbalization. It is especially tempting to do so in the case of a Newton, and I myself have done so, eulogizing him thus: Newton's generation was literate enough to listen to a recondite mathematical definition, all in words, without symbols; and the Marquise du Châtelet, woman of the world, could undertake to translate Newton's formidable *Principia* from Latin into French, ultimate ratios and all.[56] But for all the "glamor" of Newton's definition of a derivative, a college freshman of today would not easily get a passing grade in his freshmen calculus, when offering this definition as his own.

Some historians of mathematics are, overtly or covertly, "scandalized" by the fact—or at least find it incomprehensible—that persons of the stature of Kepler or Newton, although unquestionable masters of Greek mathematics in all its aspects, had peculiar difficulties with certain "fundamentals" in Analysis. Even Gauss still had them; he would have been hardly able to "define" convergence of a sequence or of a series in any manner that would have been satisfactory to twentieth-century expectations.

For my part I have learned to accept and live with this situation and I am rationalizing it in the following way.

I am accepting it as a fact that the era roughly from 1500 to 1800 which saw the rise of (modern) science was just as much an era of the rise of modern mathematics, especially of analysis, and that this newly emerging mathematics was showing some of the same symptoms of juvenility and adolescence, and of "Storm and Stress" as did the other sciences, exact and also descriptive. It is true that the new mathematics was genetically inseparable from its Greek past, which had been and remained a foundational prerequisite and a paradigmatic guide-and-beacon for it; whereas in most other science the findings of the past were aphoristically charming rather than systematically basic. There was no primer like Euclid's *Elements* anywhere outside mathematics. And yet, even in mathematics, the era of 1500–1800 bore the same marks of turbulence of youth as in the case of the sciences. Only in the early nineteenth century did mathematics bethink itself of requirements of adulthood. Only then did it, as of a sudden, turn systematically and prolongedly to the conception and needs of "mathematical rigor", which in 1500–1800 it did not have in focus at all; although it was well aware of this "mathematical rigor" in Greek mathematics and of its unique intellectual quality in (Western) Thought.

Mathematics in the West has had a rather continuous and continuing development since Thales and Pythagoras, if not already since the times of Hammurabi in Mesopotamia and of the contemporary dynast in Egypt. It has been my contention, nevertheless, that the "modern" mathematics of a Kepler and Newton, as represented by the rise of Analysis, not only exceeded Greek mathematics by volume and direct operational methodology, but was, from the first, distinctly above it by a certain hierarchical scaling and stratification.[57]

I maintain that Greek mathematics, impressive as it was and is, was, on the whole, able to perform one-step abstractions only, that is *ideations* only, that is, only immediate abstractions from reality; whereas modern mathematics performs abstractions in escalation, that is abstractions, abstractions from abstractions, abstractions from abstractions from abstractions, etc. Even the very great Plato never performed abstractions beyond one-step ideation only.

This limitation of Greek mathematics also explains why it gives the first impression of being nothing but geometry, which it simply was not. The reason being that, in the layman's "intuition", geometric abstractions are nothing but ideations, and that, conversely, ideations are overwhelmingly geometrical abstractions.

This also helps to explain why Greek mathematics did not arrive at the conception, or even notion, of a space underlying geometry in mathematics, and thus of mathematical background space underlying the physical universe; whereas analysis began to evolve the concept of space and of background space from the first, however slow the first stages of the evolution may have been. The Greeks did abstract the concept of a "general" triangle from the totality of all "special" triangles occurring in "nature" or "imagination" or "thought". Similarly, they could form the concept of a "general" quadrangle, pentagon, etc. They even succeeded in forming the conception of a general polygon, whatever the number of its sides. But the concept of a "Euclidean" space underlying their geometry was no longer accessible to their concept-formation, and the conception of a two-dimensional space E^2 which underlies the geometry of the first ten books of Euclid's *Elements*—that is of a two-dimensional space which is an object in its own right, and not just a plane in three-dimensional space—was utterly beyond their horizon.

Even more astonishing at first, and just as "fate-sealing", was the Greek failure to conceptualize our present-day notion of a real number, probably because it would have required more than mere "ideation". In fact, it would have required a succession of abstractions, as evidenced by the late nineteenth-century work of Dedekind and Cantor. And the Greeks, not having created real numbers, did not venture into any area of science in which, as it later turned out, these numbers are basically needed.

Thus, Archimedes, in his famed theory of a lever, was unable to introduce the rotational momentum $P \times L$, because this would have required the formation of a product of two real numbers. There he could not express his law as an equality

$$P_1L_1 = P_2L_2,$$

but only as a proportion

$$P_1:P_2 = L_2:L_1,$$

and this retarded the rise of mechanics by almost 2000 years.[58] Newton, however, virtually starts out with the introduction of the (translational) momentum—or "quantity of motion" as he terms it—which he defines as the product of m and v; and mechanics could, and did begin in earnest.[59] Of course, Newton was totally unable to give a "definition" of a real number, and this was, in fact, achieved only 200 years later; but such obstacles of "mathematical rigor" did not detain Newton at all. In fact, nobody else in the seventeenth century and eighteenth century was so detained.

Although it will not play a manifest part in our context, I nevertheless want to state that in addition to "space" and "real number" there is at least one more object, the vicissitudes of which sharply illumine the difference in hierarchical level between Greek mathematics and modern mathematics. And this is the mathematical notion of a function, with all its overtones and undertones, and even sidetones, if there are such. Greek mathematics never had functions in its thinking, in a "true" sense, that is, but—at first stated in Oswald Spengler's *Decline of the West*[60]—"modern" mathematics had them from the first, from its first stirring in the Middle Ages, however gropingly and inarticulately for centuries. I have stated several times since 1963[61] that the Greeks did not have functions in their thinking, and in a recent context I have put it thus.

> The Greeks did, of course, have some familiarity with categories of cognition such as "correspondence", "mapping", even "binary relation", which enter into our present-day notion of a function. But the mere occurrence of such categories does not yet make for the presence of functions. Even the occurrence of assertions or conclusions which can be readily translated or transliterated into

functional relations is not yet enough. It is still necessary, and this is decisive, that something also be "done" with those functions, or with the trains of reasoning corresponding to them, that is, that some kind of "mathematical" operations be performed with, or on, those functions, or with the trains of reasoning that "stand in" for functions. Now, it is this kind of "operational" activity, or only attitude, which it is difficult to discern in the realm of Greek thought, mathematical or other....

Furthermore, historians of logic have been recently asserting, even heatedly, that the Greeks remained creative in the field of logic even after Aristotle, and that, specifically, some principles of our "modern" propositional modes of implication were already discovered by some of the Stoics. But, here again, of propositional *functions*, that is of propositional schemes that involve "all" or "any", there is no trace.[62]

On the other hand, the effective involvement of analysis with functions is by no means predicated on the manifest textual occurrence of functions, let alone on their identification with symbols of the type $y = f(x)$. Thus, Newton's *Principia* is teeming with functions in its innermost design. It is built on them from first to last, even if the *Principia* is outwardly cast in the traditional mold of Greek non-function mathematics, and was composed about a century before mathematicians began to wonder *aloud* what functions actually are. In fact, the quantity of motion mv involves not only the multiplication of the two real numbers m,v, but, ordinarily, also the formation of the *function*

$$v(x) = \frac{dx}{dt}$$

which was already well beyond the reach of Archimedes. With some effort Archimedes might have been able to form this derivative at any one point t_0, as a slope of the tangent to a graph in the (x,t) plane. But forming this derivative at all points t "simultaneously" and envisaging the outcome as a function in t would have been a formidable task to him.[63] But even if he could manage this, the subsequent formation of the new second derivative, which occurs in Newton's law of motion,

$$\frac{d}{dt}(mv) = F$$

would have been an escalation of abstraction that would have been entirely inaccessible to his thought patterns.

Also, a close examination of the entire argument will show that these operations which Newton did perform, but the Greeks had not performed, did constantly presuppose the presence of a background space in Newton's thinking, and that on the other hand the Greek incapacity to perform such operations and the absence of a background space from their thinking are closely linked.

VII. SPACE IN NEWTON

An awareness of the need of a mathematical background space in Science manifests itself in Galileo's much quoted pronouncement that the book of nature "is written in mathematical language, and the characters are triangles, circles, and other geometric figures";[64] except that, a Galileo of today would have to say that the characters are real numbers, functions, derivatives, integrals, and other concepts of analysis. It is of no "metaphysical" consequence that Galileo still speaks instead of "triangles, circles, and other geometrical figures", just as it is of no metaphysical consequence that Newton still composed his *Principia* in an Archimedian style, ostensibly so. The architects of science in the seventeenth century, when introducing and erecting mathematical background space as the mathematical scaffolding for the new structures, were still "professing" to speak the language of Geometry while actually developing the idiom of Analysis; and some of our historians of science are taking such "professions" much too literally, *au pied de la lettre* as it were. Thus, Koyré makes a big point of stating that Galileo performed "the geometrization of space of which he was one of the foremost promoters";[65] and Koyré apparently also takes it for granted that this geometrization of space was completed by Newton and that this completion was a metaphysical high point in Newton's total achievement. To which I can only say, for my part, that Kepler, Descartes, Newton, and others, performed not a geometrization but a mathematization of the Space of Science and Astronomy, *whatever the appearances*. They did not make space "geometrical" in any sense of the past Geometry of the

Greeks, but they made it "analytical" in the sense of the Analysis in the making. That is, *whatever the appearances*, they made it "geometrical" in the sense of the coordinate geometry, that is of analytical geometry, then breaking out for just such purposes; even if the *manifest* use of coordinates in mechanics came about in the eighteenth century only.[66]

Similarly, I have previously[67] expressed my reservations against Koyré's statement that "Newton's physics, or it would be better to say, Newton's natural philosophy stands or falls with concepts of absolute time and absolute space"[68]. I strongly objected to the implication in Koyré that Newton gave a primacy to the question of the nature of space and time in a metaphysical rather than physical sense, and that, in a logical if not temporal succession, only after having developed such notions did he turn to the main task of creating a theory of motion under gravitation that would unify the three planetary laws of Kepler's and Galileo's parabolic trajectory. This, I think, is totally at variance with how a scientist's mind works, and Newton's mind must have worked. In my view, *whatever the appearances*, during Newton's creative period he was first and foremost the scientist, however much he may have been also thinking and talking of metaphysics and theology. His overwhelming urge was to develop a mathematical theory of motion under gravitation that would unify terrestial and celestial mechanics. And as a part of this undertaking and only as a part of it did he develop, mathematically, notions of space and time that would fall in with the syllogistic needs of such a theory. This was his overriding and all-deciding motivation, however much the question of the metaphysical–theological nature of space and time may have been also agitating and burrowing within his intellectuality. It misses the mark of Newton's true intellectual drive and motivation to pronounce, as done by Koyré, that Newton's physics stands *or falls* with his concepts of absolute time and absolute space; as if to allow, that if Newton's metaphysical–theological meditations about space and time would have led him to different notions with regards to these, his physical investigations might have led him to a different theory of the planetary system accordingly. This is incongruous. Newton's physics could not possibly *fall* with his conceptions of space and time, it can only stand on them. And it does stand on them, because Newton, in order to organize his theory, had arranged for his space and time to be such that his theory of gravitation, as he envisaged it in his mathematical mind, shall be able to stand on them, firmly, securely, indefinitely.

It is true that in the *Principia* Newton adduces extrasyllogistic rationalizations and metaphysical expostulations on the nature of absolute space, time, and gravitation, which he stubbornly and tenaciously holds up in various confrontations. But it is also true that these extrasyllogistic rationalizations and philosophemes are so placed in the text of the *Principia* that they could be "safely" excised from the text without causing damage to the mathematicophysical exposition of all of Newton's accomplishments in mechanics and astronomy in their consistency and syllogistic completeness. And it makes no difference to our interpretation of Newton's intellectual priorities that Newton himself would have probably heatedly denied the validity of our "psychoanalysis" of him. Einstein himself once made the much-quoted pronouncement, perhaps to a somewhat different purpose, that one should not heed what a scientist "thinks" he is doing, but, instead, appraise what it is that he is actually doing.

Koyré and others have also propounded the thesis, with almost dogmatic fervor, that Newton has performed, or at least completed, an "infinitization of space", so-called. We will heavily qualify and even contest most aspects of this thesis, excepting, of course, the incontestable fact that Euclidean space, which is the mathematical skeleton of Newton's absolute space, is infinite; which indeed it is. Otherwise, I maintain, Newton's acceptance of infinitude was guarded, reserved, reluctant.

However, Euclidean space itself was something that Newton simply had to accept, infinitude and all. The compulsion was again purely scientific–pragmatic, and not at all philosophical–theological, say. On the one hand, the Newton of the *Principia* was, in depth, far advanced in Analysis, and he very definitely had to have a mathematical background space in which to operate. But on the other hand, his Analysis was still very much rooted in the operational thought patterns of Greek mathematics, so that his background space had to be Euclidean, per force. It would have been out of operational reach for Newton to try to develop his gravitational theory for any curved space, even for a space of constant curvature, although for such a space, the three Kepler laws remain virtually the same.[69] Euclidean mathematical background space was all that Newton could handle, and this is a fact of scientific evolution that must be accepted pragmatically. In fact, it took another 200 years before mathematics advanced, by evolution, to a stage at which more intricate space forms could be handled. It would be totally inappropriate to say that Newton "wanted" to operate in an (infinite) Euclidean space, because, from some presumed philosophical or even theological conviction of his it was the best or most suitable of all possible spaces. The plain fact is that Newton was "compelled" to operate in such a space because this was the space in which he could operate within his technical mastery.

This limitation put Newton into a kind of intellectual quandary, whether he realized it or not. On the one hand, Euclidean space, by its topological nature, has a natural extent of completeness, in which it is infinite in all directions, so that the introduction of the entire infinite space into physical science seemed indicated. Yet on the other hand, in mathematico-physical expostulations Newton never needed the Euclidean space in its entire extent, but, on the contrary, in each context, only some finite part of it was required, so that the explicit introduction of the entire infinite space to begin with seems to be somehow physically "wasteful" and "inappropriate".

This quandary was not a "modern" phenomenon. The Greeks already knew about it—in their own thought patterns, that is—and it caused them many problems, and even frustrations. It may have kept Greek mathematics from ever introducing infinitude explicitly into their geometry—a fact which, as it turned out, was deathly fatal to it. Aristotle knew about it. He was very hesitant to allow actual infinitude for his *mégethos* (=magnitude). But he knew that this hesitancy has to be justified, and there is a magnificent passage in which he tries to do this [*Physica*, book 3, chapter 7, 207b 27–34]:

> This account does not, by denying an actual inexhaustable infinite of prolonged dimensionality, take away from the mathematicians their theoretical pursuits in which they neither need nor use such an infinite. Mathematical requirements are met by a figure which may be of any size without ceasing to be determinate. Any line, however short, may be divided on the same principle [*logos*] as any other however large. For the purpose of their demonstrations, therefore, it will make no difference to the mathematicians that there is no actually existing infinite magnitude.[70]

In the case of Newton we will argue in Section IX that, in his vision, the physical universe itself was not "infinite" but only "immense"; and, anticipating this, we want to add now that Newton stuck to this mere "immensity" of the Universe even in the face of a certain possibility which he dreaded might occur. According to Newton's own theory any conic section, even a hyperbolic one, was mathematically eligible to be the path of a comet, and a hyperbola in its entirety extends, of course, to infinity. Now, in a rather involved passage Newton both denies this possibility, and yet provides for it by suggesting that if such a path should occur then it might be a (finite) arc of a hyperbola which begins in the region of the stars (and presumably ends in it).[71]

I think that, from our retrospect, the dissonance between mere "immensity" of the Universe and full "infinity" of the underlying Euclidean background space created a certain tension in Newton's thought which he tried to overcome. I am making bold to speculate that Newton felt that he somehow had to draw the entire infinite Euclidean space into the cosmological setting, but in a manner that would be operationally harmless even if it would sound philosophically portentous; and it was this which contributed to his making the entire Euclidean space "absolute", that is physically and metaphysically meaningful. And immediately following the (would-be) definition of absolute space (and time) Newton makes most intriguing and arresting statements about it. But he also makes these statements not in the body of the *Principia*, but only in a "detachable" scholium, which is as beautiful metaphysically as it is inoffensive operationally. As already stated in Section II, for over two centuries scientists and philosophers were trying to find out what "absoluteness" really means. Towards the end of the nineteenth century these endeavors gradually petered out, among scientists at any rate; although some philosophers of science are still pursuing the search, even in the second half of the twentieth century.

Also, as already stated in Section II, Newton's "immensity" goes back to Copernicus, fully so. But in Copernicus there is almost no background space as yet. Still, a whiff of absolute space, but only a whiff, can be detected in Copernicus' insistence on immobilizing the sphere of the stars. If we could assume that Copernicus' "immense" Universe was somehow immersed in an infinite space that surrounded it, then the eighths sphere of the Universe would somehow already belong to the surrounding space, and making this sphere immobile would somehow be in accord with making it "absolute", in a sense. And it is not at all far-fetched to impute to Copernicus the assumption of the presence of such a space, because some ancient Pythagoreans had already something like this, according to the following famous report of Aristotle in *Physica*, 213b 22–24: "The Pythagoreans, too, held that the void exists and that breath and void enter from the infinite into the Heaven itself, which, as it were, inhales."[72]

VIII. INFINITY IN COSMOLOGY

The involvement of cosmology with infinity occurs mainly in the problem of the size of the Universe, that is, in the question whether the Universe is finite or infinite. We have already made some statements about the problem in Section III above, and they imply that the problem is one of great complexity and even intricacy. It is not even a clear-cut problem to which there "ought" to be a clear-cut answer, far from it. A mere assessment of the vicissitudes of the problem in antiquity would amply show this.[73] As J. D. North put it so aptly: "It is easy to speak of the infinite, as every theologian knows, but it is difficult to speak of it meaningfully."[74] And, in this sentence, "meaningfully" is to be interpreted as "scientifically" meaningfully. Also, North's injection of "theologian" is particularly apt, because through the length of the Middle Ages and even after, infinitude was indeed a theological concept in the main, being an attribute of the Godhead, first of all. The Old Testament, for instance, does not have a *primary* term for infinity, but it does have one for "eternity" (*'olam*), that is, infinity in time. The latter term has in the Old Testament a theological connotation, even if post-Biblically it became more secularized and acquired traits of spatiality too.[75]

There is a widespread and widely accepted thesis abroad that the Era of the Renaissance and of the Scientific Revolution, after a hard-fought prolonged struggle, brought about a transition from a "closed world" to an "infinite Universe", the first representing backwards-directed regressive Aristotelianism and the second forwards-directed progressive Modernism.[76] This is a very dubious thesis which, at least, has to be heavily qualified. The facts already adduced in Section III above show that the Eras of the Renaissance and of the Scientific Revolution have not brought about at all a consensus about the size of the Universe, and that the problem is as open today as it has ever been.

Still, the thesis of a transition from a closed world to an infinite Universe is, of course, not entirely wrong; such insinuating sweeping assertions seldom are. In fact, on the face of it, the assertion is so appealing that Oswald Spengler made it into one of his main theses in his *Decline of the West*.[77] And yet, as I see it, and as I have already stated before,[78] working scientists—in contrast to philosophers and philosophizing scientists—have always been wary, some more and some less, of declaring themselves for infinitude of space, from Thales to the present. And Aristotle's own hesitancy can be viewed as only one instance among many others, in antiquity and after. The previously quoted sentence from J. D. North gives a reason for this hesitancy, and we are going to give quotations from two other twentieth-century authors that will express the reason even more sharply.

J. L. Synge in a 1964-book on General Relativity, after having introduced the Minkowski Space–Time Universe, which is as infinite as Euclidean space, introduces a less infinite de Sitter model of a universe and he "justifies" this with the following words: "Nevertheless the de Sitter Universe is interesting in itself. It opens up new vistas, introducing us to the idea that space (a slice of space time) may be infinite, and this seems to satisfy some mental need in us, for infinity is one of those things we find difficulty in comprehending."[79]

Even more telling, and of specific consequence to us is a statement from a 1932-book by de Sitter himself, and it runs as follows.

> How the $g_{\alpha\beta}$ outside our neighborhood are, we do not know, and how they are at infinity, of either space and time, we shall never know, otherwise it would not be infinity. That is what Archimedes meant when he said that the universe could not be infinite. The universe that we know cannot be infinite, because we ourselves are finite. Infinity is not a physical, but a mathematical concept, introduced to make our equations more symmetrical and elegant.[80]

What makes this statement important to us is the magnificent phrase

"*Infinity is not a physical, but a mathematical concept.*"

We are adopting it as a *key phrase* of ours but without also adopting de Sitter's own rationale for it. This key phrase has a considerable affinity to our previous statement, made in Section II, that "in physics and cosmology a mathematical background space must be initially given under all and any circumstances", and we are merging the key phrase and this statement into the following assertion.

Key assertion. Infinity is not only a scientific conception but it occurs in other areas of human awareness too. However, wherever and whenever *infinity* does occur in science, it *is not a physical but a mathematical conception.* Any statement in science that something is infinite refers to something that cannot be described other than mathematically. In particular, in strict fact, the problem of the size of cosmological space, or the universe, or the world, etc., can properly be raised for the underlying mathematical space only. And for physical space it can be raised only to the extent to which a mathematical background space is a part of it.

Remark. This very last sentence just stated is necessitated by the crucial fact that it may and does occur that a mathematical space which is a background space of a physical system is by its own natural extent larger than the part of it needed for the mathematical scaffolding of the system which it serves.

Following a Kantian tradition most philosophers in the nineteenth century took it for granted that Space is infinite, and many scientists followed suit, or at least, they thought that they ought to follow suit. And yet, I venture to state that, in the early decades of the twentieth century, what made Einstein's General Relativity so thrilling to many was the air of ease and naturalness with which it allowed, however speculatively, for the possibility of a universe that is finite by way of being closed with respect to its spatial dimensions, that is, by being, spatially, a compact Hausdorff manifold, in strict topological parlance. And, contrary to occasional statements, these speculative innovations did not "overthrow" Euclidean–Newtonian space or anything else, but enriched the outlooks by adding different possibilities for certain suitable contexts. Nobody ceased to use the traditional Euclidean–Newtonian space in the vast areas of theoretical and practical mechanics and engineering, or in "introductory" textbooks on science on various levels of instruction. Nobody "foreswore" ever to use a Euclidean Newtonian E^3 henceforth, or did indeed cease to use it.

On the other hand, certain developments away from Euclidean space in science had already been in the making several decades before Einstein, namely since the publication in 1868 of B. Riemann's memoir on "Riemannian" manifolds; and even astronomers like Simon Newcomb and Karl Schwarzschild participated in these *avant garde* speculations.[81] Thus Newcomb, already in 1877, in an article on elliptic space, terminates with the following paragraph, which shows his awareness that what will eventually ensue from his speculations may be as innovatory as anything in man's past experience:

> It may be also remarked that there is nothing within our experience which will justify a denial of the possibility that the space in which we find ourselves may be curved in the manner here supposed. It might be claimed that the distance of the farthest visible star is but a small fraction of the greatest distance D, but nothing more. The subjective impossibility of conceiving of the relation of the most distant points in such a space does not render its existence incredible. In fact our difficulty is not unlike that which must have been felt by the first man to whom the idea of the sphericity of the earth was suggested in conceiving how, by travelling in a constant direction, he could return to the point from which he started without, during his journey, finding any sensible change in the direction of gravity.[82]

The twentieth century has been very rich in cosmological speculations, and of particular interest to our context is the fact that some of these speculations are valid for Euclidean and nonEuclidean background alike. Thus, on the one hand, it was argued by E. A. Milne and W. H. McCrea, among others, that, in a suitable interpretation, Newtonian cosmology was equivalent to relativistic cosmology.[83] On the other hand, H. Bondi in an analysis of the problem of the Olbers paradox, whose original setting is infinite Euclidean space, observed that the paradox also follows if space is "closing up on itself".[84] In the original setting of de Cheséaux in 1743, and in the later setting of Olbers in 1823, the problem is the question why at night, when only our sun is turned away from us, but all the innumerable stars in infinite space shine upon us as by day, the sky is so very much darker than by day. Now, according to Bondi's remark, if space were closed and finite, with only a finite number of stars present, the paradox would be the same, "for then we would be receiving the light from the previous periods of the universe, rather as great circles on a sphere can be continued arbitrarily far, although the area of the sphere is limited."

Thus even an extensive knowledge of phenomena in the Universe need not lead to a decision whether its background space is finite or infinite, Euclidean or other.

IX. "IMMENSITY" OF THE UNIVERSE IN NEWTON AND COPERNICUS

As already stated in Section II above, Copernicus avows in book I, chapter 6 of *De Revolutionibus* that the Universe is "immense", without any further explanation of what this is meant to be; and this is, of course, a kind of refusal, an intentional one, to commit himself as to whether the physical Universe is finite or infinite. It is tempting to take this feature in *De Revolutionibus* lightly, or even adjudge it negatively. For my part, I judge it positively, and take it to be the mark of a great scientist; because, as I have stated before, and am stating again, great scientists never "rushed" into the assertion that the Universe is infinite, as general philosophers sometimes did.

For instance, when sifting the evidence from Antiquity about the views of the Atomists from Democritus to Lucretius, there can be no doubt that the philosopher-poet Lucretius opted for infinitude of the Universe,[85] virtually in the vein of many philosophers in the eighteenth and nineteenth centuries, say. But, as I read the evidence about Democritus, who was a physicist, and a very great one too,[86] it seems to me that he is leaning towards mere "immensity", as reflected, for instance, in his view that the Milky Way "is composed of very small, tightly packed stars which seem to us joined together because of the distance of the heavens from the earth, just as if many fine grains of salt had been poured out in one place".[87] It must be admitted though, that all suchlike evidence regarding Democritus is highly uncertain as to its fine points.

But one is, of course, on much safer grounds when turning to scientists after Copernicus, and it is most illuminating to assess the views of Newton, above all. It may and must be stated, even most emphatically so, that throughout the length of Newton's *Principia*, and in all three editions of it, Newton never, but never, states that the Universe, *qua* physical Universe is infinite. On the contrary, in the main contexts in which he touches upon the problem of the size of the Universe, namely in the General Scholium to book III of the *Principia*,[88] and in section [57] of *The System of the World*,[89] he only uses the word "immense" and nothing "stronger". And this seems to be the same "immense" as occurring in Copernicus' *De Revolutionibus*, book I, chapter 6, as if to signal, that *with regard to this problem*, nothing had been changed between Copernicus and Newton. When not speaking of the Universe *qua* Universe, but of Space and Time *qua* Absolute Space and Absolute Time, the *Principia* does have once, and only just once, the word *infinity*, namely in the expression *from infinity to infinity*;[90] and Koyré comments to this as follows:

> "*From infinity to infinity* retain the same position. . . ." What does *infinity* mean in this place? Obviously not only the spatial, but also the temporal: absolute places retain *from eternity to eternity* their positions in the absolute, that is, *infinite* and *eternal* space, and it is in respect to this space that the motion of a body is defined as being absolute.[91]

But this *infinity* has nothing to do with a cosmological universe. It is a philosophico–theological term and notion that had had currency throughout all the centuries since Anaximander and Xenophanes, say. The scholasticism of the High Middle Ages was most intimately familiar with it, however much its "World" may have been "bounded" in the sense of Aristotle and Ptolemy. It is more or less the same infinity as in Lucretius' argument of the "flying dart", or, rather, "winged javelin",[92] and Lucretius may have had it from a much earlier Pythagorean vision of an "infinite" void surrounding the Heavens. It is this infinity more-or-less that Copernicus had in mind when in book I, chapter 8 he suggests that it ought to be left to philosophers of nature to quarrel over whether the World (*mundus*) is finite or infinite; whereas in book I, chapter 6, when the size of the cosmological Universe (*caelum*) is at issue Copernicus declares that it is "immense" and leaves it at that.[93]

In Aristotle a similar distinction can be made. In the *Physica* he has a "systematic" essay "On the infinite", as an object of general Cognition as it were; and there for some categories of infinitude he only denies their presence in actuality but not in potentiality.[94] But in his cosmological work "On the Heavens", when the size of the Universe is at issue Aristotle does make the statement, emphatically so, that the world is finite. He has a welter of arguments for this statement, which are very hard to follow from our retrospect. But he seems to have it in his thinking that if the world were infinite then (a) its total mass would be infinite, and, alternately, (b) its total rotational momentum in its diurnal motion would be infinite, and both alternatives seem to be preposterous to him.[95] This kind of argument from physics has been keeping generations of "modern" physicists from indulging in the presumption that the Universe is infinite, and has even led an Arthur Eddington, at the height of his career, to assert that the Universe consists of a specific finite number of elementary particles, as we have already stated in Section III above. Between Copernicus and Newton, scientists like Kepler, Galileo and Descartes, each and all of them, also balked at allowing the cosmological Universe to be infinite, outright infinite at any rate. Koyré presents their various reasonings for a non-infinitude of the Universe in a masterful and luminous fashion, almost lovingly so. But Koyré himself is a "believer" in infinitude, and so, in the summing up of the doctrine of each of the scientists, he somehow manages, in the end, to ascribe the scientist's reasoned opposition to infinitude, however well reasoned it be, to a residue of Aristotelism which he somehow was unable to overcome. On the other hand, Koyré glories in Giordano Bruno's *infinite* Universe, however convincingly Kepler argued that it was an astrophysical impossibility,[96] and however much Koyré himself, in one place, is constrained to admit that "Giordano Bruno, I regret to say, is not a very good philosopher".[97]

Returning to Newton, one might reply to our statements above that while he does not consider the infinitude of the cosmic Universe anywhere in the *Principia*, he does so nevertheless in a famous letter to Robert Bentley, one letter of a total of four.[98] However, as pointed out before,[99] while Newton does make in this letter some kind of admission that the total mass of the Universe (or perhaps the total number of separate planetary systems, it is hard to separate these two alternatives in the arguments) may be infinite, he nevertheless did not incorporate this possibility into the *Principia*. The letter was written between the publication of the first and second editions of the *Principia*, but the second edition showed no kind of after-effect of this letter at all. Furthermore, in Newton's own intent, the main thrust of his argument in the letter may have been to point out, for the first time, the phenomenon of gravitational clustering which may occur if large gravitational masses can move in large spaces,[100] and which was elaborated in the twentieth century by James Jeans and others.[101] And it is remarkable that there is indeed an allusion to gravitational clustering in the two passages mentioned above in which the "immensity" of the universe—and only its immensity—is spoken of.

Between Copernicus and Newton, that is, mainly in Galileo and Kepler, the cosmological distinction between the Universe proper and an incipient "absolute" space is not very manifest, and for a good reason. The reason being that Galileo and Kepler, between them, have virtually introduced, however primitively, the telescopic "Depth of the Universe", and thus "The Universe of Telescopic Depth". From this approach, the Universe is, pragmatically, what can be "seen", visually or otherwise, through the strongest available "telescope" of whatever kind. There is then less emphasis on an "absolute" space that might extend beyond what can be "seen" at any one time, and the "size" of the Universe becomes "indefinite".

An entirely different "indefiniteness" of the size of the Universe was claimed by Descartes for the cosmological universe of his interpretation, and his outlook has altogether a very striking feature. In his *La Géometrie* (1637), Descartes laid the foundation for the present-day structure of space as such, and the outcome is as future-oriented as anything that the seventeenth century has created. But in Descartes' physics and cosmology, spatiality is represented by, and embodied in something which he calls *extension* or *étendue*, and which hearkens back to something which the later Middle Ages called *extensio*. It is this *extension* which harbored the famed Cartesian vortices. Descartes insisted that its size is not infinite but only indefinite, and he stoutly held his ground in this matter against an ugly pressure by the Platonist Henry More that he change the verdict.[102]

But this physical *extension* of Descartes had absolutely nothing to do with the "espace à trois dimensions" of *La Géometrie*, and in no other philosophical system has the separation of Universe and mathematical space been so pronounced as in the thinking of Descartes.

And yet, as we have tried to argue before, "indefiniteness" as a characteristic of the size of the Universe may be quite ancient indeed, even early Presocratic, and already present in the thinking of Anaximander and Xenophanes more-or-less. In fact, the Greek word *ápeiron* for our *infinite*, which, since Aristotle, say, does indeed designate "infinite" in our sense, may have originally connoted "indefinite", rather than "infinite", or, at any rate, both the one and the other.[103] And such an "indefinite", whether in Descartes, or in antiquity, is a close relative to the "immense" in Copernicus and Newton, in a sense. A change-over of the meaning of ápeiron from "indefinite" (or "immense") to outright "infinite" may have taken place in Greek philosophy not only in "monistic" theories like those of Anaximander and Xenophanes, but also in "pluralistic" theories, especially in the atomistic theories of Leucippus and Democritus, Epicurus, and others. Late Greek commentators seem to take it for granted that the atomists admitted not only infinitely many atoms, but also infinitely many divers shapes of them; but, in Democritus himself, these numbers may have originally been "immense" or "indefinite" rather than outright infinite.[104]

In sum, we would like to repeat that, unlike some philosophers, scientists have been always very wary of declaring the size of the universe infinite, just as much after the Middle Ages as before; but, that in modern times, that is after the Middle Ages, a new aspect of the problem has begun to reveal itself which consisted in the separation of mathematical background spaces from Universe proper; and, that both Cusanus and Copernicus were the first to adumbrate this new concept, Cusanus more from the mathematical side, and Copernicus more from the cosmological side. It is this that separates modern science from ancient science, and not the question whether the Universe is finite or not.

APPENDIX

Maximal Universes[105]

The purpose of this appendix is to adduce a somewhat technical corroboration of our thesis that a distinction must be made between physical cosmological space and mathematical background space, and that the latter is indispensable. The cosmological space may be constructible as a cosmological feature of the Universe. The background space, however, must be posited expressly and beforehand.

We will obtain this corroboration by considering the problem whether and to what extent the knowledge of a part of a universe and of the physical events in it allows the construction or reconstruction of all of the universe with the physical events in it. We will set up a model of a universe with *seemingly* only one kind of space in it, and we will *seemingly* be able to deal with our problem nonetheless. But a closer look at the would-be reasoning will bare an oversight, and the "natural" way of overcoming the oversight will point to the introduction of an *a priori* background space after all.

We start out with an n-dimensional *connected* Hausdorff manifold S, of a *fixed* dimension n, which is meant to represent a cosmological space. Thus, ordinarily, $n = 4$, but there will be no need to specialize the value of n in our argument; and we will call S a *base space*. The "physical events" shall be represented by a family of functions on S, which we will designate by $\{f^\alpha\}$, or more explicitly by $\{f^\alpha(P)\}$, where P is a point of S, and the index α is a purely enumerative symbol which traverses an arbitrary but *fixed* index set A. Also, f^α need not at all be a function in an "ordinary" sense, that is a scaler object of the simplest kind. Rather the totality of "functions" $\{f^\alpha\}$ shall be components of any kind of objects of vectorial, tensorial, or suchlike description, and their values may very much depend on the local coordinate system given on S. All that we require is, first of all, that if another (local) coordinate system is chosen, then the values of the components $\{f^\alpha\}$ in the new coordinate system shall be obtainable from those in the first coordinate system by certain fixed rules, which, to begin with, need only satisfy the following "consistency law" and nothing else. If we transform the $\{f^\alpha\}$ from a first coordinate system into a second, and from the second into a third, then the result shall be the same as in a direct transition from the first to the third.

Now, the system of data

$$\sum = \{S; f^\alpha\} \tag{1}$$

is what we call a *universe*. Holding n and A fast, we want to introduce alternate universes, and be able to say when one universe is larger than another, and do so in such a way that no inconsistencies in the comparisons shall arise. It turns out that a safe way for being able to do this is to assume that all data are (real-)analytic and that an enlargement of one universe into another shall amount to an analytic continuation of one universe into another.

Thus, we first of all assume that the base space S of any of our universes is an analytic manifold and that the data $\{f^\alpha\}$ are such that in every analytical coordinate system, every component is analytic. Furthermore, as a condition on the nature of our data $\{f^\alpha\}$ we also expressly assume that if two sets of data on S coincide in a neighborhood of S then they are identically equal throughout S.

Furthermore, whenever we consider a homeomorphism—that is, a one-to-one mapping—of one base space into all or part of another, we expressly assume that both the mapping φ and its inverse φ^{-1} are analytic. Now, the analyticity of the data is a crucial prerequisite for the following lemma, which in its turn is a crucial prerequisite for the following lemma, which in its turn is a crucial prerequisite for the comparison of universes which we will describe immediately following it.

LEMMA. *Given three spaces S, S_1, S_2, if φ_1 is a homeomorphism of S into a part S'_1 of S_1, and φ_2 of S into a part S'_2 of S_2, and if the homeomorphism*

$$\varphi_2 \circ \varphi_1^{-1} \tag{2}$$

of S'_1 into S'_2 can be extended to an analytic mapping of all of S_1 into a part of S_2 which is not all of S_2; then the inverse homeomorphism from S'_2 to S'_1 cannot be continued to a mapping of S_2 into all of S_1.

We now hold the system (1) fast, and we consider a "variable" system

$$\sum_1 = \{S_1; f_1^\alpha\} \tag{3}$$

of the following kind. There exists a homeomorphism φ_1 of S into a *part* S'_1 of S_1, such that, on S'_1,

$$f_1^\alpha(\phi_1 P) = f^\alpha(P), \qquad P \in S, \qquad \alpha \in A. \tag{4}$$

We are adding the mapping function φ_1 to the system (2), that is we introduce the ensemble

$$\left(\sum_1 \varphi_1\right) = \{S_1; f_1^\alpha; \varphi_1\}, \tag{5}$$

and we call the ensemble (5) and (analytic) *continuation* of the system or ensemble (1).

Given a second continuation

$$\left(\sum_2 \varphi_2\right) = \{S_2; f_2^\alpha; \varphi_2\}, \tag{6}$$

of (1), we put

$$\left(\sum_1, \varphi_1\right) < \left(\sum_2, \varphi_2\right) \tag{7}$$

if the homeomorphism (2) can be continued to a homeomorphism from S_1 to a proper part of S_2.

Now, due to our analyticity assumptions stated above, relation (7) gives rise to a partial ordering of the family of all continuations of (1). Furthermore, any linearly ordered subfamily

$$\left(\sum_\lambda, \varphi_\lambda\right) = \{S_\lambda; f^\alpha_\lambda; \phi_\lambda\} \tag{8}$$

has a least upper bound. In fact, for $\lambda < \mu$, $\varphi^0_\mu \varphi^{-1}_\lambda$ is a mapping from S_λ to S_μ and this gives rise to a projective limit S_∞ which is the base in a continuation

$$\{S_\infty; f^\alpha_\infty; \varphi_\infty\} \tag{9}$$

which is the least upper bound of (6).

By a familiar reasoning,[106] we obtain the following conclusion.

THEOREM 1. *Any system* (1) *has at least one continuation which is maximal in our partial ordering.*

That is, any of our universes (1) is "maximal" in an obvious sense, or, there is at least one "extension" of it, which is maximal. But now comes the "oversight". We did not state above, but should have done so, that each space S is assumed to have a countable topological base, that is that it can be covered by a (finite, or at most) countable number of analytical coordinate systems. Now, even if we assume this to be the case for our given base space S occurring in (1) and for every other base space S_λ occurring in (8), we *cannot* gaurantee it for the base space S_∞ occurring in (9). Also, the only manageable way of guaranteeing this is to *assume* that all spaces S, S_λ are, in some sense, "subspaces" of a given one, in such a manner that S_∞ is again such a subspace; which, of course, amounts to introducing a "joint" background space that is prescribed to begin with. We leave the specific discussion of this point to the end, and for the present we are simply postulating as follows.

COUNTABILITY HYPOTHESIS. *All spaces occurring, including* S_∞, *have a countable system of coordinate neighborhoods.*

This being stipulated we are now going to formulate conditions under which the maximal continuation (9) of (1) is unique.

Definitions. (i) We call a universe (1) totally *in*homogeneous if, on the base space S, it is not possible to have two coordinate systems, around two *different* points P on S, in which the components $f^\alpha(P)$ have identical values.

(ii) We call the universe totally anisotropic if it is not possible to have two *different* coordinate systems around the *same* points, in which the components $f^\alpha(P)$ have identical values.

(iii) We call the universe totally diversified if it is both totally inhomogeneous and totally anisotropic.

With these definitions the main result of our previous paper[107] can now be stated as follows.

THEOREM 2. *Subject to the countability hypothesis, if we start with a universe* (1) *that is totally diversified, and consider only such analytic continuations* (8) *as are also totally diversified, then the maximal continuation of* (1) *is unique, and is also totally diversified.*

Finally, as for the countability hypothesis, a way of securing it is as follows. Introduce, as (generalized) background space, an analytic n-dimensional manifold B with a countable number of coordinate neighborhoods, and assume for our universe (1), and all its continuations (8), that among the "functions" f^α there is also a many-to-one analytic mapping of S into B which is locally one-to-one. This mapping makes A into a "regular" spread over B, and it is known that a connected manifold which is regularly spread out over a space with a countable number of neighborhoods retains this countability property.

NOTES

1. The Symposium was held on November 2 and 3, 1973, at the University of California at Los Angeles. My assignment was to contribute a "Commentary" to a paper of Curtis A. Wilson, entitled "Rheticus, Ravetz, and the 'Necessity' of Copernicus' Innovation", which will appear in the Proceedings of the Symposium.

   I learned a great deal from the participants of the Symposium especially from Edward Rosen and Noel Swerdlow. Regrettably, a reprint of Swerdlow's "Aristotelian planetary theory in the Renaissance", *Journal for the History of Astronomy*, 3 (1972), 36–48, reached me too late for my consideration, especially in connection with an analysis of the deferent-epicycle construction of linear harmonic motion, as presented towards the end of Section IV (of this article).
2. *Thesaurus linguae latinae*, vol. 8, cols. 13/14.
3. Lynn Thorndike, *The "Sphere" of Sacrobosco and its Commentators*, pp. 10 and 148, Chicago University Press, 1949.
4. Plato, *Timaeus*, 28b, p. 51, translated by R. G. Bury in Loeb edition.
5. Pliny, *Natural History*, vol. I, pp. 170 and 171, translated by H. Rackham in Loeb edition.
6. Pline l'ancien, *Histoire Naturelle*, livre II, p. 116, note (1). Jean Beaujeu editor, Collection des Universités de France, Paris, 1950.
7. Aristotle, *On the heavens*, translated by W. K. C. Guthrie in Loeb edition, p. xi and pp. 88 and 89. See also Leo Elders, *Aristotle's Cosmology*, pp. 40–1, Assen 1966.
8. Beaujeu, in *loc. cit.*, note 6, p. 117, note (5).
9. *Macrobius*, edited by Franciscus Eyssenhardt, p. 552, Leipzig, 1893.
10. "The origins of the Copernican Revolution", *Scientific American*, **215** (1966), 88–98. For earlier publications of his see in Edward Rosen, *Three Copernical Treatises* (3rd ed., New York; Octagon Books, 1971), The Annotated Copernicus Bibliography, pp. 299–300.
11. *Loc. cit.*, note 11.
12. *Loc. cit.*, note 1 of chapter I.
13. "Space" (by S. Bochner), in *Dictionary of the History of Ideas*, vol. 4, pp. 295–307, New York, 1973.
14. Bernard R. Goldstein, "The Arabic version of Ptolemy's 'Planetary Hypotheses'", *Transactions of the American Philosophical Society*, **57**, part 4 (1967), 3–12.
15. "Space", *loc. cit.*, note 13.
16. About Telesio, Patritius and Campanella, see Max Jammer, *Concepts of Space*, pp. 83–90, Harvard University Press, 1954. Torchbook edition, 1960. Telesio may have been the first who translated Aristotle's *topos* not by "locus" but by "spatium". See Paul Kristeller, *Eight Philosophers of the Italian Renaissance*, p. 175, 1964. Patritius, who came after Telesio even speaks of "mathematical space" (*De spatio mathematico*). See Ernst Cassirer, *Das Erkenntnisproblem in der Philosophie und Wissenschaft der neueren Zeit*, **1** (1922), 266.
17. *By Principia* we will always understand Newton's *Philosophiae naturalis principia mathematica*, and we will quote from the edition: *Sir Isaac Newton's Mathematical Principles of Natural Philosophy and his System of the World*, trans. from the Latin (1687) by Andrew Motte (1729), revised by F. Cajori (Berkeley, 1934), which we will designate by "Motte–Cajori".
18. Motte–Cajori, p. 21.
19. For a large amount of material see the recent book of Stanley L. Jaki, *The Paradox of Olbers' Paradox*, Herder & Herder, 1969.
20. Jammer, *loc. cit.*, note 16, p. 137.
21. *Ibid.*, p. 138.
22. Hannes Alfvén, *Worlds–Antiworlds*, p. 63, San Francisco and London, 1966.
23. See, for instance, A. Lichnerowicz, *Theories relativistes de la gravitation et de l'électromagnetisme*, Paris, 1955; S. W. Hawking and G. F. R. Ellis, *The Large-scaled Structure of Space-time*, Cambridge Univ. Press.
24. Nicolas Cusanus, *Of Learned Ignorance*, trans. Fr. Germain Heron, London, 1954.
25. Alexandre Koyré, *From the Closed World to the Infinite Universe*, Baltimore, 1957.
26. A. S. Eddington, *Fundamental Theory*, Cambridge, 1949; Noel B. Slater, *The Development and Meaning of Eddington's "Fundamental Theory"*, Cambridge, 1957.
27. S. Bochner, *Eclosion and Synthesis*, p. 256, New York, 1969.
28. Koyré, p. 17.
29. Koyré, chap. III.
30. To the following see S. Bochner, "The size of the Universe in Greek thought", *Scientia*, **103** (1968), 511–31; also "Infinity" in *Dictionary of the History of Ideas*.
31. "Space", section 17, *Logical Space and World of Analytical Philosophy*.
32. *Tractatus logico–philosophicus*, German edition 1918, first English translation (with German original on opposite page) 1922.
33. C. C. Gillispie, in an article on Alexandre Koyré, in *Dictionary of Scientific Biography*, vol. 7, pp. 482–91, on p. 485, near bottom of right column.
34. J. L. E. Dreyer, *A History of Astronomy from Thales to Copernicus*, p. 329, Dover, 1953.
35. Dreyer, p. 395, footnote 3.
36. Occasionally a great scientist blunders into a gross misconception and comes up with a startlingly true conclusion none the less.
37. Dreyer, p. 329.
38. Dreyer, pp. 198–9. Very similar views on the ancients are already stated in Norbert Herz, *Geschichte der Bahnbestimmungen von Planeten und Kometen*, vol. 1 (1887), part IV. "Die Breitenbewegungen des Almagest."
39. *Ibid.*
40. Dreyer, p. 199.
41. Dreyer, p. 339.
42. Dreyer, p. 342.
43. Derek J. de S. Price, "Contra-Copernicus: a critical reestimation of the mathematical planetary theory of Ptolemy, Copernicus, and Kepler", in Marshall Claggett (editor), *Critical Problems in the History of Science*, pp. 197–218, esp. p. 207, Madison, 1959.

44. Dreyer, p. 268. See also Otto Neugebauer, *The Exact Sciences in Antiquity*, 2nd ed., pp. 203–7, 1957.
45. Dreyer, p. 331. There is no satisfactory explanation why Copernicus struck out this haunting passage.
46. Dreyer, p. 201.
47. I have analyzed Newton's role in the creation—in fact though not in name—of the conception of vector for exact science in an article in 1962, which is reproduced in my book *The Role of Mathematics in the Rise of Science*, Princeton, 1966. See there pp. 191–2. Michael J. Crowe, *A History of Vector Analysis*, Notre Dame Press, 1967, on pp. 127–8, quotes from my book, approvingly, the following passage regarding Newton's comprehension of vectors. "And yet, inwardly the Euclidean Space that underlies the *Principia* is mathematically not quite the same as the Euclidean Space that underlies Greek mathematics (and physics) from Thales to Apollonius. The geometry of the Greeks emphasized congruences and similarities between figures, that is, in analytical parlance, orthogonal and homothetic transformations of the underlying space.... The Euclidean Space of the *Principia* continued to be all this, but is also something new in addition. Several significant physical entitles of the *Principia*, namely, velocities, momenta, and forces are, by mathematical structure, vectors, that is, elements of vector fields, and vectorial composition and decomposition of these entities constitute an innermost scheme of the entire theory."
48. See, for instance, Dirk Brouwer and Gerald M. Clemence, *Methods of Celestial Mechanics*, p. 73, New York, 1961.
49. Basic introductions to the theory of almost periodic functions are: H. Bohr, *Fastperiodische Funktionen*, Berlin, 1932; A. S. Besicovitch, *Almost-periodic Functions*, New York, 1958; C. Corduneanu, *Almost Periodic Functions*, New York, 1968; L. Amerio and G. Prouse, *Almost-Periodic Functions and Functional Equations*, Van Nostrand, 1971.
50. It is not hard to deduce this theorem from a general proposition of L. Amerio. See C. Corduneanu, *loc. cit.*, p. 106, Theorem 4.6. Actually, in Corduneanu the condition of separation is formulated somewhat more generally. But the difference is probably only a technical one.
51. See, for instance, the book of Carl B. Boyer, *The Concepts of the Calculus*, 1949 (reprinted as *The History of the Calculus and its Conceptual Development*, Dover, 1959), pp. 107 and 109.
52. Johannes Kepler, *Neue Astronomie* (German), translation and introduction by Max Caspar, 1929, p. 46.
53. Johannes Kepler, *Harmonies of the World*, chap. 3, sect. VIII; in *Great Books of the Western World* (R. M. Hutchins, Editors in Chief). *Encyclopedia Britannica, Inc.*, vol. 16, pp. 1019–20.
54. William Whewell, *History of the Inductive Sciences from the Earliest to the Present Time*, 3rd ed., vol. 1, p. 408, Appleton & Co., New York, 1869.
55. Motte–Cajori, p. 39.
56. *The Role of Mathematics in the Rise of Science*, p. 227.
57. *Ibid.*
58. The first systematic statement of this is in *Role of Mathematics*, pp. 48–58.
59. "The quantity of motion is the measure of the same, arising from the velocity and quantity of matter conjointly" (Motte–Cajori, p. 1). One could even say that Newton defines the quantity of motion in a rather refined way as a bilinear functional of velocity and mass, rather than in the equivalent but more simplistic way as the product of the two.
60. Vol. I, chap. 2, "The meaning of numbers".
61. *The Role of Mathematics*, chap. 6.
62. S. Bochner, "The rise of functions," *Rice University Studies*, vol. 56, no. 2 (1970), pp. 3–21, esp. p. 5.
63. To this compare, for instance, my article "Mathematics in cultural history" in *Dictionary of the History of Ideas* (1973), vol. 3, esp. p. 184.
64. G. Galilei, *Il saggiatore* (*Opere*, VI, 232).
65. A. Koyré, *From the Closed World to the Infinite Universe*, p. 99.
66. Thus Newton used coordinates extremely expertly in his *Enumeration of Lines of Third Order*; see *The Role of Mathematics*, p. 41.
67. S. Bochner, *Eclosion and Synthesis*, p. 199, New York, 1969.
68. Koyré, p. 160.
69. *Eclosion and Synthesis*, pp. 205–6.
70. Richard Hope (translation), *Aristotle's Physics*, p. 56, Lincoln, Nebraska, 1961.
71. Motte–Cajori, pp. 614–15.
72. A. S. Kirk and J. E. Raven, *The Presocratic Philosophers*, p. 152 (1957).
73. We have studied this problem in the following publication: "The size of the Universe in Greek thought", *Scientia*, **103** (1968), 511–31; *Eclosion and Synthesis*, chap. XIL; articles "Infinity" and "Space" in *Dictionary of the History of Ideas*.
74. J. D. North, *The Measure of the Universe*, p. 23, Oxford, 1965.
75. "Infinity", p. 604.
76. The classical and much admired text for this A. Koyré, *From the Closed World to the Infinite Universe*.
77. Oswald Spengler, *The Decline of the West*, vol. I, chap. 5, New York, 1926.
78. See the references in note 73.
79. J. L. Synge, *Relativity: The General Theory*, p. 257, 1969.
80. W. de Sitter, *Kosmos*, p. 113, Harvard University Press, 1932.
81. For a succinct but valuable report see North, *loc. cit.*, pp. 72–81.
82. Simon Newcomb, "Elementary theorems relating to the geometry of a space of three dimensions and of uniform pinitive curvature in the fourth dimension", *Crelle Journal*, **83** (1877), 193–219.
83. North, *loc. cit.*, chap. 8, pp. 149 ff.; H. Bondi, *Cosmology*, chap. IX, pp. 75 ff., 1952 (2nd ed. 1960).
84. Bondi, *loc. cit.*, p. 22.
85. M. Jammer, *Concepts of space*, p. 10.
86. Diogenes Laertius, IX, 41 calls him the "prince" of physicists.
87. Stanley L. Jaki, *The Paradox of Olbers Paradox*, p. 12.
88. Motte–Cajori, p. 544.
89. Motte–Cajori, p. 596.

90. Motte-Cajori, p. 9.
91. Koyré, *loc. cit.*, p. 165.
92. *De rerum naturae*, book I, line 970; Jammer, *loc. cit.*
93. It is noteworthy how perceptively Copernicus distinguishes between *mundus* and *caelum* when the distinction is of consequence; while in ordinary parlance *mundus* and *caelum* were used interchangeably. Aristotle also used *ouranos* (Heaven) and *to pan* (Universe) interchangeably, on the whole; but he would distinguish between them when the distinction was of consequence.
94. *Physica*, book 3, chaps. 4–8.
95. *Aristotle, On the Heavens*, with an English translation by W. K. C. Guthrie, book I, chaps. 5 and 6, pp. 30–53, Harvard University Press, and William Heinemann, London, 1953 (Loeb edition).
96. Koyré, *loc. cit.*, pp. 59–87.
97. *Loc. cit.*, p. 54.
98. To the following compare *Eclosion and Synthesis*, pp. 245–52.
99. *Ibid.*
100. G. R. Burbidge and E. Margaret Burbidge, "Stellar evolution", *Encyclopedia of Physics* (S. Flügge, ed.), vol. 51, p. 135 (1958).
101. See G. Gamov, *The Creation of the Universe*, 1952, p. 75.
102. *Eclosion and Synthesis*, p. 198.
103. "Infinity", p. 604.
104. *Ibid.*
105. This appendix is based on my paper "Intrinsic analytic continuation and envelopes of holomorphy", *Proceedings of the National Academy of Sciences*, **53** (1965), 904–7.
106. This is the so-called "Zorn's lemma". It began to be thus designated in the mid-1930s, but it had been formulated, very expressly, in my paper "Fortsetzung Riemannscher Flächen", *Mathematische Annalen*, **98** (1927), 406–21.
107. Quoted in note 105.

11.7.

Note on Kepler's Contribution to Mathematical Analysis

S. BOCHNER
Rice University, Houston, Texas

Extract from a letter of the author to the editor:

"... Kepler is to me, next to Archimedes, the most important precursor of Newton, inasmuch as he was, in my view, the first to apply analysis to astronomy. His *Steriometria Doliorum* was something others were on the point of doing too. But in his *New Astronomy* he 'constructs' the conic by setting up in his primitive and arcane manner, a differential equation in polar coordinates, and finding that the solution is an ellipse. This is my overall impression of his work."

COMMENTARY

Salomon Bochner
Rice University

Curtis Wilson's paper is technically so accomplished, as was also his previous paper on Kepler, that it would not be possible or even feasible for me to comment on his technical arguments, especially since I am not versed in the technical specifics of the history of astronomy. But I will make some comment, from an approach that is congenial to my way of thinking, on the actual subject matter of the papers of Wilson and Ravetz, which is not only a question of dating the *Commentariolus,* but of forming a picture of how Copernicus intellectually developed, by which challenges and stimuli of his times and by what literary means at his disposal.

Ravetz started a controversy in 1962 by asserting that Copernicus composed his *Commentariolus* even before leaving Kraków for Italy in 1496. This was opposed, by rather brief statements, by Edward Rosen in the annotated Copernicus Bibliography in the 1971 edition of his important book "Three Copernican Treatises." Now Wilson has done the same as Rosen, but very systematically with an impressive array of arguments. The decisive points at issue seem to be whether certain specific astronomical data which seem to underlie the reasoning or rationale in the *Commentariolus* were already in existence before Copernicus left for Italy, or at any rate would have been readily available to him even then. On such points of context Wilson has the better of Ravetz. However important and meaningful as the resolution of such issues may be, the problem of the true intellectual origin of the *Commentariolus* in particular, and of Copernicus' true intellectual motivation in general is not answered thereby. To put it somewhat inelegantly, this question cannot be decided exclusively by examining which "library facilities" were available to him at what times, and with what degree of attentiveness he utilized such facilities.

For my part I would not discard the possibility that Copernicus may well have started nurturing his heliocentric proclivities long before he had the specific knowledge by which to support them. Thus his dissatisfaction with the Ptolemaic assertion that the huge sphere of stars performs daily rotations need not have come about at all by the knowledge of specific astronomical detail from specific astronomical works. It may have been simply the expression of a metaphysical malaise, which was brought about by certain nascent general developments in the scientific relation between mathematical space and cosmic universe, which were destined to profoundly alter astronomy and cosmology in due course.

Let me get to the heart of the matter by commenting, from my point of view, on Copernicus' Motion in Declination, the so-called Third Motion of the Earth. Nowadays, the two motions of the Earth, which Copernicus introduced, namely the rotation around the sun, and the spinning around its axis, are vectorially independent in the underlying mathematical three-dimensional space, say in the absolute space of Newton. The superposition, or addition of these two motions in this space, is commutative, that is either of the two motions can come first. Now, Copernicus did not, or rather did not yet, have such a space, not for decisive operational purposes at any rate. For such purposes, he had only a *machina mundi,* and, in a *machina,* whenever there was a conjunction of the motions of the wheels or gears of the *machina,* there were priorities. In the present conjunction the motion along the ecliptic was deemed by Copernicus to come first, and the daily rotation was deemed to be secondary to this. But, if it was secondary, then, as in a "good" *machina,* the axis of daily rotation "ought" to be firmly affixed to the plane of the ecliptic, and thus form a fixed angle with the polar axis of the ecliptic. However, the axis of the earth does not form such a fixed axis, but remains parallel to itself, and this, in Copernicus' vision, can only come about by an extra motion of the earth, a third one, which brings this about. In this motion "the axis of the Earth describes the surface of a cone in a year, moving in an opposite direction to that of brings this about. In this motion "the axis of the Earth describes the surface of a cone in a year, moving in an opposite direction to that of the Earth's center, that is from east to west."[1]

[1]J.L.E. Dreyer, *A History of Astronomy from Thales to Kepler* (formerly titled *History of the Planetary Systems from Thales to Kepler*), (New York: Dover, 1953), p. 329.

Thus, in a sense, this third motion of Copernicus was nothing but a piece of clumsiness, a testimonial to his incapacity to add up the two motions of the Earth, the daily and the annual, simply and vectorially.[2] However, it is also a testimonial to Copernicus' greatness as astronomer that he was able to extract some very crucial astronomical yield from this piece of clumsiness, once he had installed it. He made the period of the third motion slightly less than a year, and, to quote Dreyer,

> . . . and this slight difference produces a slow backward motion of the points of intersection of the ecliptic and the equator—the precession of the equinoxes. This was now at least correctly explained as a slow motion of the earth's axis and not as hitherto as a motion of the whole celestial sphere, and this almost reconciles us to the needless third motion of the earth, which certainly had its share in the unpopularity against which the Copernican system had to do battle for a long time, as it seemed bad enough to give the earth one motion—but three![3]

But having said this in praise of Copernicus, I must backtrack and point out that even this third motion was not original with Copernicus, but was an imitation of something that Ptolemy had done in connection with the epicycles of the outer planets, in order to account for the fact, astonishing to him, that the planes of these epicycles were parallel to the ecliptic. To quote again Dreyer, from another part of this work,

> As the epicycle of an outer planet was nothing but the earth's annual orbit around the sun transferred to the planet in question, it was of course quite right that the epicycle should be parallel to the ecliptic. In thus remaining parallel to a certain

[2]Equally clumsy, and indicative of motion in a *machina mundi* rather than in a "space," is Copernicus' retention of Ptolemy's presumption that in the (epi-)cyclical motion of a celestial body it is the orbis of the body that rotates directly and not the body itself. The body is firmly affixed to the orbis.

Kepler was apparently the first to overcome all such clumsiness. He dispensed with the Third Motion of the Earth (Dreyer, p. 395), and all celestial bodies became "freely floating in space, moved by physical forces acting on them" (Arthur Koestler, *The Sleepwalkers,* (New York: Macmillan, 1959), p. 314.)

[3]Dreyer, p. 329.

> plane the epicycles did what the ancients considered an unusual thing, as they would have thought it natural that the plane of the epicycle should keep at the same angle to the radius joining the centre of the deferent to the centre of the epicycle. The hypothesis therefore demanded the introduction of a small auxiliary circle, the plane of which was perpendicular to the plane of the deferent, the center of which was in the latter plane and which revolved in the zodiacal period of the planet. If we imagine a stud on the circumference of this circle and let it slide in a slot in the epicycle, we see how the latter was kept parallel to the ecliptic.[4]

Also, may I quote the following sentence about Copernicus from another part of Dreyer:

> In his sixth book, the shortest of all, Copernicus deals with the latitudes of the planets, and this is the part of his work in which he keeps closest to Ptolemy,[5]

to which I would like to add the following observation. Unlike Kepler after him, Copernicus neither states nor shows a proper awareness of the fact that the orbit of each planet is in a plane, and that these planes are various planes in a common (three dimensional) overall space. The awareness of such a space, with such various planes in it, would perforce have carried a measure of awareness of a mathematical background space with it. Ptolemy himself did not have such a space either, but it is much easier to "forgive" this failure to Ptolemy than to Copernicus, because once a geocentric view is adopted, the planetary orbits are indeed not planar at all. But Copernicus, from his heliocentric view, was within an ace of "finding" such a space, but he did not find it. In this sense, there never was a "Copernican invention of a planetary *system*," as Derek Price, although militantly opposed to Copernicus, was somehow disposed to concede to him.[6] The glory of having created a

[4]Dreyer, pp. 198-199.

[5]Dreyer, p. 339.

[6]Derek J. de S. Price "Contra-Copernicus: A Critical Re-Estimation of the Mathematical Planetary Theory of Ptolemy, Copernicus and Kepler," in Marshall Clagett (ed.), *Critical Problems in the History of Science,* (Madison: University of Wisconsin Press, 1959), pp. 197-218, esp. p. 201.

planetary system belongs to Kepler, and perhaps Tycho Brahe before him.

May I also add that Dreyer's statement above that the ancients would have considered it unusual for the epicycle of an outer planet to remain parallel to a fixed plane in space is already to be found 20 years earlier, in 1887, in the work of Norbert Herz on the history of the determination of orbits of planets and comets, in which he says as follows.

> This uniform motion in an unchanging absolute position in space was a thought that was alien to the ancients, as can be already seen from the fact that they always counted anomalies from instantaneous apogees.[7]

The total absence of a mathematical background space from the astronomy of Ptolemy must not be held against him as an astronomer, because the much vaunted Greek mathematics did not have it either. As I have written elsewhere:

> Most conspicuous, and almost fate-sealing, was the absence from Greek thought of a general conception of space for geometry and geometrically oriented analysis. Greek mathematics did not conceive an overall space to serve as a "background space" for geometrical figures and loci. There is no such background space for the configurations and constructs in the mathematical works of Euclid, Archimedes, or Apollonius, or even in the astronomical work *Almagest* of Ptolemy. When Ptolemy designs a path of a celestial body, it lies in the astronomical universe of Ptolemy; but as a geometrical object of mathematical design and purpose, it does not lie anywhere. In Archimedes, who, in some respects was second only to Isaac Newton, the mathematical constructs were placed in some kind of metaphysical "Nowhere" from which there was "No Exit" into a mathematical "Future."
>
> Such a background space was rather slow in coming. Thus, Nicholas Copernicus did not have it yet. He was an innovator in astronomical interpretation, and not in mathematical

[7]Norbert Herz, *Geschichte der Bahnbestimmung von Planeten und Kometen,* (Leipzig 1887); part IV, Die Breitenbewegungen der Almagest (first page).

> operation. His mathematics was still largely Ptolemaic, and only a bare outline of a *mathematical* background space is discernible in his *De revolutionibus orbium coelestium*. Nevertheless, already a century before Copernicus, Nicholas of Cusa, churchman, theologian, mystic, and gifted mathematician, in Book II of his leading work *Of Learned Ignorance* (*De docta ignorantia*), adumbrated an overall space of mathematics by way of an overall mathematical framework for the space of the universe.[8]

We have observed above that, as far as mathematical background space is concerned, the Copernican Motions in Declination and in Latitude are still conceived entirely in the spirit of the *Almagest*. But in some contexts of *De revolutionibus,* anticipations of a background space are discernible.

First, a whiff of an "absolute" space, but only a whiff, can be detected in Copernicus' insistence on immobilizing the sphere of the stars. In Book I, Chapter 6, when the size of the cosmological Universe is at issue, Copernicus declares that it is "immense" and leaves it at that. Now, whatever this "immense" means—and we will talk about this soon afterwards—, it is certainly not outright "infinite." This being so, if we could assume that Copernicus' "immense" Universe were somehow immersed in an infinite space that surrounded it, then the eighth sphere of the Universe would somehow already belong to the surrounding space, and making this sphere immobile would somehow be in accord with making it "absolute." And it is not at all far-fetched to impute to Copernicus the assumption of the presence of such a space, because some ancient Pythagoreans had already conceived something like this, according to the following famous report of Aristotle in *Physica,* 213b 22-24: "The Pythagoreans, too, held that the void exists and that breath and void enter from the infinite into the Heaven itself which, as it were, inhales."

Secondly, Copernicus' insistence, in Book I, Chapter 6, on his avowal that the Universe is "immense," without any further explanation of what this is meant to be, is of course a kind of refusal, an intentional one, to commit himself as to whether the Universe is finite or infinite. It

[8]Salomon Bochner, "Space" in *Dictionary of the History of Ideas,* (New York: Scribners, 1973), Volume IV, p. 205.

is tempting to take this feature in *De revolutionibus* lightly, or even adjudge it negatively. For my part, I judge it positively, and find it very future-oriented. I adjudge it positively, because, as W. de Sitter stated it so aptly in his book *Kosmos* in 1932: "Infinity is not a physical, but a mathematical conception."[9] In fact, no matter how much you may try to define infinity for some physical object, it turns out that you attach this infinitude to some mathematical structural feature of it. And I find the Copernican "immensity" future-oriented because, as I see things, it was taken over by Newton for his *Principia* which is Newton's counterpart to Copernicus' *De revolutionibus.*

I am fully aware of the prevalent outlook that, say between Copernicus and Newton, a so-called "infinitization of Space," as Koyré has called it, has taken place, and that it somehow culminated in Newton, who was a prime architect of it. Well, I have my opinion on this outlook, which, after quoting a statement of Bishop Barnes of 1931 that "infinite space is a scandal to human thought," I previously expressed thus: "I venture to add that, beginning with Newton, scientists who seemingly acquiesced in infinitude did so not only silently, but also restlessly and even sullenly."[10]

But quite apart of the question of whether Newton was somehow involved in an infinitization of space, I find, no matter how much I go over the *Principia* of Newton, that throughout the length of the *Principia,* and in all three editions of it, Newton *never* states that the Universe, *qua* Universe, is infinite. On the contrary, in the main context, though not the only one, where he touches upon the problem of the size of the Universe, namely in the System of the World, section [57], *On the distance of the stars,* he only uses the word "immense," and nothing stronger. And this seems to be the same "immense" as occurs in Copernicus' *De revolutionibus,* Book I, Chapter 6, as if *with regard to this problem,* nothing had been changed between Copernicus and Newton. When not speaking of the Universe *qua* Universe, but of Space and Time *qua* Absolute Space and Absolute Time, the *Principia* does have once, and only just once, the expression *from infinity to infinity,* and Koyré comments upon this

[9]W. de Sitter, *Kosmos,* (Cambridge, Mass: Harvard University Press, 1932), p. 113.

[10]Salomon Bochner, *Eclosion and Synthesis,* (New York: Benjamin, 1969), p. 256.

> '*From infinity to infinity* retain the same position. . .' What does *infinity* mean in this place? Obviously not only the spatial, but also the temporal: absolute places retain *from eternity to eternity* their positions in the absolute, that is, *infinite* and *eternal* space, and it is in respect to this space that the motion of a body is defined as being absolute.[11]

But this *infinity* has nothing to do with a cosmological universe. It is a philosophico-theological term and notion that had had currency throughout all the centuries since the earliest Presocratics, especially since the era of Anaximander and Xenophanes. The scholasticism of the High Middle Ages was most intimately familiar with it, however much its "World" may have been "bounded" in the sense of Aristotle and Ptolemy. It is, more or less the same infinity as in Lucretius' argument of the "flying dart," and Lucretius may have had it from a much earlier Pythagorean vision of an "infinite" void surrounding the Heavens. It is this infinity more-or-less that Copernicus had in mind when in Book I, Chapter 8 he suggests that it ought to be left to philosophers of nature to quarrel over whether the World is finite or infinite; whereas in Book I, Chapter 6, when the size of the cosmological Universe is at issue Copernicus declares that it is "immense" and leaves it at that.

But if all this is so, then why is it that Newton somehow creates the impression of advocating infinitude, or at least acquiescing in it, while letting the cosmological Universe be only "immense"?[12] And in a

[11]Alexandre Koyre, *From the Closed World to the Infinite Universe,* (New York: Harper, 1958), p. 166.

[12]In the discussion at the Symposium I was reproved for distinguishing between *mundus* and *caelum,* since in the Middle Ages the two terms were (allegedly) used interchangeably. To which I can only retort that a residue of difference did remain, as the very manner in which Kepler uses the terms in the two contexts demonstrates. Scientists frequently use pseudo-synonyms interchangeably, when no misunderstanding can arise, but become meticulous when the residual difference in meaning is of consequence. This can even be observed in Aristotle. In *De Caelo* he frequently uses *ouranos* (Heaven) and *to pan* (the All) interchangeably. But he decidely prefers *ouranos* when he speculates on the rotation of the Heavens around the Earth, and when he demonstrates that there is only one world ("Space", p. 300). On the other hand, he uses *to pan* in *Physica,* Book 4, Chapter 5, at the end of the essay on *topos,* when raising the question whether the universe as a whole, when viewed as one comprehensive physical system, has a physical *topos* too (ibid.).

parallel feature in Copernicus, why does he say in Chapter 8 of Book I that it should be left to the physiologoi whether the world (*mundus*) is finite or infinite, but in Chapter 6, he only maintains that the heavens (*caelum*) is immense? The reason in both cases is the same. In Newton, what is infinite is the underlying Euclidean Space, and it is infinite as a mathematical object. The fact that Newton fancied to call it "absolute space" does not make the slightest difference. In Copernicus, there is an anticipation of this underlying space, but not as an operational datum, but as a semi-theological one (which Newton's absolute space still in part is), and which was very well known to the scholastics, and was already endowed in Cusanus with purely mathematical features. In fact, Cusanus had a mathematical turn of mind, and the mathematical features which I detect in his Universe are far ahead of his times. To end with a quotation, from J.D. North's important book, *The Measure of the Universe:* 'It is easy to speak of the infinite, as every theologian knows, but it is difficult to speak of it meaningfully."[13]

[13](Oxford: Clarendon Press, 1965), p. 23.

Vistas in Astronomy Vol.22, pp.19-20 0083-6656/78/0601-0019$05.00/0

KEPLER: A PERSONAL FOOTNOTE

S. Bochner

Rice University, Houston, Texas, U.S.A.

Kepler has always had a great fascination for me. As I meditate on his achievement I find more and more that he is not only the true mastermind of the Astronomical Revolution but that he is also in the vanguard of the scientific revolution at large, being just as much a leader of it as were Galileo and Descartes. In fact, for my own part, I have learned to appreciate the central "revolutionary" achievements of Galileo and Descartes by viewing them against the backdrop of Kepler's doings, which — to my perception — came ahead of theirs.

During the crucial decade 1601-1611, Kepler achieved not only his *New Astronomy*, but also his new theory of Vision and his new outlook on the mathematical theory of Conics. I learned from Vasco Ronchi that Kepler's work in Vision was just as revolutionary as his work in Astronomy, and from Charles Taylor (in a 1900 article on the conics in Kepler and Newton, in the *Proceedings of the Cambridge Philosophical Society*) I learned that his work on Conics initiated the "Law of Continuity" which a hundred years later became a leading philosopheme in Leibniz.

These three innovations may seem very disparate, and only accidentally synchronous. But from my approach they are not so, but are closely connected by flowing from the same Keplerian inspiration of envisioning a mathematico-metaphysical background space for all and anything in the Universe of Thought and Being.

In my previous article[1], I dealt with the effect of this space on the Copernican Universe.

In the case of Vision, according to Ronchi, everybody before Kepler (and his contemporary Maurolicus) based Vision, anthropocentrically, on a cone with its vertex in the eye and its opening towards the object to be visually recognized. Kepler however introduced innumerably many cones, each with its vertex at some point of the object and with its opening towards the pupil of the eye. All these vertex-points were points of the object-to-be-seen in its location in the underlying background space; and by thus decomposing the object into an aggregate of points, Kepler even cognitively anticipated the conception of an aggregate of points in the sense of Georg Cantor from the second half of the nineteenth century.

Thirdly, he manifested his Law of Continuity (Kepler spoke of "analogy") when pulling an ellipse across infinity, via a parabola, into a hyperbola, and then, by further pulling, letting the hyperbola degenerate into a pair of straight lines.

After having thus firmly grounded Astronomy on an analysis of Space and of *general orbits* and various *areas* in it, and after having dealt with Optics in a parallel way by also *dis*covering (that is *un*covering) Continuity and Infinitude in the process, Kepler took a pause from Physics for a while in order to formalize and tie together the mathematical insights gained, thus initiating the infinitesimal calculus for posterity. That Kepler chose to associate this work with a sudden concern for volumetric examination of wine casks is one of the many Kepleresque extravaganzas which even a library of books on Kepler's irrationalities, mysticism, belief in magic, etc., will not fully explain.

But this excursion into mathematics for the sake of wine casks, that is for its own sake, was only an interlude, after which Kepler returned to his beloved astronomy and succeeded in fashioning the sesquilateral law that also binds Time to Space, for all time to come; the first such permanently valid, non-trival non-obvious link between Time and Space ever.

For my own part, many of the doings of Galileo, Descartes, and even Newton, became truly intelligible when viewed against the achievements of Kepler.

For instance, I never quite understood why Newton inserted in Book I of the Principia so much about conics, until I realized, through his Lemma XXII, that they are done entirely in the spirit of Kepler (and not in the spirit of Desargues who explored the purely perspectival aspects), and that Newton simply had a compulsion to deal with *all* of Kepler's research topics, however much he (Newton) avoided mentioning the name of Kepler in such contexts.

And in the case of Descartes, I understood how he did such disparate things as (i) coordinate geometry, (ii) foreshadow the law of inertia, and (iii) introduce the subject-object duality into the philosophy of cognition. All these achievements deal with Kepler's introduction of background space. For (i), the introduction of coordinates, this is obvious. For (ii), his law of inertia simply explores and exploits the homogeneity of this space and its three-dimensional structure, whereas Galileo's terrestrial dynamics was limited to two dimensions. And finally — (iii) — the subject-object dualism in Cognition is simply the "intellectual" analog to the duality of background space and Universe in astronomy — an analogy first pointed out by Cassirer[2].

REFERENCES

(1) S. Bochner, Mathematical Background Space in Astronomy, In: *Vistas in Astronomy* (Ed. A. and P. Beer) Vol. 19, pp. 133 - 161. Pergamon Press, Oxford (1975).
(2) Ernst Cassirer, *The Individual and the Cosmos:* Univ. of Pennsylvania Press (1972).

THE EMERGENCE OF ANALYSIS IN THE RENAISSANCE AND AFTER

by Salomon Bochner

CONTENTS

I. The Key Concepts of Analysis
II. The Force of Analysis
III. Functions
IV. Real Numbers
V. Space
VI. Infinity
VII. Continuity
VIII. Post-analysis Number Theory and Algebra
References

I. THE KEY CONCEPTS OF ANALYSIS

Analysis as an independent division of mathematics—and it is the largest division by far—is a "modern" creation. Greek antiquity did not have it, nor did the Middle Ages, except for occasional gropings, mostly faint ones. But in the sixteenth and seventeenth centuries, analysis began to sprout and rise in many contexts, and its influence began to spread into the farthest precincts of mathematics. All this started very suddenly, and it was as great a revolution as any then proliferating.

The emergence of analysis in the Renaissance created a great divide in all of mathematics; there is a pre-analysis geometry and a post-analysis geometry; a pre-analysis algebra and a post-analysis algebra; a pre-analysis number thory and a post-analysis number theory, very much so; and, very importantly, a pre-analysis astronomy and a post-analysis astronomy. The first epochal manifestation of analysis—as analysis is conceived by us—was Johannes Kepler's proclamation of his Planetary Laws, especially the Law of Areas. There had been nothing like it before in an astronomy that had been a mathematically controlled science for two millennia, not even in Copernicus, whatever his astronomical innovations may have been.

Terminologically the Greeks did have analysis, in logic, in epistemology, and also in (late) mathematics, but for the most part in a contrasting parallelism with synthesis, and only rarely by itself. The verb *analyein* occurs already in Homer in the meaning of 'untie, unravel,' and a contrast between analysis (*Resolutio*) and synthesis (*Compositio*) plays a role in philosophy even today. Yet "analysis" and "analysis-synthesis" play only a secondary role in Greek mathematics, as can be gathered from the article of Mahoney [*55*], although this article is designed to present and elaborate the positive aspects of the Greek analysis. In any event, in mathematics this analysis was something procedural rather than substantive, as described in a famous passage in the work of the mathematician Pappus (third century A.D.), which is the principal ancient description known; see [*55*, p. 322]. Mahoney himself states:

> if the Greeks admitted analysis as a member of the family, they nevertheless had tried to keep it in the background, barring it from any formal appearance. Only Pappus and a few scholiasts talked about it at any length. [*56*, p. 35]

In Euclid's *Elements* there is no mention of analysis nor even any allusion to it, except for a statement on analysis-synthesis which appears after theorem XIII.5, and which, as everybody agrees, is a scholiast's interpolation. After Euclid, there are occurrences in Archimedes, Apollonius, and perhaps others, but always quite secondary ones, and from the Middle Ages there is nothing remarkable known about "analysis." But in 1591 matters came to life, terminologically and substantively, in work of the algebraist François Viète (1540-1603), who may have been stimulated by Petrus Ramus (1505-1572) (see Mahoney [*56*, pp. 11-32], *passim*). Viète put "analysis" into the title of a book (*In artem analyticam isagoge,* Introduction into the analytic art), thus creating a fashion in terminology. "Analytic" and "analysis" began to appear frequently in titles of publications, and we will adduce noteworthy instances in due course. Substantively, the aim of Viète's *isagoge* was to develop techniques for operating with algebraic symbols (which was Viète's forte), and it clearly associated "analysis" with this endeavor. This also created a fashion. Many mathematicians began to be known as "analysts"; Fermat for instance so designated himself, as a matter of course. But there were no "synthesists," probably because they would not have been very different from "classical" geometers. Besides, "analysis" was quickly exceeding the bounds within which Viète had conceived it, and it soon began to designate most of the genuinely new mathematics bursting forth all around.

Thus Isaac Newton's first leading essay on mathematics proper, which was conceived and composed in various parts between 1665 and 1669—but not published until 1711—was avowedly in analysis. Its title is "De analysi per aequationes infinitas" (*Analysis by infinite equations*), and at one point he refers to Descartes as an analyst (see [*61*, p. 222]). Yet on the title page of his great *Principia* he speaks of "mathematics," which in the sixteenth and seventeenth centuries also subsumed mathematically controlled science, especially astronomy. The full title of the work is *Philosophiae naturalis principia mathematica* (Mathematical principles of natural philosophy), and on its title page Newton identifies himself as Lucasian Professor of "Mathesis." Still, in the Scholium to the Laws of Motion, he refers to "Sir Christopher Wren, Dr. Wallis, and Mr. Huygens" as "the greatest geometers of our time" (see [*61*, p. 22]), just as, in the French language, the encomiastic designation "illustre géomètre" may refer to a distinguished mathematician of any research proclivity.

But what was this new analysis, and how did it relate to the algebra of symbols and to the infinitesimal calculus, which were such "obvious" manifestations of it? Our answer is that the operation with symbols was a kind of new technology, and that the infinitesimal calculus was a kind of automation based on it, but that analysis in a broad sense was a novel intellectual setting in which all this came to pass, and in which operational and foundational aspects were inseparable. The new setting was foreshadowed in Nicholas of Cusa (1401-1464) in the first half of the fifteenth century, but it actually formed itself in the sixteenth century, when rather suddenly, as if on a signal, mathematics began to absorb into its very texture conceptions which are the all and everything of mathematics of today, but which had been quite alien to Greek *mathematics* even at its height, as in the work of Archimedes, say. The leading such conceptions were: *Space, Infinity, Function, Continuity, Real Numbers.* These conceptions do occur, overtly or covertly, in certain areas of Greek thought, as in natural philosophy, cosmology, ontology, logic, and, very importantly, theology. But it is very difficult to locate them in Greek professional mathematics.

It is a fact, however hard to accept, that Greek geometry—Greece's pride—was a geometry without space. The geometry of Euclid did not have the "Euclidean" spaces E^2 or E^3, as objects by themselves, independent of or prior to configuration contained in them. There is, however, a separate essay on space (or rather on place, *topos*) in Aristotle's *Physics,* book 4, chapters 1-5 (compare, for instance, [*3*]), and a separate one on void, chapters 6-9. There is also, in close proximity, an essay on infinity, book 3, chapters 4-8, and on time, book 4, chapters 10-13. But there is nothing syllogistic about infinity in Greek mathematics, or about continuity, and these have been the intellectual's hallmarks of Time, from Homer to Newton and beyond (see below, section VII).

Our key conceptions, although distinct from each other, cannot be kept apart in meaningful contexts. Aristotle knew this too. He knew that they come close together in the conception of motion, and he made great efforts to comprehend their interaction in the second half of the *Physics*, books 5-8. He was especially intrigued by the puzzles of Zeno of Elea (dichotomy, Achilles and the tortoise, the flying arrow, the race-course) and made great efforts to resolve them [*25*]. But there is absolutely no mention of the puzzles in Greek mathematics, which apparently did not profit at all from philosophical developments towards elucidating them. For instance, there is nothing evidential in the extant corpus of Greek writings to suggest that Archimedes was interested in the puzzles, or even that he knew about them or knew who Zeno of Elea had been. A reader may recoil in disbelief from the supposition that Archimedes could have been totally unaware of such developments in philosophy and of their potential import for mathematics, but the possibility must be faced. The great reluctance to entertain such a "negative" supposition about an Archimedes stems from a presumption that Greek professional mathematics was still close to philosophy, closer than it is today. But this presumption is not justified. It is true that Eudoxus of Cnidus, a great mathematician and member of Plato's Academy, could apparently lecture on Plato's Theory of Forms, and (according to Aristotle, *Nichomachean Ethics*, book X) was even an expert on problems of ethics. But, in Plato's dialogue "Theaetetus," Theodorus of Cyrene—a distinguished mathematician, teacher to Plato and Theaetetus—is reluctant to participate in the discussion of the nature of Knowledge, pleading a lack of competence in the field. Also, it is true that the system of Definitions, Postulates, and Common Notions, on which Euclid's *Elements* is built, was originally debated by philosophers and mathematicians alike (see [*42*, pp. 114-136]). But there is little evidence for such a continued "cooperation" after Euclid; and, very importantly, Archimedes, for instance, showed almost no regard for principles and niceties of the system, in sharp contrast to Poincaré, Hilbert, L.E.J. Brouwer, or Hermann Weyl of our times (see the thorough-going investigation [*38*] in its entirety, and especially a critique of Archimedes on p. 57). It is misleading to judge the relation between philosophy and professional mathematics in antiquity by Plato's aphorism that God is a Geometer and his conceit of intervening in matters of professional mathematics (*Republic*, vii).

The key concepts of analysis are in their roots common to mathematics and other knowledge, general or particular, and, to a considerable extent, analysis had to wrench them by a very slow process of adaptation out of their extra-mathematical contexts and fit them to its peculiar molds. It is this genetic universality of the key conceptions of analysis that made modern mathematics so very efficacious in interpreting nature and knowledge, much more so than Greek mathematics ever was or could be. The "tools" of Greek mathematics were by origin narrowly mathematical. Those of modern mathematics, however, are genetically universal; and the more universal by origin, the more effective they are in mathematics after becoming attuned to its needs and purposes.

II. THE FORCE OF ANALYSIS

It is our thesis that the era of the emergence of analysis is virtually coextensive with the Era of the Scientific Revolution in its full duration; that the emergence of analysis was an all-dominant outcome of the Revolution; and that the one-ness of analysis instills a feature of one-ness into the Scientific Revolution, whatever its diversity.

The Scientific Revolution spans the three centuries 1500-1800 (see [*39*]), even the four centuries 1400-1800 (compare [*24*]). There is nothing uncertain about the terminal date 1800; in particular, in our outlook, we would not replace it by an earlier date, although the eighteenth century is a renowned age by itself, the Age of Enlightenment (see [*28*]). But the beginning date of the Scientific Revolution is flexible. The century 1400-1500 was not as "revolutionary" as the following one, although it brought about two separate achievements both seriously altering our key conception of space. It systematically initiated linear perspective in painting (see [*37*]), and, in forwards-directed pronouncements of Nicholas of Cusa, it inititated "modern" insights into the spatial fabric of the universe [*49* and *17*].

It is an essential part of my thesis that the three centuries 1500-1800 in the rise of modern mathematics "genetically" correspond to the three centuries 500-200 B.C. in the rise of Greek mathematics, and the four-century span 1400-1800 similarly corresponds to 600-200 B.C., but that this genetic correspondence breaks down after the terminal dates, totally and irretrievably.

In the century 300-200 B.C., just before the terminal date, Greek mathematics was in its fullest vigor. The century began with Euclid's *Elements,* ended with the *Conics* of Apollonius, and culminated with the works of Archimedes, who was one of the very greatest mathematicians ever. The century could also boast of the great astronomer Aristarchus of Samos, who was even a heliocentrist (according to a statement of Archimedes), and of the polymath Eratosthenes of Cyrene, eminent mathematician (Sieve of Eratosthenes) and even more eminent geographer. But, as I stated elsewhere [*12*], after 200 B.C., "unexpectedly and inexplicably, as if on a signal," the upward trend of mathematics proper, that is, of mathematics as "research" mathematics, came almost to a halt. The development of mathematics "began to level off, to lose its impetus, and then to falter." An inexorable downward trend set in, which even the imposing figure of a Pappus (third century A.D.) could not stem; and around A.D. 500, that is, around the time of the fall of Rome, came the final extinction of Greek mathematics in its own phase.

To an extent, the decline of Greek mathematics was part of the so-called decline of ancient civilization, many features of which began to manifest themselves in the second century B.C. But (see [*12*]) the decline of mathematics from the heights of the third century B.C. was too large, too sudden, and too incongruous to be thus fully explained, and an additional cause peculiar to mathematics must have been operative. I maintained (in [*12*]) that, from first to last, the intellectual base of Greek mathematics was altogether too narrow and, above all, too rigidly fixed in its intellectual narrowness to support an ever heavier mathematical edifice; so that at a certain critical stage the edifice ceased to be firmly grounded. It gradually became unstable, fractured, decadent.

A glance at the fortunes of astronomy will corroborate this outlook. The mathematically controlled astronomy of the Greeks did continue to thrive after 200 B.C. Around 150 B.C. Hipparchus made the breathtaking discovery of the precession of the equinoxes, and three centuries later, around A.D. 150, Claudius Ptolemy composed his *Almagest,* which, by mathematical formalism, held sway over astronomy not only through the length of the medieval Islamic Era, but still in the structural framework of Copernicus's *De Revolutionibus,* however "revolutionary" its astronomical content may have been. This perdurance of the *Almagest* is due not so much to its greatness as to the general stagnation of the Greek mathematics employed. As soon as, after Copernicus, the breath of the new analysis blew into astronomy, it produced the unprecedented Planetary Laws of Kepler, and his new Theory of Vision to boot (see below, section V). Also, not long afterwards, whatever is worth remembering of the work of the Late Hellenistic mathematician Pappus was enhanced by dosages of analysis—directly in the n-line locus problem of Descartes, and indirectly in the Perspective of Kepler and Desargues.

After that, until 1800, analysis was in a state of prolonged adolescent growth, as it were, whence comes an impression of freshness, robustness, and inimitable prowess; and the progress of this growing-up can be gauged by a quick assessment of several works which carry "analysis" in the title.

First, we have already noted Newton's work of 1665-1669 on "analysis by infinite equations." It deals with an array of infinite processes, mainly with infinite series for various definite and indefinite integrals, and successive applications of his method of approximation to solutions of polynomial equations.

Second, in 1748, came Euler's most popular treatise among his many popular ones, under the title "Introduction into analysis of infinitude" (*Introductio in analysin infinitorum*), and it is already a veritable grab-bag of analysis of today. It contains, among other things, an exposition of infinite series, including those for e^x, cos x, sin x, and the relation $e^{ix} = \cos x + i \sin x$; curves and surfaces investigated with the aid of their equations; an algebraic theory of elimination; and, mark! a chapter on the Zeta function and its relation to the prime number theorem, as well as a chapter on *partitio numerorum* (see [*71*, p. 169]).

Next, four decades after Euler's "Analysis," in 1786, came Lagrange's *Mécanique Analytique* (Analytical Mechanics), which was a new kind of title. Whatever Lagrange himself and his immediate contempories may have associated with this "analytique," in retrospect it does *not* connote a contrast to something else, like synthetical, practical, or pragmatic, but simply the presentation of mechanics in the spirit, ambience, and style of Lagrange's "today." It almost means "modern" mechanics, mechanics of "today." And it was a long "today," because Lagrange's treatise was a paradigm for textbooks in colleges and universities until World War I. Also only ten years after the *Mécanique Analytique,* in 1795, Monge began to publish his *Feuilles d'analyse* (that is, "Notes" in analysis), which later became the famed *Application d'analyse a la géométrie,* which promulgated a new kind of geometry, an avant garde geometry based on analysis, not at all replacing the "classical" geometry, but becoming a novel field alongside the latter.

Lagrange's *Mechanics* came precisely a hundred years after Newton's *Principia,* and by contents it is, in a sense, a translation of the *Principia* from the ostensibly Archimedean *mise-en-scène* into an up-to-date setting of analysis. But the *Principia* was its basis, foundation, and fount of inspiration, so that in innermost substance the *Principia* must have anticipated the "analytical" character of the *Mechanics.* Such was indeed the case. We will verify that, from our outlook, the *Principia* is simply soaked in analysis, which is of the same substance as in Lagrange's *Mechanics,* and which had been in the making from the first. Nevertheless, syllogistically, at the time of the *Principia,* analysis was still in a state of juvenile un-readiness, which the Archimedean setting of the *Principia* masks and counteracts in one. I argued this already (in [*17*]), and the emphasis there was on the fact that Kepler's analysis was in an even earlier state of infancy, but that this state of infancy in no wise impaired its power and efficacy.

Even Lagrange's work on "Analytic Funtions" (*Fonctions analytiques*) is syllogistically still very wobbly. In fact, the maturing of analysis (and of modern mathematics as a whole) into a syllogistically acceptable doctrine began only after 1800, and the process extended till 1900 and beyond.

The era of the Scientific Revolution contains diverse and disparate developments, but the extramathematical universality of the key concepts of analysis suggests that there ought to be internal features of unity within the diversity, and there are indeed. For instance, Kepler calls his optical work "astronomiae pars optica," thus insisting on a link with his astronomy. Kepler himself does not suggest what the link is, and Alexandre Koyré in his *Révolution Astronomique* [*50*], a splendid book about Kepler by-and-large, does not enter into optics at all. But there is indeed such a link, namely through the key conception of Space, as we will see in section V. I will even (*pace* Descartesians) suggest such a link in Descartes, namely a link between his geometry and his body-and-soul problem. Here again I found nothing about it in Koyré, although he wrote two important books about Descartes [*47* and *48*], in addition to expounding his work in mechanics. But I did find encouraging corroboration in the basic essay of Ernst Cassirer [*28*], especially in its last chapter.

In the eighteenth century analysis and mechanics were almost inseparable in their growth, but after 1800 they began to separate, although in a very amicable way; and, what is more remarkable, a concern for a syllogistically rigorous foundation of analysis began to spread, however gradually, into ever wider circles of creative mathematicians. Analogous concerns sprang up in virtually all academic fields. In historical sciences, for instance, a concern for rigorous foundation resulted in the emergence of a so-called "higher criticism" (see [*9*]). There seems to be no satisfactory explanation for this "universal" development.

Similarly, in the unfolding of the key concepts of analysis since 1800 there has been an overall difference between the nineteenth and twentieth centuries, which again appears to have analogs in most other academic departments. The difference is that gradually a certain "secularization" of the key concepts, followed by a certain "flattening-out," has come about. Space, for instance, in the hands of Riemann became something intricately lofty in two separate grandiose constructions, namely construction of a Riemann surface in complex analysis—or, rather, to Riemann, in algebraic geometry—and of a general Hausdorff manifold with or without a Riemannian metric. These are structures as lofty as cathedrals. But a secularized class of spaces nowadays, indispensable, yet very "flat," and, to me "vulgar," are the various linear vector spaces, which are mushrooming all over mathematics with no end in sight. Also, space in cosmology and relativity is highly "structured," and is becoming ever more so. On the other hand, almost any aggregate of mathematical objects, or alternately any "point-set" can be viewed as a space, even if no kind of structure is postulated to begin with.

Or consider continuity. After Cantor and Dedekind had finally settled the nature of real numbers and of the linear *continuum* in one, the nature of continuous *functions* followed next. They were explained as continuous mappings from one topological space into another, and until around 1935 topology was dominated by the quest to elucidate the implications. It was all very exciting. But the sobering "secularization" set in, when "topological structure" of a space was made something very general, subject to some minimal requirements only. Any point-set whatsoever could be given a topological structure without having one, namely the so-called discrete structure. Finally, infinity, which had been highly theological throughout, and still was an awe-inspiring conception in the nineteenth century, became quite vulgarized in this sense. For instance, any linearly ordered set whatsoever can be enlarged by a fictitious last element, which is irreverently called infinite, and denoted by $+\infty$, say; or it can be deprived of it, which is equally irreverent.

III. FUNCTIONS

The emphasis on operation with symbols in Viète and Fermat undoubtedly had an impact on the rise of functions and of real numbers, because the simplest kinds of functions are those given by algebraic expressions in terms of symbols, and real numbers are the simplest "constant" coefficients that are elements of an algebraic field or ring. But functions and real numbers also had "analytical" roots of their own, which were not intergrown with symbols, reaching back into the sixteenth century and earlier.

Outwardly, functions unfolded rather slowly. According to Moritz Cantor's *History of Mathematics*, the systematic conception of functions began only near 1700 with Leibniz, who also coined the name "function." But, syllogistically, it was a rather crude "approximation" to our $y = f(x)$ of today; in the eighteenth century there were several attempts to do better, but progress was slow. Even Lagrange's book, *Fonctions analytiques,* which was composed toward the very end of the eighteenth century, is not, or not yet, what an unwary reader today would expect. It does have power series and their usual properties. But to Lagrange an analytic function was one appropriate for *his* analysis, and this was the analysis involved in his Analytical Mechanics; thus he was not "programmed" to distinguish between a function which is analytic in our present-day sense and one that is, say, continuous and piecewise analytic only. As a matter of fact, Lagrange knew that in the mechanics of continuous media (vibrating strings, etc.), a "general" solution can be obtained approximately by replacing continuously distributed masses by finite systems of mass points, and he may have therefore taken it for granted that a (uniform) limit of such piecewise analytic functions (in our sense) is analytic (in his sense) too. Cauchy was probably the first successor to Lagrange who could clearly perceive the distinction. Lagrange and Cauchy were in the same French tradition, and chronologically not too far apart. But Lagrange was active before 1800 and Cauchy after 1800, and the difference between them is a telling manifestation of the change-over around 1800. To Lagrange a mathematical function, however "general" and "abstract," is somehow always an orbit from mechanics; but to Cauchy any path or orbit from mechanics, any mechanics, is a mathematical function, as a matter of course.

Unlike power series expansions, trigonometric expansions are suitable for very "general" functions. This was emphatically proclaimed by Fourier, who was active after Lagrange and before Cauchy, and who asserted that "any" function admits a trigonometric expansion, even if it is "absolument arbitraire." This provoked Dirichlet into defining, in 1837, a general function $y = f(x)$ essentially as an "arbitrary" association of a suitable value y with any value x; and the priority of Dirichlet in this matter has been "institutionalized" by A. Pringsheim [*64*] in the German Encyclopedia of Mathematics. But in another volume of the Encyclopedia, H. Burkhardt [*23*], author of an article which is a fundamental work in the "early" history of Fourier Analysis, states that the twentieth-century conception of a general function originated in the remarkable book (1838) of Antoine Augustin Cournot, *Recherches sur les principes mathématiques des richesses* (Investigation of the mathematical principles of the theory of wealths). The "Calculus" work [*62*] by A. Ostrowski, which has competent historical notes, agrees with Burkhardt. It should be stated though, that, in a serious vein, a "function" cannot be really defined at all, because any description of it as a "correspondence" or "association," or even "ordered binary relation," is logically a vicious circle.

After this very brief survey of the *overt* career of functions since their formal introduction by Leibniz in 1700, we shall turn to their *covert* presence *before* 1700.

I find that, covertly, functions are in the very center of Newton's *Principia*, which appeared in 1686 but had been long in the making. Newton's primary *concept* is "quantity of motion." If he had had formulas, then for the motion of a particle on the line, $x = \varphi(t)$, the quantity of motion would be

$$(1) \qquad p = mv \quad , \quad v = \frac{dx}{dt} = \frac{d\varphi}{dt} \ .$$

And Newton's primary *fact* is that the rate of change of the quantity of motion is (equal to, or is determined by, or, perhaps even determines) the total force acting on the mass point, that is, in terms of (1),

$$(2) \qquad \frac{dp}{dt} \equiv \frac{d(mv)}{dt} = F \ .$$

In the case of a brusquely acting force, one might also have

$$\Delta p \equiv p(t + 0) - p(t - 0) = F(t) \ .$$

There is nothing like all this in Greek mathematics, in any kind of verbal circumlocution of these formulas. We might concede that a point function $\varphi(t)$ is reminiscent of a Greek curve, and that a derivative $\frac{d\varphi}{dt}$ at a point is reminiscent of the tangent to a curve at a point. But gathering up the derivatives at various points into a new function $v(t)$ is not reminiscent of anything occurring in Greek mathematics anywhere, and the formation of the second derivate of $\varphi(t)$, which occurs in (2) when writing it in Euler's form

$$m \frac{d^2x}{dt^2} = F , \tag{3}$$

is far outside anything Greek thought has ever conceived. (We note that Euler's form (3) was very important in the creation of the theory of elliptic operators. But Newton's form (2) occurs in most theories of physics after Euler, in the Hamilton-Jacobi theory of Mechanical Systems with Restraints; in the theories of Gas Dynamics of Maxwell and Boltzman; in the theory of Relativity; and in Quantum Mechanics.)

But even the concession that the derivative $\frac{d\varphi}{dt}$ is reminiscent of a Greek tangent to a curve goes too far. From what is known, the Greeks envisaged tangents to curves that are convex, at least locally, and only to these, so that their tangents were supporting lines, at least locally (see [*12*, p. 184]). They apparently never introduced a tangent which crosses the curve at the point of tangency, although some of the curves known to them had such tangents. In fact it was one of Fermat's major achievements to have become aware of such a possibility. In 1640 he envisaged, as a part of his pioneering work on maxima and minima of functions, the problem "of investigating by the (analytic) art the points of inflection at which the curvature changes from convex to concave, and conversely" (see [*56*, p. 201]). Fermat thus created the concepts of a "point of inflection" of a curve, and of the "(point-by-point) curvature" of the curve, which Frenet and others took up systematically only in the nineteenth century.

This achievement of Fermat belongs to the "theory of functions" irrespective of the geometric setting, and earlier achievements affecting functions also appeared in settings that were not overtly "analytic." Functions in our sense are discernible already in the fourteenth century—as a species of graph and other—in the work of Nicole Oresme (1325-1382); and the renowned mathematician-turned-historian Oswald Spengler was so impressed with them that he made them into a litmus-like indicator by which to distinguish present-day civilization from ancient ones.

In the fifteenth century Regiomontanus created a field in function theory by freeing trigonometry from its subordination to geometry and astronomy and by beginning (but only beginning) to view sin x and cos x as functions on the x-axis subject to the familiar addition theorems as functional equations. Later on the logarithmic and exponential functions were also drawn into this context.

Simon Stevin shortly before 1600, and Luca Valerio in 1604, began to consider rather *general* functions when standardizing and streamlining the formation of "Riemann-Darboux Sums" for the computation of volumes, which for *special* functions already occur in the masterful work of Archimedes on Conoids and Spheroids. Thus Carl Boyer comments on Luca Valerio: "this geometrical reasoning is strikingly similar to that presented in many present-day elementary textbooks on the calculus" [*21*, p. 105]. I fully concur, except that I would have said "analytical" rather than "geometrical" reasoning. And all this was done without Viète symbolism, which, in similar contexts, does not occur before the work of Fermat.

In denying the creation of functions to the Greeks, I do not mean to say that they were not familiar with categories of cognition such as "correspondence," "dependence," "mapping," even "binary relation," without which "functions" cannot be formally conceived, but only that they did not perform mathematically controlled operations with them, in symbols or in words. Loosely conceived functions become mathematical objects in earnest when they are subjected to manipulations which are recognizably mathematical, in symbols or verbal circumlocutions. It is this that the Greeks did not achieve.

IV. REAL NUMBERS

The Greeks could not or would not form the product of two general real numbers. Archimedes did not form it. The Renaissance quickly did it, and the consequences are all around us.

As is done even today, the Greeks represented real numbers by the (lengths) of rectilinear segments in the plane, and when Greek mathematics was at its height it had a (vague) notion that there are as many real numbers as there are (incongruent) segments. The addition of real numbers was obvious: segment was followed by segment. But for a product of two segments, the factor segments invoked the image of a rectangle whose sides they were, and the Greeks could not or would not "convert" the rectangle quantitatively into a new segment. This blockage created a stand-off for two thousand years.

On an early page of his *Geometry* (1637), Descartes performed the conversion as follows: by similar triangles, the equation

$$a : b = c : d \tag{4}$$

determines any of the four magnitudes in terms of the other three. But now *standardize* one of them as a unit, say $d = 1$. Then

$$a : b = c : 1$$

performs the multiplication of b,c, thus $a = bc$. Zeuthen, in his important work [*75*], celebrates this calculation as a crucial achievement, and Mahoney [*56*, p. 44], sees in this "bold" move a triumph of the analytic art.

The fact of the matter is that all this had been achieved—standardized unit and all—long before Viète, let alone Fermat and Descartes, in the famed *Algebra* of Raffael Bombelli (1530- after 1572), in a part of it that was published in 1929 by Ettore Bortolotti [*19*], but had been unedited till then. In one of the theorems, by putting $b = 1$ in (4) Bombelli performs the division $a = c/d$, and in another theorem he expressly forms for any real number a, its successive powers $a^2, a^3, a^4, \ldots$, for any exponent wanted.

Any history of mathematics quotes Bombelli, but only Bourbaki [*20*] reports this capital achievement of his. All the others praise him for his anticipation of the fact that for any real numbers a,b the expression $(a + b\sqrt{-1})^{1/3} + (a - b\sqrt{-1})^{1/3}$ is real-valued. This is a striking enough achievement, but in doing this Bombelli was working in the area of many other contemporary mathematicians. In shaping real numbers into a semiring he was ahead of them.

But why was this so important? And what were the consequences? The importance and consequences were that with multiplication of real numbers at his command Isaac Newton could resume where Archimedes had left off nineteen centuries before.

Of course in everyday life the Greeks could and did multiply numbers; carpenters, masons, builders, and engineers could not have done without it. What they could not or would not do was to multiply the numerical values of two physical quantities of heterogeneous physical provenance and thereby obtain the numerical value of the quantity of a third physical species. As I have maintained from the first [7, p. 182], the crucial evidence for this is Archimedes' own formulation of his law of the lever in equilibrium. He expressed it by the proportion $L_1 : L_2 = P_2 : P_1$ where L_1, L_2 are the lengths of the two arms of the lever, and P_1 and P_2 are the magnitudes of the weights suspended from them. But he was apparently unable to express it, by introducing the product $L \cdot P$, in the form

$$L_1 \cdot P_1 = L_2 \cdot P_2, \tag{5}$$

let alone to create the conception of the "rotational momentum" $L \cdot P$ in the process, and to interpret (5) as the sameness of the momentum for the two sides of the lever.

But this is precisely the manner in which Isaac Newton did proceed. He did form a product of heterogeneous factors, namely his quantity of motion

$$p = mv \tag{6}$$

and he did state that in the case of equilibrium it is constant in time. He fully knew that he broke new ground, and he described the product (6), entirely without symbols:

> *The quantity of motion is the measure of the same, arising from the velocity and quantity of matter conjointly.*
>
> The motion of the whole is the sum of the motions of all the parts; and therefore in a

body double in quantity, with equal velocity, the motion is double; with twice the velocity, it is quadruple. [*60*, p. 1]

Note that Newton is on the verge of describing the product (6) as a bilinear functional in (m, v), which is further evidence for my contention that the notion of function was fermenting inside the *Principia*.

Many developments initiated by Newton were quite slow in unfolding (see [*7*, pp. 346-348]). Thus, only the nineteenth century extended the process of multiplication involved in (6) to physical magnitudes in general, creating such objects as

$$[\text{energy}] = [M \cdot L^2 \cdot T^{-2}]$$

and, say,

$$[\text{``action''}] = [M \cdot L^2 \cdot T^{-1}],$$

which is the dimension of Planck's universal constant h. A first essay on this subject, short but systematic, was included in the *Théorie de la chaleur* of J. Fourier, and a much more systematic account was given by Clerk Maxwell and Fleeming Jenkins in 1863 (see [*7*, pp. 211-212]).

In the case of energy, Leibniz introduced a *vis viva* by mv^2, which is twice our kinetic energy, and historians of science are unappeasably puzzled by the fact that Newton himself did not introduce any kind of energy at all. Even one grotesque solution of the would-be puzzle has its audience (see [*7*, pp. 79-80]). The explanation of the "puzzle" is simply that Newton did not need the concept of energy operationally, and not needing it, he did not form it. In traditional physics the concept of energy arises in two ways: It arises in mechanics when variational principles are invoked, but Newton did not resort to such principles. And it is indispensable in physics proper, that is in theories of heat and electrodynamics, but Newton never dealt systematically with physics proper. Besides, quantity of motion, that is momentum, which was Newton's central concept of mechanics, is mathematically more subtle than energy, because it is a vector whereas energy is a scalar. As I have stated previously—and found corroborated in M.J. Crowe [*33*, pp. 127-128]—Newton was the actual creator of the conception of vector for exact science. "Several significant physical entities of the *Principia*, namely velocities, momenta, and forces are, by mathematical structure, vectors, that is elements of vector fields, and vectorial composition and decomposition of these entities constitute our universal scheme of the entire theory" [*7*, p. 192].

But Newton did not introduce tensors, and so, in keeping with the general remark made above, only the nineteenth century was ready to create them.

We have to return to the Greeks for two observations regarding multiplication: First, the Greeks did not arrive at the theorem—and they would not have been able to formulate it—that any natural number is a (unique) product of primes, although they had all the "lemmas" at hand for doing so.

Even the Scientific Revolution itself was not yet ready for it, and the theorem surfaced only in 1801 in the "Disquisitiones Arithmeticae" of C.F. Gauss (see [*7*, p. 216] and [*15*, p. 828]). Second, there is nowhere in Greek mathematics an anticipation of the group operation $a \cdot b = c$; and, for my part, I would say that groups appear only in the work of Euler as groups of motion in 3-space, in his mechanics of rigid bodies. (But see also the book of Wussing [*74*].)

Finally, the difficulty of multiplication was not the only obstacle that kept the Greeks from arriving at real numbers. There was also the fact that an escalation of abstractions would have been involved. Starting from natural numbers, one must introduce the "cone" of positive fractions, extend it to the field of rational numbers by introducing zero and negative numbers, and last extend this to the field of real numbers by completion in the order topology. Now, as I have frequently stated before (see for instance [*7*, pp. 51 ff]) the Greeks hardly ever went beyond mere ideations—that is, one-step abstractions, in and out of mathematics, and a veritable scale of abstractions, as involved in the formation of real numbers, was beyond their reach. And may I add that when in modern mathematics Cardano and Bombelli in the sixteenth century ventured the first sallies into the realm of complex numbers, they did so long before Viète symbolism became operative. Descartes, however, in the seventeenth century, who mastered the symbolism as expertly as any of his contemporaries, was one of those who still rejected complex numbers, coining the pejorative appellation "imaginary" for them.

V. SPACE

In our outlook, nothing separates ancient and modern mathematics more decisively and emphatically than the concept of space. Greek mathematics was a mathematics without space, all of it, but post-medieval mathematics was a mathematics with and by space from the very first. Mathematical space is a hallmark of analysis, and analysis was launched by it. Space was not an appurtenance of Greek mathematics, geometry or other, but all geometry since 1600 has been molded in space and thus imbued with the spirit of analysis, the geometry of Desargues (1593-1662) just as much as the geometry of Descartes (1596-1650), the so-called "synthetic" geometry of the nineteenth century just as much as the avowedly "analytical" one.

The advent of mathematical space was heralded in the fifteenth century by the emergence of linear perspective in painting in the work of Alberti (1404-1472) (which was followed by beginnings of projective and descriptive geometry in the sixteenth century), and by the emergence of a mathematically conceived spatial structure of the universe in the theology of Nicholas of Cusa (1401-1464).

In the middle of the sixteenth century there was a sudden change of course in Italian Renaissance Philosophy. It is hard to account for (see Kristeller [*51*]). The best known representative of the "New Directions" was Giordano Bruno (1540-1600), whose spectacular cosmological speculations became widely known. For our purposes it is more significant that in the work of Patritius (1529-1579) there appears the expression "mathematical space" in a chapter heading *De spatio mathematico,* and also the parallel expression "physical space." I cannot imagine Archimedes announcing a lecture or writing an essay "On mathematical space" or "On space in mathematics."

For us, the paramount achievement involving the rise of space in Renaissance thought is the scientific effort of Johannes Kepler (1571-1630): his momentous Planetary Laws, his New Theory of Vision, and also his purely mathematical analysis of volumes (compare [*13*, *17*, and *16*]).

And now for some elaborations of the above.

The Greeks did have space in physics, cosmology, and perhaps also theology, but not in logical, ontological, or psychological perception, and above all not in mathematics, where a modern reader would expect it first of all. There is no *pre-existent* mathematical background space for the configurations and constructs in the mathematical works of Euclid, Archimedes, or Apollonius of Perga, nor in the astronomical work of Ptolemy. In these works, as soon as a mathematical configuration or construction is envisaged, an ambient space is evoked, but there is no Euclidean space E^2 or E^3 as a mathematical object in its own right, independent of configurations in it. When Ptolemy's *Almagest* designs a path of a celestial body, then the path lies in the astronomical universe of Ptolemy, but as a geometrical object of mathematical design and purpose it does not lie anywhere. In his work on the lever and on hydrostatic equilibrium, Archimedes does not have anything remotely resembling Newton's Space in the *Principia,* "absolute" or other.

We will offer an explanation why there is no theorizing on perspective in the arts in Greek works, extant or known by title. Aristotle, for instance, who wrote something about almost everything, apparently never had a treatise on this subject matter. The explanation is simple. Perspective, whatever it be experientially and esthetically, is mathematically a mapping of E^3 into E^2, and the Greeks, not having created these spaces, could not create such a mapping either. But the fifteenth century obviously did already have such spaces, and linear perspective could and did get under way. I think that my opinion is in accord with, or at least not at variance with, insights of the art historian Erwin Panofsky, who made searching studies about the rise of perspective in the West (see [*37*, chapter XI] and [*63*]). But most art historians (and other scholars) cannot believe and will not concede that the Greeks had not had linear perspective at all. For instance, S.Y. Edgerton's very knowledgeable book [*37*], from which I have quoted, is entitled "The Renaissance *Re*discovery of Linear Perspective," as if it were reasonable to assume that a linear perspective had been known before, that is, in antiquity. But there is no evidence to support such an assumption.

In the early part of the fifteenth century, Cusanus, or Nicholas of Cusa, the theologian-philosopher and gifted mathematician mentioned above, has many intriguing philosophemes on Space and Universe in his famous work "On Learned Ignorance" (*De docta ignorantia*). In his cosmological meditations, Cusanus has something that Ptolemy definitely did not have, namely the first outlines of an abstract mathematical background space in addition to and interpenetrating the concrete physical and cosmological space (see [*17*], and [*47*, pp. 5-27]). In Cusanus, and also in others, the universe itself is sometimes called *machina mundi* (the world machine), an expression which occurs already in the poem "De rerum natura" of the great Roman poet Lucretius (first century B.C.). Cusanus makes the following two statements:

(A) The world is neither finite nor infinite, or, what is virtually the same, it is both finite and infinite.

(B) The *machina mundi* has its center everywhere, and thus nowhere, and its circumference is nowhere.

Statement (A) will be discussed in the next section; statement (B) suggests to me a homogeneous manifold whose center, due to the homogeneity, is everywhere and thus, in a sense, nowhere. And statements (A) and (B) taken together suggest the Bolyai-Lobachevsky geometry in the unit ball $y_1^2 + y_2^2 + y_3^2 < 1$ of the E^3. In the non-Euclidean metric, this unit ball is homogeneous relative to a transitive group of motions, so that there can be no "absolute" center, and it has no circumference because it is open. It is infinite in its own metric, but finite in the metric of the ambient E^3.

These forward directed insights of Cusanus were adumbrations only, but unique ones; and until Kepler almost nothing like them occurred for over a century. It is true that before Kepler there was Giordano Bruno. But Bruno was unmathematical and his speculations have no nuance of mathematical persuasion, whereas Cusanus and Kepler were real mathematicians, while Copernicus was primarily an astronomer.

Copernicus decomposed the motion of the earth into a yearly translation around the sun and a daily spinning around the axis; and in Kepler's subsequent outlook the two motions are in effect represented by two vectorially independent paths in an underlying E^3, so that, in suitable coordinates $\{x,y,z\}$ the location of a point of the body of the Earth at time t is representable by the formulas

$$
\begin{aligned}
x(t) &= r\sin\theta\cos(\varphi + nt) + \xi(t)\,,\\
y(t) &= r\sin\theta\sin(\varphi + nt) + \eta(t)\,,\\
z(t) &= r\cos\theta + \zeta(t)\,,
\end{aligned}
\tag{7}
$$

where r is the distance of the point of the body of the earth from the center of the earth, (θ,φ) are its latitude and longitude, n is a constant depending on the normalization of time, and $\{\xi(t), \eta(t), \zeta(t)\}$ are the coordinates of the center point of the earth at time t.

In Copernicus there is no such underlying mathematical paradigm for the compound motion of the earth. As I have described it in [*16* and *17*], the Copernican *World Machine* was, vaguely, an arrangement of cycles or epicycles in the shape of wheels, rods, and gears, in which the movements of the parts of the machine had to be centrally controlled and hierarchically ordered. Nothing could be "loose" and "free-floating," and no two submovements could be "independent" and "commutative." Also, the celestial body itself was firmly affixed to a spot on the hierarchically last *orbis,* and it is this orbis which moved (as did other wheels), not the body itself. Even Copernicus never decided whether an orbis was a circular hoop or a spherical shell. Kepler very knowingly swept away the hoops and shells, and let the planet traverse a path in E^3, as envisioned today.

The outlook of Copernicus led him to a peculiar construction. Since, in his machine, the daily and yearly motions of the earth were hierarchically ordered, therefore, *prima facie,* under whichever ordering, the axis of daily rotation "ought" to be firmly affixed to the plane of the ecliptic in which the yearly motion takes place. But it is not so affixed, since in E^3 the axis of the earth remains parallel to itself. Copernicus accounted for this by introducing a third motion, a Motion of Declination, in which "the axis of the Earth describes the surface of a cone in a year, moving in an opposite direction to that of the Earth's center, that is from east to west" [*35*, pp. 140-141]. This Third Motion had no appeal. Even before Kepler it was rejected by Christoph Rothmann and Giordano Bruno [*58*, p. 199], although Copernicus was able to make magnificent use of it. He made the period of the Third Motion slightly less than a year, and explained the precession of the equinoxes thereby.

Furthermore, unlike Kepler after him, Copernicus does not, because he cannot, show an awareness of the fact that, due to his heliocentric hypothesis, the orbit of each planet is in a (mathematical) plane of its own, and that these planes are various planes in a common three-dimensional (mathematical) space. It seems that Copernicus expressly introduces only the plane of the ecliptic, and if he also speaks of the *line* of intersection of the planetary orbit with the plane of the ecliptic then he apparently refers only to the line joining the two nodes of the planetary orbit without envisaging a plane of the planetary orbit too.

Kepler's "post-Copernican" work began in earnest with his *New Astronomy*. In it he immediately takes it for granted and expressly posits that any planetary orbit—which he frequently calls an "excenter"—lies in a (mathematical) plane. If he had not taken the plane for granted, how could he have debated for so long what kind of "oval" the orbit of Mars actually is? Or, even more pertinently, how could Kepler have arrived at his tremendous *Law* that the radius vector from the focal center of the orbit to the planet sweeps out equal areas in equal times, unless he took it for granted that the various areas are "flat" areas in one and the same plane. Copernicus did not even begin to conceive such a law because he was still far removed from the thought patterns of space and analysis that the inception of such a law presupposes. The great originality of the Law of Areas is best attested to by the fact that Isaac Newton made his version of it the very first theorem of the *Principia,* although for reasons best known to himself Newton did not choose to quote Kepler in the context.

Kepler's planetary laws are not the only evidence for his awareness of the importance of space in science. Other evidence is a new spatial setting, a revolutionary one, in his optical Theory of Vision, which came into being virtually simultaneously with his *New Astronomy* and which he even called "the optical part of astronomy (*astronomiae pars optica*)." The change of spatial setting was as follows. Before Kepler—with one exception to which we will soon turn—the light rays between the eye and the object seen by the eye were thought to move in a cone which had its vertex in the pupil and the base on the object. Kepler, however, changed and reversed the setting. In the process of seeing there are many cones, narrow ones. Each point of the object seen is the vertex of a cone, and all cones thus arising have their bases on the eyeball. Thus Kepler views the object as an aggregate of points, à la Cantor, and it is obviously a pointset in a pre-existent mathematical space. Before Kepler, the cone had the same "space-less" nature and setting as in Apollonius. But Kepler's cones are "modern" cones, in a "modern" pre-existent space. The great originality of Kepler's innovation is emphasized in Vasco Ronchi [*67* and *68*], and a valuable account is also in Alistair Crombie [*32*].

The one exception to the pre-Kepler conception was the mathematician Francesco Maurolico (1494-1575) whose work showed that he already recognized the reversal of cones [*57*]. But Ronchi feels certain that Maurolico's work was unknown to Kepler when he worked out his optics. As stated at the beginning of this section, in the second half of the sixteenth century there was among philosophers a sudden awakening of interest in the role of mathematics in science whose suddenness is hard to explain. Thus the present-day historian P.O. Kristeller says:

> The thinkers whom we shall discuss in the three remaining chapters were all active during the second half of the sixteenth century, and what separates them from those we

have considered so far is not merely the passing of a few decades, but the emergence of a completely different intellectual atmosphere. [*51*, p. 91]

Apart from observing that the first half of the sixteenth century was the era of the great religious revolution that goes under the name of Reformation, Kristeller makes no real attempt to trace back this intellectual change-over to tangible roots.

Kepler also demonstrated penetrating insights into the "analytical" role of space in purely mathematical contexts; I will adduce two instances. First, with skill and imagination, he computed the volumes of a large number of solid bodies, many of which were of his own devising, by viewing them as infinite series of infinitely thin laminas, or of infinitesimal wedge-shaped segments radiating from the axis, or of other types of vertical or oblique sections (see Boyer [*21*, p. 108, and also pp. 110-118] for similar exploits).

Second, in the context of his optical work, Kepler introduced into the theory of conics a certain "space-related" way of seeing things which may appear commonplace today but was unprecedented in its time, just as many a present cliché was unexpectedly novel when Shakespeare uttered it first; and Shakespeare and Kepler were contemporaries. Kepler introduced foci of conics (which Apollonius did not have) and infinitely distant points in space, and he interconnected various conics in the following way: A circle, in which the foci coincide, can be stretched into an ellipse, and then into a parabola, in which one focus, the "blind" one, is at infinity; and this process can be reversed. Similarly, from a pair of intersecting lines, which is a "degenerate" hyperbola in which the foci and vertices all coincide, one passes through an infinity of hyperbolas to the parabola.

Kepler summarized these findings by saying that the conics are related by "analogy." Seemingly independently, over two centuries later Jean-Victor Poncelet stated a "Law of Continuity" in projective geometry which subsumes Kepler's finding, and Charles Taylor [*72*], a strong advocate of Kepler's originality in the field, demands that the Law of Continuity be credited to him by priority. He maintains even more emphatically that Newton's "essay" on conics, which is incorporated in the *Principia* (see [*60*, pp. 76-108]), and in which Kepler is not mentioned, is a legacy of Kepler's. In this assertion Taylor may be right because the style and diction of the "essay," especially the manner in which it speaks of points at infinity, are all entirely Keplerian.

Taylor [*72*] also assigns to Kepler statements in Desargues that are frequently assumed to be original with Desargues. This, I think, has to be qualified. Desargues's theorem on triangles in perspective (published in Abraham Bosse) has a "message" about space in geometry for which I find no precedent in Kepler. The theorem states that if two triangles in E^3 are in perspective—that is, if the three lines joining corresponding vertices meet at a point, finite or infinite—then the three finite or infinite points at which corresponding sides of the triangle intersect lie in a straight line, finite or infinite. The theorem also holds for two triangles in a plane, but as Hilbert has shown [*44*], the plane must be imbeddable in a three-dimensional space in the following sense: There is a system of Euclid-like axioms for a 3-space for which the theorem holds throughout, but which is such that if one omits those axioms which refer to three-dimensionality then the theorem may fail (Non-Desarguean Geometry). I cannot imagine the Greeks or the medieval schoolmen conceiving a theorem in two dimensions whose proof requires that the 2-space be imbeddable in a 3-space in which a three-dimensional version of the theorem is demonstrable; and I do not think that Kepler had such anticipations either.

The same book of Hilbert has another feature that quite directly highlights the supremacy of analysis over geometry in mathematics today. Hilbert introduces a system of axioms that *implicitly* defines objects like points, lines, and planes, and in order to verify the consistency of the axioms he *explicitly* defines the objects using various kinds of numbers whose existence and consistency is taken as known. But these numbers are in the last instance grounded in analysis, whose consistency is thus taken for granted, so that altogether geometry is built on analysis. Archimedes would have been perplexed. By his entire mathematical upbringing, it is geometry that would have a claim for being assumed consistent, with analysis, whatever that be, to be founded on it.

We now turn to Descartes. His Geometry (*La Géométrie,* 1637) is etched on the E^3 of today and is thus entirely "modern," its only flaws resulting from Descartes's uncertainty about negative numbers. But in his physical science Descartes has a species of space, called *extensio* or *étendue*, in which the structure of E^3 manifests itself so little that commentators are provoked into remarking that it is "still medieval." And yet the sharp separation of the two spaces in Descartes was in keeping with an irreversible development then in progress. Beginning with Kepler, with anticipations in Cusanus and Copernicus, every physical (or cosmological) system had two spaces in its fabric, whether the author of the system realized it or not: an "objective" space of the *Physis* of the system, and a "subjective" mathematical background space; the latter space being "subjective" because all objects of mathematics—however much suggested by the external world—are nevertheless *internally* conceived, *internally* created, and *inwardly* structured (see [7, p. 47]). Usually these two spaces are interpenetrating and intergrown to some extent. But Descartes, not being a "mathematical" physicist at all, and leaning towards neat, crisp separations, kept the (would-be) mathematical (background) space virtually out of direct physical settings, and displayed it mainly as the E^3 of his *Geometry.*

We note in passing that the emergence of a "subjective" mathematical background space in contrast to the "objective" space of physical manifestations is only a special aspect of the emergence of an attitude of "subjectivity" in contrast to one of "objectivity" in a vast realm of intellectuality during the Renaissance which has become a leading permanent distinction between "ancient" and "modern." Descartes's revolution in philosophy was a signal event in this context; and, to my way of thinking, even the body-soul dualism in Descartes is an instance of the permanent opposition between objective and subjective, which has thus come into being. A key reference to this entire subject-matter is *The Individual and the Cosmos in Renaissance Philosophy,* by Ernst Cassirer [*29*], especially chapter IV, "The subject-object problem in the philosophy of the Renaissance." I found further enlightenment in the even more difficult book of Georg Lukács, "The Theory of the Novel," but it would exceed the plan of the present article to unravel its pertinence to our context.

Isaac Newton undoubtedly had our E^3 in his thinking when he started the very first theorem of the *Principia* with his version of the Law of Areas (wording it thus: "the areas which revolving bodies describe by radii drawn to an immovable center of force do lie in the same immovable planes, and are proportional to the time in which they are described"), and when in the converse to the Law, which is the next theorem, the center is even "moving forwards with an uniform rectilinear motion" [*60*, pp. 40-43]. On the other hand, in book III, p. 419, Newton states the hypothesis "that the center of the system of the world is immovable," which suggests a gravitational space of the kind that eventually the Theory of General Relativity would explicate. Yet, as if fusing the two spaces into one, Newton solemnly introduces a single space, which he terms *absolute*, obviously endowing it even with physico-metaphysical features beyond the call of operational necessity. This space failed, although many physicists tried to make sense of it. But the failure did not affect the course of physics. As Jammer put it:

> It is interesting to note how little the actual progress of the science of mechanics was affected by general considerations concerning the nature of absolute space. Among the great French writers on mechanics, Lagrange, Laplace and Poisson, none of them was much interested in the problem of absolute space. They all accepted the idea as a working hypothesis without worrying about its theoretical justification. In reading the introductions to their works, one discovers that they felt that science could very well dispense with general considerations about absolute space. [*45*, pp. 137-138]

Also, in the eighteenth century, Diderot and D'Alembert, in the article "Espace" of their *Encyclopédie*, say with regard to the existence of absolute space: "cette question obscure est inutile à la Géométrie et à la Physique."

But all such negative statements about Newton's absolute space must be tempered by the observation that Newton refers to this space only in the so-called "Scholia" of the *Principia,* and not in the operational body of the

Principia itself, and that the total omission of the Scholia from the work would in no wise affect its operational integrity or its scientific standing.

After Newton, in the eighteenth century there was no major overt development involving space as a key concept of analysis. In the nineteenth century, however, space made world-wide "headlines" through non-Euclidean and Riemannian geometry, and in the twentieth century through the General Theory of Relativity. None of the other key concepts received this kind of publicity even in scientific circles, except that Georg Cantor's Theory of Aggregates affected each of them to an extent.

The *Encyclopedic Almanac 1970* of *The New York Times* (on p. 458, under *Landmarks of Science*, which are listed chronologically) assigns to the years 1825-1826 the "foundation of basic concepts of non-Euclidean geometry" by Lobachevsky and Bolyai, and (on p. 471, under *Men of Science and Mathematics*) it states that Bernhard Riemann "developed non-Euclidean system of geometry representing elliptic space," obviously referring to his renowned paper [*65*], which was composed and orally presented in 1856 and published posthumously in 1867. A comparison of these two listings cannot but suggest to the reader, general or mathematical—or rather confirm him in a belief already vaguely held—that the achievement of Bolyai-Lobachevsky was *the* landmark event—undoubtedly instantly hailed as such—and that the achievement of Riemann, however important, was only a kind of follow-up to it.

But this is not how developments unfolded, in the realm of mathematics at any rate. The work of Bolyai and Lobachevsky lay half-dormant for many years. Mathematicians like Jacobi, Hamilton, Poncelet, Möbius, Plücker, and others in the first half of the nineteenth century showed almost no awareness of it [*18*]. There is absolutely no allusion to it in [*65*], or anywhere else in Riemann. It was first incorporated into a textbook by R. Baltzer in 1867, the same year in which Riemann's memoir was published, and only then did things begin to happen very briskly. In 1868, E. Beltrami linked up the hyperbolic geometry of Bolyai-Lobachevsky with the elliptic geometry of Riemann, by assimilating the style of the hyperbolic geometry to Riemann's. In 1871, Felix Klein made the distinction, which was subtle for that period, between Riemann's elliptic line-element on the sphere, as Riemann himself had introduced it, and the same line element on projective space, which arises from the sphere by identifying pairs of antipodal points. And in 1877 the American astronomer Simon Newcomb, referring back to Riemann, in an article (written in English) in the then very prestigious *Crelle Journal* [*59*], was obviously groping for this projective space in three dimensions as a possible model for our universe, thus anticipating by forty years Albert Einstein's proposal that the universe might be a relativistic compact space of the type of the sphere. It is worth quoting the last paragraph of Newcomb's article:

It may be also remarked that there is nothing within our experience which will justify a denial of the possibility that the space in which we find ourselves may be curved in the manner here supposed. It might be claimed that the distance of the farthest visible star is but a small fraction of the greatest distance D, but nothing more. The subjective impossibility of conceiving of the relation of the most distant points in such a space does not render its existence incredible. In fact our difficulty is not unlike that which must have been felt by the first man to whom the idea of the sphericity of the earth was suggested in conceiving how, by travelling in a constant direction, he could return to the point from which he started without, during his journey, finding any sensible change in the direction of gravity.

But why did Riemann gain the attention of other mathematicians so much more readily than Bolyai and Lobachevsky? Because the latter, although playing a new game in geometry, were nevertheless playing it on the traditional turf of Euclid and Archimedes. They may have revolted against Euclid, but they were fighting the revolt with the traditional Greek weaponry. And in a certain sense they were not even the first to play on Euclid's own field a game different from his. The geometry of perspectivity, of Kepler and Desargues, of Monge, Lazar Carnot, and Poncelet had already been "new" in this sense. Furthermore, a kind of non-Euclidean geometry was involved in Lagrange's mechanics of finite systems of mass points with constraints. As I have put it: The emergence of general coordinates and free parameters in the eighteenth century was intellectually a prelude to the rise of multi-dimensional geometries, Euclidean and non-Euclidean, in the nineteenth century [*7*, p. 201].

But Riemann's turf was a radically new one. It was the most general Hausdorff space as a topological substratum, the like of which Bolyai and Lobachevsky did not have in their vision at all. And on this substratum Riemann built a new kind of geometric structure. He took the theorem of Pythagoras, which is the pinnacle of attainment of book I of Euclid's Elements, and turned it into a presumptive (infinitesimal) building stone of the structure. This turning of a conclusion into an assumption set the pattern for many similar "inversions" in the twentieth century, but none was as profoundly original as Riemann's own.

The aim and achievement of Bolyai and Lobachevsky, who were interested in Euclid's axiom on parallels and hardly anything else, was the erection of axiomatics, and methodologically their work culminated in Hilbert's book [*44*], which, for the working mathematician, is a textbook in applied axiomatics, as it were. But Riemann, during his relatively short span of mathematical activity (he did not start publishing very early and died before reaching forty), showed no interest in axiomatics, mathematical logic, or anything else "foundational," nor in any kind of "abstract" algebra, but only in the promotion of analysis, in the very broad sense of the term in which it is used in the present paper. In this aim he succeeded as had nobody since Pythagoras, even Euler not excepted.

Half a century after its publication, Riemann's memoir led, via C.G. Ricci, to Einstein's General Theory of Relativity, as it came about (with Einstein aficionados maintaining that he would have found the theory anyway, Ricci or no Ricci); and we note that the theory is providing a vindication of the thesis that there is no physical "foreground" space without a mathematical "background" space. Thus the treatise of Hawking and Ellis [*40*] firmly posits a four-dimensional Hausdorff manifold globally, with physical data imposed on it. And not only a final dissolution of the physical universe but also the initial creation of it are nothing but "singularities" in the physical data, with the mathematical substratum itself being "intact."

Albert Einstein himself would probably have hesitated to agree to a "dualistic" distinction between physical and mathematical space in a universe as he envisaged it. He probably would have tried to argue that in any creation of the universe—be it big-bang or peaceful—the instantly evolving physical space ought to co-evolve a mathematical background space too, by a kind of "analytic continuation," as it were. But it would be very difficult to find a specific "model" for the would-be universe in which to rationalize such an "evolution." Even "mythological" cosmogonies knew this. In Plato's dialogue *Timaeus*, Space (*chora*) is a pre-existent datum, a feature of Necessity (*Ananke*), whereas Time (*chronos*), a "moving likeness of Eternity," came into existence along with the Heaven (*Ouranos*) (see [*31*, p. 102]).

H.P. Robertson (1903-1961), a talented cosmologist, tried to popularize a slogan of his, that "Geometry is a part of Physics." The slogan is all right, as long as one realizes that the physics to which geometry is to be subordinated presupposes a mathematical background space on which to be erected.

A clear-cut, though not intended corroboration of our thesis can be found in Alfvén [*2*, p. 68], and I have already adduced it in [*17*, p. 137]. In *Worlds - Antiworlds*, Alfvén is concerned with the creation of a megagalaxy out of previously given ambiplasma, and, guided by a hydrodynamical paradigm that had been set up by James Jeans decades before, he makes the following assumption: "The starting point in our model is an ambiplasma, which fills the huge sphere. Its density is uniform throughout." This huge primordial blob of ambiplasma is obviously inert, but something somehow sets it in motion, and a universe emerges with a cosmological space-setting peculiar to itself. But before such an "individual" cosmological space can be called into being there must at first be, in a mathematical "somewhere," a "neutral" space in which the "sphere"-like blob of ambiplasma is located. This neutral space is apparently a Euclidean background space, because in such a space it is easy to verify mathematically the demand that the ambiplasma shall have a density, and uniformly the same; or it would have to be another kind of "neutral" space in which spheres and point densities can be readily depicted.

This section began with the introduction of linear perspective in painting and will be terminated with one or two observations on the abandonment of perspective in painting about a century or so ago. I have already commented on this situation very briefly in [*7*, p. 250 and *13*, p. 302], and having since then found my outlook corroborated by R.L. Delevoy [*34*] (see also [*37*, chapter XI), I now venture to be a little more explicit. The grand name in the abandonment of perspective—the Riemann of the movement, as it were—is Paul Cézanne. He did not "destroy" linear perspective, as is sometimes asserted, but replaced it by a richer and more variegated one, just as Riemann in creating a general manifold did not "destroy" Euclidean space but replaced it by a rich variety of "space forms," as Felix Klein termed them. As I have gathered from reading Loran [*53*]—whether this was precisely Loran's judgment or not—what Cézanne did, perhaps more by inspired drive than by calculated design, was, in effect, to retain the principle of perspectivity, but to replace the strict linear perspective with only one vanishing point by a much more richly structured perspective in which there are several "regional" vanishing points towards which lines of vision can and do severally converge. Analogously, by Riemann's express prescription, a manifold is a union of several overlapping neighborhoods, each Euclidean. Just as even before Riemann there were non-Euclidean, or less-than-Euclidean structures on a Euclidean substratum, so also before Cézanne there were painters who, without violating linear perspective, were straining to become dissociated from it. A prominent such predecessor of Cézanne, who is singled out as such somewhere in Delevoy [*34*], was Dominique Ingres.

After Cézanne, an actual break with perspective began to manifest itself in Cubism, and in Picasso's *Guernica* not only is there no perspective left, but there is only the flattest lamina of space left at all; this development is in accord with the statement made towards the end of section II, that in the twentieth century, space, like other key concepts of analysis, has been "secularized" and "flattened-out" in various ways. Finally, however superficial the observation may be, the pointillism of Seurat and others, which appeared between Cézanne and Picasso, reminds me of the lesson taught by Georg Cantor (a contemporary of Seurat), that any space, however tightly organized, is an aggregate of points first of all.

In art, as opposed to mathematics, there is still a presumption that linear perspective, however painters may have rebelled against it, is the one and only way for the human eye actually to "see" things in space. The late art historian Erwin Panofsky, in a very erudite early essay, raised doubts whether the optical situation is really so, unchangeably; and even some of his most admiring students and friends have never since forgiven him this "heresy." A precedent from mathematics ought to warn against being too dogmatic in such contexts. G. Helmholtz, also a great student of human vision, once aimed at proving that man's ambient space must be Euclidean if it is sufficiently homogeneous to human vision. Whereupon Friedrich Schur [*69*] pointed out (see also [*36*]) that in *any* space of constant curvature, positive, zero, or negative, at any point, and perpendicular to any direction, there is a totally geodesic hyperplane E^{n-1}; and conversely, that any two such hyperplanes can be transformed into each other by a motion. Since for negative curvature, geodesics emanating from the same point never meet again, it is certainly impossible by *vision* to separate Euclidean space from a hyperbolic one.

VI. INFINITY

In Greek natural philosophy the meaning of infinity was a major problem from first to last, from the earliest pre-Socratics in the sixth century B.C. to the last "Commentators" in the fifth century A.D. Anaximander of Miletus, a younger contemporary of Thales, had an intellectual brush with infinity, not much more, and this was enough to immortalize him in the annals of philosophy, any philosophy of the West.

But Greek professional mathematics, although vaguely familiar with the concept of infinity, shied from any direct face-to-face encounter with it, even in contexts which, from our retrospect, would cry out for at least a mention of it, if not for an outright involvement with it. For instance, Euclid has the theorem (9,20) that *"prime numbers are more than any assigned multitude of prime numbers."* Wording and proof of the theorem are flawless; the proof is ingenious and the same as today's. But what is missing, in retrospect at least, is the paraphrase of this assertion: that *the number of prime numbers is infinite*. No such straight paraphrase appears to be found anywhere in extant Greek writings, although the content of the paraphrase must have been in the forefront of the mind of Eratosthenes when he constructed his famous "sieve." Thus Greek mathematics shied away from the infinitely *large*; and in the case of the "infinitely *small*," the mighty Archimedes apparently never "saw" that the position of the tangent to a curve at a given point is the "limit" of positions of chords through the given point and a movable second point, as the movable point converges to the given one. Yet Archimedes was not at all far from such an insight; while Euclid defines a tangent as a supporting line to the curve in its entire extent, Archimedes in his book *On Spirals* demands the property only locally, as a matter of course.

But this limitation changed visibly with the earliest advent of the Renaissance; infinity in mathematics and in mathematically oriented science began to manifest itself directly and openly, and in various aspects.

The clearest and most unmistakably "modern" manifestation of infinity occurred when John Wallis in 1656 introduced the perduring symbol "∞" for an infinitely large number and started operating with it, faultily and perhaps even recklessly, but determinedly, as with any other symbol or number. This was far from a front-ranking achievement of the century. But it was modernism vibrating. Archimedes knew—in his own thought patterns as clearly and sharply as we in ours—that $1 + \frac{1}{n}$ tends to 1 as n tends to infinity; but he did not have the elation of being able to express this by

$$\lim_{n = \infty} (1 + \frac{1}{n}) = 1$$

or

$$\lim_{n \to \infty} (1 + \frac{1}{n}) = 1 ,$$

and he could articulate the statement and its proof only by a mass of words which would weary even his most ardent admirer of today.

Not all the limitations of Greek mathematics in comprehending the infinitely small were limitations of Greek intellectuality in its entirety. Thus, Aristotle made a strenuous effort in the second half of his *Physics* to shed light on the nature of the linear continuum of real numbers, especially by examining the precise meaning of concepts like together (*háma*), apart (*chóris*), in contact (*haptómenon*), between (*metaxý*), in succession (*ephexés*), contiguous (*echómenon*), continuous (*synechés*); but Greek professional mathematics became in no wise involved in the effort, as if it were totally unaware of it. Mathematics could not or would not recognize that the problem involved, although "masquerading" as physics or natural philosophy, was in fact a profoundly mathematical one. There is no evidence that the Middle Ages recognized this either; for instance, there certainly is nothing in this direction in the extensive commentary on the *Physics* by Thomas Aquinas (1225-1274). The first to recognize the mathematical nature of the problem was Gregory St. Vincent. In a large mathematical work which appeared in 1647 (but was apparently done twenty years before), he put Zeno's puzzle, Achilles and the Tortoise, into a mathematical setting, and thus resolved the puzzle mathematically (see [*55*, pp. 79-80]).

The puzzle maintains, against all evidence, that in a race between the quick-footed Achilles and the slow-moving tortoise, if the tortoise has any head start at all, then Achilles cannot overtake him, ever. By the time Achilles has reached the starting point of the tortoise, the latter has moved on by a certain distance. When Achilles covers that distance, the tortoise has gained a further distance, etc. Thus the tortoise always remains ahead. Aristotle maintains that the puzzle is wrong. He knows that the refutation of the puzzle turns on using the structural properties of the linear continuum and of uniform motion, and he talks endlessly about it all. But it is almost impossible to say whether and how the actual refutation of the puzzle by Aristotle comes about. Gregory St. Vincent, however, sets up the obvious geometric series involved in bringing Achilles and the Tortoise abreast of each other and verifies that the series has a finite sum, which he calls the limit of the series (see [*46*, p. 437]). Having done this he takes it for granted that the puzzle is refuted thereby. His contemporary readers applauded his solution, totally unconcerned about the "foundational" question of whether there really is a point on the line of pursuit which the sum of the geometric series represents. The seventeenth and even eighteenth centuries were not worried, not yet, by such questions of rigor. It is worth observing that only in 1760 did Father Gerdil "simplify" Gregory's argument by noticing that if the starting distance is one league and Achilles runs ten times faster than the Tortoise, and x is the distance covered by the Tortoise before they meet, then x is a solution of the equation $10x = 1 + x$.

The slowness with which leading analytic conceptions crystallized even in the fast-moving seventeenth century can be gauged by the following comparison. Gregory, in his *Opus geometricum* of 1647, says that the

> terminus of a progression is the end of the series to which the progression does not attain, even if continued to infinity, but to which it can approach more closely than by any given interval [*46*, p. 437],

which is a rather clear definition of the sum of an infinite series. But the concepts of the limit of a function at a point, and of the derivative of a function at a point, are deeply embedded in the fabric of Newton's *Principia* of 1686 and are defined there by the following statements, which are hardly an advance on Gregory's formulation of forty years before.

> Quantities, and the ratios of quantities, which in any finite time converge continually to equality, and before the end of that time approach nearer to each other than by any given difference, become ultimately equal.
>
> For those ultimate ratios with which quantities vanish are not truly the ratios of ultimate quantities, but limits towards which the ratios of quantities, decreasing without a limit, do always converge; and to which they approach nearer than by any given difference, but never go beyond, nor in effect attain to, until the quantities have diminished *in infinitum*. [*60*, pp. 29 and 39]

Returning from the infinitely small to the infinitely large, with which this section actually began, we note that there is one area of the infinitely large to which the age of the Scientific Revolution did not contribute anything constructive. Galileo observed something that had been sporadically noticed since antiquity: that an infinite set can be in one-to-one correspondence with a proper part of itself—or, rather, he adhered to a vaguely shared consensus that because of such a possibility infinite sets do not "in actuality" exist. Galileo's observation was that two line segments of *unequal* length can be put into a point-by-point correspondence by a perspectivity; and, what is a little subtler, that the set of integers $\{n\}$, $n \geq 1$, can be put into a one-to-one correspondence with its subset $\{n^2\}$ (see [*46*, p. 993]). It is remarkable that until the middle of the nineteenth century such observations were discouraging rather than provocative. A possible explanation for this is that transfinite numbers somehow belong to, or at least are contiguous to "foundations" of mathematics and that the mathematicians of the seventeenth and eighteenth centuries were fully occupied with first amassing "substantive" subject matter, for which to lay "foundations" afterwards.

Also, there were the misgivings of philosophers to overcome. Aristotle's *Physics* has a series of arguments against the existence of the infinitely large, one of them being that if there were an infinitely large body (*soma*), then any finite body when added to it would be "annihilated" by it. Georg Cantor himself interpreted this argument to be a reasoned opposition against the relation $n + a = a + n = a$, which holds if a is any infinite *cardinal* number and n is a finite one. And he adds that even for an infinite *ordinal* number α, one has $n + \alpha = \alpha$, but that $\alpha + n > \alpha$. Cantor apparently considers that to be a sufficient refutation of Aristotle because he adds, as if in support of the refutation, that if a finite ordinal number has the temerity to range itself in front of an infinite one, then it is being absorbed by the latter, but if it has the modesty of placing itself in the rear of the infinite one, it is saved (see [*26*, p. 176], also [*11*, p. 612]). Even gods nod at times.

Aristotle's argument as stated above has some application even today. The "classical" principle from thermodynamics that the total energy of a closed physical system is constant becomes meaningless if the total energy is infinite, and a physicist who strongly believes in the principle might therefore hesitate to advocate a cosmological theory of the universe in which the total energy need not be finite. Aristotle already argues, in thought patterns of his, that an infinite universe would mean an infinite total mass and an infinite rotational momentum (his universe performed a diurnal rotation), both of which he considered too absurd to contemplate (*De caelo,* book I, chapters 5 and 6). The statement that the total energy of the universe, or of any closed physical system, is finite, or that it is infinite, is a mathematical statement of the kind that can be articulated only if the system is presented by a mathematical paradigm in a mathematical setting; because energy, as usually conceived, is a non-negative real-valued additive set function, so that a ring of sets, preferably of Borel subsets of a manifold M, must be given before any statement can be made. And the total energy is finite or infinite depending on whether the least upper bound of the set function is finite or infinite. Thus, barring an unusual structure of the ring of sets on which the energy is defined and finitely-valued, the total energy will also be finite if the manifold M is compact; unless the mass-distribution has singularities S, in which case, however, $M - S$ will usually cease to be compact.

It was this kind of intricacy that caused Willem de Sitter, the eminent early student of general relativity, to declare that *"Infinity is not a physical, but a mathematical concept."* This is a very important *"key phrase"* [*17*, p.

152] about the role of infinitude in physics and cosmology especially with regard to the problem of the size of the universe, to which we will turn next. But before taking up the problem, we ought to quote the context from which the key phrase is taken, because we are not adopting the particulars of de Sitter's rationale:

> How the $g_{\alpha\beta}$ outside our neighborhood are, we do not know, and how they are at infinity, of either space and time, we shall never know, otherwise it would not be infinity. That is what Archimedes meant when he said that the universe could not be infinite. The universe that we know cannot be infinite, because we ourselves are finite. Infinity is not a physical, but a mathematical concept, introduced to make our equations more symmetrical and elegant. [*70*, p. 113]

I have dealt with the problem of the size of the universe extensively in [*8*, *9*, chapter 14, *11*, and *17*], and I will restrict myself now to the basic interactions between this problem and the emergence of analysis in the Renaissance and after. The fact of the matter is that the only persons loudly and flatly to proclaim the infinitude of the universe were *non*-scientists like Giordano Bruno (1548-1600) and, long before him, Hasdai Crescas (1340-1411) (see [*45*, p. 79]), and that scientists proper were very circumspect in asserting the infinity of the universe even if they came close to it.

Thus, as we have already observed in section V, Cusanus held that the universe is both finite and infinite. Copernicus, a century later, was even more restrained. In "On the Revolutions of the Heavenly Spheres" [*30*, book I, chapter 6], he states that the heaven (*Caelum*) is "immense," without ever amplifying what "immense" means. Further on, in chapter 8 of book I, he touches upon the question whether the world (*mundus*) is finite or infinite, but evades making a decision by declaring that this is the kind of question which should be left to *physiologoi* (obviously: philosophers of nature) to argue about. Johannes Kepler, in a stirring polemic against Bruno, maintains with astrophysical arguments that the universe is finite [*49*, pp. 58-87]. Alexander Koyré, a "secret" admirer of Kepler, presents Kepler's case with admirable objectivity and sympathy, and makes him out to be a better philosopher than Bruno. But not being able to control his own hankering after "infinitization," Koyré terminates the eulogy of Kepler with the following "rebuke":

> All that is not new, nor specific to Kepler; it is the traditional teaching of Aristotelian scholasticism. Thus we have to admit that Johannes Kepler, the great and truly revolutionary thinker, was, nevertheless, bound by tradition. In his conception of being, of motion, though not of science, Kepler, in the last Analysis remains an Aristotelian. [*49*, p. 87]

Galileo could not make up his mind whether the universe is finite or infinite [*49*, pp. 95-99]. Descartes, however, with his customary incisiveness, came to a wonderful compromise decision. He called God infinite and only Him, and everything else was either finite or indefinite: thus his "extension," which was his space of physics and cosmology, and which he identified with matter, was to him indefinite, but not infinite; and he would not budge from this position even under threat of intellectual "excommunication" by the British Platonist Henry More, a partisan of Descartes turned adversary [*49*, pp. 101-124]. More maintained in correspondence with Descartes that the world, being God's creation, is as infinite as God himself, and that to declare it to be only indefinitely large is an unpardonable blasphemy. But Descartes would not yield; and sudden death relieved him of further acrimony.

Finally we come to Isaac Newton, who is supposed to have brought the infinitization of the universe to its completion, primarily in the *Principia*, I presume. But I do not find it there. In book III of the *Principia*, entitled "The system of the world," the main characterization of the size of the universe, which occurs a number of times, is the same as in Copernicus, as if it were taken from there, namely that it is "immense." Now, in the General Scholium of the book, on p. 544, the term "infinite" does occur, but unmistakably as an attribute of God, in the sentence: "the Supreme God is a Being eternal, infinite, absolutely perfect." Similarly, when infinity occurs in the Scholium of book I, it is in a *temporal* sense (see [*49*, p. 166]); and, when it occurs in the "essay" on conics which is incorporated in the *Principia* (see [*60*, pp. 76-108])— which we have described in the preceding section—then it refers only to infinite points of mathematical space, and not at all to points in physical or cosmological space. I do not mean to say, and it would be incorrect to assert, that these are the only occurrences of infinitude in the works and correspondence of Newton. In correspondence with Richard Bentley, Newton makes some kind of admission, reluctantly, that the gravitational universe may be infinite [*49*, pp. 178-179]. But I do wish to emphasize that in the *Principia* themselves, which went through three editions in forty years, the matter of the infinity of the universe is handled extremely gingerly.

Having adduced all this evidence against the presumption that the seventeenth century made the universe infinite, we still have to explain how, in spite of all such evidence, the impression does prevail that an irreversible infinitization of some very tangible kind did take place. From our outlook the explanation is very simple. Something did happen, something very momentous indeed. It was the creation and emergence of Euclidean space, E^3 and E^2, as an object in its own mathematical right, and as an indispensable background space for many other contexts; first, as background space for geometry itself, which had never had a ready-made background space before, and then, conjointly with this, as background space for any kind of "perspective" in the arts, for astronomy and cosmology, for mechanics, terrestrial and celestial, and for physics, inorganic and also organic (as in Kepler's theory of vision). And this Euclidean space, which thus intruded itself as background space into everything, is of course infinite, by whatever mathematical criteria of infinitude there are. The theory of relativity added other space forms as candidates for background space, some of which are compact and thus "finite." But the ordinary Euclidean background space was in no wise displaced thereby. Classical mechanics, and most parts of classical physics in their classical Euclidean setting, continue to be the basis of college and university education for purposes of science and technology.

Now, an infinite mathematical background space is of course not the same as an infinite universe. But, beginning with Giordano Bruno, philosophers and philosophizing scientists have been tending to equate the two, sometimes even to the detriment of their philosophical systems. We will conclude this section by adducing three instances of such "detrimental," or at least "unnecessary," identification.

As pointed out above, Newton in book III of the *Principia* appraised the size of the universe as "immense" and not as "infinite." In a letter to Richard Bentley, when pressed hard to come out for infinitude of the total gravitational mass of the universe, Newton mused that this could come about only by a suitable gravitational clustering of the masses, which, however, he could "not think explicable by mere natural causes," but would be forced to ascribe "to the counsel and contrivance of a voluntary Agent" [*49*, p. 185]. Yet his absolute space is a metaphysical (and theological) feature of his universe, and in its true role it is the Euclidean background space which is indispensable to his analysis, although it is burdened with extra-mathematical attributes as unnecessary as they are obscure.

Next, in his *Critique of Pure Reason,* Immanuel Kant introduced *a priori* intuitions of space and time which are fascinating by their originality, whether they be accepted or not. But, quite unnecessarily, he identified this space of intuitive perception with the mathematical background space of Euclid's geometry, thus causing embarrassment to his followers after non-Euclidean geometry came to the fore. As we have seen at the end of the preceding section, spaces of any constant curvature are indeed not separable from Euclidean space in many respects.

Finally, let us adduce an instance from philosophy in the twentieth century. The *Tractatus logico-philosophicus* of the linguo-philosopher Ludwig Wittgenstein—which originally appeared in German in 1918—introduces a "logical space" (*logischer Raum*) and a "world" or "universe" (*Welt*). The logical space is a kind of background space to the world, but it is hardly recognizable as such, being a kind of aggregate or congeries of logical entities like "facts," "atomic facts," "states of affairs," "propositions," etc. Also, in a certain sense the logical space seems to be more substantive than the world, inasmuch as the constituents of the world are only some kind of "pictorial" representation of the constituents of the logical space. Yet, quite unnecessarily, Wittgenstein asserts that the logical space is "infinite";

this is simply a standard philosophers' assertion since Giordano Bruno, and nothing else (for further details see [*13*, pp. 304-305]).

VII. CONTINUITY

As elaborated in previous contexts ([*10*] and the second half of [*15*]), continuity was highly non-mathematical by origin, extending, frequently under one synonym or another, into vast areas of cognition, knowledge, and belief. It had an early affiliation with time and eternity, sharing their mystique and perennial appeal. Its non-mathematical roots were older, deeper, and stronger than those of any other key concept of analysis, and although the mathematization was partially begun already in antiquity, it proceeded rather slowly even through the length of the Scientific Revolution. Only in the course of the nineteenth century was continuity finally fitted to the exigencies of working mathematics, and after this process had come to an end, the consequences became immeasurable; most of topology since the late nineteenth century has been one of the consequences, in a sense.

The Greek word for "continuous" is *synechés*. It is an all-purpose word, which can be used on any level of abstraction, and its basic meaning is "to keep, or hold together" (which is also, as it happens, the basic meaning of the root of the Latin form "con-tinuous"). It occurs already in Homer, once in the *Iliad,* and once in the *Odyssey*. Ulysses relates that after one of his adventures was all over, he and his companions slept three days and three nights "continuously" (*synechés*). It is remarkable that according to the Oxford English Dictionary the first verbal form pertaining to "continuous" in the English language was also associated with time. It is the word *continual* (in time), and it occurs, already before Chaucer, in the phrase "great exercise of body and continual travail of the spirit," in one of the so-called English Prose Treatises of the hermit Richard Rolle of Hampole (1290-1349). The word "continuous" itself gained currency only in the seventeenth century, and the antonyms to it have been not only "discontinuous," "discrete," and suchlike, but also "atomic," "particulate," and "monadic."

A philosophico-mathematical conception of continuity pervades the work of G.W. Leibniz (1646-1716), who was a mathematician, philosopher, and logician all in one. He affirmed the presence, in many contexts, of something he called Law of Continuity *(Loi de Continuité; lex continui).* It runs through his entire metaphysics and science, and also seems to involve mathematical continuous functions, but in rather broad settings and applications. It is not easy to give a fully satisfactory presentation of this Law of Continuity because Leibniz himself did not present it in a systematic study of its own, but reverted to it in various contexts, presenting each time some of its aspects pertinent to the context. But if I extract from his various pronouncements what might be applicable and relevant to mathematics, I am led to crediting him with the following two insights:

A. First, Leibniz somehow conceived, broadly and systematically, the continuity of a *function* $y = f(x)$ in a rather general setting. This was original with him. Whatever continuity had been anticipated by the Greeks referred only to the element of continuity that is involved in the structure of the *linear continuum* (of real numbers). And while Aristotle tried to approach the problem of the linear continuum front-face (see the preceding section), Greek professional mathematics dealt with it only indirectly by creating the Eudoxian theory of proportions (Euclid, book 5), and conjointly with it the Archimedean process of exhaustion for the computation of areas, volumes, and similar objects.

B. Second, Leibniz made a statement or statements that can be interpreted to imply that the solutions of functional equations governing physics and cosmology depend on initial data in a continuous manner. This means that the equations are "stable," in the sense that "when the essential determinations of one being approximate those of another, as a consequence, all the properties of the former should also gradually approximate those of the latter" [*10*, p. 498]. This of course is part of the vision and outlook of the Age of Enlightenment: that we live in an orderly, sensible, rational world governed by reason, perhaps even the best possible world attainable. The nineteenth century cast mathematical doubts on this expectation, and the twentieth century proved it deathly wrong; but of this there is more later on.

The above features of continuity in Leibniz, although not mathematically operational immediately, became so eventually. There is nothing similar in Newton's *Principia* except for aspects of continuity in the description of limits and derivatives cited in the preceding chapter. There is also a metaphysical circumlocution of continuity in Newton's description of his absolute time which runs as follows:

> Absolute, true, and mathematical time, of itself, and from its own nature, flows equably [Latin: *aequabiliter*] without relation to anything eternal, and by another name is called duration. [*60*, p. 6]

This "equably" is a kind of circumlocution for "uniformly" or "continuously," so that we again have a link between time and continuity, as we have had since Homer.

Surprisingly, Immanuel Kant, after linking his *a priori* space to Euclid's geometry, links his *a priori* time to (the obviously Greek) enumerative numbers, which constitute a discrete set, as if he had never outgrown an attachment to antiquarian Greek mathematical conceptions of his early schooling. There have been, though, occasional attempts, sometimes implied ones, to "quantize" time by introducing minimal time lapses in atomic theories of various ages: in Epicurean philosophy, in medieval Islamic philosophy [*10*, p. 494]), and even in twentieth-century quantum physics; but they were not serious enough to engage our (or anybody's) attention.

Going back from the era of Newton to the earliest days of the seventeenth century, we recall from section V that Kepler's procedure of merging various types of conics into one comprehensive family by adding to E^2 points at infinity was broadened by Poncelet, in the first half of the nineteenth century, into something which he also called *Law of Continuity,* a law which by intent was quite different from that of Leibniz.

Going even further back, into the last decades of the sixteenth century, I must emphasize that François Viète and Simon Stevin, especially the latter, were vigorously recommending and propagating the use of decimal fractions for daily and theoretical purposes [*22*, pp. 347 ff]. In the twentieth century nothing brings the "continuity property" of the linear system of real numbers more intuitively to light and life than their representation by (infinite) decimal expansions. But this was not the immediate effect, and only in the second half of the nineteenth century were the real numbers syllogistically secured. That this development was almost unbelievably protracted can be seen from the following: in the very first theorem of his *Elements,* Euclid took it for granted that a circular arc joining a point inside the circle with a point outside the circle meets its circumference somewhere. Virtually the same syllogistic gap was still present in C.F. Gauss, when, in the proofs of the theorem that a polynomial with complex coefficients has a complex root, he took it for granted that a polynomial of odd degree with real coefficients has a real root. But soon afterwards, in 1817, Bolzano published an essay under the self-explanatory title: *Purely analytical proof of the proposition that between any two values which yield an opposite result, there lies at least one root of the equation.* In a sense, Archimedes had already laid bare the syllogistic gap [*41*, chapter V] when, making a virtue of necessity, he advertised a "principle" which he called *neusis,* and which, somewhat anachronistically, can be stated as the following broad proposition: if in a plane there are given two curves and the half-lines emanating from a fixed point, and if two of the half-lines intercept intervals of different length between the two curves, then any interval of intermediate length can be intercepted by a half-line of intermediate position.

Bernhard Riemann was a great elucidator of the role of continuity, and of continuity versus discontinuity, in the structure of functions, especially functions that actually occur in analysis. And these elucidations were near the center of his achievements. Most intriguing of all is his conception of a removable singularity, which he formed for holomorphic functions in one complex variable; actually Riemann speaks not of a "singularity" but of a "discontinuity" (*unstetigkeit*), and he does not call it "removable" but "liftable" (*hebbar*), both of which when taken together suggest something rather gentle. And, while most significant in the case of holomorphic functions of one complex variable, a "removable singularity" can of course also occur in other contexts. In the same memoir, which is Riemann's doctoral dissertation, he also constructs his Riemann surfaces for holomorphic functions, and the heart of the construction is the introduction of certain points of ramification, and then of local parameters to fit the neighborhoods of such points. A holomorphic function on the Riemann surface, when projected down to the base space, may become many-valued in the neighborhood of a point of ramification and thus cease to be "regular" there, so that, in a sense, the Riemann surface also "removes" certain isolated singularities of the function.

Next, Riemann's grand memoir on trigonometric series has as a "preamble" his theory of the Riemann integral on an interval. This theory was something radically new. Before it, "everybody," that is, Cauchy, Dirichlet, and others started with defining the definite integral for a function which is continuous to begin with, but which afterwards is permitted to have certain isolated discontinuities or singularities. Riemann, however, starts out only with the demand that the function be bounded, then sets up a precise condition for the function to have an integral, and envisages the class of functions that are integrable in this sense. There was nothing like it before; that is, nobody before him introduced, with syllogistic precision, a class of functions to suit a pre-conceived purpose. In the creation of the integral there is no mention of continuity at all. Only after the class of functions has been secured does Riemann show by an example that an integrable function may have discontinuities everywhere dense. Riemann was not yet "ready" to formulate Lebesgue's criterion that a bounded function is Riemann integrable if and only if the set of its discontinuities has outer measure zero; but in style and spirit he was not far from it.

Finally, in his seminal work on shock waves [*66*, pp. 156-178], Riemann pioneered in exhibiting for certain partial differential equations solutions with rather violent discontinuities, and he coined the strong term *compression thrust* (*Verdichtungsstoss*) in the context. All Riemann's work was exceptionally future-oriented, but this essay was perhaps more so than any other. The paper appeared in 1860. In 1877 E.B. Christoffel added some elaborations, and in 1887 a French mathematician, P.H. Hugoniot, rediscovered the results, apparently independently, so that sometimes a Riemann-Hugoniot theory is referred to. In 1903 the distinguished analyst Jacques Hadamard, in a book on propagation of waves, incorporated a detailed account of such discontinuous solutions, which represent shock waves. Whatever the original appeal of Hadamard's book may have been, during World War II it was suddenly closely studied by some members of the Manhattan Project for the construction of an atomic bomb.

Whether Riemann, in the nineteenth century, realized it or not, his memoir on the possibility of shock waves was a rebuttal of Leibniz's presumption that the mathematical functional equations which govern the universe depend continuously on initial values each time, and thus also on intermediate values, and therefore remain "stable" indefinitely, as the Age of Enlightenment was intellectually conditioned to envisage the situation. And it was left to the twentieth century to turn the speculative possibility into a physical reality. Riemann gives the impression of having been a gentle person himself. But by *Zeitgeist* he was a contemporary of Karl Marx and other propagators of violence, and being one of the most far-sighted mathematicians ever, he could not but foresee that "discontinuities," even violent ones, would be the need, demand, and fate of the future, for good or ill (see [*14*]).

The mathematical conception of continuity as syllogistically formulated in the course of the nineteenth century has been an indispensable leaven in the growth of mathematics ever since. It is a fact, though, that these syllogistic formulations have not directly affected age-old conceptions of continuity in some descriptive sciences, especially in earth and life sciences, in which presumptions of continuity have been of the very essence from the first. This may help to explain why a new generation of investigators spearheaded by René Thom and C.H. Waddington [*73*] has been introducing radically new mathematical paradigms into these disciplines.

It is of course obvious that continuity is the backbone of any evolutionary theory. But continuity had been present, in a descriptive non-evolutionary manner, long before Darwin, in the conception of a Great Chain of Being, which had been perceived by Aristotle, and was advocated with special ardor by Leibniz. The standard reference for this subject matter is Lovejoy [*54*], whom I have cited before [*10*, pp. 498-500].

Aristotle merely said that "Nature passes little by little from things lifeless to animal life, so that, by continuity, it is impossible to present the exact lines of demarcation, or to determine to which of the two groups intermediate forms belong." But the Great Chain of Being in Leibniz can be briefly described thus: "All the order of natural beings form but a single chain in which the various classes, like so many rings, are so closely linked one to another that it is impossible for the senses or the imagination to determine precisely the point at which one ends and the next begins." Leibniz also equated the adjective "great" in the Great Chain variously with "maximal," "optimal," "perfect," "complete," and "*continuous*," thus subordinating it to his *Law of Continuity*. Leibniz obviously shut his eyes, and perhaps even his mind, to the very simple verity, which was clearly recognized by the redoubtable Dr. Samuel Johnson, that a Great Chain of Being, however "complete," cannot be "continuous" because each individual "link" in the chain is isolated, meaning that it has a neighbor that is hierarchically above it, and another that is hierarchically below it.

This kind of disparity between physical actuality and mathematical possibility began to be much more significant and pronounced after evolutionary outlooks began to demand everybody's attention [*10*, *59*, and *1*]. In 1837 the leading historian of science, William Whewell, in his *History of the Inductive Sciences*, coined the name *uniformitarianism* for hypotheses of continuity peculiar to geology and *catastrophism* for its contrary. For geology, and also biology, and even linguistics, uniformitarianism asserts that there has been a certain continuity of evolution since the formation of the Earth. But no satisfactory definition of this continuity has been agreed upon. The indecision is largely a substantive one, in that the demand of continuity may mean different things purely geologically. But it is also a fact that mathematically uniformitarianism, in almost any of its versions, is a compound of continuity proper, strict constancy, and even cyclicity, and that there is no "special" mathematical conception of continuity by which to bring about some kind of "synthesis" or conflation of the various aspects of uniformitarianism by sheer force of mathematical intervention, as it came about in the seventeenth century when Newton's definitions of momentum and force, based on those of velocity and acceleration, made an end to uncertainties and inconsistencies in the work of Galileo and others.

Even in general philosophy a similar phenomenon can be observed. I have described (in [*10* and *15*]) how the eminent American philosopher Charles Sanders Peirce—whose basic academic training was in mathematics with subsequent durable achievements in algebra of logic—strove to formulate a universal law of continuity, which he termed *Synechism*, for the governance of the universe; but he was not able to arrive at such a universal law, strenuous efforts notwithstanding. Peirce knew quite well that there are ineluctable discontinuities in the world, and he even adumbrated the fundamental discontinuity inherent to the world picture of quantum physics. But he was also a philosopher, very much so, and a philosopher has to have an integrated system of Thought and thoughts, and there can be no such system without a measure of continuity of some kind of another. Peirce was thoroughly familiar with the Dedekind-Cantor continuity of his time, and he probably also knew a good part of the theory of real variables of his time. But apparently none of his labors would assist and guide him in his quest for a principle of continuity on which to erect a comprehensive philosophical structure of his own.

VIII. POST-ANALYSIS NUMBER THEORY AND ALGEBRA

I have stated repeatedly that in the case of astronomy, geometry, optics, and also mechanics the irruption of background space and of other appurtenances of analysis into their texture took place rather suddenly around 1600, and that the effect was instantaneous and unmistakable.

In the case of number theory, I am not sufficiently familiar with the texture of the subject matter to be able to judge what it was specifically that made the number theory of Fermat and John Wallis different from that of Euclid and Diophantus, and how to attribute the differences, such as they were, to an upsurge of analysis. My cursory reading would suggest that, although there palpably was a change of outlook, spirit, and intellectual motivation, still the achievements of Fermat and Wallis themselves might conceivably have come about in the idiom and with the resources of antiquity (although the formulation of the Fermat conjecture regarding $x^n + y^n = z^n$ for "all" and therefore arbitrary large natural numbers n might have strained the perceptions of a Diophantus to the limit).

But in the eighteenth century a number theory of a genuinely *analytical* affiliation and orientation began to take shape. The Goldbach conjecture and the Waring problem may sound deceptively pre-analytical, but their solutions, complete or partial, have thus far been swathed in analysis.

Furthermore, in the eighteenth century an extremely recondite arithmetical problem has been posed, however indirectly, which can be stated only within analysis itself. One of Euler's finest triumphs is the formula, $k \geq 1$,

$$\Sigma_{n=1}^{\infty} \frac{1}{n^{2k}} = (-1)^{k-1} \frac{B_{2k}}{(2k)!} 2^{2k-1} \pi^{2k}, \tag{8}$$

in which the B_{2k} are rational numbers, so-called Bernoulli numbers, which are given by

$$\frac{t}{e^t - 1} = \Sigma_{k=0}^{\infty} \frac{B_k}{k!} t^k ,$$

see, for instance, [*46*, pp. 449-452]). There is absolutely nothing corresponding to (8) known for the odd-powered sums

$$\Sigma_{n=1}^{\infty} \frac{1}{n^{2k+1}} , \quad k = 1,2, \ldots , \tag{9}$$

and particularly irritating is the "limiting" case $k = 0$. In this case the series (9) does not converge. But, as already found by Euler, the limit

$$\gamma = \lim_{n \to \infty} \left(\frac{1}{1} + \frac{1}{2} + \ldots + \frac{1}{n} - \log n \right)$$

exists, and is a positive fractional number, approximately $\gamma = 0.577218 \ldots$. It is embarrassing to have to say that nothing arithmetical is known about this number, not even whether it is rational or irrational. There is certainly a problem for some future generations to attend to.

Of pivotal consequence to the future of number theory was another achievement of Euler, his creation of the function

$$\Gamma(s) = \int_0^{\infty} e^{-t} t^{s-1} dt \tag{10}$$

as a holomorphic extension of the factorial function $n!$ from the domain of integers to the domain of complex numbers, the extension being

$$(11) \qquad \Gamma(n+1) = n! \quad , \quad \Gamma(s+1) = s\Gamma(s) .$$

Now, this function $\Gamma(s)$ appears, without any number-theoretic rhyme or reason, in the Riemann function equation

$$(12) \qquad \Gamma(\frac{s}{2})\pi^{-\frac{s}{2}} \zeta(s) = \Gamma(\frac{1-s}{2})\pi^{-\frac{1-s}{2}} \zeta(1-s) ,$$

that is

$$(13) \qquad \Delta(s)\,\zeta(s) = \Delta(1-s)\,\zeta(1-s) ,$$

where

$$(14) \qquad \Delta(s) = \Gamma(\frac{s}{2})\pi^{-\frac{s}{2}} ,$$

and the renowned "Zeta function"

$$(15) \qquad \zeta(s) = \Sigma_1^\infty \frac{1}{n^s}$$

admits the Euler product

$$(16) \qquad \zeta(s) = \Pi_p \left(1 - \frac{1}{p^s}\right)^{-1} ,$$

the index p running over all primes.

Presumably, the two relations (12) and (16) between them hold the key to the secret of the Riemann hypothesis; and this places the gamma function, as a participant in (12), into the very midst of the intriguing number-theoretic setting that the two relations create and represent. But from looking at the integral (10) directly, and surveying the properties of the gamma function as emanating from it, one would never suspect it of being involved in such a highly charged context. Except for being outwardly involved in (12), and in a plethora of other functional equations of a similar structure—which were devised by Dirichlet, Dedekind, Erich Hecke, Carl Ludwig Siegel, Max Koecher, and others—the gamma function is really a humdrum function with very "predictable" properties. As G.H. Hardy said casually, he was never held up in his research by not being able to decide a specific question relating to the behavior of the gamma function. And I have never heard it said that in developing the standard theory of the gamma function a piece of syllogism can be modified or illuminated by a number-theoretic argument. Thus Euler's gamma function is an opaque piece of "hard-rock" analysis indispensably lodged among sparkling "jewels" of number theory.

Another gift from analysis to number theory, even grander, was the very mechanism by which the Riemann functional equation and then numerous other functional equations were established, at first indirectly and then ever more directly. The gift was a piece of Fourier analysis, which the twentieth century has named the Poisson summation formula, but which was introduced into number theory in the first half of the nineteenth century by Lejeune Dirichlet, who may have learned it while studying in Paris as a young man from Joseph Fourier himself. Dirichlet's own application of the Poisson summation formula was to the theory of Gaussian Sums (see [*4*, pp. 146-187; also *6*]). An exposition of the theory of Gaussian Sums in Hecke [*43*, pp. 235-248] features a contribution of Cauchy, who applied his "residue formula" to it. As already stated before [*15*, pp. 832-833], the Poisson summation formula and Cauchy's integral and residue formulas are two different aspects of a comprehensive broad-gauged duality formula which lies athwart most of analysis.

The Poisson summation formula does not establish the Riemann-type functional equations themselves, but only "intermediate" theta relations which were the creation of Jacobi; and theta relations lead to functional equations by way of the so-called Mellin Transform. The Poisson summation formula did grow with its application to analytic number theory and is being gradually "identified" with it. Hecke's application of the formula to situations in E^n (in his memoirs, articles 7, 8, 12, and 14 in [*43*]) was n-dimensional Fourier analysis of high order, and the part of it represented in Landau [*52*] is only a foretaste of it (see also the Appendix in [*5*]). After that, extensions of the formula to non-Euclidean settings, in work by André Weil and also by Atle Selberg (Selberg trace formula) have even "algebrified" the formula to an extent. The broadest settings of the formula and of correspondingly broad functional equations have been initiated in the doctoral thesis of J.T. Tate (see [*27*, pp. 305-347]). But when we come down to the original Hecke cases, the gamma function remains the *deus ex machina* as before.

We may conclude with a few remarks on algebra itself. According to Otto Neugebauer, algebra started quite early in the West; what the Babylonians had before the Greeks was mainly algebra rather than geometry. Similarly, after the Middle Ages, algebra resumed with the algebra of polynomials of third or fourth degree long before analysis started properly around 1600. And yet, after this promising beginning, as if deferring its development to that of analysis, the algebra of polynominals stood virtually still for about two-and-a-half centuries, and after preliminaries by Lagrange and Gauss prominently started up again in the nineteenth century with the work of Abel and Galois. Even though to a twentieth-century algebraist the work of Galois is apparently lucid and comprehensible, in the nineteenth century it was reputed to be obscure and recondite. Until after World War I there were very few textbooks in algebra from which the Galois theory could be conveniently learned; but until books like van der Waerden's *Algebra* began to appear, a very accessible text from which an analyst could learn it was a chapter in Picard's *Traité d'Analyse*.

While the syllogistic formalization of analysis was started in the early nineteenth century, the corresponding formalization of algebra was a hundred years behind (only after World War I did a textbook in algebra dare to introduce concepts like fields, rings, ideals, etc., in relatively early chapters). This observation was made to me around 1930 by the leading analyst C. Carathéodory, who also had intended to put it in print in some publication *à l'occasion*, but was prevented from doing so by a fellow editor of the planned publication, a young algebraist of the Emmy Noether school who was outraged by this kind of judgment. Yet Carathéodory was right. For instance, when E. Steinitz in a memoir in 1910 introduced the concept of a characteristic for a general commutative field and showed that it can be 0 or a positive prime, this was considered something of a milestone. But this construction was hardly more innovative than a Weierstrass continuous function which is nowhere differentiable, let alone a Cantor non-constant monotone continuous function which has a zero derivative almost everywhere, both of which were discovered long before.

I think it can be said that during the three centuries 1615-1915 analysis was blazing the trails, with algebra somehow following along. After 1915 algebra was beginning to catch up, and in the second half of the twentieth century it has even happened that certain partial algebraizations of analysis have taken place, at least ostensibly. But this last phenomenon belongs not to the "emergence" of analysis but to its "maturing," and so we need not concern ourselves with it.

REFERENCES

[*1*] ALBRITTON, C.C., JR., ed., *The Fabric of Geology* (Stanford, 1963).

[*2*] ALFVÉN, HANNES, *Worlds-Antiworlds* (San Francisco and London, 1960).

[*3*] *Aristotle's Physics*, translated by Richard Hope (Lincoln, Nebraska, 1961).

[*4*] BACHMANN, PAUL, *Die Analytische Zahlentheorie* (Leipzig, 1921).

[*5*] BOCHNER, SALOMON, "Some properties of modular relations," *Annals of Mathematics* **53** (1951): 332-363.

[*6*] ———, "Remarks on Gaussian sums and Tauberian theorems," *Journal of the Indian Mathematical Society* **15** (1951): 97-104.

[*7*] ———, *The Role of Mathematics in the Rise of Science* (Princeton, 1966).

[*8*] ———, "The size of the Universe in Greek thought," *Scientia* **103** (1968): 511-531.

[*9*] ———, *Eclosion and Synthesis* (New York, 1969).

[*10*] ———, "Continuity and Discontinuity in Nature and Knowledge," *Dictionary of the History of Ideas* (New York, 1973), vol. I, pp. 492-504.

[*11*] ———, "Infinity," ibid., vol. II, pp. 604-617.

[*12*] ———, "Mathematics in Cultural History," ibid., vol. III, pp. 177-185.

[*13*] ———, "Space," ibid., vol. IV, pp. 295-307.

[*14*] ———, "Singularities and Discontinuities," *Complex Analysis 1972, Rice University Studies* **59**, No. 2 (1973): 21-40.

[*15*] ———, "Mathematical Reflections," *American Mathematical Monthly* **81** (1974): 827-852.

[*16*] ———, "Commentary (to an article of Curtis A. Wilson)," in *The Copernican Achievement,* ed. by Robert S. Westman (Berkeley, 1975), pp. 40-48.

[*17*] ———, "Mathematical Background Space in Astronomy and Cosmology," *Vistas in Astronomy* **19** (1976): 133-161.

[*18*] BONOLA, ROBERTO, *Non-Euclidean Geometry*, translated by H.S. Carslaw (Dover, 1955).

[*19*] BORTOLOTTI, ETTORE, *L'Algebra, opera di Rafael Bombelli de Bologna* (Bologna, 1929).

[*20*] BOURBAKI, NICHOLAS, *Eléments d'Histoire des Mathématiques,* 2nd edition (Paris, 1969).

[*21*] BOYER, CARL B., *The History of the Calculus* (New York, 1959).

[*22*] ———, *A History of Mathematics* (New York, 1968).

[*23*] BURKHARDT, H., article in *Encyklopädie der Mathematischen Wissenschaften,* vol. 2.1.2, p. 968.

[*24*] BUTTERFIELD, HERBERT, *The Origins of Modern Science 1300-1800* (New York, 1957).

[*25*] CAJORI, F., "The History of Zeno's arguments on Motion," *American Mathematical Monthly* **22** (1915): 1-6, 39-47, 77-82, 109-115, 143-149, 179-186, 215-220, 253-258, and 352-357.

[*26*] CANTOR, GEORG, *Gesammelte Abhandlungen* (Berlin, 1932).

[*27*] CASSELS, J.W., and A. FRÖHLICH, *Algebraic Number Theory* (Washington, 1967).

[*28*] CASSIRER, ERNST, *The Philosophy of the Enlightenment* (Boston, 1955).

[*29*] ———, *The Individual and the Cosmos in Renaissance Philosophy* (Philadelphia, 1972).

[*30*] COPERNICUS, NICOLAUS, "On the Revolutions of the Heavenly Spheres," *Great Books of the Western World*, R.M. Hutchins, editor in chief, vol. 16 (Chicago, 1952).

[*31*] CORNFORD, F.M., *Plato's Cosmology* (New York, 1957).

[32] CROMBIE, A.C., "Kepler: de modo visionis," *Mélanges Alexandre Koyré* (Paris, 1964), vol. I, pp. 135-172.
[33] CROWE, M.J., *A History of Vector Analysis* (Notre Dame, 1967).
[34] DELEVOY, R.L., *Directions of the 20th Century Arts,* translated by Stuart Gilbert (Geneva, 1965).
[35] DREYER, J.L.E., *A History of Astronomy from Thales to Copernicus* (Dover, 1953).
[36] DUSCHEK, A., and W. MAYER, *Lehrbuch der Differentialgeometrie,* vol. II (Leipzig, 1930).
[37] EDGERTON, S.Y., JR., *The Renaissance Rediscovery of Linear Perspective* (New York 1975).
[38] VON FRITZ, KURT, "Die *ARXAI* in der griechischen Mathematik," *Archiv für Begriffsgeschichte* **1** (1955): 13-103.
[39] HALL, A.R., *The Scientific Revolution 1500-1800* (London, 1954).
[40] HAWKING, S.W., and G.F.R. ELLIS, *The Large-scale Structure of Space-time* (Cambridge, 1973).
[41] HEATH, THOMAS L., *The Works of Archimedes* (Cambridge, 1897).
[42] ———, *The Thirteen Books of Euclid's Elements,* vol. I (Cambridge, 1908).
[43] HECKE, ERICH, *Mathematische Werke* (Göttingen, 1959).
[44] HILBERT, DAVID, *Foundations of Geometry*, translated by E.J. Townsend, 2nd edition (Chicago, 1921).
[45] JAMMER, MAX, *Concepts of Space* (New York, 1960).
[46] KLINE, MORRIS, *Mathematical Thought from Ancient to Modern Times* (New York, 1972).
[47] KOYRÉ, ALEXANDRE, *Essai sur l'idée de Dieu et les preuves de son existence chez Descartes* (Paris, 1922).
[48] ———, *Entretiens sur Descartes* (New York, 1944).
[49] ———, *From the Closed World to the Infinite Universe* (New York, 1957).
[50] ———, *The Astronomical Revolution*, translated by R.E.W. Maddison (Paris, 1973).
[51] KRISTELLER, P.O., *Eight Philosophers of the Italian Renaissance* (Stanford, 1964).
[52] LANDAU, EDMUND, *Einführung in die elementare und analytische theorie der algebraischen Zahlen und Ideale* (Leipzig, 1918).
[53] LORAN, EARLE, *Cézanne's Composition* (Berkeley, 1963).
[54] LOVEJOY, A.O., *The Great Chain of Being* (New York, 1960).
[55] MAHONEY, M.S., "Another Look at Greek Geometrical Analysis," *Archive for History of Exact Sciences* **5** (1968): 318-348.
[56] ———, *The Mathematical Career of Pierre de Fermat (1601-1665)* (Princeton, 1973).

[57] MAUROLICO, FRANCESCO, *The "Photismi de Lumine" of Maurolycus*, translated and edited by Henry Crew (New York, 1940).

[58] MICHEL, P.H., *The Cosmos of Giordano Bruno*, translated by R.E.W. Maddison (Paris, 1973).

[59] NEWCOMB, SIMON, "Elementary theorems relating to the geometry of a space of three dimensions and of uniform primitive curvature in the fourth dimension," *Crelle Journal* **83** (1877): 193-219.

[60] NEWTON, ISAAC, *Sir Isaac Newton's Mathematical Principles of Natural Philosophy* ("Motte-Cajori" translation), edited by Florian Cajori (Berkeley, 1962).

[61] ———, *The Mathematical Papers of Isaac Newton*, edited by D.T. Whiteside, vol. II, 1667-1676 (Cambridge, 1968).

[62] OSTROWSKI, ALEXANDER, *Vorlesungen über Differential- und Integralrechnung* (Basel, 1945).

[63] PANOFSKY, ERWIN, *Renaissance and Renascences in Western Art* (Stockholm, 1960).

[64] PRINGSHEIM, ADOLF, article in *Encyklopädie der Mathematischen Wissenschaften*, vol. 2.1.1, p. 7.

[65] RIEMANN, BERNHARD, "Ueber die Hypothesen, welche der Geometrie zu Grunde liegen," *Gesammelte Mathematische Werke*, edited by Heinrich Weber, 2nd edition (Leipzig, 1892), pp. 272-287.

[66] ———, "Ueber die Fortpflanzung ebener Luftwellen von endlicher Schwingungsweite," ibid., pp. 156-175.

[67] RONCHI, VASCO, *Optics, the Science of Vision*, translated and edited by Edward Rosen (New York, 1957).

[68] ———, *The Nature of Light*, translated and edited by V. Barocas (Cambridge, Mass., 1970).

[69] SCHUR, FRIEDRICH, "Ueber den Zusammenhang der Räume Konstanten Krümmungsmasses mit den projektiven Räumen," *Mathematische Annalen* **27** (1886): 537-567.

[70] DE SITTER, WILLEM, *Kosmos* (Cambridge, Mass., 1932).

[71] STRUIK, D.J., *A Concise History of Mathematics* (New York, 1948).

[72] TAYLOR, CHARLES, The geometry of Kepler and Newton," *Cambridge Philosophical Society,* Transaction **18** (1900): 197-219.

[73] WADDINGTON, C.H., (ed.), *Towards a theoretical biology, 1. Prolegomena* (Edinburgh, 1968).

[74] WUSSING, HANS, *Die Genesis des abstrakten Gruppenbegriffes* (Berlin, 1969).

[75] ZEUTHEN, H.G., "Geschichte der Mathematik im XVI. und XVII. Jahrhundert," *Abhandlungen zur Geschichte der Mathematischen Wissenschaften* **17** (1903).

RICE UNIVERSITY

FOURIER SERIES CAME FIRST

SALOMON BOCHNER

In an illuminating note on Carleson's theorem for Fourier series, Halmos (1978) starts out with the following sentence: "It is a historical misfortune ... that Fourier series were discovered before convergence" (this MONTHLY, volume 85, p. 33).

This sentence gave me a jolt. As I have always known—or believed that I knew—it was Fourier series that in effect "created" all modes of convergence of functions (and I mean of functions and not of mere numbers) and their principal consequences, so that a discovery of convergence before Fourier series would have been evolution running in reverse, much of the way.

After power series in Abel and others before him, the first major result about convergence of a series of functions was Dirichlet's theorem about convergence of Fourier series of functions that are piecewise monotone (1829) [2]. The most perduring key result about summability is Fejér's theorem on $(C,1)$-summability of Fourier series of continuous functions (1904)[6]. In between, Du Bois-Reymond (1876) [5] constructed a continuous function whose Fourier series does not converge everywhere, at a time when "counterexamples" were still at a premium; and C. Jordan (1893) [11] introduced the very important class of functions of bounded variation, apparently in order to express Dirichlet's convergent test more "elegantly" (1894) [12].

In 1907, Lebesgue integration and convergence in L^2-norm were put on the mathematical map by the Riesz–Fischer theorem [7], [16]; it was undoubtedly provoked by Fourier series, even if its application to general orthogonal expansion immediately captured the headlines. Fourier series were again of the essence in the 1927 result of Marcel Riesz, for $p>1$, on L^p-convergence of L^p-Fourier series and the existence of conjugate functions [17]. The result in effect demonstrated that a Banach space can have a projection of norm greater than 1, even into a subspace spanned by a subset of a basis of the whole space.

Stochastic modes of convergence of probability distributions came to life when interpreted as modes of convergence of their Fourier transforms [14], [1]. When Norbert Wiener created his "Wiener process" (1923) [14], he modeled it on the body of Fourier series, which, as it happens, is not even the best way of modeling it. Also, Gelfand's creation of the theory of normed rings (1939) [9], was triggered by an attempt to disengage Wiener's Tauberian theorems (1932) [20] from their special setting amidst Fourier series and Fourier integrals.

Gauss had to work hard to derive his reciprocity law for Gaussian sums (1811) [8]. However, Dirichlet did it smoothly after him with the Poisson summation formula for Fourier series (1835) [3]; and he established a new tier in the edifice of number theory, using as building stones his Dirichlet series, which are almost periodic objects of Fourier analysis [2]. Erich Hecke, Dirichlet's counterpart in the twentieth century, in a string of papers 1917–1920 [10], which created Riemann–Hecke functional equations for various zeta functions in algebraic number fields, bested eminent Fourier analysts of his generation by operating intricately with the Mellin transform on the body of the Poisson summation formula in several variables. The intricacy can be savored by reading the expository book of Landau (1918) [13], which does not even deal with Dirichlet characters, let alone Hecke's general quasi-characters. Decades later the role of the algebraically conceived idèles, which had been introduced for another purpose, was considerably enhanced when they were used by Tate (1950) **[18]** to relieve the intricacy in Hecke's work.

Riemann created his integral, the first of its kind, within his memoir on trigonometric series (1856, posthumously published 1867) **[15]**; and the underlying concept of a general function had been anticipated in Dirichlet's above-mentioned work on Fourier series (1829).

Georg Cantor, while still rather young, noticed that Riemann assumes without proof that if, for real A_n, α_n,

$$\lim_{n\to\infty} A_n \cos(nx+\alpha_n)=0 \qquad (*)$$

for all x in $-\infty<x<\infty$, then $A_n \to 0$. He set out to fill in the gap and he succeeded; and he even proved that the hypothesis (*) can be dispensed with on certain "exceptional" sets. Actually the problem is a natural one for Lebesgue integration. It can then be easily seen that it suffices to assume (*) on some Lebesgue set of positive measure, even for

$$\lim_{n\to\infty} A_n \cos(\lambda_n x+\alpha_n)=0$$

if λ_n are any real exponents going to ∞; compare for instance **[21]**. But not having Lebesgue integration, Cantor had to make do with what he had. And he went about this business in such a clumsy way that he stumbled full-face into his glorious theory of sets. This is the honest truth (see Georg Cantor, *Gesammelte Abhandlungen*, edited by Ernst Zermelo, Berlin, 1932, p. 102, last four lines).

Yes, Fourier series did indeed come first.

References

1. S. Bochner, Harmonic Analysis and the Theory of Probability, Berkeley, 1955.

2. P. G. L. Dirichlet, Sur la convergence des séries trigonométriques qui servent à representer une fonction arbitraire entre des limites données, J. Reine Angew. Math., 4 (1829) 157–169; reproduced in **[4]**, vol. 1, p. 117.

3. ______, Sur l'usage des intégrales définies dans la sommation des series finies ou infinies, J. Reine Angew. Math., 17 (1837) 57–67; **[4]**, vol. 1, 257 (this is an extract from, but easier reading than, an earlier paper (1835), reproduced in **[4]**, vol. 1, p. 238).

4. ______, Werke, ed. by L. Kronecker, 2 volumes, 1889.

5. P. Du Bois-Reymond, Untersuchungen über die Convergenz und Divergenz der Fourierschen Darstellungsformeln, Abhandlungen, Akad. München, 12 (1876) 1–103.

6. L. Fejér, Untersuchungen über Fouriersche Reihen, Math. Ann., 58 (1904) 501–569.

7. E. Fischer, Sur la convergence en moyenne, C. R. Acad. Sci. Paris, 144 (1907) 1022–1024.

8. C. F. Gauss, Summatio quarumdam serierum singularium. Best accessible in: Carl Friedrich Gauss' *Untersuchungen über höhere arithmetik*, German ed. by H. Maser, 1889, p. 463–495.

9. I. Gelfand, Theory of normed rings I, II, III, C. R. (Doklady) Acad. Sci. URSS (N.S.), 23 (1939) 1430–1432, 25 (1939) 570–572, 573–574.

10. E. Hecke, Mathematische Werke, 1939, especially nos. 7, 8, 9, 13, 14.

11. C. Jordan, Cours d'analyse, 2nd ed., vol. 1, 1893, p. 54.

12. ______, vol. 2, 1894, p. 233.

13. E. Landau, Einführung in die elementare und analytische theorie der algebraischen Zahlen und der Ideale, 2nd ed. 1927.

14. M. Loève, Probability theory, several editions.

15. B. Riemann, Mathematische Werke, 1892, 227–271.

16. F. Riesz, Über orthogonale Funktionensysteme, Göttinger Nachrichten, 1907, 116–122.

17. M. Riesz, Sur les fonctions conjuguées, Math. Z., 27 (1927) 218–244.

18. J. T. Tate, Fourier Analysis in Number Fields and Hecke's Zeta-Functions, doctoral thesis 1950, reproduced in J. W. Cassels and A. Fröhlich, Algebraic Number Theory, Academic Press, London; Thompson Book Co., New York, 1967, Chap. 15.

19. Norbert Wiener, Differential Space, J. Mathematical Phys., 2 (1923) 131–174, reproduced in Selected Papers of Norbert Wiener, MIT Press, 1965.

20. ______, Tauberian theorems, Ann. of Math., 33 (1932) 1–100.

21. A. Zymund, Trigonometrical Series, 1st ed. Warsaw, 1935, p. 267–268; 2nd ed., Cambridge, 1959, p. 316.

DEPARTMENT OF MATHEMATICS, RICE UNIVERSITY, HOUSTON, TX 77001.

RICE UNIVERSITY STUDIES

Vol. 65, No. 3 Summer 1979

EINSTEIN BETWEEN CENTURIES

INTRODUCTION 1

PART I. NINETEENTH- AND TWENTIETH-CENTURY BACKGROUND 4

1. Transportation, Communication, Illumination 4
2. The Age of the Earth and of the Universe 7
3. Probing the Galactic Universe 9
4. Space and Structure 11
5. Atoms, Continuity 15
6. Discontinuity versus Continuity 18
7. Everything Has a History 22
8. Pessimists 27
9. The Irrational versus the Rational 29
10. Greatness of the Nineteenth Century 34

Notes to Part I 35

PART II. A LECTURE ABOUT EINSTEIN 38

Notes to Part II 53

INTRODUCTION

This essay has two parts.

Part II reproduces in substance a lecture, "Albert Einstein, how he was and what he did," given on the Rice University campus on March 14, 1979, in commemoration of Einstein's hundredth birthday. It is an analysis of Einstein's achievements in physics, terrestrial and cosmological; of his mode of originality; of his impact both on physics and on philosophy; and, importantly, of his resistance to certain innovations in physics that came after his. The lecture harmonizes certain features of Einstein's intellectual constitution, scientific and general, by viewing him as a personage with intellectual roots in and flowering fully in the ambience of the nineteenth century. It is true that Einstein's high creativity falls into the period 1900-1920. But—as was stipulated in the lecture,. and will be observed throughout the essay—our nineteenth century is deemed not to end in 1900 but to extend till after World War I, so that our twentieth century (proper) begins only then, that is, not much before 1920. This is not at all an *ad hoc* postulate. This manner of separating the two centuries has been burrowing in my thinking for a long time and has surfaced, insinuatingly and demandingly, in the present context. Still, Einstein did enjoy enormous applause and a raptly listening audience in the "true" twentieth century, so that he was indeed situated "between centuries," as is signified in the title of the essay.

Part I creates a setting for Part II, but has an identity of its own. It offers divers observations on developments in the nineteenth and twentieth centuries that widen out Einstein's topics in physics into themes in intellectuality at large. The physical topic of general relativity subsumes itself under a comprehensive theme of "Space and (geometric) Structure"; and the subject matter of quantum theory, "old" and "new," that deeply influenced Einstein's career from first to last, is part of a universal theme of "Comparisons and Contrasts between the Continuous and the Discontinuous (or the Discrete)." The theme of Space and Structure has been, in my view, highly creative since the end of the Middle Ages, and the theme of Discontinuity versus Continuity clearly goes back to classical antiquity. But in the present context we will concentrate on developments since 1800, and my presentations will of themselves justify this restriction. Also, I will round out Part I with some observations that may not visibly bear directly on Part II.

Our leading theme in Part I will be Continuity and Discontinuity, because it gives rise to a significant distinction between the two centuries, the nineteenth and twentieth. Nineteenth-century physics created a major theory on this theme, the statistical theory of matter. But in this theory as everywhere else, the century gave considerable presumptive preference to Continuity over Discontinuity. As will be explained in Part II, however, in the twentieth century, after Einstein had peaked, physics took a sharp turn towards giving recognition to discontinuity, much to Einstein's abiding but futile chagrin. This turn towards discontinuity was a momentous development, gloriously forward-directed, and the twentieth century can take pride in this rare mark of superiority of strength over the nineteenth. It is an aim of Part I to extend this complex of presumptions and insights from the confines of physics into open areas of intellection, showing how the Continuous and the Discontinuous are becoming evermore equipoised. Immanuel Velikovsky's *Worlds in Collision* (1950) endeavored to identify large-scale discontinuities in very recent behavior of our planetary system. Academic professionals were scandalized by this work, but it subsumes under our general theme nonetheless.

We will not start Part I with a blunt assault on my primary theme. Rather I will ease into it after presenting striking instances of nineteenth century originality. That the nineteenth century had an originality *sui generis* readily suggests itself by the following: Carl Becker in a small but unforgettable book, *The Heavenly City of Eighteenth Century Philosophers* (1932), could propose that, on the whole, the intellectual attitudes of that century were medieval rather than modern, Voltaire or no Voltaire.[1] But no such pull-back of the nineteenth century to a previous one is on record, or is likely to be proposed in earnest. At any rate we will find that the nineteenth century was the first to overcome certain impedimenta, technological and scientific, that had persisted since Antiquity, right through the Middle Ages, the Renaissance, the Scientific Revolution, and the Age of Enlightenment.

For mathematics I have anticipated this in a previous essay, "The Emergence of Analysis in the Renaissance and After."[2] We there found that, in the age of the Renaissance, mathematics came under the rejuvenating dominance of suddenly emergent Analysis; that "the three centuries 1500-1800 in the rise of modern mathematics 'genetically' correspond to the three centuries 500-200 B.C. in the rise of Greek mathematics, . . . but that this genetic correspondence breaks down after the terminal dates, totally and irretrievably."[3] After 200 B.C., Greek mathematics began to slide down, slowly but inexorably, towards an extinction in its own phase; whereas in modern mathematics after 1800 phenomena of "firstness" came to the fore analogous to those in fields of science and technology. In the present context I will not give to mathematics a precedence of description, but let it find its turn in due course.

Among the differences between the nineteenth and twentieth centuries there is a "sociological" one, which can be presented immediately now, however elusive and contestable it be. The difference is that the twentieth century has not produced anybody of the truly universal standing and recognition of an Einstein, Freud, Tolstoy, Darwin, or even Karl Marx. To be sure, in the twentieth century there are counterparts to them who have profoundly influenced our world, lives, and fortunes. But their acclaim and recognition come more from professional confrères than, as in the nineteenth century, from the public at large. Heisenberg and Schrödinger, the creators of the new quantum theory, have received a good deal of applause. But how many people know that Heisenberg, at least in his youth, played the piano almost like a stage artist, probably much better than Einstein his fiddle; or that Schrödinger struggled with the problem of the role of physics in biology, something Einstein never was concerned with? Kurt Gödel created in 1931 a law in logic[4] that would do honor to Aristotle (we will have occasion to invoke it towards the end of Part II). But who cares to know whether there is a baby photograph of Gödel around?

In sum, it seems to me that Einstein was the last representative of a species of folk heroes that is presently extinct, or at least invisible. And I do not quite know why this should so be, unless I accept Spengler's pronouncement that, in the West, the twentieth century would be epigonic to the nineteenth.

This one difference between the two centuries is perhaps more striking than profound; but there will be substantive ones too.

NOTES

1. See my article "Mathematics in Cultural History," *Dictionary of the History of Ideas* (New York, 1973), vol. III, p. 179.

2. In "History of Analysis," *Rice University Studies* **64**, nos. 2 & 3 (Spring-Summer 1978): 11-56.

3. Ibid., p. 15.

4. I mean the so-called "incompleteness theorem," that if a formal logical system is expressible within the arithmetical language ordinarily taught in secondary schools, then it includes assertions which can be neither proven nor disproven. See Andrzej Mostowski, *Sentences undecidable in formalized arithmetic; an exposition of the theory of Kurt Gödel* (Amsterdam, 1952).

PART I

NINETEENTH- AND TWENTIETH-CENTURY BACKGROUND

In 1803 the Reverend Samuel Miller, Trustee of Princeton University, published a *Brief Retrospect of the Eighteenth Century* that, even though an intended second volume never appeared, was anything but brief. In the judgment of the critic Gilbert Chinard, Miller's survey found that the eighteenth century was, among other things, "the age of literary honors and of international memberships in learned societies, and the age of literary and scientific intercourse, for while in all preceding ages literary men were in great measure 'insulated,' increased facilities in transportation have enabled them to travel and to communicate freely."[1] Our own retrospect of the nineteenth century will indeed be very brief. It will begin with pointing out how incomparably *more* innovative travel and communication since 1800 have become, and what the consequences of this have been; and this will set the tone for much that will follow.

1. TRANSPORTATION, COMMUNICATION, ILLUMINATION

When in 1812 Napoleon and his *grande armée* set out to conquer Russia, their basic mode of travel was the same as that of Caesar and his legions when setting out to conquer Gaul: on foot, on horseback, and in wagons. By now the horses had shoes, the riders had spurs, and some of the wagons were spring-cushioned coaches. But horse-drawn vehicles they were. And when retreating from General Kutusov, Napoleon traveled the same way as Darius when fleeing from Alexander the Great. Also, Napoleon's overland postal service was not much different from Darius'. For urgent pithy messages he might have conceivably still used beacons of fire as did Aeschylus in his *Agamemnon*. Papal conclaves are still signaling by columns of smoke.

But in the middle of the nineteenth century the American Civil War was suddenly fought by railroad and Morse telegraph; at the end of the (extended) century, in World War I, military planes met in dogfights; and already, years before, there were wireless communications ship-to-shore and ship-to-ship. After that, the twentieth century made transport by plane and communication by satellite a way of life.

In the wake of such strides in transportation, the nineteenth century did not limit itself to the creation of "learned" societies for a "select" few as Samuel Miller boasted for the eighteenth century, but there sprang up, all over, various academic and professional societies, whose membership comprised all those engaged in a certain kind of activity. Societies like these began to hold annual, or even more frequent, meetings to which the *total* membership was invited; and, very significantly, towards the end of the nineteenth century, international congresses for ever-larger numbers of participants began to be organized, sometimes with telling results. For instance, in 1904 Henri Poincaré proclaimed his Relativity Principle not in Paris, where he was domiciled, or even in France, or even in Europe, but in the U.S.A., at the World's Fair in St. Louis, at something called the International Congress for Arts and Sciences.

In the twentieth century, scientific meetings and conferences—local, regional, national, and international—have become a rage, and scientific collaboration, even outright research collaboration, by long-distance telephone is a common occurrence.

Furthermore, the history of the rise of Socialism, Communism, Nihilism, and Anarchy in the nineteenth century is replete with accounts of various travels of partisans, either among themselves or to formal gatherings. These travels became increasingly frequent as the century progressed, and "Internationals" and "Congresses" proliferated. Karl Marx made his true debut in 1864, at a meeting of the so-called First International, whose full name was The International Working Men's Association. It was set up in London in 1864, began as a joint affair of the British and French Trade Unions, and subsequently organized Congresses, Conferences, and Sections in various European cities like London, Geneva, Lausanne, Brussels, and the Hague.[2] Marx and Marxism would hardly have come to influence and power without these propagandistic outlets.

The crowning event of "The Rise of Communism by Travel Facilities" came in 1917, during World War I, when Germany's General Erich Ludendorff outwitted himself monumentally by transporting Lenin from Switzerland to Leningrad, then St. Petersburg, in a sealed railroad car right across Germany, with the avowed aim of overthrowing the czarist regime.

In the twentieth century ease of travel has vulgarized life. Even before 1890, Jane Harrison, author of *Prolegomena to the Study of Greek Religion* (1903), an academic nurse to a generation of lustrous classicists, braved great difficulties when going to Athens for topographic studies of its ancient precinct and for visiting the birthplace of Zeus in a cave of the lonely island of Crete. Nowadays these places are overrun by organized tourist groups. The Tahiti of today is not the Tahiti of Matisse. And an American, fleeing an island off New England for an island off Spain, winds up with the same Coca Cola bottle in hand.

Quick travel and communication call for adequate nighttime illumination, and it was an important modernism when in 1784 Aimé Argand put a glass chimney around the wick of the oil lamp.[3] Nevertheless, in its beginning, the nineteenth century was still very close to age-old candlelight, flickering as ever. Balzac (1799-1850), for instance, wrote still by candlelight, working through many a long night. But the century spent most of its productive years by cheerful kerosene lamps and ever-shining gas burners, and many bulky scholarly works bear the hallmark of this effective innovation.

It is surprising to find, day in and day out, how many of these nineteenth-century "gaslight" works either do not have any adequate successor works in the twentieth century, or, at least, refuse to be superseded by them; and so they have to be reprinted, in ever increasing numbers, for active research and reference activities. There are very few reference books in continuing use that were written before 1800, but astonishingly many from the nineteenth century are still flourishing. Moritz Cantor's voluminous *History of Mathematics* is a reference work in active use, and so is Eduard Zeller's comprehensive *History of Greek Philosophy*. In 1874 Alexander Schmidt, a sickly Prussian schoolmaster in Koenigsberg, who apparently had never set foot in an English-speaking land, compiled a Shakespeare dictionary that is still on American reference shelves. The British "Liddell-Scott," in an up-dating, is *the* dictionary of classical Greek all over the world. Emile Littré compiled not only his renowned *Dictionnaire de la langue française,* but, being a physician by training, he also compiled an indispensible many-volume edition of the Greek Hippocratic corpus, with translation, commentary, and essays. With all that, he also spread the gospel of Auguste Comte's positivism among "machine" politicians, making himself one of them. The grand *Oxford English Dictionary* appeared. And there is always the eleventh edition of the *Encyclopedia Britannica* (1911), challenging anybody to dare turn it to pulp with which to produce a would-be better edition of the work.

In the twentieth century, many-volumed academic works, scientific and humanistic, tend to become "corporate" or "collective," that is, to become collections of contributions by *soi-disant* specialists. In the sciences, physical and social, even research work and its publication are done with increasing frequency by two or more authors jointly. In the nineteenth century people were clannish and socially gregarious, but did "brain work" individually, by themselves. In the twentieth century they are members of Margaret Mead's nuclear family and David Riesner's lonely crowd, but do their brain work jointly.

In consequence of this development, standards for originality of achievement by an individual may become harder to apply. Einstein himself became involved in a minor identity problem resulting from joint authorship. Around 1913, on the road to the theory of general relativity, he and the mathematician Marcel Grossmann had written a few joint papers on the mathematics that was prerequisite for the eventual theory. It is reported that in later years Einstein insisted that the joint work was only a convenience to him, and that he could have managed without it.

Jointly authored research in mathematics, physics, and other sciences began to multiply after the late Victorian era, and it frequently came into being in the coffee houses in which academics congregated. Europe before World War I was a vast peninsula of the Eurasian landmass teeming with coffee houses. Mathematicians, for instance, did a good deal of joint work there, but some others did not. Jean Paul Sartre and Simone de Beauvoir, the longest-surviving fossils of the coffee-house era, have not been known to cooperate literally, either between themselves or with others.

2. THE AGE OF THE EARTH AND OF THE UNIVERSE

A drastic change of outlook between the two ends of the nineteenth century took place with regard to the age of the earth, affecting the questions of the chronology of the earth and of the scientific standing of the Biblical account of creation as given in *Genesis*.

Around A.D. 1800, it was still presumed that the time that had elapsed since the last flood, Noah's Flood—the occurrence of which, in one form or another, was somehow taken for granted—was around six thousand years, as implied in the Biblical account that had been composed around 700 B.C.[4] It was already taken for granted, though, by church fathers and their Jewish counterparts, that creation itself had taken place appreciably earlier, the six days of creation having been six epochs of events and not just six days in our conventional sense. Even so, around 1800 the *total* age of the earth was not thought to be very great. The last generally known chronological enumeration before 1800 had been made in 1778, in Georges de Buffon's *Epochs of Nature*, in which he distinguished several epochs: from the creation of the earth as a hot nebula to its consolidation, hence to its cooling-off, hence to the beginning of organic life, hence to the creation of man, and hence to the present. With all these various epochs to account for, the total age of the earth ran only into one hundred thousand years or so.[5]

But at the other end of the century, in 1906, a calculation was made by which the age of the earth was estimated to be at least a *billion* years old. Radioactive decay was a recent discovery then; and the physicist B. B. Boltwood, an exceedingly quiet American, was apparently the first to suggest, in correspondence with Ernest Rutherford, that the element lead in the earth had arisen by decay from uranium. At least a billion years would have been needed for the total process to come about; Boltwood thus hinted indirectly at a minimum age for the earth.[6]

Between Buffon and Boltwood there were of course estimates running into millions of years "only." The first paragraph of the National Research Council's study, "The Age of the Earth," runs thus:

> At the beginning of the present century the problem of the age of the Earth was envisaged as requiring the reconciliation of three independent estimates, all of the same order of magnitude. These estimates were G. H. Darwin's, of 57 million years, based on the separation of the moon from the earth; Lord Kelvin's, of 20-40 millions, based on the secular cooling of the globe; and Joly's, of 80-90 millions, based on the rate of accumulation of sodium in the world-ocean. To these should be added Helmholtz's estimate of 22 million years, based on the source of the sun's heat and its probable duration.

These intermediate estimates came very late in the nineteenth century, in the last third of it, and apparently only after the publication of Charles Darwin's *Origin of Species*. This work finally liberated the average scientist from the necessity of comparing the scientific account of the age of the universe with the Biblical one. Until then, the geologist somehow had to take a stand with regard to the Biblical account, and it was more admissible to oppose it than to ignore it. The requirement that science be answerable to religion was one of the longest-lasting survivals of antiquity into modern times. The nineteenth century finally succeeded in loosening scientific ties to theology that the Renaissance, and even the mighty Scientific Revolution after it, had not. Since Darwin's *Origin*, anybody who still considered himself scientifically accountable to the Bible has been called a Fundamentalist.

Helmholtz, and also some of the others, were wondering not only how old the earth is, but also how long life on it, which obviously depends on a steady supply of heat, is likely to continue. Buffon had already made an estimate. He greatly underestimated the heat coming from the sun, and took into account mainly the heat that reaches the habitable earth by propagation from the hot center of the earth. This is little enough. In addition to that, Buffon's calculations, which came before J. Fourier's great treatise on the propagation of heat, were inexpert. Even from his own premises, Buffon's estimate for the future of life on earth was shorter than it might have been.

Helmholtz and his contemporaries knew, of course, that the energy reaching us from the sun is the decisive figure in the calculations, and the question before them was how the sun's heat can maintain itself, and for how long. Only two large-scale sources for production of heat suggested themselves to them, chemical combustion and mechanical motion. They examined both, the mechanical motion being that of the sun itself, of meteorites, and of other cosmic matter attracted by the sun. Their hypotheses allowed for millions of years only, both in the past and for the future life of the earth,[7] but paleontological and related estimates called for billions of years instead.

The microphysics of the twentieth century finally brought relief. Close to 1940 it was suggested that the production of energy in the sun is an atomic process converting hydrogen into helium; such a process does allow for billions of years, past and future. A great irritant of the nineteenth century was removed.

Otherwise, the twentieth century has raised Boltwood's estimate of over a billion years for the age of the earth to several billion, which is not much of a change, and big-bang theories of creation have assessed the age of the universe to be ten to fifteen billion years.

The preceding account owes much to a conversation with the Rice geologist John Allan Stewart Adams.

3. PROBING THE GALACTIC UNIVERSE

William Herschel in the eighteenth century felt that all the pursuit of astronomy required was bigger and better telescopes, and still in 1857, G. F. Bond of Harvard opined "that it was simply a matter of finding one or two hundred thousand dollars."[8] Actual developments proved the problem was not so simple.

The nineteenth century created spectroscopy[9] and laid the foundation for radio astronomy. These are powerful searchlights for the exploration of the telescopic depth of the universe that had been bared in Galileo's *Starry Messenger*. The nineteenth century readied both devices, and began to test spectroscopy seriously. The twentieth century has been developing and using both of them to the hilt. They complement each other to such an extent that radio-astronomy is forgiven if, somewhat irreverently, it has taken to calling its "big dish" a radio-telescope.

These devices derive their power from living on Maxwell's vast electromagnetic spectrum that stretches from frequency 0 to frequency ∞—a discovery that was as great a hallmark of the Victorian Age as any. Spectroscopy resides in the visible part of the spectrum, Newton's domain, and radio waves emanate from its nethermost part, the realm of Heinrich Hertz. The uppermost part is also sending messages, unsolicited ones, via cosmic rays, but they have not yet been unscrambled.

Spectroscopy can identify the composition of matter, say its chemical composition on earth, in the solar system, and anywhere beyond. That is, it can say what stars are made of in defiance of Auguste Comte's *obiter dictum*, "the field of positive philosophy lies wholly within the limits of our solar system, the study of the universe being inaccessible in any positive sense."[10] The discoveries came about gradually at first, then the pace quickened. The eighteenth century was in awe of Newton's spectral resolution of sunlight, but did nothing to continue the work. Punctually in 1800, however, William Herschel discovered the infrared; in 1801 J. W. Ritter discovered the ultraviolet; and in 1802 Wollaston was the first to discover in Newton's spectrum certain dark lines in fixed positions, the so-called Fraunhofer lines.[11]

Spectra, and thin lines in the spectra—the lines being sometimes dark and sometimes bright—can be produced by letting any kind of light, from any luminous source, terrestrial or celestial, pass through a crystal. It was soon sensed, however inarticulately at first, that the coloration of the spectra and the lines in them, especially the latter, transmit information about the chemistry of the light sources and of the media that the light traverses before reaching the refractive crystals. This is what spectroscopy is all about.

A first high point was reached around 1860 by Gustav R. Kirchhoff, partly in collaboration with R. W. Bunsen. Gathering up preceding developments, he greatly emphasized the possibility of applications "to the chemistry and structure of the suns and stars. His treatment was more complete than that of any of his predecessors, and made a much greater impression at the time on the scientific world."[12] Acting on this precept, the British astronomer William Huggins decided in 1863/4 that certain nebulae, which until then (that is, by mere telescopic examination) were deemed indistinguishable, were partly clusters of stars and partly luminous gas.[13]

The creation of the chemistry of celestial matter was only half of what spectroscopy contributed to the comprehension and exploration of the universe. Another half has come about in the way I will elaborate next, and it was again Huggins who was in the vanguard of developments.

By the time J. von Fraunhofer examined his lines in 1815, there was a general acceptance of the undulatory theory of light of Thomas Young. It was part of Young's theory that the length of the wave varies with the color of the light, increasing from violet to red, and Fraunhofer gave precise figures for the waves of his lines. In 1841 the Austrian physicist Christian Doppler announced the principle—now generally known by his name—that if a source of light, or of any other wave of the same mathematical structure, is in motion relative to an observer, then the period of the waves as received by the observer will be different from the period of the waves as emitted by the source: if the source moves away from the observer the wave becomes longer, if it moves towards the observer the wave becomes shorter. Also, soon after Doppler's work, "the French physicist Armand Hippolyte Louis Fizeau pointed out that to detect the [Doppler-] effect in light, one should not worry about overall color, but should measure the exact position of the spectral lines and note their shift."[14] Here again Huggins scored a first. In 1868 he was able to detect a small shift in the spectral lines of the bright star Sirius and to show that it is moving away from the sun. His figures for the velocity of relative motion were not right, largely because he could train his telescope on the star only visually and not yet photographically. Eventually he pioneered in attaching a photographic lens too.

What Huggins observed in 1868 in the spectrum of Sirius is nowadays called a "red shift"; and this phenomenon and Huggins's conclusion from it were only small tokens of a big thing to come in the twentieth century. Edwin Powell Hubble and others trained huge telescopes equipped with the latest spectroscopic devices on distant nebulae and galaxies outside our own, and the red shift was observed everywhere and any time. Thus the conclusion imposed itself that our universe as a whole is expanding, and this momentous insight of the twentieth century harks back to something that first happened in the 1850s and the 1860s, in the works of Kirchhoff and Huggins.

Even before Hubble's astronomical discoveries, Albert Einstein had encountered the possibility of an expanding universe from purely theoretical considerations and found it puzzling. Hubble's observational confirmation was welcome, but a philosophical sense of wonderment lingers on.

Radio astronomy is largely a creation of the twentieth century. But an attempt, even if an abortive one, to detect solar waves was made around 1890 by the English physicist Oliver Joseph Lodge (1851-1940), a pioneer in radio communication.[15]

4. SPACE AND STRUCTURE

What follows is less spectacular but more substantive than what has preceded.

"Everybody" knows that in the early nineteenth century an age-wrinkled Euclidean geometry passed scepter and diadem to the young and promising non-Euclidean geometry of Bolyai-Lobachevsky. No revisionist historian can gainsay this, but I venture to qualify it.

The horse-drawn vehicles of Napoleon, which I mentioned in section 1, were not the same as those of Julius Caesar, except that horse-drawn they remained. So also, the Euclidean geometry of Gauss, Monge, and Poncelet in the beginning of the nineteenth century was not the same as that of Euclid, Archimedes, and Apollonius in the third century B.C. But recognizably Euclidean it remained, until Bolyai and Lobachevsky made it recognizably non-Euclidean.

Theirs was only a first start on the road to a new geometry. Nowadays geometry-at-large, especially the geometry in science, is something that happens in space, and space is something for geometry to live in. But it was not so for Bolyai and Lobachevsky. For them, as for Euclid, geometry was something to house axioms (and postulates), and axioms were something on which to erect geometry. Still, in the end, something very avant-garde came of it.

The second start came when, in the middle of the nineteenth century, Bernhard Riemann, ignoring Bolyai-Lobachevsky, introduced a general Hausdorff manifold as an underlying variety; and this being given, he imposed on it a certain geometric structure, named "Riemannian metric (or structure)" after him. It cannot be over-emphasized that the generality of the underlying variety upon which the geometrical structure is imposed is of the essence. It is this that first begat Einstein's General Relativity and Cosmology and still lures the masses to him. In the case of Bolyai-Lobachevsky the underlying variety remained the same, namely a Euclidean expanse, and only the imposed structure was different; Einstein hardly used it.

Like all great art, Riemann's constructs are "representational" and "abstract" at once. They are representational as intuited, but are swaddled in a layer of mathematics that makes them appear to be abstract. And they are indeed so only by appearance. It is this fusion of representation and abstraction that appeals to the "sophisticates" of our day, be they ever so defiantly apologetic about not understanding the mathematics of it at all.

The geometrical axiomatization *à la* Bolyai-Lobachevsky can also be updated and refined, as we will see. But in comparison with Riemann's vertical edifices, it leads to a vast flatness of abstraction, both monotonous and productive, as flatlands can be.

In order to grasp what the nineteenth century did in this matter, and then what the twentieth century made of it, we must go back to the Greeks; go back to them and accuse them of having brought forth the most freakish kind of geometry possible (freakish from our retrospect, of course). When compared to Riemann's geometry, they had nothing "structural," not only no overall Space Structure, but even no underlying so-called Euclidean Space, on which conceivably to erect a structure. As I have stated repeatedly before,[16] nowhere in their thinking did the Greeks have a background space—that is, a Euclidean E^2 or E^3—as a mathematical object in its own right, prior to configurations to place in it. For instance, it is unthinkable to find a record of Archimedes having given a lecture on "Some properties of Space." Ptolemy did not use the concept either. A planetary path of his was of course lying in his cosmos, and a grand conception it was, but as a mathematical construct it was lying in some kind of indistinct metaphysical Nowhere. This continued throughout the Middle Ages. Although there was an adumbration of background space in the work of Nicholas of Cusa in the first half of the fifteenth century, the concept was still lacking in the thought patterns of Copernicus in the first half of the sixteenth century. Concurrently, there was no mathematically oriented theory of gravitation put forward during that long stretch of time.

A break did finally come, at the very end of the Renaissance in the first years of the seventeenth century, and the protagonist was Johannes Kepler. Suddenly and imperiously he introduced mathematical background space into science—any science—and, in effect, into cognition at large, mathematical and other. I have first stated and elaborated this in detail for the case of astronomy and cosmology,[17] and afterwards for virtually all of mathematics.[18] I hope eventually to gather up all Kepler's achievements into one portmanteau, even his explanation of the retinal image of human vision that is second only to his planetary laws, his pride of prides.

Kepler's planetary laws formed half or even two-thirds of Newton's law of gravitation. I would even say nine-tenths of gravitation, if Kepler had derived the laws tolerably well. But he did not. As it was, he only half-guessed them, and Newton after him had to provide a derivation. In the context of the derivation, in the *Principia* as a whole, Newton imposed on Euclidean Space a certain structure, nowadays called *affine* structure, by making the space the carrier, at each point, of a vector field, by which to identify velocities, momenta, and forces.[19] Before Newton, Galileo, for all the innovations of his *Starry Messenger*, made no contribution to the creation of mathematical background space. Therefore, his gravitational findings are only terrestrial, and his pro-Copernicus argument requires the lengthy *Dialogue Concerning the Two Chief World Systems* of his most mature years, whereas Kepler got his justification of Copernicus out of the way in his slim post-graduate *Cosmographic Mystery*.

Descartes, between Kepler and Newton, imposed on Euclidean Space a structure prerequisite for the unfolding of Analysis through symbolism and infinitesimal calculus. He himself contributed more to the establishment of the method than to its unfolding; uncannily he made his *La Géometrie* an integral part of his *Discours de la méthode*. As will be emphasized in Part II, in the context of gravitation, this "method" went into action well over a century after Newton, in the work of S. D. Poisson at the beginning of the nineteenth century. Poisson firmly introduced the gravitational potential that changed gravitation from a distance action into a proximity action. After another century this insight of Poisson's guided Einstein through the Riemannian maze towards the goal of a gravitation of his own.

Aristotle insists that everything in Nature and Knowledge runs in contrasts and contrarieties. In addition to the vertically cresting space category of Riemann, we have, by contrast, the horizontally undulating space of Georg Cantor, of set theory fame, who was led to generalize the traditional concept of space until it extenuated into being a general "point set," that is, a general (Cantorian) aggregate, the elements of which are viewed as "points." This conception penetrated into all of mathematics. But it also had a great attraction for the "axiomatics" that was an end-product of Bolyai-Lobachevsky developments, and for the mathematical logic of the Boolean algebra type, also a creation of the nineteenth century. This novel context of mathematical science is the "flatland" previously alluded to.

It is a vast savannah, and it must be vast because it is the breeding and nurturing ground of all the automation, computerization, containerization, etc., that is engulfing us ever more pervasively, changing ever more rapidly by the week, day, and hour. Air conditioning, Xeroxing, even food packaging —they are all domiciled on the same twentieth-century flatland, with Henry Ford's assembly line being perhaps in the vanguard; and artifacts like refrigerators, freezers, and washing machines also occupy a corner of it.

There were harbingers of these modern innovations in the nineteenth century, such as standardizing of clothes, shoes, etc., to fit various customers, rather than having them custom-made individually. Thus the department store is in a sense only a nineteenth-century forerunner of the twentieth-century supermarket and superdrugstore, which now flourish with it, side by side.

The contrariety just sketched between "vertical" and "horizontal" space in science and technology is not haphazard and inconsequential in the details adduced. It also meaningfully occurs in the representational arts, especially painting; and there I can point to an art book wholly devoted to this subject matter for an implied corroboration of my outlook.[20] Having stated my views before,[21] I will here only hint at the high points, and quote passages from two books not previously cited. The Riemann of modern painting is Paul Cézanne, and our first quotation is as follows:

> it was necessary to abandon (as Cézanne had done) the postulates of Euclidean geometry. This involved the adoption of Riemann's topological schemas, schemas which Cézanne had intuitively lit on in his analysis of the phenomenon of extension in the visible world. The significance of this orientation can best be grasped if we bear in mind the fact that, as against Euclidean geometry based on the notion of distance, Riemann's topology, based on surfaces and elastic curves, tended to show that all fundamental relations can be expressed without recourse to any instrument of measurement. Since matter is extensible and distortable, a circle can become an ellipse or a sphere a convex surface without the loss, in either case, of its initial qualities. Ruling out the notions of rigidity, right angles, straight lines and distance (hence of projective coordinates), this topology deals with forms that can be manipulated and reshaped at will. This explains why the many versions of *Montagne Sainte-Victoire* differ from each other far more than all the replicas and variants of a classical "Adoration of the Magi."[22]

The second quotation is from a book by Kurt Badt:

> Cézanne did not plan space as an empty shell into which he later imported objects, and placed them in a certain relationship to each other, expressed through diminishing size, diminishing clarity and changing color; on the contrary, his essential method of reproducing space was to show solids as convex, as detached and separate from each other or as partially covering each other, in other words by making the plasticity of things manifest everywhere, and by making them overlap.[23]

This sentence of Badt's is reminiscent of the description of a locally Euclidean manifold, which is given by overlapping neighborhoods or by adjoining (convex) polyhedral simplices.

The "flatland" that is contrapuntal to Cézanne's "plateau," together with what is built on the flatland, is the land of Picasso the cubist and much of what followed after him, Picasso's flattest production being his "Guernica." There is even a counterpart to Georg Cantor, creator of the point set; it is not of the same intellectual dimension, but a counterpart: Georges Seurat and his pointillism.

A flattening-out also took place in the geometry stimulated by Bolyai-Lobachevsky. In Euclid, and still very much in Bolyai-Lobachevsky, points, lines, and planes were conceived and defined intuitively, and linked up by two kinds of axioms: geometrical postulates, and general "common notions." In

1900 David Hilbert, in a landmark book, *Foundation of Geometry*, "de-intuitivized" the setting. Points, lines, and planes were not defined descriptively at all, but simply "posited" and linked up by a single kind of axiom from which all relations between them were deductively obtained.[24]

After Hilbert, Ernst Zermelo even flattened out the hierarchical relation between a general aggregate and its elements. Aggregates and their elements became one family of objects, but with a formalized binary relation between them, $a \in b$, read: a is an element of b, subject to suitable assumptions.

Finally, during the last years of the (extended) nineteenth century, the linguo-philosopher Ludwig Wittgenstein, in his *Tractatus* (*logico-philosophicus*), brought together the flattened spaces of (Riemann-) Cantor and (Bolyai-Lobachevsky-) Hilbert in a "logical space" of his own. It is a congeries of logical entities like "facts," "atomic facts," "states of affairs," "propositions," etc., and serves as a (mathematico-) ontological background space to a metaphysical "space" or "universe" of reality. The internal contrasts in this construction are comparable to those in a contemporary museum exhibiting paintings on the flattest of background spaces side-by-side with huge sculptures bulging and thrusting out in all directions.

The Edwardian Era, also called the Belle Epoque, was a matrix of such contrasts, as set down in studies on history of art and culture, and also in Barbara Tuchman's *Proud Tower* and similar works.

5. ATOMS, CONTINUITY

Atoms and Continuity are conceptions from natural philosophy that are thoroughly Greek, pre-Socratic or even older. Also, Atomism was frequently representative of Discontinuity, especially to Aristotle, so that Atom and Continuity were even twin conceptions, of sorts. They were talked about throughout the centuries and millennia, but only the nineteenth century brought them to productive life and even put them at center stage. The Renaissance and the Scientific Revolution had not been equal to this task.

Atom is a Greek word (*átomos*, uncuttable). Atomism was created by Leucippus and Democritus in the fifth century B.C., and it has been agitating Western thought unceasingly since. Yet between classical antiquity and A.D. 1800 atomism was aphoristic; after 1800 it became systematic. Before 1800 it was mostly "talk"; immediately after 1800 it became "fact." Hardly anything that was said between Democritus' time and 1800 was more arresting than what had been said by Democritus himself.

What Democritus said was not just that matter is granulated, which is a trivial observation, or even that matter is not always infinitely subdivisible, which even Aristotle, an opponent of atomism, apparently admitted,[25] but that (i) matter consists of microscopic, that is imperceptible, particles, which (ii) are separated by a void between them, the existence of the void being just as much a part of the assertion as that of the atoms. Furthermore, (iii) the particles are constantly in motion, (iv) all physical events are some kind of "interactions" between the atoms across the void, and (v) different interactions come about by the fact that different atoms have different structural features that dovetail in different combinations.

These postulates, which are also postulates of the basic physical science of today, have been talked about incessantly since Democritus; but only the nineteenth century, starting immediately after 1800, began to implement them by specific statements which can be adduced in a college textbook of today.

The significance of Atomism was recognized by Aristotle even as he steadfastly opposed it (mainly because he "abhorred" its void). He couched his opposition in terms of the highest respect. After Aristotle, the philosophical system of Epicurus, which had an ethical leitmotif and a social orientation, was built around a physical doctrine of atomism that was obviously derived from Democritus, even if, as tradition has it, Epicurus emphatically denied this and claimed the originality of its philosophical creation for himself. After Epicurus, Lucretius saw fit poetically to enshrine the atomism of Epicurus in his *De rerum natura*, thereby making it immortal.[26]

In the Middle Ages, Islamic philosophy at its height permeated its theological doctrine of the Kalam with a far-reaching atomism, mostly though not exclusively of Hellenic origin.[27] The Latin Middle Ages also speculated on versions of atomism whose objects may have been compounds of the "hard" atoms of Democritus and Epicurus with a "softer" kind of "natural minima" perhaps descended from Aristotle. Giordano Bruno, in the second half of the sixteenth century, also propagated a kind of atomism of little scientific value, harking back to medieval conceptions. In the seventeenth and eighteenth centuries the mechanical theory of a finite system of masses suggested an atomistic outlook on the constitution of matter. Philosophically, a systematic elaboration of this outlook was produced in the second half of the eighteenth century by R. G. Boscovitch; it was of no serious consequence to science, but is prized by some philosophers. Also, in the seventeenth century, Father Pierre Gassendi fused a staunch belief in traditional atomism with a kind of Radical Liberalism to such an extent that victorious Communists of the twentieth century canonized Gassendi to Early Sainthood of Marxism. René Descartes, however, proclaimed a doctrine of the constitution of matter that has no relation to atomism at all.

All this discussion of early theories of atomism becomes quaint, antiquarian, and nostalgic when compared to what happened after 1800. First, the classical theory of chemical compounds was put on a satisfactory basis by the introduction of the twin concepts of atom and molecule, with Dalton, Avogadro, and Prout in the lead, a molecule being defined as a compound of atoms of the same or different elements. This entire development culminated in the Mendeleev table of 1868, and it is noteworthy that already in 1815 William Prout surmised that all elements are built up of hydrogen.

Second, the theory of electrolysis, as developed by Galvani, Volta, Davy, Faraday, etc., led to the introduction of ions, that is, electrically charged atoms, called cations if the charge is positive and anions if the charge is negative. After that, towards the end of the century a pure indivisible unit of negative electrical charge, called the electron, began to take shape in scientific thinking, and at the very end of the century, in November 1900, Max Planck presented his "quantum" of energy. What followed is the subject matter of Part II of this book.

That was not all. There was still a further atomistic conception, one that was nearest to Democritus' own, and it is the subject matter of the theory of statistical mechanics, or, alternately, of the Kinetic Theory of Gases, two theories that are closely related even if not quite synonymous. This theory is a powerful mathematical paradigm for the study of gaseous and similar matter, which can be viewed as a large assemblage of atomlike particles that move with uniform velocity in all possible directions, collide with each other, and are reflected from walls of containers. Except for a solitary anticipation in the eighteenth century by Daniel Bernoulli, the theory is entirely a creation of the nineteenth century.

Einstein's "juvenilia" were a rediscovery of the elements of this theory on his own.

We now turn to continuity.

Continuity is one of the oldest categories of knowledge, perhaps as old as rationality itself. It is basic, sprawling, and all-pervading. A form of continuity, under one name or another, occurs in mythology and theology, history and philosophy, science and mathematics, psychology and anthropology. Without continuity there can be no awareness of one's inner self or outer setting, or of the all-connecting sensation of time. The Greeks somehow knew all this very early. They had a standard word for it (*synechês*), which was firmly planted in the language and was dominant over occasional synonyms. The word occurs in Homer, and already there it shows signs of becoming an abstract notion.[28]

Continuity occurs in our present-day ordinary dictionary meaning in the ontological poem by Parmenides and, similarly, in the company of related terms, in Aristotle's search for a quantitative paradigm for space and time in the second half of his *Physics*. But it does not occur *anywhere* in the corpus of Greek professional mathematical writing in an intended technical mathematical meaning like that of today (the "today" beginning with the early nineteenth century).

If it were not for the Eudoxian theory of proportions, which in its beautiful clumsiness is a substitute for certain aspects of continuity, I would pronounce that Greek mathematics was a mathematics without space and without continuity either.

Between antiquity and 1800, Kepler introduced into the theory of conics a Law of Continuity, named by him the Law of Analogy, which was later greatly elaborated by Poncelet in the nineteenth century. Leibniz proposed for all of natural philosophy, not only for mathematics, a Law of Continuity under this express name, which the American philosopher Charles Sanders Peirce in the nineteenth century tried, but failed, to duplicate into a much more ambitious law of *Synechism*. And Newton took recourse to a version of continuity in characterizing absolute time as flowing equably (Latin: *aequabiliter*) without relation to anything external. But when the great mathematicians of the eighteenth century tried to give a specific definition for a continuous mathematical function, they flapped about badly.

In the first half of the nineteenth century Cauchy inaugurated a definition of a continuous function that is still in use in beginners' college texts; and in the second half of the century Dedekind and Cantor finally gave a satisfactory characterization of the linear continuum that Aristotle, to his eternal glory, was passionately searching for. After that, the newly constituted discipline of topology took over. Soon after World War I—that is, at the proper beginning of the twentieth century—it came up with a definition of a continuous function that is "the real thing," however premature it is even today to tax the average freshman with it.

And if our freshman, coming to college full of hopeful expectation, is disenchanted with the technical dryness of the definition of continuity offered to him, then let him consider that this is the only definition of it there is in the wide world of knowledge; or rather that mathematics is the only segment of human knowledge in which a definition of continuity can be rigorously formulated at all. Also, in mathematical contexts continuity is sharply distinguished from neighboring concepts like uniformity, steadiness, constancy, etc., all of which have, in mathematical contexts, definitions of their own. Outside of mathematics, however, in vast reaches of applications, continuity is ambiguously conceived and loosely applied, and mergers, fusions, and crossings with neighboring concepts are unavoidable and perplexing.

To restate my thesis: mathematics is the only academic discipline in which the conception of continuity can be called upon to give a rigorous account of itself, and in which it is being bridled, mastered, and made "rationally" operative.

6. DISCONTINUITY VERSUS CONTINUITY

In virtually all areas of knowledge, Western intellection since 1800 has been active against the background of a multifaceted contrariety between Continuity and Discontinuity, with the nineteenth century tending, in varying degrees, to favor continuity over discontinuity, and the twentieth century tending toward neutrality. For mathematics I have sketched the course of the contest in a previous essay in these *Studies*;[29] for physics the contrariety will be accounted for in the second half of Part II; in the present section we will cast brief looks, really brief ones, into other parts of the landscape of knowledge.

In the social sciences and humanities the preference for continuity amounted to a veritable discovery of continuity. This was closely allied to a discovery of the version of time in the course of which developments evolve continuously and in which "everything there is has a history," as will be substantiated in the next section. But even in science the bias towards continuity was unmistakable.

Since Newton and Huygens there had been a debate over whether light is corpuscular (discrete) or undulatory (continuous). In the very beginning of the nineteenth century, Thomas Young, for whatever objective reason, decreed it to be continuous (only), and so it remained virtually till 1905, the year of a crucial paper by Einstein. Similarly, in electromagnetism, "field" theories tried to smother "particle" (that is, electron) theories out of existence for the major part of the century.

Chemistry, for all its devotion to atoms and molecules, accepted the fiction of thermodynamics, that for many "macroscopic" statements, a continuous distribution of masses is appropriate. The century developed the theory of mechanics of continuous media in various states of aggregation, and in standard mathematical paradigms mechanical matter was not only distributed continuously, but even had a finite density point-by-point. Not only was the true atomistic composition of matter ignored, but Lord Kelvin even tried to envisage single atoms in the form of continuously spread-out hydrodynamical vortices of a kind conceived by Helmholtz. The resulting "vortex-atoms" turned out to be something of an embarrassment, and they have virtually been banished from present-day textbooks. Even historical surveys by professional physicists mention them only reluctantly. But there is no reason to be quite so bashful about it; that is the way the nineteenth century functioned.

Even a physicist of the stature of Helmholtz was not able to be even-handed in the debate between Discrete and Continuous. He pioneered in the discovery, for which he is justly famous, that physiologically the ear is "attuned" to receive only a discrete set of tones, but he showed a proclivity for the presumption that acoustics is a theory of waves, that is a field theory, without bestowing on the tone levels the same theoretical importance.

In the geology and biology of the nineteenth century, the doctrine of (historical) evolution not only asserted that various states of development were following upon each other in a chronological succession, but also that the entire process was, and has been, a continuous one. The Continuity of geology and biology is hard to define; it has many facets and connotations, and it goes under various names. It is frequently called "uniformity," and the affirmation of it is the presumption of "uniformitarianism." The notion of uniformitarianism is perplexingly difficult to define, or to apply, or both. Also, sometimes, but only sometimes, it is meant to be the antithesis to "catastrophism," but the descriptions of uniformitarianism do not necessarily suggest this opposition. It is even difficult to make out, unequivocally, whether the requirement of uniformitarianism is an internal part of the doctrine of organic evolution, that is, inseparable from "organicity,"[30] or something externally superimposed on it. In spite of all such difficulties, of which the nineteenth century was well aware, uniformitarianism won out, and the twentieth century is doggedly holding on to it. As I stated in the introduction, Immanuel Velikovsky, a very knowledgeable outsider, capitalized on these difficulties in his best-seller *Worlds in Collision*. The professionals were enraged; they have not yet forgiven him for bringing the problem so crudely into the open, and apparently do not intend to.

In social science, a deep-seated presumption of continuity, from which the nineteenth century never quite emancipated itself, was the doctrine of Francis Bacon, from the early seventeenth century, that all science advances gradually and unerringly from observation to experimentation to theoretization, and that there are ways in science of deciding between right and wrong in such a manner that an experimenting or observing scientist may collect his facts indubitably "objectively" before forming a theoretical opinion of their meaning. The twentieth century has been reacting quite militantly to this doctrine. Alexandre Koyré, for instance, opposed it almost to a fault.

The presumption of continuity was, and is, closely linked to a presumption of inevitable progress. R. G. Collingwood quotes the following absurdity from an 1880 book, *The Nineteenth Century—A History*, by a certain Robert Mackenzie:

> Human history is a record of progress—a record of accumulating knowledge and increasing wisdom, of continual advancement from a lower to a higher platform of intelligence and well-being. Each generation passes on to the next the treasures which it inherited, beneficially modified by its own experience, enlarged by the fruits of all the victories which itself has gained. The rate of this progress . . . is irregular and even fitful . . . but the stagnation is only apparent. . . . The nineteenth century has witnessed progress rapid beyond all precedent, for it has witnessed the overthrow of the barriers which prevented progress. . . . Despotism thwarts and frustrates the forces by which providence has provided for the progress of man; liberty secures for these forces their natural scope and exercise. . . . The growth of man's well-being, rescued from the mischievous tampering of self-willed princes, is left now to the beneficient regulation of great providential laws.[31]

A subtler form of continuity, which I have termed "continuism,"[32] assumes that any event of today is indeed a reaction to an event that must have taken place yesterday, but that the reaction may also be a negative one. Today's event may concur with yesterday's event and carry it forward, or it may disagree with it, and oppose it with something different. Continuism can be tedious if overdone or practiced lopsidedly. This applies to some extent to the work of the eminent historian of science Pierre Duhem in the waning nineteenth century. He was quite prolific, and even after he died in 1916 at the age of 55, his wife published five more volumes of his *Système du monde*. He was mainly a medievalist participating in the Great Rehabilitation of the Middle Ages that had been vigorously inaugurated by the nineteenth century and that has become a large-scale academic industry in the twentieth century. As I said before, Duhem

> forges what seems to be an unbroken chain of human links, from Thales to Galileo, clear across the entire Middle Ages, without omitting a single decade, or even a single year of them. He does not naively whiten out all the darknesses of the Middle Ages; but to Duhem they only indicate a certain lowering of the level of intellectuality, and not a chasmal rupture in the substance of the flooring.[33]

Duhem's intellectual commitment to continuism barred him from carrying a historical accounting of events in the Middle Ages beyond the axis from Copernicus to Kepler and Galileo. As I have been maintaining in previous essays, the eruption of Analysis in the era of Kepler constituted a large-scale discontinuity and a great divide in the unfolding of mathematics, with consequences in the various areas of knowledge, scientific and other. No continuist narration of actions and reactions can account for it. Even in the circumscribed field of Italian Renaissance philosophy there was a sudden change of direction in the middle of the sixteenth century that cannot be explained.[34]

It is a truism that the Greek historian Herodotus was rather episodic in his presentation of political developments, whereas his successor Thucydides was powerfully continuist (in a good sense) in linking cause and effect in the arena of politics. Also Herodotus was rather expansive in his geographical, social, and related asides, whereas Thucydides was almost single-mindedly intent on unfolding his political narration. In view of this, the outgoing nineteenth century tended condescendingly to reduce Herodotus to the standing of a mere story-teller, with the unconcealed suspicion that most of his stories are only fables anyway; some bold historians might even have liked to read Herodotus out of the guild of historians because of his many asides, and to enthrone Thucydides as the Father of History instead. But in the twentieth century archeology has been triumphantly vindicating the fables of Herodotus; and the linking up of geographic, economic, and social circumstances with political developments has become a new trend.

The nineteenth century exegete of ancient literary texts was prone—even at his best—to "emend," "rearrange," or "reinterpret" the statements of the texts, in order to arrive at more "continuous" and more "natural," and thus more probable, courses of developments than those offered by the texts as they are. Frequently the twentieth century finds good causes for rejecting or at least modifying the well-meant removal of such discontinuities.

For instance, the book *Leviticus* of the Old Testament is a large collection of elaborate priestly ordinances, alien to our sensibility, composed in a highly formalized priestly lingo, which, as everybody agrees, was formed rather late, about the time, say, of Ezekiel, the last of the great prophets. But *Leviticus* also insists that these ordinances were handed down to Moses quite early in the history of Israel, at the Sinaitic revelation. Enlightened scholars of the nineteenth century were inclined to take the reference to Moses very lightly and to conclude that not only the language but also the substance of the so-called Priestly Code was a late construction of post-Exilic times. But in the twentieth century William Foxwell Albright, an eminent American student of the Old Testament, has devoted many studies to the proposition that the substance of the Priestly Code is indeed as ancient as the Bible maintains, going back to early stages of the communal organization of the tribes of Israel.[35]

Finally, I have to dwell briefly on the fact that the nineteenth century was also a century that proposed the most rigid scientific determinism ever claimed. It is the statement of Laplace that any person possessing complete knowledge of the state of the universe at any one given moment would be able to predict the state of the universe for all times to come.[36] There is nothing more intolerant of discontinuities than this approach; and we will find in Part II that the spirit of this intolerance was still operative in Einstein's rejection of the Bohr-Born-Heisenberg world-picture of twentieth century physics.

May I state that there was an equally intolerant *social* determinism and that it was most eloquently expressed in the tenets of Karl Marx and Marxism. However revolutionary Karl Marx's expectations may have been, according to his doctrine the revolution between the classes was nevertheless historically inevitable, and the outcome was firmly predetermined. Dialectical Materialism is a fundamentally historical conception of society and even of nature. "According to this conception," said Engels towards the end of the nineteenth century, "Socialism was no longer an accidental discovery of this or that ingenious brain, but the necessary outcome of the struggle between historically developed classes."[37] This belief in social determinism did not win out without a struggle. An opposite to it, namely social *in*determinism, was embodied in Proudhon's Anarchism, and Marx had to strain all his intellectual resources to combat it.[38]

7. EVERYTHING HAS A HISTORY

The nineteenth century hardly initiated the writing of history of or in any academic field, but it intensified it to an unprecedented degree in every field.

In mathematics, in the latter half of the nineteenth century Moritz Cantor wrote a history that remains fundamental. In 1840, M. Chasles wrote a history of geometry; I. Todhunter wrote first a history of probability (1865), then a history of *Mathematical Theories of Attraction and Figure of Earth from Newton to Laplace* (1873), and then a history of the *Theory of Elasticity from Galileo to the Present* (1893).

Astronomy is, next to mathematics, the oldest "exact" science there is, yet histories of it had been very rare and sparse. As if to make up for this, the nineteenth century produced a spate of them.

At the beginning of the century, J. B. Delambre wrote an indispensable history of astronomy in six volumes (1817-1827). About the same time, P. S. Laplace wrote a somewhat popular *Précis de l'Historie de l'astronomie* (1821), and even before that J. S. Bailly composed an even more popular one in several volumes (1775-1782).

In medicine, in contrast to astronomy, a notable historical work had already been done in 1696 by the Swiss physician Daniel Leclerc. But a true beginning was made only around 1800 with a multivolume work by Kurt Polykarp Sprengel; of this work the twentieth-century historian of medicine Fielding H. Garrison said that it "has been the great source book for facts and footnotes for all subsequent investigators." In 1790 a journal in the history of medicine was begun under the title *Archiv fuer die Geschichte der Arzneykunde.*

A large-scale history of physics was compiled in 1860 by F. Rosenberger. It had been preceded in 1838 by a history of optics by S. Wilde, and even much earlier by histories of electricity (1770) and of light (1771) by Joseph Priestley. In chemistry two fundamental works by H. Kopp and by F. Hoefer were published in the 1840s. In geology, a systematic full-scale history came only in 1899, and it was the *Geschichte der Geologie* by K. A. Zittel. That such a book should appear relatively late is surprising, seeing that geology in the nineteenth century had an intrinsic historical orientation from the outset.

Special mention should be made of two works that, in their separate ways, were most influential. The many-sided Cambridge don William Whewell produced in 1837 a *History of the Inductive Sciences*, which in a sense created the genre of "History of Science" in our present-day sense. The second work appeared in the last decades of the Victorian era, a "critical" history of mechanics by Ernst Mach, which greatly enhanced the role of mechanics within physics as a whole. Above all, this was Einstein's lifelong *vade-mecum* in matters philosophical. His self-indoctrination with the philosophy of Mach may have created the intellectual barrier that kept Einstein from entering the twentieth century of physics during the long second half of his career.

In the history of law, there appeared relatively early in the century two compendious works from Germany, one on German law by Karl Friedrich Eichhorn (*Deutsche Staats- und Rechtsgeschichte*, 4 volumes, 1808-1823), and another on Roman law, by Friedrich Karl von Savigny (*Geschichte des Römischen Rechts im Mittelalter*, 6 volumes, 1815-1831). They were part of the rehabilitation of the Middle Ages mentioned in the preceding section, and of a growing estimation, and sometimes overestimation, of medieval achievements. For many scholars, and even generations of scholars, to vindicate the Middle Ages became, and continues to be, a consummation devoutly to be wished. Many books on aspects of life in the Middle Ages began to sound like by-lines of contemporary American correspondents abroad. Many books on popular themes can be written in this way, the story of Abélard and Héloïse being probably the most popular.

Present-day theories of automation, communication, and information involve abstract situations of linguistics, and such situations began to be shaped in the nineteenth century within its evolutionary outlooks. For instance, the introductory paragraph of a leading work on the *The Discovery of Language* runs thus:

> Until the close of the eighteenth century, European linguistic science had advanced but little beyond the knowledge of linguistics achieved by the Greeks and Romans. It is true that little by little some of the elements had appeared which were necessary for development beyond the limitations and errors of the ancients, but as yet they had no influence. The soil was prepared, but it was not until the nineteenth century that growth was destined to spring from it.[39]

This statement exaggerates in one respect. Another book on linguistics, a recent one, adduces an overwhelming amount of evidence for successful linguistic studies even before the beginning of the nineteenth century.[40] Thus, an impressive achievement was the 1770 publication by J. Sajnovics of a *Proof that the Languages of the Hungarians and the Lapps Are One and the Same*, followed in 1799 by a study of S. Gyarmathi—apparently a much more systematic one—which deals with "the historical kinship of Hungarian and Finnish."[41] Already Leibniz, in the *first* half of the eighteenth century, "was one of the first to posit historical relations between Finnish and Hungarian."[42]

What definitely belong to the nineteenth century are the highly spectacular feats of deciphering hieroglyphs and cuneiform writing. And if the nineteenth century did not succeed in reading Etruscan inscriptions, then it need not be ashamed of itself, because the twentieth century has not much advanced in this task either.

Books on the history of linguistics hardly mention feats that astound the impressionable layman. What they count mostly as the truly innovatory and unprecedented achievement of the nineteenth century is the emergence and systematic pursuit of *comparative* grammar, which officially began in the late eighteenth century with a famous statement by Sir William Jones, a British judge in India. Jones had studied a wide variety of languages, and in 1786 he delivered his famous *Third Anniversary Discourse*, which reads in part:

> The Sanskrit language, whatever be its antiquity, is of a wonderful structure; more perfect than the Greek, more copious than the Latin, and more exquisitely refined than either, yet bearing to both of them a stronger affinity, both in the roots of verbs and in the forms of grammar, than could possibly have been produced by accident; so strong indeed, that no philologer could examine them all three, without believing them to have sprung from some common source, which, perhaps, no longer exists. There is a similar reason, though

not quite so forcible, for supposing that both the Gothic and the Celtic, though blended with a very different idiom, had the same origin with the Sanskrit; and the Old Persian might be added to the same family.[43]

The expression "comparative grammar" was coined in 1808 in Friedrich von Schlegel's *Über die Sprache und Weisheit der Inder* (On the Language and Wisdom of the Hindus), with the following send-off: "Comparative grammar will give us entirely new information on the genealogy of languages, in exactly the same way in which comparative anatomy has thrown light upon natural history."[44] The ensuing findings about genealogical proximities and distances between languages must have been thrilling to contemporaries, even if the twentieth century has become blasé about such statements. But even today, word-compounds like Indo-European and Finno-Ugric speak volumes of history; and even today it can be astonishing to hear that Old Norse and Old Albanian are much more consanguineous than, say, Hebrew and Greek, even if central passages in the Gospels bear tell-tale marks of being translations from Hebrew-Aramaic originals.

In very few cases of historically oriented studies has there been such a difference in knowledge between A.D. 1800 and A.D. 1900 as in the understanding of the rise, spread, and distribution of spoken and written languages, especially the written. The subsequent developments in the twentieth century seem to concentrate on *spoken* language and languages, as against the *written* ones. This accords with the fact that until the nineteenth century the study of linguistics was preponderantly historical. In the twentieth century it is historical and analytical in equal measure.

Next, the pedagogy (this is a formal name for education) of today is laced with psychology. This link was forged systematically in the early decades of the nineteenth century by the all-purpose philosopher Johann Friedrich Herbart, even before psychology was an academic field in its own right, independent of philosophy. Herbart was not only also a psychologist, as most philosophers at the time still were automatically, but he was also a lifelong devotee of pedagogy, which he included in his psychology, thus making pedagogy into a kind of applied psychology. He was successful in this although the psychology current at his time was not sufficiently articulated to be adequate for the purpose.[45] It was Herbart who became the founder of "scientific" pedagogy—albeit by turning it into an applied psychology in a sense—and not the celebrated Pestalozzi, his forerunner, for whom the theoretical basis of education consisted of the maxim that "Love in the classroom conquers all." A recent history of education fondly remembers Herbart thus:

The great advance made by Herbart—and its importance can hardly be exaggerated—is that he based his educational views on a new and fully worked out psychology. He is therefore the forerunner of the whole subsequent school of educational reformers who had held that their first requirement was an independent study of the workings of the human

> mind. . . . For the first time an educational prophet saw the necessity of deriving his theories from a coherent psychology.[46]

In psychology proper, Herbart jumped ahead of the twentieth century and conceived of "thresholds of consciousness," "inhibitions," and "repressions." His inhibitions and repressions referred only to ideas and not yet to fears and desires; even so, his adumbrations were enough to secure for him several mentions in the standard biography of Sigmund Freud.[47] By searching diligently enough, one could probably find that any proposition by Herbart had been conceived before him, by Kant, Leibniz, Locke, Aquinas, Aristotle, and so forth. But this finding would be of no consequence: every one of Freud's major maxims had once been somebody's aphorism long before Freud. Also, Herbart's notions were sufficiently free from metaphysical encumbrances to be noticed by Freud, whereas previous ideas were very much a part of the philosophical framework.

Finally, the everyday vocabulary of social science—though not all of it—is also a creation of the nineteenth century.

Aristotle even pioneered in saying that man is a "political animal." But the Greeks did not understand our concept of "state," only something like "the totality of the citizenry." The Greeks could not have made the pronouncement that "the state will wither away." Our concept of state began with Machiavelli.

The Greeks did not have our notion of a "constitution," only of something in the nature of "the totality of laws." They could not have said that something is "unconstitutional," only that it is unlawful or illegal. Nor did they have our conception of "international law." In the Roman Empire, conceptions like "cosmopolitanism" and "international law" would have referred to the Roman Empire in its total extent, and perhaps also to its spheres of influence, but not to something truly worldwide. Such conceptions emerged only in the seventeenth and eighteenth century.

Next, the nineteenth century created the conception of socialism. Although the term occurs in print as early as 1803, it took several decades before it became knowingly associated with reforms of social order and assertions of human rights.[48] Soon afterwards the term *anarchy* was consciously created by Proudhon, around 1840, and anarchism was apparently from the first intended to be opposed to socialism. Communism, however, not only as a principle of joint property, but even as a way of life, was already fully known to the Greeks. Plato advocated it in his *Republic*, but with a very pronounced authoritarian slant; nothing corresponding to our present-day "leftish" notion of civil rights was involved.

But the slogan of "the exploitation of man by man" was coined in the early nineteenth century; it was created, well before Marx, by Saint-Armand Bazard (1791-1832).[49] Marx institutionalized the term and concept of Alienation; but it had been used, earlier in the nineteenth century, by Hegel.

"*Sociologie*" was coined in 1839 by Comte. It was apparently intended to replace in Comte's usage the earlier expression "*physique sociale*," which had been created by Henri de Saint-Simon (1760-1823), then used in 1835 in a book title by A. Quetelet, the creator of the "average man" (*homme moyen*), and also used by Comte himself. In English, "sociology" began to be used in book titles in the 1870s both in Great Britain and the United States. In the 1890s, Chicago University may have had the first professor of sociology so titled.

> Except for the academic field of Politics, if Sociology is not just an objective representation of social facts, but a program of prying into the meanings of social facts; if it is not just the history of ethical systems, of cultic practices, of mythological beliefs, etc., but an attempt and effort to find systematic techniques for interpreting ethical systems, cultic practices, mythological beliefs, etc.; then sociology has begun around and after 1800, and is one more of the really great achievements of the century.[50]

8. PESSIMISTS

The nineteenth century had pessimists and prophets of doom. So did earlier centuries. But in the nineteenth century they were radically different, a new breed.

Even before the century began, in 1792, Thomas Robert Malthus, a parson among many, warned mankind that it will starve itself out of food unless it controls the size of its population. The leading aspects and arguments of the Malthusian Principle may have been worked out by various population-economists long before Malthus.[51] But it was Malthus who caught the attention of the public, by casting a pall that refuses to dissipate. In fact, the world seems to be running out not only of food, but of natural resources altogether. Karl Marx, being a humorless optimist and eschatologist, and unrelenting when in opposition, hated Malthus, and, as was his wont in intellectual confrontations, heaped scorn on him. To Marx, the proletariat was fated to win, albeit by violence; a betterment of the world was sure to come. No Malthus would change that.

Next, in the middle of the nineteenth century, culminating in 1865, the thermodynamicist Rudolf Clausius created his second law of thermodynamics and applied it to the universe as a whole. The outcome was that the available energy of the universe is unceasingly transforming itself into uncreative uniform heat, thus leading the universe inexorably to a "heat death, in which the world shall be reduced to one vast temperature equilibrium in which nothing really new ever happens. There will be nothing left but a drab uniformity out of which we can expect only minor and insignificant local fluctuations."[52]

This heat death would take millions, even billions of years to come about. But intellectuals have been frightened by the prospect of it nonetheless. Thus

Henry Adams wrote "A letter to American Teachers of History" about it, and quite recently an economist wrote a large economic treatise on the theme of the Second Law of Thermodynamics.[53]

Clausius coined the term *entropy* by which to quantify his law. It designates a real-valued quantity measuring, in a thermodynamic process, when it is completed, the extent to which the amount of heat has increased at the expense of other available energy. Thus the entropy of the universe in its entirety is on the increase. But only for the universe in its entirety. In select parts of the universe it may on occasion decrease. Such a decrease in a part of the universe must be compensated for somewhere else by an increase at least as large. In the kinetic theory of gases, itself part of so-called statistical mechanics, heat is generated by the "irregular" motion of microscopic particles. If one interprets this irregular motion as "disorder," then the entropy measures the degree of this disorder, the higher the entropy the larger the disorder. And the second law of thermodynamics states that in the universe as a whole, the total disorder is constantly on the increase.

On the other hand,

> Living organisms represent systems that are highly "organized," that is, they display less randomness than the materials from which they are "built"; and it is therefore justifiable to say that a *decrease* in entropy is involved in their building.[54]

By what we have noted above, this decrease in entropy must be compensated for by an increase somewhere else. In fact, "the small local decrease in entropy represented in the building of the organism is coupled with a much larger increase in the entropy of the universe."[55]

A very discouraging consequence arises if one equates disorder with the "pollution" of everyday parlance. In this interpretation, the total pollution of the universe can only increase, and any attempt at clearing up an environment results in an even worse pollution in some layer outside the given environment. Thus, terrestrial nuclear waste cannot be literally disposed of. At best it can be transported into another part of the universe, with some additional pollution as "transportation compensation" arising. There is one dubious theoretical way out. Even if one admits that entropy is constituted by disorder, it is still very difficult to say what "disorder" in entropy really is.[56]

Finally, at the end of the nineteenth century, there appeared, before the end of World War I, Volume I of *The Decline of the West*, by the "prophet of doom" Oswald Spengler. Leaving aside Spengler's insistence on the cyclicity of history, but focusing on his comments on tangible phenomena, one sees that his analysis, whether retrodictive or predictive, cannot be simply ignored. His retrodictive interpretation of differences between Greek and modern mathematics is most pertinent; and before 1920 he predicted for the second half of the twentieth century the kind of civic lawlessness and social fragmentation that we are undergoing nowadays.

Malthus, Clausius, and Spengler were social or physical scientists. In between were the literary pessimists.

Between Malthus and Clausius there was Arthur Schopenhauer, a philosopher-pessimist in the manner of Heracleitus. According to him, human life not only *may* contain more misery than happiness but this *must* be so, on philosophical grounds. There is only one progress, progress towards more misery.[57]

Between Clausius and Spengler there was Lev Tolstoy with his pessimistically moralistic *Kreutzer Sonata*, and above all, Friedrich Nietzsche, a brooding philosopher of ambiguous outlooks, whom Spengler viewed as one of his two mentors (the other being Goethe).

In the twentieth century, successor pessimists to all the above were, among others, Franz Kafka, the Jean Paul Sartre of *Nausea*, and the Sigmund Freud of *Discontent in Civilization*. Also (and this will be a prime lesson in Part II), much of what can be taken for general pessimism was "sublimated" into twentieth-century physics. More about this in Part II itself.

9. THE IRRATIONAL VERSUS THE RATIONAL

After World War I, the anti-democratic forces in post-war Germany that eventually brought forth Nazism rallied intellectually under the flag of a contrariety of Irrationality versus Rationality. They did so from an overwhelming Teutonic fear of, distrust of, contempt of, and horror of what the German intelligentsia, and especially the superintelligentsia, called "rationality," and they gave way to a headlong abandonment into and passion for what they called "irrationality," with a religio-mystical flavor entirely of its own. The despised and feared and to-be-exterminated rationality was, according to the Germans, a mark of Frenchmen, Jews, Freemasons, and misguided Anglo-Saxons for whom rationality and the profit motive were inseparable.

A recent study by an expert author seems to suggest, if I interpret the author correctly, that Werner Heisenberg was compelled and impelled towards inventing his matrix mechanics in response to an all-encompassing "ecological" and cultural pressure for overcoming determinism, causality, "mechanical" rationality, etc., in favor of indeterminism, uncertainty *qua* a-causality, freedom from stifling and un-German "rationality," etc.[58] The study is extremely well documented, and for my part I can only rely on memory of my personal experiences and impressions, having lived through the period in Germany already as an adult. I do not even have any kind of diaries by which to refresh my memory. I nevertheless venture to make some observations on the intellectual climate during the so-called Weimar Period or Weimar Culture, as the setting in Germany between World War I and the ascent of Hitler has come to be known.

My first overall observation is that things were very bad, and the academic atmosphere in Germany was oppressive indeed, and growing more so. But I simply do not remember at all that mathematicians and physicists and other scientists were the choice targets of this oppressiveness, and that they acted and re-acted as such. In particular, young and aspiring mathematicians and physicists did in no way consider themselves as being worse off than students of classics, or of other humanistic pursuits, or even of German history and literature, than which there was nothing more sacred in the land. Things were bad, but not along such a scientific/humanistic line of division.

I do not mean to absolve the bearers of Weimar Culture of their very many and very stubborn sins of omission and commission. I agree with the present-day German historian Fritz Fischer in his thesis—which he stoutly defends against all critics and detractors—that the World War of 1914-1918 was brought about, almost inevitably, by the German bourgeoisie in all its layers, and foremost by its intellectuals; by uncontrollable dreams of glory of globe-spanning world domination, physically, materially, and also intellectually. And when the dreams were all shattered, most of the nation, especially the educated classes, and most especially the university crowds, were seeking for scapegoats all over the landscape, not only in political areas but also in intellectual ones. And a climate of oppressiveness was developed and sustained.

All this is true. But to cap these insights by the conclusion that these movements of intellectual obfuscation were somehow responsible for the rise and creation of the theories of quantum mechanics and of wave mechanics does not sound right to me. It simply did not happen this way.

My second general observation is that there were two very separate phenomena involving discontinuities and that they must not be conflated, as is sometimes regrettably done. Both are large-scale phenomena of Western intellectuality, both worldwide, and not necessarily of German origin, however much Weimar Germany may have cultivated them to excess. These several phenomena do overlap, and the Weimar intellectuals themselves did conflate them too. But from a historical distance of a half-century they ought to be kept apart and traced to their divers origins.

1. The first phenomenon, or movement, a most "international" one, was, and still is, a struggle to grasp the interrelations, differences, and (in-)consistencies of some seemingly related or contrasting cognitive concepts, axioms, and procedures, like causality, determinism, functional (inter-)dependencies, mathematical certainty, statistical certainty, certainty as a limit of statistical probability, etc.

This struggle may have had antecedents in antiquity and the Middle Ages, but it really began after the Renaissance with the British empiricists and was institutionalized by David Hume with his affirmation that "causality," whatever that may be, is at best "association" and nothing more "binding." Kant made one of philosophy's greatest efforts to save causality by shoring it up with Apriorism. But hardly anybody believed that Kant's was the last word in the matter. Even the relatively secondary question, whether and to what extent causality is co-extensive with determinism, refuses to come to rest. And the widespread identification of causality and/or determinism with the "certainty" and "predictability" of a Newtonian mechanical point system was in the twentieth century almost too childish to debate, even if thundering superintellectuals of the Weimar Culture sometimes took this identification for granted, as did also, regrettably, many intellectuals of the "enlightened" West.

The "experiential" or "empirical" association of Hume leads only too naturally to such intellectual constructs as Mill's inductive logic in Britain, and pragmatism and instrumentalism in the United States; or, at least, it cannot be neatly separated from them. And even the element of "probabilism" in the great theories of statistical mechanics and physics of the nineteenth century has a certain link to it.

In sum and in short, the profession of a-causality in the broadest sense was in no wise a Teutonic institution, however much the Weimar Culture may have pushed it to extremes.

2. A rather different phenomenon, which the Germans had a greater share in creating, was the quest for "irrationality." It was more popular than the quest for a-causality, even though only its academic formulations by various Germans have become generally known.

Of course in the heady outpouring, verbal or written, of a muddle-headed or mischievous German on the (pseudo-) intellectual war-path, the two phenomena, that is, irrationality and a-causality, frequently merged, but in a serious study today they ought to be kept apart. Furthermore, when the outpouring came in a solemn address, or an article-to-such-a-purpose by a German professor, then one still must consider the difference between delivering oneself of such a declaration or declarations, and actually applying the lesson from it in one's day-by-day routine work in the lab or classroom, or behind a desk with a pile of books on it. The latter application occurred much more rarely than so-called documentations might suggest.

That irrationality is really something very different from a-causality in its main manifestations can be seen from the following statement about it by an expert:

> Future historians will, I believe, recognize in this preoccupation with the surd element the governing impulse of our time, the *daimon* or Zeitgeist which in different guises has haunted minds as various as Nietzsche, Bergson, Heidegger in philosophy; Jung in psychology; Sorel, Pareto, Spengler in political theory; Yeats, Lawrence, Joyce, Kafka, Sartre in literature; Picasso and the surrealists in painting.

The expert is E. R. Dodds, a prominent British classicist, author of the book *The Greeks and the Irrational*; the aphorism above is a footnote to his article

"Plato and the Irrational," which is Chapter VII in his book *The Ancient Concept of Progress and Other Essays on Greek Literature and Beliefs* (1973). This author's own work on irrationality deals with a very special aspect of it, namely with its manifestations in Greek religion, but the author seems very knowledgeable about the conception in general.

It is noteworthy that in Dodds's list the Germans do not even predominate. The only true Germans on it are Heidegger, Nietzsche, and Spengler, because Jung is culturally Swiss, and this is not the same as being German. And among non-Germans Bergson, I think, was a much more pronounced and influential irrationalist than is sometimes conceded. I remember that around the time when Hitler was close to coming to power, some "Nazi intellectuals" were disappointed that Bergson was a Jew, because they would have liked to put him into their Living Hall of Fame (he lived till 1938).[59]

3. I also question the widespread view that anybody in the Weimar period who quoted something meaningful from Spengler with approval was necessarily an irrationalist like Spengler. This does not follow. It is wrong to posit that everybody during the Weimar Culture who was fascinated by Spengler and was carried away into quoting him was necessarily a grave-digger of this culture. In this sense I must put in a kind word for Ernst Troeltsch (1863-1923), who is sometimes mistaken for an unmitigated irrationalist. Troeltsch is the author of the monumental work in two volumes *The Social Teaching of the Christian Churches*; of the large historiographic work *Der Historismus und Seine Probleme* (Tübingen, 1922); of the still very readable article on the German Enlightenment, "Die Aufklärung," in *Real-Encyclopädie für protestantische Theologie und Kirche*, Third Edition, vol. II (1897), pp. 225-241; and of some articles to the effect that the outbreak of the (First) World War was due to some kind of "breakdown of communications," as one would put it today, rather than to diabolic intent on either side. During my student days at the University of Berlin (I enrolled in the autumn of 1918) Troeltsch used to give, every term, I think, a lecture two hours a week (Saturdays, 10-12, with an intermission) which bore the title "Introduction to Philosophy" and was in fact an introduction to his current interests in philosophy. During one of the lecture hours he simply presented, in an agitated manner, a "book review" of volume I of Spengler. As I remember the book review, it was not at all an extolling of irrationality or a-causality, but simply the exposition of a new approach to the "morphology of world history," as Spengler himself had called the book originally. (As I heard it at the time, it was the publisher who suggested the title *Decline* [Untergang] *of the West*, simply as an advertising gimmick.) It soon became mandatory to discuss Spengler endlessly. It was just a fashion, and as a fashion it was not nearly so widespread as had been twenty years before in Paris the compulsion to discuss Bergson.

Also it must be taken into account that Troeltsch died relatively soon after the outburst of the Spengler mania, and perhaps had he lived longer, he might have had reasons and opportunities for reconsidering his stand on various implications of Spenglerism.

4. One must not make too much of the fact that during the Weimar period mathematics and physics were taken to task for being intellectually too presumptuous and too demanding of attention, and that some mathematicians cringingly admitted guilt. To this charge I can only say that during the Weimar period mathematics stood generally in very high regard, and that this kind of ambivalence of attitude towards mathematics is as old as academic mathematics itself, the word *academic* to be taken literally (that is, since the times of the Platonic Academy itself).

In the early nineteenth century, the German *Naturphilosophen* (nature-philosophers) had a deep intellectual suspicion of mathematics. Yet, from around 1800 until today every great university of the West has had a great mathematics department. In the seventeenth century Francis Bacon, a paragon of modernism, was contemptuous of mathematics. But it was a great century in mathematics nevertheless, and hardly any of Bacon's general followers echoed his attitude towards mathematics. In the thirteenth century Roger Bacon was a noisy advocate of mathematics, and he was so noisy about it that there must have been many others who were indifferent to it or worse. Around 1200, Fibonacci made such a splash with his *Liber abaci* that at the court of Frederick the Second in Sicily he was dined and wined, figuratively speaking. On the other hand, the important work of Oresme and others from the fourteenth century was all but forgotten after 1600, and was only re-discovered in the nineteenth century as a part of the Great Rehabilitation of the Middle Ages then commencing. Ambivalence towards mathematics is as old as mathematics itself.

5. Finally we come to the suggestion that Heisenberg's matrix mechanics, with the consequent uncertainty-relation, and thus the entire physics since, came about by a soul-anguished desire to respond to the (allegedly) all-pervading quests for a-causality, indeterminism, irrationality, etc. This suggestion is untenable.

The matrix-mechanics was a genuine development in physics-for-physics' sake, and it was spawned not in Germany but in Copenhagen. Only a part of the German sense of exhilaration over this achievement was pride in Heisenberg's being a German. And several among the leading co-architects of the new theory were as un-irrational as they could possibly be; nobody who ever met Dirac or Fermi could conceivably suspect them of allowing themselves to be tools in a drive towards legitimizing irrationality of any kind. Schrödinger, creator of the theory of wave mechanics, although an Austrian, may have had some tendencies towards Teutonic irrationality; but he acknowledged a measure of indebtedness to Louis de Broglie, who not only is as rational as only a certain caste of Frenchmen can be, but also was a staunch ally of

Einstein in his battle to eliminate any kind of Heisenberg uncertainty or indeterminacy altogether.

It is noteworthy that Russian textbooks on quantum mechanics, especially on the more advanced quantum *field* theory, speak not of a Heisenberg "uncertainty"-relation but simply of a "Heisenberg Law," which reigns in the "microscopic" realm of elementary particles, and they seem to have no difficulties at all. That is, they seem to have no difficulties in harmonizing this with the "macroscopic" Marx-Engels law to the effect that in the realm of economics and class relations everything is deterministic, and is in fact determined by dialectical materialism.

An undergraduate today is not disconcerted by the "Heisenberg Law." He is no more disconcerted by the statement that p and q cannot be observed with a joint precision more than $\frac{h}{2\pi}$ than by the statement that light of frequency ν travels in units of energy $h\nu$. Both these statements assert the existence, or rather presence, of certain "discontinuities" in physical nature, and our *Zeitgeist* is simply willing to accept them. It would be carrying things too far to assert that it was the Weimar Culture that mightily promoted this universal movement that distinguishes the twentieth century from the preceding nineteenth century, and that the bearers of the Weimar Culture brought this about from "sinister" motives too.

10. GREATNESS OF THE NINETEENTH CENTURY

I will list some noteworthy achievements from the span A.D. 1850-1870. It is a ribbon across the center of the extended nineteenth century 1800-1920, of the same width as the interval between the two World Wars in the twentieth century.

1850s	Gustav R. Kirchhoff, with the partial cooperation of Robert M. Bunsen, founds spectroscopy
1850s	Theodor Mommsen, *Roman History*
1851	Herman Melville, *Moby Dick*
1854	Bernhard Riemann's lectures on (Riemannian) geometry, published posthumously 1868
1857	Gustave Flaubert, *Madame Bovary*
1859	Charles Darwin, *The Origin of Species*
1860	Jacob Burckhardt, *The Civilization of the Renaissance in Italy*
1860	Louis Pasteur becomes "Father of Bacteriology"

1860	Bernhard Riemann initiates the mathematical theory of (dreaded) shockwaves
1860	Gustav T. Fechner and Ernst H. Weber create the Weber-Fechner Law (Response = logarithm of stimulus)
1862	Victor Hugo, *Les Miserables*
1862	Abraham Lincoln, *Emancipation Proclamation*
1864	James Clerk Maxwell, theory of electro-magnetic radiation
1864	Angelo Secchi begins spectroscopic analysis of sun and stars
1865	Rudolf J. E. Clausius, entropy and its Law
1865	Friedrich J. Kekulé, ring structure of organic compounds (benzene ring)
1865	Joseph Lister introduces antiseptic surgery
1865	Gregor Mendel, Law of Inheritance
1865-1869	Lev Tolstoy, *War and Peace*
1865-1870	Paul Cézanne begins to exhibit
1866	Feodor Dostoevski, *Crime and Punishment*
1868	Karl Marx, *Das Kapital*
1868	William Huggins uses red-shift to prove that Sirius is moving away from the sun
1869	Dimitri Mendeleev, *Periodic Table*
1870	Heinrich von Stephan conceives the Universal Postal Union, the first permanent international organization

[I wish I could include Verdi's *Aida* (1871) and, above all, Bizet's *Carmen* (1875).]

NOTES

1. Gilbert Chinard, "Progress and Perfectibility in Samuel Miller's Intellectual History," *Studies in Intellectual History* (Baltimore, 1953).
2. G. D. H. Cole, *A History of Socialist Thought* (London, 1954), vol. II.
3. N. Lukiesh, *Artificial Light, its Influence upon Civilization* (New York, 1920).
4. C. C. Gillispie, *Genesis and Geology* (Cambridge, Mass., 1951), pp. 42-43.

5. Francis C. Haber, *The Age of the World, Moses to Darwin* (Baltimore, 1959), p. 118.

6. See appropriate passages in "The Age of the Earth," *Bulletin of the National Research Council* **80** (June 1931); see also Lawrence Badash, *Rutherford and Boltwood, Letters on Radioactivity* (New Haven, 1969), p. 17.

7. In 1871, in a lecture given twice, Helmholtz gave 22 million years for the past of the earth, and 17 million for the future ("On the origin of the Planetary System," *Popular Lectures on Scientific Subjects*, Second Series, p. 182). His main argument had already been stated in 1854, but without specific figures ("On the interaction of natural forces," *Popular Lectures on Scientific Subjects* [London, 1873], pp. 153-196).

8. J. D. North, *The Measure of the Universe* (Oxford, 1965), p. 8.

9. There was one very isolated occurrence in the middle of the eighteenth century. See Edmund Whittaker, *A History of the Theories of Aether and Electricity*, vol. I, *The Classical Theories* (Harper Torchbook, 1960), p. 367.

10. Ibid.

11. Ibid., pp. 99-100 and 367.

12. Ibid., p. 371.

13. See the article on Huggins in *Dictionary of Scientific Biography*, C. C. Gillispie, editor, vol. 6, p. 592.

14. Isaac Asimov, *The Universe, From Flat Earth to Quasar* (New York, 1966), p. 68.

15. Ibid.

16. Most recently in "The Emergence of Analysis in the Renaissance and After," *Rice University Studies* **64**, nos. 2 & 3 (Spring-Summer 1978): 25ff.

17. Salomon Bochner, "Mathematical Background Space in Astronomy and Cosmology," *Vistas in Astronomy* **19** (1975): 133-161.

18. In "Emergence of Analysis," cited in note 16.

19. Bochner, *The Role of Mathematics in the Rise of Science* (Princeton, 1966), pp. 191-192; see also, "Mathematical Background Space" (cited in note 17), p. 160, top.

20. Robert L. Delevoy, *Directions of the 20th Century Arts*, translated by Stuart Gilbert (Geneva, 1965).

21. "Emergence of Analysis," p. 36.

22. Delevoy (see note 20), p. 74 (modified).

23. Kurt Badt, *The Art of Cézanne* (Berkeley, Calif., 1965), p. 161.

24. To the above see my article "Space" in *Dictionary of the History of Ideas* (1973), vol. 4, pp. 295-307, especially p. 305.

25. Andrew G. van Melsen, *From Atomos to Atom* (Pittsburgh, 1952).

26. On this subject and following, see my article, "Continuity and Discontinuity in Nature and Knowledge," *Dictionary of the History of Ideas* (New York, 1973), vol. I, pp. 492ff.

27. Probably the most authoritative account is in H. A. Wolfson, *The Philosophy of the Kalam* (Cambridge, Mass., 1976); but see also Majid Fakhry, *Islamic Occasionalism and its Critique of Averroës and Aquinas* (London and New York, 1958).

28. See "Continuity and Discontinuity"; and "The Emergence of Analysis."

29. "Singularities and Discontinuities," *Rice University Studies* **59**, no. 2 (1973): 21-40.

30. The most penetrating study of this problem known to me is François Jacob, *The Logic of Life, A History of Heredity*, translated by B. C. Spillman (Pantheon Books, 1973).

31. R. G. Collingwood, *The Idea of History* (New York, 1956), p. 145.

32. *The Role of Mathematics*, cited in note 19; see index under "continuism."

33. Ibid., p. 117.

34. "Mathematical Background Space," and "Emergence of Analysis."

35. William Foxwell Albright, *From the Stone Age to Christianity* (Anchor Book, 1957).

36. Alfred Stern, *Philosophy of History and the Problem of Values* (The Hague, 1962), p. 15.

37. Ibid.

38. Leonard I. Krimerman and Lewis Perry, editors, *Patterns of Anarchy* (Garden City, N.Y., 1966); David E. Apter and James Joll, editors, *Anarchism Today* (London, 1971).

39. Holger Pederson, *The Discovery of Language* (Cambridge, Mass., 1931).

40. R. H. Robins, *A Short History of Linguistics* (Bloomington, Indiana, 1967).

41. Ibid., p. 170.

42. Ibid., p. 168.

43. Ibid., p. 134; Thomas H. Sebeok, ed., *Portraits of Linguists: A Biographical Source Book in the History of Western Linguistics, 1746-1963* (Bloomington and London, 1966), vol. I, p. 5.

44. Pederson, *Discovery*, p. 19.

45. Luella Cole, *A History of Education, Socrates to Montessori* (New York, 1950), p. 495.

46. A. D. C. Peterson, *A Hundred Years of Education* (London, 1952), p. 78.

47. Ernst Jones, *The Life and Work of Sigmund Freud* (New York, 1953), vol. I, pp. 281, 366, and 372-376.

48. G. D. H. Cole, *A History of Socialist Thought* (New York, 1953), vol. I, pp. 1-2.

49. Heinz Maus, *A Short History of Sociology* (New York, 1956), p. 11.

50. Albion W. Small, *Origins of Sociology* (Chicago, 1924, reprinted New York, 1967), p. 37.

51. Joseph A. Schumpeter, *History of Economic Analysis* (New York, 1954), pp. 254-255.

52. Charles Lincoln Van Doren, *The Idea of Progress* (New York, 1967), p. 155.

53. Nicholas Georgescu-Roegen, *The Entropy and the Economic Process* (Cambridge, Mass., 1971).

54. Harold F. Blum, *Time's Arrow and Evolution*, second edition (Princeton, 1955), p. 94.

55. Ibid.

56. P. W. Bridgman, *The Nature of Thermodynamics* (Cambridge, Mass., 1940), pp. 166ff.

57. Frederick Copleston, *Arthur Schopenhauer: Philosopher of Pessimism* (London, 1947), p. 28.

58. Paul Forman, "Weimar Culture, Causality and Quantum Theory, 1918-1927. Adaptation of German Physicists and Mathematicians to a Hostile Intellectual Environment," *Historical Studies in the Physical Sciences*, ed. by R. McCormmach (Philadelphia, 1971), vol. 3, pp. 1-116.

59. Henri Bergson taught that in addition to "rational" time that can be measured by the clock there is an "irrational" duration that cannot be quantified.

PART II

A LECTURE ABOUT EINSTEIN

Albert Einstein[1] was universally lionized during his lifetime, perhaps more so than any scientist in modern times.

It is hard to describe the storm that broke loose around him in Berlin, where I also lived, after World War I, in late 1919, when his theory of relativity began to be talked about, everywhere, anywhere, anytime, in academic halls and in beer halls. The waves of the storm spread quickly all over the world. Newspapers talked about him and reported on his movements for years on end.

Nothing detracted from his standing. In Berlin it was taken for granted that his mannerisms were all publicity devices. His wearing long hair was jocularly ascribed to a competition against the similarly coiffed playwright Gerhard Hauptmann, 1912 Nobel Prize winner for plays on social themes. Around 1926 an older mathematician who had observed him over many years remarked to me that Einstein was his own best and most watchful publicity agent. But none of this was intended to impugn his greatness, nor did it affect the adulation of him.

In this country the *New York Times* began to report on him in 1920.

In 1926, Mrs. Josepha Wheeler, daughter of the eminent American astronomer Simon Newcomb, solicited from Einstein, then in Berlin, a statement about her father's work to be deposited with his papers in the Library of Congress. Einstein did respond with a statement, praising Newcomb's work towards explaining deviations of planetary orbits from their elliptical Keplerian paths due to gravitational interference from other planets.[2]

In 1929 Einstein wrote a paper on Unified Field Theory, one of several attempts, none of enduring outcome, to force all of physics into the mathematico-physical mold of his theory of gravitation. As I remember it, the American newsmen in Berlin burned up the transatlantic cables transmitting the entire text of the paper to the United States, intricate formulas and all. This account is from my memory. But it is a documented fact that a trustee of Wesleyan University purchased the manuscript directly for deposit at the Olin Library at Wesleyan.

Abraham Flexner, the first director of the Institute for Advanced Study in Princeton, liked to regale young people with stories of his exploits. I was sometimes present, being a member not of his Institute but of Princeton University. Sometime in 1934 he told us how he persuaded the Bamberger Family of Newark, New Jersey, to come across with a donation to found the Institute. It was in 1931, and nobody was willing to make money commitments. Finally the patriarch of the clan blurted out: give us Einstein and you can have your Institute.

And I do remember that probably late in 1931, newspapers in Germany reported that henceforth Einstein would be dividing his time between his position in Berlin and a lush academic sinecure in Princeton, U.S.A.

When late in 1933 he was sailing from Europe for permanent residence here, at the pilot stop in New York harbor two trustees of the Institute took him off the steamer in a tugboat for quick processing in the port.

Finally, during World War II, at a war bond rally in Kansas in 1944, a manuscript copy of Einstein's 1905 paper on special relativity was sold for six million dollars and deposited in the Library of Congress.

Why Einstein should have received such public attention and esteem is, I think, an intriguing problem in mass psychology. After all, Einstein did not write a treatise like Darwin's *Origin of Species*—which, as I remember it, the contemporary *New York Times* called a book in a century—or, if this be too controversial, like Newton's *Principia* (that is, *Mathematical Principles of Natural Philosophy*), the like of which has not been created before or since. For instance, the *Principia* predicted that some day a projectile would be launched with a sufficiently high velocity to make it an earth satellite. This was seriously spoken of in the eighteenth century, even though it came true only in the middle of the twentieth century. I do not think that there is anything like this in the writings of Einstein. In a splendid paper of 1916 Einstein recognized the existence of *stimulated emission* that decades later became a pillar of the theory of lasers. But the paper had no predictive features.

Newton also wrote a book called *Opticks* which, according to Marjorie Hope Nicolson, inspired sages and poets through the length of the eighteenth century and beyond.[3] Einstein wrote nothing of the kind. The *Opticks* is based on the resolution of white light into the visible spectrum from red to violet, together with related experiments. Newton performed all his experimental feats himself, as did other great physicists after him. Maxwell, for instance, however eminent a theoretician, was also the first professor of *Experimental* Physics at Cambridge University. But Einstein never performed an experiment in his life. Einstein spent twenty-two years at the Institute for Advanced Study, but he never formed a school of modern physics, whereas Niels Bohr directed such a school of world renown.

Einstein never coined an original name for anything. He created the elementary particle "photon," calling it a "quantum of light." The beguiling name *photon* was coined by somebody else. Even his "principle of relativity" was not his by name, but Poincaré's. Heisenberg, at the age of twenty-four or twenty-five, proclaimed an "Uncertainty Principle" that scared the daylights out of some philosophers. Einstein never came up with anything so suggestive. I know only one aphorism of his worth remembering: "As far as the laws of mathematics refer to reality they are not certain, and as far as they are certain they do not refer to reality." Interesting as it may be, it does not have half the punch of the cognate saying that "mathematics is the field of knowledge in which nobody knows what it is that he is talking about," which I think is due to Bertrand Russell.

An awareness of the fact that Einstein was, after all, a product of a period is beginning to seep through. The National Academy of Sciences in Washington, D.C., has recently decided to erect an out-of-doors monument to Einstein. I have no quarrel with this at all. Einstein became not only an American citizen, but also an American property. After World War II many immigrant celebrities started dashing about all over Europe, usually ending up in Switzerland, where at that time they could get creature comforts approximating those in the United States. Einstein, however, stayed put in Princeton, officially at any rate, and the community, in all its strata, was very appreciative of this. When he died in the Princeton Hospital, right at home, a blanket of sadness spread over the town. Thus a monument to him is well deserved. But the prospectus of the Academy soliciting contributions towards the erection of the monument is surprisingly circumspect. It bills him as the greatest theoretical physicist of the twentieth century, as if shunning a comparison with a nineteenth-century physicist like Maxwell, or with a twentieth-century scientist other than a physicist—say a Sigmund Freud, who would be formidable competition indeed.

Still, with all possible qualifications duly noted, Einstein was a person of really great impact, not only on physics but also on philosophy, as even a schematic presentation of his achievements ought to put beyond doubt.

Einstein's span of high creativity extended from 1900 till 1920, when he was forty-one, or, if one wants to include the Bose-Einstein statistics, till 1924 when he was forty-five. This seems typically to be the approximate productive span for persons of his caliber. Maxwell died at forty-eight, and Riemann, but for whom there would be no Einstein gravitation, at less than forty. Gauss managed to lay the foundation for Riemann's pre-Einstein memoir when he was approaching fifty. Newton published his *Principia* at forty-five, and his contemporary Spinoza died at forty-five. Galileo crested at the age of forty- six with his *Starry Messenger*, which created the telescopic depth of the universe. His contemporary Johannes Kepler terminated his high creativity with his third planetary law at about forty-seven, and Galileo's and Kepler's contemporary William Shakespeare retired from London to Stratford at forty-eight. Before

Shakespeare, Giordano Bruno was incarcerated by the Inquisition at the age of forty-four; and long before him Thomas Aquinas amassed a shelf of writings amidst an active career in church policy before dying at the age of forty-eight or -nine. There were notable exceptions though: Miguél de Cervantes, a contemporary of Shakespeare, was already fifty-eight when the publication of *Don Quixote* plunged him into immortality. Also, Immanuel Kant of Königsberg was about fifty-eight when the publication of the *Critique of Pure Reason* made him a successor—some would say *the* successor—to Plato of Athens.

Einstein's creative span from 1900 till 1920 is neatly divided in two halves by the First Solvay Congress of 1911. These Congresses, founded by the manufacturer of sodium carbonate Ernest Solvay, meeting in Brussels, and having about twenty-five participants each, are resplendent with famous names, and the first was particularly star-studded. Einstein, at thirty-two, was the youngest participant, and James Jeans, at thirty-four, the next youngest. These two "juniors" were charged with reporting on the bad state of health of the Dulong-Petit Law about specific heat, with Jeans presenting the patient's past history in the nineteenth century when ministered to by Maxwell and Boltzmann, and Einstein explaining the fancy remedy, quanta and all, prescribed by himself in 1907. It leaps at the reader of the Proceedings of the Congress that luminaries like Lorentz, Poincaré, Planck, Nernst, and W. Wien were profoundly impressed with Einstein, and were so indulgent with him. He talked a great deal, and butted in at will. And the luminaries beamed at him, at whatever he said. In return, Einstein was quite deferential in letters to them (less so in letters about them).

Poincaré died soon after, which made it the easier for Einstein to outshine the others within a decade, shining them out of existence, as it were. In a way, they did it to themselves. Only two years after the Congress, in 1913, Planck, speaking for the German delegation to the Congress, told some Prussian authorities that Einstein was a physicist in a century, and *must* be brought to Berlin. He was made both a salaried member of the Prussian Academy of Sciences with the privilege of lecturing at the University (the two buildings are, or were, close by) and the director of a separate research Institute of Physics whose creation was hastened by his appointment. All this at the age of barely thirty-five, in the spring days of 1914 before the outbreak of World War I, even before Einstein was ready to present his *pièce de résistance*, the theory of General Relativity and Gravitation, which, incidentally, as some rumor had it, Planck abhorred.

Einstein did indeed lecture at the university, making it a platform for the propagation of the faith in relativity and himself. I attended such a lecture in 1919, sixty years ago, and I vividly remember part of what he said, and certainly how he said it. From time to time, in any lecture or debate, he would look up above the audience, as if actually communing with the powers above, but letting the audience be witness to it.

In content, what I remember most about the lecture was a sharp upgrading of the stature of Michael Faraday. Nowadays there is a Newton-Faraday-Maxwell axis, and I think that Einstein contributed greatly to its creation. At the beginning of the century Faraday was not included in such an exalted company. Even in the late 1930s an Oxford don, a humanist visiting the Institute for Advanced Study at Princeton, told me how enlightening it was for him to learn in Princeton of the great esteem that Faraday enjoyed in the U.S.A. Einstein in his lecture gave Faraday much credit for the creation of the electromagnetic field, even if the lines of force in the field had the material quality of strings or even cables for him.

But now for Einstein's real achievements.

In 1905, at twenty-six, he published, in one and the same volume of the *Annalen der Physik*, three papers, each now enshrined in fame. In the first paper he introduced, but for the name, the universal *photon* as an elementary particle much as it is understood today. Its only rival was the electron, which, although long known, had been institutionalized as a kind of elementary particle in our present-day sense about ten years before. The most notable point of the paper is that Einstein satisfactorily explained the way in which these two rivals interact in the so-called photoelectric effect. It was this paper that brought Einstein the Nobel Prize in 1921; at least, no other achievement is singled out in the citation. Using the technique of this paper, Einstein began in 1907 to repair the Dulong-Petit law. He carried the repair only halfway, Einstein himself pointing out an imperfection. Debye and others completed the job.

In the second paper of the same volume of the *Annalen*, and in some follow-up papers, Einstein gave a satisfactory theoretical account of Brownian motion, which is the random motion of particles suspended in liquids. The magnitude of this achievement is not dimmed by the fact that, independently of Einstein, Marian Smoluchowski[4] in Cracow did something similar at the same time, and that subsequent theories of Brownian motion, some quite elaborate, were on a different level of mathematical expertise.

The third paper in the volume was the noisily celebrated special theory of relativity—universally celebrated, that is, till 1953, when Edmund Whittaker, in a masterful history of physics, rudely stripped the paper of celebrity and reduced it in rank to a glorified footnote to what Whittaker called "the theory of relativity of Poincaré and Lorentz"; but in the matter of general relativity and gravitation he left Einstein's standing intact, offering, in fact, an excellent sketch of the theory. The anguish of Einstein partisans notwithstanding, most physicists are tending to acknowledge that Whittaker is not entirely wrong, at least in the sense that there were several originators of the theory and that Einstein was a relative late-comer.

For my part, I do assign to Einstein an outstanding share in the discovery of special relativity, because Einstein's paper contained a philosophical blockbuster that had a very salutary effect when it exploded, somewhat belatedly in 1919, on the agora of philosophers and bystanders. It was the philosophical question of what it means to say that two events at different locations in space are synchronous, and how one "verifies" this, the matter of verification being of course the heart of the question.

Surprisingly, philosophers did not resent the intrusion onto their turf by a professional outsider. Many of them were exhilarated, on the Continent at any rate. None dared to spurn it, "lightweight" as it was when compared to customary "heavy" stuff like ethics, aesthetics, mind-and-body, existence of God, transcendental knowledge, etc. Some of them might even sneak Einstein's question into their so-called "Introduction" to philosophy, a grab-bag of what a budding philosopher ought to become familiar with. Einstein's kind of question was not quite unprecedented. Immanuel Kant once asked whether it is meaningful to say that a treasure is buried in the Sahara in a location inaccessible to humans. But these were marginal questions in exotic settings. Nothing can be more common than wanting to know whether two events in different cities have occurred at the same time, however, and, if they have, wanting to establish this in a manner that will stand up in court.

I can cite a witness to the impact of the question. It is Sir Karl Popper, widely known for his book *The Open Society and Its Enemies.* I remember reading in an autobiographical account of his that it was Einstein's special relativity, especially the question of synchronicity, that made him into the philosopher he has become, except that it suited his philosophical temper to feature not "verifiability" but "falsifiability," as a criterion for the validity of any theory, that is.

Formally, special relativity is an alternate to Newtonian mechanics, but different from it. It contains a positive constant c, and Newtonian mechanics arises as a limiting case by letting c go to infinity. By pedigree, however, special relativity was an outgrowth of and even a part of electrodynamics. Lorentz and Poincaré so viewed it, and they included the problem of synchronicity somewhere in their context, as a problem of physics. But Einstein was the first to make it clearly into a problem of knowledge in general. I find that Einstein always thought in this manner. This was his *modus operandi* throughout his entire career, scientific and other, usually for the good, but not always. It made him extremely apt to be successful for the present, but less enduring as a prophet. We will return to this distinction later.

Einstein was what I venture to call an "essentialist." He had the drive and capacity to single out, in any scientific situation under purview, what is really *essential* to the context, not only neutrally, but also in the hierarchical order of its components, and in its scope.

Einstein arrived at the universal photon in this way. The name *photon* was introduced much later, by G. N. Lewis in 1926. As I have already stated, Einstein himself spoke of a "quantum of light," because he sifted its existence out of Planck's quantum theoretic explanation of black body radiation. In fact, in the 1916 paper mentioned before, by penetrating through to essentials, Einstein even gave the simplest proof yet known for Planck's radiation formula from Planck's own premises.

Or, consider the famous formula $E = mc^2$. It was actually found by Poincaré, but presented by him without derivation or explanation. Einstein, however, after proving it, gave it the universal interpretation that if a physical system, any system, loses energy, then it also loses inertial mass. The amount of that loss is E/c^2, with the implied suggestion to a later generation that, conversely, by atomic or nuclear disintegration of mass a corresponding amount of energy can be liberated. This is precisely what happens in atomic fission or thermonuclear fusion. Because of this universal interpretation, which was heavily emphasized in the so-called *Smyth Report*,[5] the formula has been firmly attached to Einstein, irrespective of its origin in Poincaré, who had stated it in an electrodynamic context.

Even Einstein's gravitation theory, his most original creation ever, was not independent of an essentialist provenance, in Einstein's own pedigree for it, at any rate. He packaged it inside something he called general theory of relativity; this in turn was an essentialist elaboration of something called special theory of relativity, which was designed to feature the fact that light velocity is maximal and constant. And this special theory in turn was an essentialist shell in which to house something that had been created by Poincaré, a principle of relativity, expressly so named by him. This principle had its debut right in this country, on September 24, 1904, when Poincaré presented it at a Congress of Arts and Sciences assembled at the World Exposition in St. Louis. As if this route were not devious enough, on the road from the principle of relativity to gravitation Einstein built a major way station called the Principle of Equivalence, and a secondary one, a Principle of Covariance, both mandatory stopovers; the stopover at Equivalence involved an eerie plunge in the so-called Einstein Elevator.

All this sounds very involved. Nevertheless, the magnitude of Einstein's achievement is such that it can be savored from outside without entering into a maze of technicalities at all.

Gravitation is the oldest and most ubiquitous phenomenon, and it is always accessible to observation. Objects are falling and moving all the time and in all places. But explaining gravitation has always been a forbidding problem. Even today the innermost structure of gravitation is more arcane than anything else in physics, and if you overhear a physicist say that something is only *probably* so, then more likely than not he is talking about gravitation, or something involving gravitation. Thus, some physicists say incessantly that there "probably" are gravitons (elementary corpuscular particles of gravitational *force*), as if by saying this often enough the gravitons would indeed show up suddenly.

Aristotle proposed a theory in which each body in the universe has its natural place, and if it is not located in it, it seeks to reach it. There was no effective rival theory till the end of the sixteenth century, although dissatisfaction with Aristotle's theory had been on the increase since the late Middle Ages. In the early seventeenth century Galileo offered a satisfactory explanation, but only for terrestrial events. His paths of falling bodies were even parallel, as if the earth were flat. At the same time Kepler was seeking to explain gravity's cosmic manifestations. With his powerful intuition he half-guessed the three planetary laws, but an underlying rationale eluded him. Newton then found one, on a comprehensive scale.

The simplicity and grandeur of Newton's theory fascinated lay intellectuals like Voltaire and unemotional philosophers like Kant. But professionals in astronomy and mechanics had cause to view it soberly, even disenchantedly. When trying to regulate the motions of planets, moons, and comets by the new theory, they encountered difficulty after difficulty. As soon as one difficulty was seemingly over, a new one demanded attention. Therefore there were no discouraging side-glances at Einstein when—after some attempts by others—he started from the very roots, building up something very different, even if it was in close proximity to Newton's ideas.

From our retrospect, the working mathematics needed for any mathematically controlled theory of gravitation was in Newton's time still in its infancy, and there was only one space form available in which to create it, the Euclidean. That was all that Newton had to work with, and he made the most of it. Euclidean space happens to be infinite by well-intuited mathematical notions of infinitude. But this does not mean, as some philosopher-historians volubly proclaim, that Newton was intellectually motivated by a metaphysico-theological drive for infinitude of space. Even the fact that Newton viewed his Euclidean space as "absolute" has no such implication for me. If closed Riemannian space had been mathematically accessible to Newton and he had used such a space instead of the standard Euclidean type, then philosopher-historians would have undoubtedly adjusted his alleged metaphysico-theological drive accordingly.

From motives that nobody can fathom, after 1911 (and even earlier) Einstein was driven to erect a theory of gravitation of his own, a genuinely new one. By chance and circumstances he fell in with a body of elaborate mathematics that had come into being long after Newton, in the course of the nineteenth century. As if by a predestined harmony, it was highly fit to articulate Einstein's desiderata from physics and give suitable responses to them. It was the so-called Riemannian geometry, that is, the multidimensional differential geometry mainly of Riemann, but indispensably also of Christoffel,

Beltrami, Ricci, and Levi-Civita. The mathematician Marcel Grossmann, an old acquaintance of Einstein's and a faculty colleague at Zurich, was Einstein's mentor and even co-author in a few papers. But the final theory, and above all the inspiration, drive, and passion of purpose, were wholly Einstein's. At any rate, there are no reports of Grossmann having had the sense of a brush with immortality followed by failure due to Einstein's ill-will. Einstein did exhaust his store of mathematical knowledge, but any physicist has the privilege of doing so with any mathematician. Einstein was after big game, and he applied whatever hunting laws were legitimate.

When mathematicians became aware of this activity in a highly charged mathematical area, their curiosity was aroused, and David Hilbert, the greatest among them, set out to rob Einstein of his prey. In defense, or rather counter-attack, Einstein tore the center piece out of his kill, left the denuded carcass for mathematicians to argue over, dashed off into the promising field of cosmology, and very quickly inaugurated a novel kind of cosmological speculation of which he became the father or godfather for the rest of the century. This was a glorious evasive tactic, and it was Einstein's all-overshadowing enterprise, uniquely his, although mathematicians hesitated to follow Einstein, and even physicists were skeptical. I heard it said that Niels Bohr warned his followers to stay out of cosmology until problems of microphysics have been solved. This is like saying that one should defer space exploration until urban slums have been cleared.

Again, the effect of these developments on philosophy was incalculable. Since Giordano Bruno in the sixteenth century, the space problem involved in astronomy and cosmology was philosophically beset by an utterly sterile yes-or-no question: whether space is finite or infinite and also what infinitude of space means cognitively. Some philosophers of science are clinging to this question even today. Aristotle was naively condemned for finitude, and Bruno was even more naively hailed for infinitude. Those in the know, the true cognoscenti, were mostly hedging. Copernicus hedged—and this was one of his finest features; and so did Galileo, from indecision. Kepler inclined towards finitude, but less from siding with Aristotle than from disapproving of Bruno, whose brand of infinitude he considered astrophysically reckless, and rightly so. Descartes created his own wrinkle and hedged by making his so-called extension (*extension, étendue*) neither finite nor infinite but indefinite, which, in its inventiveness, went back right to Anaximander. Newton in the *Principia* was fully hedging, avoiding a decision throughout the three editions. When the rector Richard Bentley in a pesky letter tried to force Newton's hand by making him agree that a universe of finite matter might suffer gravitational collapse, Newton hemmed and hawed and counter-suggested that such a universe might remain stable by arranging itself in what is nowadays called gravitational clustering. And Immanuel Kant weaseled out of the situation some of the time by devising an antinomy that if the world is infinite it is finite, and if finite infinite. Among medieval schoolmen such antinomies must have been quite common. Any trained schoolman could probably make up one on any contrariety offered. Kant himself used several.

To my surprise, Ludwig Wittgenstein was quite conventional in this matter. The logical space of his *Tractatus* is, by parenthetic remark, infinite. Commentators notwithstanding, this infinitude serves no purpose, but is the standard philosophical presumption since Giordano Bruno.

Einstein changed all this by abolishing the traditional presumption that the universe is the infinite Euclidean space or a finite portion of it. Instead he approached the outlook of Riemann that almost any topological manifold should be eligible to begin with. A compact manifold, for instance, offers a neat compromise between finite and infinite, and Einstein gave attention to the simplest such case, namely, the case of a sphere. Also, it is not necessary to elect one fixed space and adhere to it. One may envisage a family of spaces and examine how observational and speculative data are compatible with all of them or some of them. As a consequence, even the problem whether the universe had an origin in time, in a cosmically meaningful gauge of time, or has been in an eternal state of existence, appeared in a very new light. All in all, as I remember it, one felt one was finally free after centuries of constriction in a straitjacket of presumptions.

Of course, curved spaces were introduced in the nineteenth century, long before Einstein. However, the renowned hyperbolic space of Bolyai-Lobachevsky is topologically not different from Euclidean space and made no immediate impression on physics, or even on mathematics. But the elliptic space of Riemann did draw a quick response. Soon after Riemann's memoir became known in English, Simon Newcomb, whom I have previously mentioned, in an 1877 paper in the prestigious *Crelle Journal* speculated on how elliptic and even projective space would be detectable by human or astronomical experience. Also, in 1900 an astronomer, Karl Schwarzschild, went into greater quantitative detail. Yet these papers were of no enduring consequence, because they dealt with the same physics as before, only in different spatial frameworks. But in 1916 the same Schwarzschild, in response to the latest work of Einstein, constructed a relativistic space-time model whose importance is such that Schwarzschild's name, I think, currently occurs in cosmology more frequently than Einstein's. Regrettably, he died soon after, at the age of forty-three.

A similar early contribution to relativity was made by Willem de Sitter. In addition to scoring specific achievements, he also wrote in 1932 a small general book called *Kosmos*. In it he makes a statement that should be emblazoned on all books in physics or astronomy. It is the statement that "Infinity is not a physical, but a mathematical concept."[6] That is, whenever you state in exact science that some physical magnitude is infinite, then, without exception, you refer to a quantity in an appropriate mathematical skeleton of the physical context. Without a mathematical skeleton, however primitive it be, no such statement can be made. This in itself does away with many a version of the query whether the universe is finite or infinite.

In comparison to Einstein's gravitation, Newton's appears to be simplistic. But it is wrong to say that Einstein "overthrew" Newton. Einstein used mathematical resources, readied by others, that would have been inaccessible to Newton in his wildest dreams. Newton was still so ingenuous in the ways of analysis that he took it for granted, half-apologetically, that his gravitational force was an action-at-a-distance, not knowing that the simple analytical device of introducing the gravitational potential turns it into a proximity-action, instantly. In fact, the gravitational potential as a dominant concept revealed itself only in the early nineteenth century in the work of Poisson. Einstein himself stated very expressly that, *mathematically*, what he did was to replace Poisson's one-component-object by a more elaborate object of ten components.

Einstein's advance over Newton was evolutionary, and nothing but evolutionary. He overthrew Newton as little as did birds overthrow reptiles out of whom they evolved. In fact, reptiles continue to flourish alongside birds, as does Newton's gravitation alongside Einstein's, for instance, in all engineering mechanics and physics in our time. The theory and production of high velocity particles prerequires special relativity, but particle accelerators in which they are generated are architecturally built on Newtonian engineering mechanics exclusively. As Niels Bohr put it in general terms: "No more is it likely that the fundamental concepts of the classical theories will ever become superfluous for the description of physical experience."[7]

Einstein did not overthrow Maxwell either. In fact, it can be argued that special relativity was an apex achievement evolving almost rectilinearly out of Maxwell's electromagnetism, and all but completed it.

Nor can it be maintained that Einstein was superior to Maxwell in an overall sense. It is true that, except for his study of the ring of Saturn, Maxwell did not contribute directly to the study of gravitation in which Newton and Einstein excelled. But he did perform the extraordinary feat, to which there is no parallel, of fusing optics with electromagnetism. Optics, Aristotle's *hé optiké*, was the most venerable body of knowledge there was, and everybody who was somebody, literally everybody, felt bound to have an outlook on it—physical, philosophical, even inspirational—in late antiquity, the Middle Ages, the seventeenth century, the eighteenth century, etc. Electromagnetism, however, William Gilbert notwithstanding, was an upstart field; the Leyden jar was the perennial toy for adults. Yet Maxwell fused these two fields into one, or, at any rate, consummated their fusion. In the process he created, with posthumous assistance from Hertz and others, the awe-inspiring electromagnetic spectrum from end to end, from frequency 0 to frequency ∞. To appraise its importance one need only consider what would be left of laboratory physics, red-shift cosmology, and radio-astronomy if it were not available. To offset this achievement Newton at least wrote his two tremendous works, the *Principia* and the *Opticks*, on which Voltaire could compose a beguiling commentary. Einstein wrote nothing to induce Arthur Koestler to include him among his sleepwalkers.

Any survey of Einstein's scientific career, however sketchy, must also take into account activities in his later years, after the high creativity had exhausted itself, but the glory and the presence remained. His scientific later years lasted at least three decades, beginning not later than 1925, although he continued to bask in glory for another ten years or so before gradually becoming an "institution" resting on past achievements.

Einstein's later years had a regrettably negative scientific cast. If, as I have maintained all along in this essay, the nineteenth century lasted from the French Revolution and Napoleon till the end of World War I, then, in physics, Einstein was a perfect embodiment of the last two decades of it, shining brilliantly forth into the twentieth century that came afterwards, but not being able to see into it himself, as if the light that he threw into it flashed back at him, blinding his vision. In this sense Einstein was a person of the present, of the lived present and the escapist present, and it was this that made him, especially in this country, a smash hit of the 1920s.

But he was not a prophet; he did not have the passion and vision, selflessness and calling, to be one. Between 1925 and 1927, a crew of post-teenagers, headed by Heisenberg, Schrödinger, and Dirac, seized the helm of the ship of quantum theory, set out into uncharted waters, and, to the amazement of many, started making discovery after discovery, not only of islands, but even of continents, and of new navigational principles too. This was the so-called new quantum theory. Mathematically, the "classical" equations of mechanics and electrodynamics that had emerged from the work of Newton, Euler, Lagrange, and Maxwell were radically re-interpreted, and the effect on the open mind was compelling and irresistible.

Einstein, amazingly, resisted and remained aloof. He soon became antagonistic to the new discoveries, and gradually more rejectionist. He could not or would not perceive that a momentous development in a new phase of modernity was in the making. He refused to legitimize it, in whole or in any part, and stuck to his refusal to the end, pleas by personal friends notwithstanding. There were no overt public demonstrations, no excommunications of heretics, just a hard standoff, and an agreement to disagree reluctantly maintained.

There were debates, though. Einstein, being unbeatable in debates, scored on points, sometimes markedly. And like a medieval schoolman, he was content with this kind of victory and the applause of his retinue. But, before the God of Physics, his were illusory victories. The new quantum theory was couched in a new idiom, subdivided into dialects. Its grammar is not yet settled, even today, to the satisfaction of all the faithful. This bespeaks strength, not feebleness. For instance, analysis was a new idiom for mathematically controlled science. It suddenly sprang up close to 1600, and immediately came up with Kepler's remarkable planetary laws. But it went through the pangs of birth, childhood, and adolescence for over two centuries before reaching the adulthood of rigor and consistency in the nineteenth century, and at first it had many more vulnerabilities than anything that Einstein could adduce against the innovations of his day. One can only pity a twentieth-century student of Kepler bent on making twentieth-century sense out of the ramblings of Kepler's *New Astronomy*. Yet the achievements of Kepler are here to stay. Einstein should have known that a debate or two will not topple something destined to stay, but, alas, he did not. Whether he could not or would not know it is hard to determine, and does not make a great difference anyway.

There were specific findings to which Einstein and a small band of other physicists were dialectically opposed, as some still are. The primary one is Heisenberg's Uncertainty Principle taken in conjunction with Born's probability doctrine. It states that in the case of an elementary particle, say an electron, its position and its momentum (that is, velocity) cannot be, both of them, observed with sharp precision at the same time. If, say, the position has been fixed absolutely, then the momentum is entirely random; joint approximations of the two have thresholds beyond which they cannot be improved, every threshold involving Planck's constant of action; and if one of the two has been observed to fall within a certain range, then the value of the other has possibilities that are given by associated probability distributions.[8]

These assertions about limitations of microphysical observability produced in Einstein and a handful of other physicists an emotional reaction as if a sacrament hallowing the concepts of space and time had been defiled. Also, pragmatically Einstein must have instantly perceived that this was the end of his dream of a Unified Field Theory *on his terms*, although for years on end he would go through the motions of trying to find one; and it was perhaps this that he would not ever forgive. While only some scientists balked, many more philosophers were dismayed. Until 1927, when Heisenberg proclaimed his Uncertainty Principle, professional philosophers took little notice of quantum theory, of the "old" theory and even the "new" one. But this apparent contravention of the customary presumption of determinism in the laws and prescriptions of physics took them by very unpleasant surprise and jolted them into attention. It was one thing say, to oppose Hegelianism to Kantianism, old or "neo-"; but it was quite another thing to call nineteenth century determinism into question. An Einstein could afford just to stand aside and pout. The philosophers could not. They had to get into the scrimmage and try to explain things to themselves and others. For instance, Popper wrote a lengthy article about it, and the distinguished philosopher Ernst Cassirer, author of a string of important works, added one last book to his list, to deal with determinism and indeterminism in physics. From what I have seen, these philosophers' cogitations have not been overly illuminating so far.

From my viewpoint, their lack of success does not reflect on the new physics or the struggling philosophers, nor does it justify Einstein's aloofness. The new quantum theory is finally a creation of the twentieth century proper. It expresses the vicissitudes and anxieties, the fears and uncertainties of our age much more faithfully than did the old theory, and professional philosophy will have to find the drive and the idiom by which and in which to articulate this.

The nineteenth century knew a great deal about discontinuities, and it spiced the naiveté of the eighteenth century by inserting "additives" of discontinuities, some good-sized, into its philosophemes about continuity all around. But these added discontinuities were somehow felt to be undesirable, even if ineluctable, and continuity remained the paramount wish and expectation, even more intensely than in the eighteenth century. In fact, academics have been wont to proclaim that the nineteenth century was *the* century of continuity. Although Darwin's continuist evolution had to be salvaged from immobilizing rigidity by an injection of potent mutationist discontinuity, overall continuity nonetheless remained paramount. Bernhard Riemann, one of the gentlest of men, in a paper of 1860 inaugurated thought patterns and even words of violence that only twentieth century technology was able to follow up. He did not have "shock wave" in his vocabulary, but he had "compression wave" (*Verdichtungswelle*) and "compression thrust" (*Verdichtungsstoss*); yet the world around him took little notice of his anticipation. Karl Marx envisaged a violent socioeconomic upheaval, that is, discontinuity, to come. But it was sociopolitically predetermined and bound to be followed by a future free of such discontinuities. And Marx bitterly fought anarchists, that is, advocates of political discontinuity as a way of life.

But as the twentieth century wanes, discontinuities are gradually advancing to a level equal in day-to-day importance to that of continuities. They are becoming severally and jointly a way of life and thought throughout, and not only in select areas. Einstein had no qualms about imputing to elementary particles two co-existent aspects, a continuous undulatory aspect and a discontinuous corpuscular aspect. And he had no objection to, but productively cooperated in the large nineteenth-century statistical theory of matter that takes into account the difficulties of truly precise observations and the mathematical inconvenience of dealing rigorously with a mechanical system of many particles. But he balked more than half of his academic life at the Born-Heisenberg statistical theory, because its statistical aspect is not marginally approximative but centrally legislative.

And yet, already decades before Born-Heisenberg, the American philosopher Charles Sanders Peirce (1839-1914) perceived, from pure thinking as it were, the possibility of an "absolute chance," that is, of a genuine indeterminacy as opposed to an indeterminacy arising merely from "ignorance," be the ignorance genuine or presumptive. Peirce even installed the minor Greek goddess *Tyche* (chance) to be in charge of this "Tychism."[9]

Niels Bohr, only six years younger than Einstein, proposed to come to grips with the perplexities of twentieth-century physics by the use of a new rule of comprehension, a tricky one. He called it the principle of complementarity. By this principle there are near-contradictory aspects in the process of verifying phenomena. Yet, however contradictory, the aspects complement each other. If an experiment illumines one aspect it necessarily obscures the other. On translating from experimentation to cogitation it follows that there are theories that are both contradictory and complementary.[10] As if to corroborate this, there are physicists who find that relativity theory and quantum theory are of this kind.

This sort of thinking was outside Einstein's horizon. Before World War I certain intellectual circles hopefully expected that a common language of communication, a true Esperanto, would cure many if not all ills of the world; and Einstein's heart was arrested in this expectation. But it is a vain expectation. In our century there is a built-in possibility for ambiguity and uncertainty that cannot be eliminated. Gödel's theorem, when boldly transferred from Logic to Life, already points to this, whether Gödel himself would approve of such a transfer or not. Suppose the United States and Russia would agree, in the purest of intentions, to draw up a detailed comprehensive document of coexistence in the most precise language possible. By Gödel's theorem no precaution of wording can prevent situations from arising about which it could not be decided whether they are or are not subsumable under the provisions of the document. An equally bold transfer of the principle of complementarity suggests that subsumability in all cases could be insured by composing the treaty in two versions, English and Russian, but at a price: The price is that in some instances conflicting interpretations might arise, either outright contradictory, or differing so much in degrees of clarity as to amount to a contradiction. Albert Einstein, physicist *sans pareil*, could not or would not see what was quintessential in the workaday doings of a younger generation of physicists all around him.

Bertrand Russell, with whom Einstein was in rapport, was not far different from him. Russell's great disciple Ludwig Wittgenstein did have a sense of something new in the offing. His *Tractatus* was done at the same time as Einstein's work on General Relativity, and fits snugly into the outgoing phase of the extended nineteenth century. After that, Wittgenstein began to grope for something radically different, towards entering into the twentieth century proper, as I see it; this is shown by the difference of style in his *Philosophical Investigations*, and in the *Brown* and *Blue Books* before. But he could not make it. His intellectual resources were too limited and too exhausted for such a radical reorientation.

Kafka and other existentialists, with the Sigmund Freud of around 1930 among them, did show an awareness of a rising Discontent in Civilization, but they were too continuity-bound, too conditioned and too overwhelmed by the past effusions of Schopenhauer, Kierkegaard, and Nietzsche, to recognize the twentieth-century novelty of it all.

Even Arnold Toynbee, when he tried to extrapolate from the past into the future, did not have things in focus. Too much eschatological mysticism clouded his vision. But his rival prognosticator Oswald Spengler, author of *The Decline of the West*, clearly described the woes of our "imperial presidency," for instance, and he clearly foresaw the oncoming struggle between the West—which he calls the Faustian culture—and a resurgent Russia. What is more, he insisted before 1918 that in the end it will be not the Lenins but the Solzhenitsyns that will oppose us.

The contrast between the nineteenth and twentieth centuries can be epitomized by a changing of the guard in *Central Europe*, and I mean Central Europe literally. In the extension of the nineteenth century there were Freud, Einstein, Spengler. Their counterparts in the twentieth century proper were Kafka, Heisenberg, Gödel. Overwhelming as the nineteenth century is in this confrontation—and it was an overwhelming century indeed—the twentieth century does have a very distinctive composition and orientation of its own, a radically different one. How the twenty-first century will compare with the two preceding ones will be for the twenty-first to decide.

NOTES

1. Albert Einstein was born March 14, 1879, in Ulm, Germany, and died April 15, 1955, in Princeton, New Jersey. He was cremated and his ashes were scattered, except for the brain, which has been preserved.

After an irregular attendance at high schools in Munich, Germany, and Aurau, Switzerland, he was enrolled at the reputed Zurich Institute of Technology ETH (Eidgenössische Technische Hochschule) from July 1896 till August 1900, graduating from its Division IV with a diploma for teaching mathematics and physics. After temporary teaching jobs he became on June 23, 1902, a clerk at the Patent Office in Bern, from which position he resigned October 15, 1909, to become an associate professor (Extraordinarius) at the University of Zurich. Until joining this university, while still with the Patent Office, he was also a *Privatdozent* at the University of Bern. He left the University of Zurich in the summer of 1911, to become professor at the German University of Prague. He stayed only one year, 1911/1912, leaving to become full professor (Ordinarius) at the ETH, upon the recommendations of Marie Curie and Henri Poincaré. This stay was also brief, and from April 1, 1914, till the middle of March 1933 he was a salaried member of the Prussian Academy of Sciences in Berlin. In October 1933 he reached his final academic destination. He joined the Institute for Advanced Study at Princeton, New Jersey, as one of its faculty members, becoming a resident Emeritus in 1946.

These data are taken from Max Flückiger, *Albert Einstein in Bern* (Bern, 1974), and the comprehensive biography of Einstein: Ronald W. Clark, *Einstein, the Life and Times* (New York and Cleveland: World Publishing Company, 1971).

Some data in the text have been taken from the bibliography in P. A. Schilpp, *Albert Einstein, Philosopher-Scientist* (Evanston, Ill., 1949); from Jagdish Mehra, *The Solvay Conferences on Physics* (Dordrecht and Boston, 1975); and from A. Eucken, *Die Theorie der Strahlung und der Quanten* (Halle, 1914), which is a detailed report on the First Solvay Conference.

2. *Science* **49** (1929): 248-249. Einstein observes that, whatever Newcomb achieved, he could of course not explain the irregularity of the perihelion of Mercury because only General Relativity could cope with this anomaly. In an assessment of Newcomb's achievements this was an odd observation to make, since Newcomb died in 1909, years before there was a theory of General Relativity. Newton could not have explained it either.

3. Marjorie Hope Nicolson, *Newton Demands the Muse: Newton's "Opticks" and the Eighteenth-Century Poets* (Princeton, 1946). See also the "Preface" by I. Bernard Cohen in the edition of the *Opticks* by Dover Publications (New York, 1952). This edition also has a valuable introduction by Sir Edmund Whittaker.

4. Died at age forty-five.

5. H. D. Smyth, *Atomic Energy for Military Power* (Washington, D.C., 1945). This book, a best-seller, was, in effect, a government publication.

6. W. de Sitter, *Kosmos* (Cambridge, Mass., 1932), p. 113.

7. Niels Bohr, *Atomic Theory and the Description of Nature* (Cambridge, 1934), p. 16.

8. Ibid.

9. Salomon Bochner, "Mathematical Reflections," *The American Mathematical Monthly* **81** (1974): 827-857, especially p. 850.

10. Bohr, *Atomic Theory*, pp. 1-20.

Salomon Bochner—Vita

Personal Data
Born 20 August 1899, Cracow, Austria-Hungary (now Poland)
U.S. Citizen
Widower, 1 child

Education
University of Berlin, Ph.D., 1921

Area of Research
Analysis

Professional Positions
University of Munich, Lecturer, 1927–1932

Princeton University, Associate, 1933–1934

Princeton University, Assistant Professor, 1934–1939

Princeton University, Associate Professor, 1939–1946

Princeton University, Professor, 1946–1959

Princeton University, Henry Burchard Fine Professor, 1959–1968

Rice University, E. O. Lovett Professor, 1968–

Member (part time), Institute for Advanced Study, 1945–1948

Consultant, Los Alamos Project, Princeton, 1951

Harvard Lecturer, second term, 1946–1947

University of California, Berkeley, Visiting Professor
(Department of Statistics), second term, 1952–1953

Invited Addresses
Keynote Speaker, AAAS Symposium: the Role of Mathematics in the Development of Science, 1971.

Honors
Fellow, International Education Board, Copenhagen, Oxford, Cambridge, 1925–1926

A symposium in honor of Salomon Bochner, Princeton University, 1969

Leroy P. Steele Prize, January 1979

Professional Activities
Vice-President, American Mathematical Society, three years

Colloquium Lecturer, American Mathematical Society

Consulting Editor, McGraw-Hill Encyclopedia of Science and Technology

Editorial Board, Dictionary of the History of Ideas

Member, National Academy of Sciences

University Activities
Chairman, Department of Mathematics, Rice University, 1969–1976

THESES SUPERVISED BY S. BOCHNER

1934 John L. Barnes ON THE LAPLACE-STIELTJES TRANSFORMATION

1936 Israel Halperin ADJOINTS AND CLOSURES OF LINEAR DIFFERENTIAL OPERATORS

1939 Andrew Sobczyk PROJECTIONS IN MINKOWSKI AND BANACH SPACES

1939 Robert Greenwood HANKEL SERIES

1939 F. Joachim Weyl ANALYTIC CURVES

1940 John M.H.Olmsted LEBESQUE THEORY ON A BOOLEAN ALGEBRA

1942 Paco Lagerstrom MEASURE AND INTEGRAL IN PARTIALLY ORDERED SPACES

1944 Charles Dolph NON-LINEAR INTEGRAL EQUATIONS OF THE HAMMERSTEIN TYPE

1947 Samuel Karlin INDEPENDENT FUNCTIONS

1948 Bernard R. Gelbaum EXPANSIONS IN BANACH SPACES

1948 Gilbert Hunt ON STATIONARY STOCHASTIC PROCESSES

1948 Harry E. Rauch GENERALIZATIONS OF SOME CLASSIC THEOREMS TO THE CASE OF FUNCTIONS OF SEVERAL VARIABLES

1949 Min-Teh Cheng ON THE UNIQUENESS THEOREM OF MULTIPLE TRIGONOMETRICAL SERIES

1949 Newton Seymour Hawley HOMOGENEOUS AND COMPLEX MANIFOLDS

1950 E. Calabi ISOMETRIC COMPLEX ANALYTIC IMBEDDING OF KÄHLER MANIFOLDS

1950 Gerard Washnitzer A DIRICHLET PRINCIPLE FOR ANALYTIC FUNCTIONS OF SEVERAL COMPLEX VARIABLES

1951 Walter M. Gilbert COMPLETELY MONOTONIC FUNCTIONS OF SEVERAL VARIABLES

1951 Melvin Hausner DIRICHLET'S PRINCIPLE AND GENERALIZED BOUNDARY VALUES

1951 J.H.Sampson THE DIRICHLET PROBLEM IN THE LARGE

1952 Arthur C. Allen SOME THEOREMS ON HARMONIC FUNCTIONS AND RELATED TOPICS

1952 Solomon Leader SET FUNCTION SPACES AND ORTHOGONAL EXPANSIONS OF FINITELY ADDITIVE SET FUNCTIONS

1953 Carl Herz BESSEL FUNCTIONS OF MATRIX ARGUMENT

1954 Sigurdur Helgason BANACH ALGEBRAS AND ALMOST PERIODIC FUNCTIONS

1954 H. Scarf DIFFERENTIAL OPERATORS ON MANIFOLDS, AND APPLICATIONS TO STOCHASTIC PROCESSES

1955 Robert C. Gunning A CLASSIFICATION OF FACTORS OF AUTOMORPHY

1956 John W. Woll HOMOGENEOUS STOCHASTIC PROCESSES

1958 Harry Furstenberg PREDICTION THEORY

1958 Oscar Rothaus DOMAINS OF POSITIVITY

1961 Richard A. Askey MEAN CONVERGENCE OF ORTHOGONAL SERIES AND CONJUGATE SERIES

1962 Burton Randol GENERALIZED ZETA-FUNCTIONS

1963 Stephen Scheinberg HARDY SPACES AND BOUNDARY PROBLEMS IN ONE COMPLEX VARIABLE

1963 William Veech ALMOST AUTOMORPHIC FUNCTIONS

1965 Anthony W. Knapp DISTAL FUNCTIONS ON ABELIAN GROUPS

1966 Roger Lee Cooke RESULTS ON THE STRUCTURE AND APPLICATIONS OF ABSTRACT-VALUED FUNCTIONS

1967 Jeff Cheeger COMPARISON AND FINITENESS THEOREMS FOR RIEMANNIAN MANIFOLDS

Acknowledgments

The American Mathematical Society gratefully acknowledges the kindness of these institutions in granting the following permissions:

Accademia Nazionale Dei Lincei:

Su un teorema di Frobenius Pern Le Funzioni di Jacobi, Rendiconti di questa Accademia, vol. IX, series VIII, fasc. 6, 1950, pp. 327–331.

ACTA Scientiarum Mathematicarum (Szeged):

Lattice points and Fourier expansions, Acta Sci. Math. 12 (1950), 1–15.

Akademie der Wissenschaften zu Göttingen:

Abstrakte Funktionen und die Besselsche Ungleichung Nachrichten von der Gesellschaft der Wissenschaften zu Göttingen, Math. Phys. (1933), 178–180.

American Council of Learned Societies:

Bochner, Salomon, *Dictionary of Scientific Biography*, vol. 17, supp. II, pp. 88–90.

American Scientist:

Role of Mathematics in the Rise of Mechanics, American Scientist 50 (1962), 294–311.

Annali di Matematica:

Theorem of Morera in several variables, Annali di Matematica, serie IV, Tomo XXXIV, 1953.

Vector fields on Riemannian spaces with boundary, Annali di Matematica Pura ed Applicata, serie IV, Tomo LIII, 1961.

Bayerische Akademie der Wissenschaften:

Über die Struktur von Fourierreihen fastperiodischer Funktionen, Sitzungsberichte der Bayerischen Akademie der Wisenschaften, 1928, pp. 181–190.

Berliner Mathematische Gesellschaft E.V.:

Über Fourierreihen von fastperiodischen Funktionen, Sitzungsberichte BMG 26. Jahrgang.

B. G. Teubner, GMBH:

Orthogonale Systeme analytischer und harmonischer Funktionen, Jahresbericht der Deutschen Mathematiker-Vereinigung Bd. 30, 1921, pp. 100–101.

Beitrag zur absoluten Konergenz fastperiodischer Fourierreihen, Deutsche Mathematiker, Vereiningung, Jahresbericht, Band 39, 1930, pp. 52–54.

Kurven Endlicher Länge, Deutsche Mathematiker, Vereiningung, Jahresbericht, Band 40, 1931, pp. 245–249.

Calcutta Mathematical Society:

General analytical setting for the general limit theory of probability, Golden Jubilee Commemoration volume (1958–1959), pp. 111–128.

Canadian Mathematical Society:

A new viewpoint in differential geometry, Canad. J. Math. 3 (1951), 460–470.

Centre National de la Recherche Scientifique-Publications:

Systémes des équations aux dérivées partielles qui se rattachent aux formules de Green, La Theorie des Equations aux Dérivées Partielles, CNRS (1956), 19–27.

Duke University Press:

Fourier series of L_2-Functions, Duke Math. J. 16 (1949), 579–583.

F. Centrale des Reviues Dunod-Gauthier Villars:

Function rings and analytic measures, J. Math. Pures Appl. (1963), 209–221.

Weak solutions of linear partial differential equations, J. Math. Pures Appl. 35 (1956), 194–202.

Sur les fonctions presque periodiques de Bohr, Anal. Math., 1–3.

Friends of Fondren Library, Rice University:

Frontispiece appearing in Part 4, from the cover of The Flyleaf, 33 (1983).

Fundamenta Mathematicae:

Integration von Funktionen, Deren Werte die Elemente Eines Vektorraumes Sind, Fund. Math. (1933), 262–276.

Eine Bemerkung zum Satz von Fubini, Fund. Math. (1933), 277–280.

Absolute-Additive Abstrakte Mengenfunktionen, Fund. Math. (1933), 211–213.

Hungarian Academy of Sciences:

Über Faktorenfolgen Für Fouriersche Reihe, Acta Litterarum Ac Scientiaru, Tom IV, Fasc. III, vol. 28, 1929, pp. 125–129.

Indian Mathematical Society:

On compact complex manifolds, J. Indian Math. Soc. 11 (1947), 1–21.

Remarks on Gaussian sums and Tauberian theorems, J. Indian Math. Soc. 15 (1951), 97–104.

Algebraic and linear dependence of automorphic functions in several variable, J. Indian Math. Soc. 16 (1952), 1–6.

Connection between functional equations and modular relations, J. Indian Math. Soc. 16 (1952), 99–102.

Hartogs' theorem in complex spaces with singularities, J. Indian Math. Soc. 16 (1952), 138–146.

Theorems on analytic continuation which occur in the study of Riemann's functional equation, J. Indian Math. Soc. 21 (1957), 127–147.

Institut Mittage-Leffler:

Hauptlösungen von Differenzenglichungen, Acta Math. 51 (1927), 1–21.

Abstrakte fastperiodisch Funktionen, Acta Math. 61 (1933), 150–184.

Fastperiodische Lösungen der Wellwngleichung, Acta Math. 62 (1933), 227–237.

Institute of Mathematical Statistics:

A limit theorem for the periodogram, Ann. Math. Stat. 29 (1958).

International Philosophical Quarterly:

Aristotle's physics and todays physics, Internat. Philos. Quart. 4 (1964), 217–244.

History of Science Society, ISIS:

The significance of some basic mathematical conceptions for physics, ISIS 54 (1963), 179–205.

Johns Hopkins University Press:

Bounded analytic functions in several variables and multiple Laplace integrals, 59 (1937), 732–738.

Some theorems of quasi-analyticity for functions of several variables, Amer. J. Math. 61 (1939), 303–328.

Gauss summability of trigonometric integrals, Amer. J. Math. 71 (1949), 50–59.

London Mathematical Society:

Properties of Fourier series of almost periodic functions, Proc. London Math. Soc. (2) 26 (1926), 432–452.

Remarks on the theorems of Picard-Landau and Picard-Schottky, J. London Math. Soc. 1 (1925), 100–103.

Properties of Fourier series of almost periodic functions, J. London Math. Soc. 1 (1926), 132–133.

Notes on two theorems of Norbert Weiner, J. London Math. Soc. 1 (1926), 240–244.

Remarks on the integration of almost periodic functions, J. London Math. Soc. 8 (1933), 250–254.

Homogeneous systems of differential equations with almost periodic coefficients, J. London Math. Soc. 8 (1933), 279–288.

Remark on the Euclidean algorithm, J. London Math. Soc. 9 (1933).

An extension of a Tauberian theorem on series with positive terms, J. London Math. Soc. 9 (1933), 141–148.

On a theorem of M. Riesz for Fourier series, J. London Math. Soc. 14 (1939), 62–73.

Properties of Fourier series of almost periodic functions, Proc. London Math. Soc (2) 26 (1926), 432–452.

Lunds Universitets Matematiska Institution

Bessel functions and modular relations of higher type and hyperbolic differential equations, Communications du Seminaire mathematique de L'universite de Lund Tome Supplementaire, 1952, pp. 12–20.

MacMillan Publishing Company:

Space, Dictionary of the History of Ideas (Philip P. Wiener, ed.), vol. IV, pp. 295–307.

Continuity and discontinuity in nature and knowledge, Dictionary of the History of Ideas (Philip P. Wiener, ed.), vol. I, pp. 492–504.

Symmetry and asymmetry, Dictionary of the History of Ideas (Philip P. Wiener, ed.), vol. IV, 345–353.

Mathematics in cultural history, Dictionary of the History of Ideas (Philip P. Wiener, ed.), vol. III, pp. 177–185.

Infinity, Dictionary of the History of Ideas (Philip P. Wiener, ed.), vol. II.

McGraw-Hill Publishing Company:

Mathematics, McGraw-Hill Encyclopedia of Science and Technology, 1987, pp. 175–180.

Mathematical Association of America:

Fourier series came first, Amer. Math. Monthly 86 (1979), 197–199.

Mathematical reflections, Amer. Math. Monthly 81 (1974), 827–852.

National Academy Press:

John Von Neumann 1903–1957: *A biographical memoir*, Biographical Memoirs, vol. 32, National Academy Press, Washington, DC, 1958.

General almost automorphy, Proc. Nat. Acad. Sci. U.S.A. 72 (1975), 3815–3818.

Oxford University Press:

Summation over lattice points in K-space, Quart. J. Math. 19 (1948), 238–248.

Summations over lattice points in K-space (*A supplemental note*), Quart. J. Math. Oxford Ser. (2) 1 (1950), 80.

Princeton University Press:

Green's formula and analytic continuation, Ann. of Math. Stud. 33 (1954), 1–14.

Contributions to the theory of Riemann surfaces (L. Alfors et al., eds.), Princeton University Press, Princeton, NJ, 1953, pp. 189–201.

(*frontispiece* #3) *from Problems in Analysis: A symposium in honor of Salomon Bochner*, Princeton University Press, Princeton, NJ, 1970.

Plato and the one-ness of knowledge, The Role of Mathematics in the Rise of Science, Princeton University Press, Princeton, NJ, 1966.

Structure of complex spaces, Contributions to the theory of Riemann surfaces, Princeton University Press, Princeton, NJ 1953, 189–201.

Inversion formulae and unitary transformations, Ann. of Math. 35 (1934), 111–115.

Distribution functions and positive-definite functions, Ann. of Math. 35 (1934), 252–257.

On compact solutions of operational-differential equations. I, Ann. of Math. 36 (1935), 255–291.

Summation of derived Fourier Series—An application to Fourier expansions on compact Lie groups, Ann. of Math. 37 (1936), 345–356.

A converse of Poisson's theorem in the theory of probability, Ann. of Math. 37 (1936), 816–822.

Singularities of composite functions in several variables, Ann. of Math. 38 (1937), 293–302.

A theorem on analytic continuation of functions in several variables, Ann. of Math. 39 (1938), 14–19.

Closure theorems for translations, Ann. of Math. 39 (1938), 287–300.

Linear functionals on certain spaces of abstractly-valued functions, Ann. of Math. 39 (1938), 913–944.

Additive set functions on groups, Ann. of Math. 40 (1939), 769–799.

Finitely additive integral, Ann. of Math. 41 (1940), 495–504.

Hadamard's theorem for Dirichlet series, Ann. of Math. 41 (1940), 711–714.

Additive set functions and vector lattices, Ann. of Math. 42 (1940), 316–324.

Hilbert distances and positive definite functions, Ann. of Math. 42 (1941), 647–656.

On a theorem of Tannaka and Krein, Ann. of Math. 43 (1942), 56–58.

Absolutely convergent Fourier expansions for non-commutative normed rings, Ann. of Math. 43 (1942), 409–418.

Analytic and meromorphic continuation by means of Green's formula, Ann. of Math. 44 (1943), 652–673.

Group invariance of Cauchy's formula in several variables, Ann. of Math. 45 (1944), 686–707.

Boundary values of analytic functions in several variables and of almost periodic functions, Ann. of Math. 45 (1944), 708–722.

Compact groups of differentiable transformations, Ann. of Math. 46 (1945), 372–381.

Groups of differentiable and real or complex analytic transformations, Ann. of Math. 46 (1945), 685–694.

Formal Lie groups, Ann. of Math. 47 (1946), 192–201.

Linear partial differential equations with constant coefficients, Ann. of Math. 47 (1946), 202–212.

Locally compact groups of differential transformations, Ann. of Math. 47 (1946), 639–653.

Distributive order-preserving operations in partially ordered vector sets, Ann. of Math. 48 (1947), 168–179.

Groups on analytic manifolds, Ann. of Math. 48 (1947), 659–669.

Stochastic processes, Ann. of Math. 48 (1942), 1014–1061.

Curvature and Betti numbers, Ann. of Math. 48 (1948), 379–390.

On the localization property for multiple Fourier series, Ann. of Math. 49 (1948), 966–978.

Curvature and Betti numbers. II, Ann. of Math. 50 (1949), 77–93.

Quasi-analytic functions, Laplace operator, positive kernels, Ann. of Math. 51 (1950), 68–91.

Euler-Poincaré characteristic for locally homogeneous and complex spaces, Ann. of Math. 515 (1950), 241–261.

Vector fields on complex and real manifolds, Ann. of Math. 52 (1950), 642–649.

Some properties of modular relations, Ann. of Math. 53 (1951), 332–363.

Tensor fields with finite bases, Ann. of Math. 53 (1951), 400–411.

Tensor fields with finite bases, Ann. of Math. 56 (1952), 504–519.

Zeta functions and Green's functions for linear partial differential operators of elliptic type with constant coefficients, Ann. of Math. 57 (1953), 32–56.

Length of random paths on general homogeneous spaces, Ann. of Math. 57 (1953), 309–319.

Complex spaces with singularities, Ann. of Math. 57 (1953), 490–516.

Partial ordering in the theory of Martingales, Ann. of Math. 62 (1955), 162–169.

On Riemann's functional equation, Ann. of Math. 63 (1956), 336–360.

On Riemann's functional equation with multiple gamma factors, Ann. of Math. 67 (1958), 29–41.

A special integral transformation in Euclidean space, Ann. of Math. 68 (1958), 150–158.

Strong law of large numbers for sequences of almost periodic functions, Ann. of Math. 68 (1959), 718–732.

Indefinite linear pseudogroups of transformations, Ann. of Math. 75 (1962), 93–104.

Contributions to Fourier analysis, Ann. of Math. 25 (1950), 3–45.

Analytic functions with almost periodic coefficients, Ann. of Math. (2) 35 (1934), 152–161.

Rice University Press:

The rise of functions, Complex Analysis II: Proceedings of the Conference on Complex Analysis, Rice University, 26–29 March 1969, Rice Univ. Stud., vol. 56, Rice Univ. Press, Houston, TX, 1970, pp. 3–23.

Singularities and discontinuities, Proceedings of the Converence on Complex Analysis, Rice University, March 1972, Rice Univ. Stud., vol. 59, Rice Univ. Press, Houston, TX, 1973.

The emergence of analysis in the Renaissance and after, Rice Univ. Stud., vol. 64, Rice Univ. Press, Houston, TX, 1978, pp. 11–56.

Einstein between centuries, Rice Univ. Stud., vol. 65, Rice Univ. Press, Houston, TX, 1979, pp. 1–54.

Seminario Matematico:

Curvature and Betti numbers in real and complex vector bundles, Rendiconti del Seminario Matematico 15 (1955–56), 225–253.

S. Hirzel Verlag Stuttgart:

Über einen arithmetischen Satz der Röntgenkristallographie, Probleme Der Modernen Physik, 1928, pp. 88–94.

Springer-Verlag:

Über orthogonale Systeme analytischer Funktionen, Math. Z. 14 (1922), 180–207.

Konvergenzsatze für Fourierreihen grenzperiodischer Funktionen, Math. Z. 27 (1928), 187–211.

Ein Satz über lineare Operationen, Math. Z. 29 (1929), 737–744.

Über Stürm-Liouvillesche Polynomsysteme, Math. Z. 29 (1929), 730–736.

Allgemeine lineare Differenzengleichungen mit asymptotish konstanten Koeffizienten, Math. Z. 33 (1931), 426–450.

Ein Konvergenzsatz für mehrvariablige Fouriersche Integrale, Math. Z. 34 (1931), 440–447.

Limitierung mehrfacher Folgen nach dem Verfahren der arithmetischen Mittel, Math. Z. 35 (1932), 122–126.

Ein Satz von Landau und Ikehara, Math. Z. 37 (1933), 1–9.

Simultankonvergenz von mehrfachen Fourierschen Reihen und Integralen, Math. Z. 50 (1933), 1–6.

Green-Goursat theorem, Math. Z. 63 (1955), 230–242.

Beitrage zue Theorie der fastperiodischen Funktionen, I. *Teil. Funktionen einer Variablen*, Math. Ann. 96 (1926), 119–147.

Beitrage zue Theorie der fastperiodishen Funktionen, II. *Teil. Funktionen Mehrerer Variablen*, Math. Ann. 96 (1926), 383–409.

Fortsetzung Riemannscher Flachen, Math. Ann. 98 (1928), 406–421.

Darstellungen reellvariabler und analytischer Funktionen durch verallgemeinerte Fourier und Laplace-Integrale, Math. Ann. 97 (1927), 632–662.

Der Dimensionsbegriff für Punktmengen in kartesischen Räumen, Math. Ann. 102 (1930), 768–784.

Über gewisse Differential und allgemeinere Gleichungen, deren lösungen fastperiodisch sind I. *Teil. Der Existenzsatz*, Math. Ann. 102 (1929), 489–504.

Über gewisse Differential und allgemeinere Gleichungen, deren lösungen fastperiodisch sind. II. *Teil. Der Beschränktheitrassatz*, Math. Ann. 103 (1930), 588–597.

Über gewisse Differential und allgemeinere Gleichungen, deren lösungen fastperiodisch sind. III. *Teil. Systeme von Gleichungen*, Math. Ann. 104 (1931), 579–587.

Die Poissonsche Summationsformel in mehreren Veränderlichen, Math. Ann. 106 (1932), 56–63.

Monotone Funktionen, Stieljessche Integrale und harmonische Analyse, Math. Ann. 108 (1933), 378–410.

Hartogs' theorem in Euclidean space and a related Cousin theorem on the torus, Contributions to Function Theory, Tata Institute of Fundamental Research, 1960.

Potential-theoretic approach to analytic Bohr almost periodic functions in half-planes, Math. Ann. 150 (1963), 150–155.

A formal approach to stochastic stability, Zeitschrift für Wahrscheinlichkeitstheorie und Verwandte Geibiete 31 (1975), pp. 187–198.

Positivity of the heat kernel for ultraspherical polynomials and similar functions, Arch. Rational Mech. Anal. 70 (1979), 211–217.

Tôhoku Mathematical Journal:

Some general convergence theorems, Tôhoku Math. J. 42 (1936), 191–194.

The University of California Press:

Stationarity, boundedness, almost periodicity of random-valued functions, Proceedings of the Third Berkeley Symposium on Mathematical Statistics and Probability, The University of California Press, Berkeley, CA, pp. 7–27.

University of Maryland Book Center:

Sturm-Liouville and heat equations whose eigen-functions are ultraspherical polynomials or associated Bessel functions, Proceedings of the Conference on Differential Equations (J. B. Diaz and L. E. Payne, eds.).

The University of Michigan Press:

Functions in one complex variable as viewed from the theory of functions in several variables, Lectures on Functions of a Complex Variable (Wilfred Kaplan et al., eds.), The University of Michigan Press, Ann Arbor, MI, pp. 315–333.

Walter de Gruyter & Company:

Über eine Klasse singulärer Integralgleichungen, Sonderausgabe Aus Den Sitzungsberichten Der Preussischen Akademie Der Wissenschaften, Phys-Math Klasse, 1930, XXII.

Umkehrsätze für allgemeine Limitierungsverfahren, Sonderausgabe Aus Den Sitzungsberichten Der Preussischen Akademie Der Wissenschaften, Phys-Math Klasse, 1933, III.

Spektralzerlegung linearer Scharen unitärer Operatoren, Sopnderausgabe Aus Den Sitzungsberichten Der Preussischen Akademie Der Wissenschasften, Phys-Math Klasse, 1933, IX.